T0214504

Lecture Notes in Computer Science 11356

Commenced Publication in 1973
Founding and Former Series Editors:
Gerhard Goos, Juris Hartmanis, and Jan van Leeuwen

Editorial Board

More information about this series at http://www.springer.com/series/7410

Debrup Chakraborty · Tetsu Iwata (Eds.)

Progress in Cryptology – INDOCRYPT 2018

19th International Conference on Cryptology in India
New Delhi, India, December 9–12, 2018
Proceedings

 Springer

Editors
Debrup Chakraborty ⓘD
Indian Statistical Institute
Kolkata, India

Tetsu Iwata ⓘD
Nagoya University
Nagoya, Japan

ISSN 0302-9743   ISSN 1611-3349   (electronic)
Lecture Notes in Computer Science
ISBN 978-3-030-05377-2   ISBN 978-3-030-05378-9   (eBook)
https://doi.org/10.1007/978-3-030-05378-9

Library of Congress Control Number: 2018962936

LNCS Sublibrary: SL4 – Security and Cryptology

This Springer imprint is published by the registered company Springer Nature Switzerland AG
The registered company address is: Gewerbestrasse 11, 6330 Cham, Switzerland

Preface

INDOCRYPT 2018, the 19th edition of the International Conference on Cryptology in India, was held during December 9–12, 2018, in India Habitat Center, New Delhi. Indocrypt is organized under the aegis of the Cryptology Research Society of India (CRSI). It began in 2000 under the leadership of Prof. Bimal Roy of the Indian Statistical Institute, Kolkata, and since then this annual event has gained its place among prestigious cryptology conferences and is considered as the leading Indian conference for cryptology. In the past, the conference took place in various cities of India: Kolkata (2000, 2006, 2012, 2016), Chennai (2001, 2004, 2007, 2011, 2017), Hyderabad (2002, 2010), New Delhi (2003, 2009, 2014), Bangalore (2005, 2015), Kharagpur (2008), and Mumbai (2013).

INDOCRYPT 2018 attracted 60 submissions from 14 different countries. Out of these 60 submissions, papers that were withdrawn before the submission deadline and those submitted after the submission deadline were not reviewed, and after the review process, 20 papers were accepted for inclusion in the program. All the papers that satisfied the submission guidelines were reviewed by at least three reviewers. Submissions of the Program Committee members were reviewed by at least four reviewers. The individual review phase was followed by a discussion phase that generated additional comments from the Program Committee members and the external reviewers. A total of 44 Program Committee members and 48 external reviewers took part in the process of reviewing and the subsequent discussions. We take this opportunity to thank the Program Committee members and the external reviewers for their tremendous job in selecting the current program. The submissions and reviews were managed using the "Web Submission and Review Software" written and maintained by Shai Halevi. We thank him for providing us the software.

The proceedings include the revised versions of the 20 contributed papers. Revisions were not checked by the Program Committee members and the authors bear the full responsibility for the contents of the respective papers. In addition to the 20 papers, the program included three invited talks. Gilles Van Assche gave a talk about "On dec(k) Functions," Takahiro Matsuda spoke on "Public Key Encryption Secure Against Related Randomness Attacks," and Mridul Nandi's talk was about "How to Make a Single-Key Beyond Birthday Secure Nonce-Based MAC." The abstracts of the invited talks are also included in these proceedings.

We would like to thank the general chairs, Dr. Anu Khosla and Prof. Brishbhan Singh Panwar, and the organizing chairs, Prof. Shri Kant and Dr. Indivar Gupta, along with the Organizing Committee comprising members of Sharda University and SAG DRDO for making the conference a success. Finally, we would like to thank all the authors who submitted their work to INDOCRYPT 2018, and we also would like to thank all the participants. Without their support and enthusiasm, the conference would not have succeeded.

December 2018   Debrup Chakraborty
                                                          Tetsu Iwata

INDOCRYPT 2018

The 19th International Conference on Cryptology in India

New Delhi, India
December 9–12, 2018

General Chairs

Anu Khosla — Scientific Analysis Group, DRDO, Delhi, India
Brishbhan Singh Panwar — Sharda University, India

Program Chairs

Debrup Chakraborty — Indian Statistical Institute, Kolkata, India
Tetsu Iwata — Nagoya University, Japan

Organizing Chairs

Indivar Gupta — Scientific Analysis Group, DRDO, Delhi, India
Shri Kant — Sharda University, India

Program Committee

Diego Aranha — University of Campinas, Brazil and Aarhus University, Denmark
Shi Bai — Florida Atlantic University, USA
Subhadeep Banik — EPFL, Switzerland
Lejla Batina — Radboud University, The Netherlands
Rishiraj Bhattacharyya — NISER, India
Christina Boura — University of Versailles and Inria, France
Debrup Chakraborty — Indian Statistical Institute, Kolkata, India
Sanjit Chatterjee — Indian Institute of Science, Bangalore, India
Geoffroy Couteau — Karlsruher Institut für Technologie, Germany
Pooya Farshim — CNRS and ENS, France
Shay Gueron — University of Haifa, Israel
Divya Gupta — Microsoft Research India, India
Indivar Gupta — SAG, DRDO, Delhi, India
Gottfried Herold — ENS de Lyon, France
Viet Tung Hoang — Florida State University, USA
Takanori Isobe — University of Hyogo, Japan
Tetsu Iwata — Nagoya University, Japan
Elena Kirshanova — ENS Lyon, France

Shanta Laishram	Indian Statistical Institute, Delhi, India
Patrick Longa	Microsoft Research, Redmond, USA
Atul Luykx	Visa Research, USA
Subhamoy Maitra	Indian Statistical Institute, Kolkata, India
Hemanta K. Maji	Purdue University, USA
Bart Mennink	Radboud University, The Netherlands
Kazuhiko Minematsu	NEC Corporation, Japan
Debdeep Mukhopadhyay	IIT Kharagpur, India
Mridul Nandi	Indian Statistical Institute, Kolkata, India
Khoa Nguyen	NTU, Singapore
Ryo Nishimaki	NTT, Japan
Raphael Phan	Multimedia University, Malaysia
Manoj Prabhakaran	Indian Institute of Technology, Bombay, India
Somindu C. Ramanna	Indian Institute of Technology, Kharagpur, India
Francisco Rodriguez-Henriquez	CINVESTAV-IPN, Mexico
Adeline Roux-Langlois	University of Rennes, CNRS, IRISA, France
Jacob Schuldt	AIST, Japan
Peter Schwabe	Radboud University, The Netherlands
Francois-Xavier Standaert	UCL, Belgium
Siwei Sun	Chinese Academy of Sciences, China
Atsushi Takayasu	University of Tokyo, Japan
Srinivas Vivek	IIIT Bangalore, India
Shota Yamada	AIST, Japan
Kazuki Yoneyama	Ibaraki University, Japan
Yu Yu	Shanghai Jiao Tong University, China
Vassilis Zikas	University of Edinburgh, UK

External Reviewers

Nuttapong Attrapadung	Manoj Kumar
Arnab Bag	Iraklis Leontiadis
Balthazar Bauer	Shun Li
Sai Lakshmi Bhavana	Fuchun Lin
Avik Chakraborti	Fukang Liu
Bishwajit Chakraborty	Alice Pellet–Mary
Joan Daemen	Ryutaroh Matsumoto
Prem Laxman Das	Nicky Mouha
Martianus Frederic Ezerman	Fabrice Mouhartem
Chun Guo	Pierrick Méaux
Jian Guo	Tapas Pandit
Muhammad Ishaq	Christophe Petit
Matthias Kannwischer	Shravan K. Parshuram Puria
Louiza Khati	Yogachandran Rahulamathavan

Joost Renes
Yusuke Sakai
Palash Sarkar
Akash Shah
Danping Shi
Bhupendra Singh
Ben Smith
Shifeng Sun
Sharwan Kumar Tiwari
Yiannis Tselekounis

Alexandre Wallet
Weijia Wang
Xiao Wang
Yuyu Wang
Yohei Watanabe
Weiqiang Wen
Masaya Yasuda
Thomas Zacharias
Bin Zhang
Juanyang Zhang

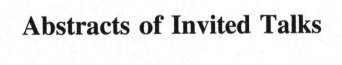

Abstracts of Invited Talks

On dec(k) Functions

Gilles Van Assche

STMicroelectronics, Diegem, Belgium

Cryptographic objects with input and output extension properties are very convenient in numerous situations. With the duplex construction, we defined a cryptographic object that can return a digest on a growing sequence of strings, with an incremental cost, i.e., without the need to process again the entire sequence [2]. Similarly, the Farfalle construction builds a keyed cryptographic function with an extendable input and able to return an output of arbitrary length [1]. It supports for sequences of strings as input and a specific incremental property, namely that computing $F(Y \circ X)$ costs only the processing of Y if $F(X)$ was previously computed. Clearly, duplex and Farfalle are not the only way to build functions with such properties, and the construction should be decoupled from the input-output signature.

For this purpose, we propose the name *dec function* for a function that takes a sequence of input strings and returns a digest of arbitrary length and that can be computed incrementally. Here, "dec" stands for *Doubly-Extendable Cryptographic*. Note that a dec function is a particular case of extendable-output function (XOF), as a XOF is not required to accept growing inputs at an incremental cost. Likewise, we propose the name *deck function*, with an additional "k" for *Keyed*, for a keyed function with the same incremental properties and whose output is a pseudorandom string of arbitrary length.

In this talk, I will explain the purpose of dec(k) functions, from transcript hashing to authenticated encryption, and how to implement them. On this last point, I will relate them to the duplex and full-state keyed duplex constructions, as well as to the Strobe protocol framework [2, 4, 5]. Then, I will explore the permutation-based Farfalle construction as a way to build an efficient deck function from permutation components [1]. Finally, I will detail the recent Xoodoo permutation, its cryptographic properties and the deck function Xoofff built on top of it [3].

References

1. Bertoni, G., Daemen, J., Hoffert, S., Peeters, M., Van Assche, G., Van Keer, R.: Farfalle: parallel permutation-based cryptography. IACR Trans. Symmetric Cryptol. **2017**(4), 1–38 (2017)
2. Bertoni, G., Daemen, J., Peeters, M., Van Assche, G.: Duplexing the sponge: single-pass authenticated encryption and other applications. In: Miri, A., Vaudenay, S. (eds.) SAC 2011. LNCS, vol. 7118, pp. 320–337. Springer, Heidelberg (2012)
3. Daemen, J., Hoffert, S., Van Assche, G., Van Keer, R.: Xoodoo cookbook. IACR Cryptol. ePrint Arch. **2018**, 767 (2018)

4. Daemen, J., Mennink, B., Van Assche, G.: Full-state keyed duplex with built-in multi-user support. In: Takagi, T., Peyrin, T. (eds.) ASIACRYPT 2017. LNCS, vol. 10625, pp. 606–637. Springer, Cham (2017)
5. Hamburg, M.: The STROBE protocol framework. IACR Cryptol. ePrint Archive **2017**, 3 (2017)

Public Key Encryption Secure Against Related Randomness Attacks

Takahiro Matsuda

National Institute of Advanced Industrial Science and Technology (AIST),
Tokyo, Japan
t-matsuda@aist.go.jp

Abstract. Most cryptographic primitives are designed under the assumption that perfect (uniform) randomness is available. Unfortunately, however, random number generators (RNGs) are notoriously hard to implement and test, and we have seen many examples of the failures of RNGs in practice. Motivated by the challenge of designing public key encryption secure under randomness failure, Paterson, Schuldt, and Sibborn (PKC 2014) introduced a security notion called *related randomness attack (RRA) security*. This notion captures security against adversaries that are allowed to control the randomness used in the encryption scheme, but still requires that messages encrypted under an honestly generated public key remain hidden, given that certain restrictions are placed on the adversaries' queries. RRA security is one of the promising security notions that allows us to hedge against randomness failures in the usage of public key encryption. In this talk, I will give a brief survey of the topic, in particular the formalizations, existing results, and techniques used for achieving RRA security.

How to Make a Single-Key Beyond Birthday Secure Nonce-Based MAC

Mridul Nandi

Indian Statistical Institute, Kolkata
mridul.nandi@gmail.com

Abstract. At CRYPTO 2016, Cogliati and Seurin [1] have proposed a highly secure nonce-based MAC called Encrypted Wegman-Carter with Davies-Meyer (EWCDM) construction, as $E_{K_2}(E_{K_1}(N) \oplus N \oplus H_{K_h}(M))$ for a nonce N and a message M. This construction achieves roughly $2^{2n/3}$ bit MAC security with the assumption that E is a PRP secure n-bit block cipher and H is an almost xor universal n-bit hash function. Note that EWCDM requires three keys; two block cipher keys K_1 and K_2 and one hash key K_h. Thus, it is natural to ask that whether one can achieve the similar security in the case of using less number of keys. In fact, proving BBB security of single-keyed EDM ($E_{K_1}(E_{K_1}(N) \oplus N)$), is a highly complicated task as evident from [2] and it is not clear at all how to build on this result to prove the MAC security of EWCDM construction with $K_1 = K_2$. Moreover, Cogliati and Seurin, in their proof of single-keyed EDM [2], have also stated that

"For now, we have been unable to extend the current (already cumbersome) counting used for the proof of the single-permutation EDM construction to the more complicated case of single-key EWCDM."

In this talk, I will discuss a recent design - Decrypted Wegman-Carter with Davies-Meyer (DWCDM) construction - which is structurally very similar to its predecessor EWCDM except that the outer encryption call is replaced by decryption. The biggest advantage of DWCDM is that we can make a truly single key MAC: the two block cipher calls can use the same block cipher key $K = K_1 = K_2$. Moreover, we can derive the hash key as $K_h = E_K(1)$, as long as $|K_h| = n$. Whether we use encryption or decryption in the outer layer makes a huge difference; using the decryption instead enables us to apply an extended version of the mirror theory by Patarin to the security analysis of the construction. DWCDM is secure beyond the birthday bound, roughly up to $2^{2n/3}$ MAC queries and 2^n verification queries against nonce-respecting adversaries when nonce is a $2n/3$ bits string. I will also describe how this construction can be further improved in two directions. We extend the nonce space to as large as the set of all $n - 1$ bits. Moreover, the security bound can be extended against $2^{3n/4}$ MAC queries. The details of a part of this talk can be found in [3].

Keywords: EDM · EWCDM · Mirror theory · Extended mirror theory · H-Coefficient

References

1. Cogliati, B., Seurin, Y.: EWCDM: an efficient, beyond-birthday secure, nonce-misuse resistant MAC. In: Robshaw, M., Katz, J. (eds.) CRYPTO 2016. LNCS, vol. 9814, pp. 121–149. Springer, Heidelberg (2016)
2. Cogliati, B., Seurin, Y.: Analysis of the single-permutation encrypted Davies-Meyer construction. Des. Codes Cryptography 2018 (2018, to appear)
3. Datta, N., Dutta, A., Nandi, M., Yasuda, K.: Encrypt or decrypt? To make a single-key beyond birthday secure nonce-based MAC. In: Shacham, H., Boldyreva, A. (eds.) CRYPTO 2018. LNCS, vol. 10991, pp. 631–661. Springer, Cham (2018)

Contents

Asymmetric Key Cryptography and Cryptanalysis

Symmetric Key Cryptanalysis

Theory

Secure Computations and Protocols

Outsourced Computation and
Searchable Encryption

Revisiting Single-Server Algorithms for Outsourcing Modular Exponentiation

Jothi Rangasamy$^{(\boxtimes)}$ and Lakshmi Kuppusamy

Society for Electronic Transactions and Security (SETS), Chennai, India
{jothiram,lakshdev}@setsindia.net

Abstract. We investigate the problem of securely outsourcing modular exponentiations to a single, malicious computational resource. We revisit recently proposed schemes using single server and analyse them against two fundamental security properties, namely privacy of inputs and verifiability of outputs. Interestingly, we observe that the chosen schemes do not appear to meet both the security properties. In fact we present a simple polynomial-time attack on each algorithm, allowing the malicious server either to recover a secret input or to convincingly fool the client with wrong outputs. Then we provide a fix to the identified problem in the **ExpSOS** scheme. With our fix and without pre-processing, the improved scheme becomes the best to-date outsourcing scheme for single-server case. Finally we present the *first precomputation-free single-server algorithm*, π**ExpSOS** for *simultaneous* exponentiations, thereby solving an important problem formulated in [6].

1 Introduction

The problem of securely offloading cryptographic computations from a (comparatively) weak device to a more powerful device has been considered since many years [1] but the need for such a solution has been increasing rapidly [8–11]. A low-cost RFID tag is a natural example as it has limited computing resources but will benefit from running cryptographic protocols [18]. Proliferation of the usage of mobile applications adds one more scenario wherein outsourcing resource-consuming cryptographic tasks to a third-party is desirable.

The growing utilisation of cloud services such as Dropbox, Google and Amazon Cloud Drives has raised concerns about the availability and integrity of the data being handled and stored. By using cryptographic primitives such as provable data possession [2] and proofs of retrievability [3], these service providers could convince their clients that the actual data given by clients has been retrieved entirely. However, during this process, the clients have to engage in performing computationally-intensive operations to verify the claims of their storage providers and this is not practically viable for many devices in use today. Among complex cryptographic operations, modular exponentiation is invariably the predominant and core operation; that is, to compute $u^a \bmod p$ with a variable base u, a variable exponent a and a prime or RSA modulus p. In this paper,

© Springer Nature Switzerland AG 2018
D. Chakraborty and T. Iwata (Eds.): INDOCRYPT 2018, LNCS 11356, pp. 3–20, 2018.
https://doi.org/10.1007/978-3-030-05378-9_1

our goal is to inspect the recently proposed single-server outsourcing algorithms for modular exponentiation.

1.1 Related Work

The problem of secure delegation of crypto computations to (untrusted) helpers has been considered in various contexts [1,8–11,17,24]. In particular, the idea of secure delegation of modular exponentiation can be attributed to the work of Schnorr [21,22] as he was the first to propose speeding up modular exponentiations in cryptography. However it has not received formal treatment until Hohenberger and Lysyanskaya [12] developed a formal security framework for secure outsourcing of cryptographic computations to untrusted servers in 2005. In secure outsourcing scenario, preserving the secrecy of the inputs and/or outputs is vital. Hence, in Hohenberger-Lysyanskaya formalism, secrecy is the first notion an outsourcing algorithm should aim to satisfy. The second security notion, namely verifiability addresses the correctness of the output of the powerful helper/server. Hohenberger and Lysyanskaya also presented a scheme for outsourcing modular exponentiation to two non-colluding servers. This approach was improved in [6] and further in [15] with better verifiability results.

Designing an efficient algorithm using single untrusted (cloud) server for securely outsourcing (multi-)modular exponentiation has been a perennial problem. Towards solving this, Wang et al. presented (at ESORICS 2014) an efficient protocol to outsource modular exponentiation to a *single* untrusted server [23]. They also presented a generic protocol for outsourcing multi-exponentiations to a single server. However Chevalier et al. [7] presented a lattice-based attack on Wang et al. scheme recovering the secret exponent. Independently, Kiraz and Uzunkol designed an outsourcing scheme but requires an additional sub algorithm [14]. In 2017, Cai et al. [5] proposed a new scheme using redundant inputs but with increased communication complexity undesirably. Recently, Li et al. [16] have come up with a novel approach of using logical divisions twice for the given inputs. Then Zhou et al. [25] proposed a new scheme, which they call **ExpSOS**, using special ring structure of $\mathbb{Z}_N$ with the goal of eliminating the (de facto) preprocessing and achieving near-full verifiability simultaneously. To the best of our knowledge, **ExpSOS** is the only single-server algorithm which does not require resource-demanding pre-processing techniques such as *Rand* in [4].

1.2 Our Contributions

Contributions of the paper is two-fold. We first present practical attacks on three recent single-server algorithms and then we resolve the issue with the **ExpSOS** scheme, making it the best to-date scheme for this purpose.

First we show that the **CExp** scheme due to Li et al. [16] is unfortunately *zero* verifiable, instead of the authors' claim of satisfying the full verifiability. We demonstrate how to manipulate the outputs so that the delegator can be tricked by 100% into accepting false outputs.

Secondly, we note that the **SgExp** scheme proposed by Cai *et al.* [5] does not provide the claimed verifiability guarantee. The idea behind this tricky attack is to classify the queries to the untrusted (cloud) server into two categories because exponents in a set of queries need to be powers of two. This makes the scheme unfortunately *totally* unverifiable, instead of having the verifiability probability $1 - \frac{1}{n^2}$ which is ≈ 0.99996 for $n = 160$. The result is applicable to the **SmExp** scheme proposed by Cai *et al.* for the case of simultaneous exponentiation.

Thirdly, we describe an attack on the **ExpSOS** scheme of Zhou *et al.* [25] on its second invocation with the same secret exponent. The scheme will leak the exponent if it is used again, invalidating the claimed secrecy guarantee. The demonstrated attack extends to the other versions of the **ExpSOS** scheme in [25].

Our last and main contribution are new single-server algorithms for secure outsourcing of single and simultaneous modular exponentiations. For single exponentiation case, our algorithm is obtained by modifying **ExpSOS**. The modified scheme, which we call **MExpSOS** becomes the most efficient and simple scheme available in the related literature in addition to eliminating the *memory requirement* and the substantial *computational cost* of the precomputation step. Finally, we present an elegant algorithm, π**ExpSOS** extending **MExpSOS** for simultaneous multiplications. The π**ExpSOS** algorithm is near error-free and preprocessing-free and hence is the first of its kind in the literature. Our observations are summarised in Table 1.

Table 1. Single server based outsourcing algorithms and their properties

Algorithm	Secrecy	Verifiability probability	Pre-processing required
SgExp, SmExp [5]	YES	0	YES
CExp [16]	YES	0	YES
Kiraz–Uzunkol [14]	YES	< 1	YES
ExpSOS [25]	No	≈ 1	No
MExpSOS, π**ExpSOS**	YES	≈ 1	No

Outline. The paper is organised as follows. The Hohenberger-Lysyanskaya security model is recalled in Sect. 2. In Sect. 3 we observe weaknesses in recent single-server based outsourcing schemes. In Sect. 4 we propose a re-designed **ExpSOS** scheme and prove that it satisfies security notions: secrecy and verifiability and Sect. 5 concludes the paper.

2   Security Definitions

The first formal treatment for the problem of outsourcing cryptographic computations from a weak client to a powerful server was due to Hohenberger and Lysyanskaya [12]. The security model is useful in checking the privacy, efficiency and verifiability probability when outsourcing the task. In this section, we reproduce the security definitions of the Hohenberger-Lysyanskaya framework.

ADVERSARIAL BEHAVIOUR. Assume that an algorithm Alg is run by two parties: a computationally weak and trusted party C (i.e., a client) who invokes a computationally powerful and untrusted party U through oracle queries. An outsource-secure implementation of an algorithm $\mathsf{Alg} = C^U$ is specified by (C, U). where C carries out the tasks by invoking U.

Hohenberger and Lysyanskaya modelled adversary $A = (\mathcal{E}, U')$ and its behaviour in two parts: (i) the adversarial environment $\mathcal{E}$ simulated to send/submit inputs to Alg; (ii) a malicious oracle U' simulated to mimic U. $\mathcal{E}$ and U' can establish a direct communication channel only before agreeing on a joint initial strategy after which the only way they can communicate is by passing the messages through a channel re-directed/monitored by C.

INPUT/OUTPUT SPECIFICATIONS. The following are the forms of information the algorithm's input/output may have:

Secret information possessed by C;

Protected information known to both C and $\mathcal{E}$ but unknown to U'. This protected information is categorized depending upon the honest or adversarial generation of inputs;

Unprotected information known to C, $\mathcal{E}$ and U'.

Definition 1 (Algorithm with IO-outsource). *The outsource algorithm* Alg *obeys the input/output specification if it accepts five inputs and produces three outputs. The honest entity generates the first three inputs and the last two inputs are generated by the environment $\mathcal{E}$. The first three inputs can be further classified based on the information about them available to the adversary $A = (\mathcal{E}, U')$. The first input is the honest, secret input which is unknown to both $\mathcal{E}$ and U'. The second input is the honest, protected input which may be known by $\mathcal{E}$, but is protected from U'. The third input is the honest, unprotected input which may be known by both $\mathcal{E}$ and U'. The fourth input is the adversarial, protected input which may be known by $\mathcal{E}$, but is protected from U'. The fifth input is the adversarial, protected input which may be known by $\mathcal{E}$, but is protected from U'. Similarly, the first, second and third outputs are called secret, protected and unprotected outputs respectively.*

2.1 Outsource-Security Definitions

The following are the security requirements an outsource algorithm should satisfy:

– *Secrecy.* It should be ensured that the malicious environment $\mathcal{E}$ should not learn secret inputs and outputs of the algorithm Alg, although there exist a joint initial strategy between $\mathcal{E}$ and the oracle U'. In the formal definition, it is assumed that a simulator $\mathcal{S}_1$ exists without having access to the secret inputs and simulates the view of $\mathcal{E}$.

- *verifiability.* The malicious oracle $\mathcal{U}'$ does not gain any knowledge about the inputs to Alg even when it mimics the behaviour of $\mathcal{U}$ to $\mathcal{C}$. In the formal definition it is assumed that a simulator $\mathcal{S}_2$ exists without having access to the secret/protected inputs and simulates the view of $\mathcal{U}'$.

The following Definitions 2, 3, 4 and 5 are reproduced from [12].

Definition 2 (Outsource-security). *[12] A pair of algorithms $(\mathcal{C},\mathcal{U})$ is said to be an outsource-secure implementation of an algorithm* Alg *with IO-outsource if:*

Correctness. $\mathcal{C}^{\mathcal{U}}$ *is a correct implementation of* Alg.
Security. *For all probabilistic polynomial-time adversaries $\mathcal{A} = (\mathcal{E},\mathcal{U}')$, there exist probabilistic expected polynomial-time simulators $(\mathcal{S}_1,\mathcal{S}_2)$ such that the following pairs of random variables are computationally indistinguishable.*

Pair One. *($\mathcal{E}$ learns nothing): $EVIEW_{real} \sim EVIEW_{ideal}$.*

The real process: This process proceeds in rounds. Assume that the honestly generated inputs are chosen by a process I. The view that the adversarial environment obtains by participating in the following process:

$$EVIEW_{real}^i = \{\left(istate^i, x_{hs}^i, x_{hp}^i, x_{hu}^i\right) \leftarrow I\left(1^k, istate^{i-1}\right);$$
$$\left(estate^i, j^i, x_{ap}^i, x_{au}^i, stop^i\right) \leftarrow E\left(1^k, EVIEW_{real}^{i-1}, x_{hp}^i, x_{hu}^i\right);$$
$$\left(tstate^i, ustate^i, y_s^i, y_p^i, y_u^i\right) \leftarrow \mathcal{C}^{\mathcal{U}'(ustate^{i-1})}\left(tstate^{i-1}, x_{hs}^{j^i}, x_{hp}^{j^i}, x_{hu}^{j^i}, x_{ap}^i, x_{au}^i\right):$$
$$\left(estate^i, y_p^i, y_u^i\right)\}$$

$$EVIEW_{real} = EVIEW_{real}^i \text{ if } stop^i = TRUE.$$

In round i, The adversarial environment does not have access to the honest inputs $(x_{hs}^i, x_{hp}^i, x_{hu}^i)$ that are picked using an honest, stateful process I. The environment based on its view from last round, chooses the value of its $estate_i$ variable that is used to recall what it did next time it is invoked. Then, among the previously generated honest inputs, the environment chooses a input vector $(x_{hs}^{j^i}, x_{hp}^{j^i}, x_{hu}^{j^i})$ to give it to $\mathcal{C}^{\mathcal{U}'}$. Observe that the environment can specify the index j^i of the inputs but not the values. The environment also chooses the adversarial protected and unprotected input x_{ap}^i and x_{au}^i respectively. It also chooses the boolean variable $stop^i$ that determines whether round i is the last round in this process.
Then, $\mathcal{C}^{\mathcal{U}'}$ is run on inputs $(tstate^{i-1}, x_{hs}^{j^i}, x_{hp}^{j^i}, x_{hu}^{j^i}, x_{ap}^i, x_{au}^i)$ where $tstate^{i-1}$ is $\mathcal{C}$'s previously saved state. The algorithm produces a new state $tstate^i$ for $\mathcal{C}$ along with the secret y_s^i, protected y_p^i and unprotected y_u^i outputs. The oracle $\mathcal{U}'$ is given $ustate^{i-1}$ as input and the current state in saved in $ustate^i$. The view of the real process in round i consists of $estate^i$, and the values y_p^i and y_u^i. The overall view of the environment in the real process is just its view in the last round.

The ideal process:

$$EVIEW^i_{ideal} = \{\left(istate^i, x^i_{hs}, x^i_{hp}, x^i_{hu}\right) \leftarrow I\left(1^k, istate^{i-1}\right);$$

$$\left(estate^i, j^i, x^i_{ap}, x^i_{au}, stop^i\right) \leftarrow E\left(1^k, EVIEW^{i-1}_{ideal}, x^i_{hp}, x^i_{hu}\right);$$

$$\left(astate^i, y^i_s, y^i_p, y^i_u\right) \leftarrow \mathsf{Alg}\left(astate^{i-1}, x^{j^i}_{hs}, x^{j^i}_{hp}, x^{j^i}_{hu}, x^i_{ap}, x^i_{au}\right);$$

$$\left(sstate^i, ustate^i, Y^i_p, Y^i_u, replace^i\right) \leftarrow \mathcal{S}^{\mathcal{U}'(ustate^{i-1})}_1\left(sstate^{i-1}, x^{j^i}_{hp}, x^{j^i}_{hu}, x^i_{ap}, x^i_{au}, y^i_p, y^i_u\right);$$

$$\left(z^i_p, z^i_u\right) = replace^i\left(Y^i_p, Y^i_u\right) + \left(1 - replace^i\right)\left(y^i_p, y^i_u\right):$$

$$\left(estate^i, z^i_p, z^i_u\right)\}$$

$$EVIEW_{ideal} = EVIEW^i_{ideal} \text{ if } stop^i = TRUE.$$

This process also proceeds in rounds. The secret input x^i_{hs} is hidden from the stateful simulator $\mathcal{S}_1$. But, the non-secret inputs produced by the algorithm that is run on all inputs of round i is given to $\mathcal{S}_1$. Now, $\mathcal{S}_1$ decides whether to output the values (y^i_p, y^i_u) generated by the algorithm Alg or replace them with some other values (Y^i_p, Y^i_u). This replacement is captured using the indicator variable $replace^i \in \{0,1\}$. The simulator is allowed to query the oracle $\mathcal{U}'$ which saves its state as in the real experiment.

Pair two *($\mathcal{U}'$ learns nothing):* $UVIEW_{real} \sim UVIEW_{ideal}$.

The view that the untrusted entity $\mathcal{U}'$ obtains by participating in the real process is described in pair one. $UVIEW_{real} = ustate^i \text{ if } stop^i = TRUE$. The ideal process:

$$UVIEW^i_{ideal} = \{\left(istate^i, x^i_{hs}, x^i_{hp}, x^i_{hu}\right) \leftarrow I\left(1^k, istate^{i-1}\right);$$

$$\left(estate^i, j^i, x^i_{ap}, x^i_{au}, stop^i\right) \leftarrow E\left(1^k, estate^{i-1}, x^i_{hp}, x^i_{hu}, y^{i-1}_p, y^{i-1}_u\right);$$

$$\left(astate^i, y^i_s, y^i_p, y^i_u\right) \leftarrow \mathsf{Alg}\left(astate^{i-1}, x^{j^i}_{hs}, x^{j^i}_{hp}, x^{j^i}_{hu}, x^i_{ap}, x^i_{au}\right);$$

$$\left(sstate^i, ustate^i\right) \leftarrow \mathcal{S}^{\mathcal{U}'(ustate^{i-1})}_2\left(sstate^{i-1}, x^{j^i}_{hu}, x^i_{au}\right);$$

$$\left(ustate^i\right)\}$$

$$UVIEW_{ideal} = UVIEW^i_{ideal} \text{ if } stop^i = TRUE.$$

In the ideal process, the stateful simulator $\mathcal{S}_2$ is given with only the unprotected inputs (x^i_{hu}, x^i_{au}), queries $\mathcal{U}'$. As before, $\mathcal{U}'$ may maintain state.

Definition 3 (α−efficient, secure outsourcing). *[12] A pair of algorithms $(\mathcal{C}, \mathcal{U})$ is said to be an α−efficient implementation of an algorithm Alg if $(\mathcal{C}, \mathcal{U})$ is an outsource secure implementation of algorithm Alg and for all inputs x, the running time of $\mathcal{C}$ is $\leq$ an α− multiplicative factor of the running time of $\mathsf{Alg}(x)$*

Definition 4 (β−verifiable, secure outsourcing). *[12] A pair of algorithms $(\mathcal{C}, \mathcal{U})$ is a β−verifiable implementation of an algorithm Alg if $(\mathcal{C}, \mathcal{U})$ is an outsource secure implementation of algorithm Alg and for all inputs x, if $\mathcal{U}'$ deviates from its advertised functionality during the execution of $\mathcal{C}^{\mathcal{U}'}(x)$, $\mathcal{C}$ will detect the error with probability $\geq \beta$*

Definition 5 ((α, β)–outsource-security). *[12] A pair of algorithms $(\mathcal{C}, \mathcal{U})$ is said to be an (α, β)–outsource-secure implementation of an algorithm* Alg *if they are both α–efficient and β–checkable.*

3   On Recent Single-Server Outsourcing Schemes

This section presents security issues with three single-server algorithms proposed in 2017; They are (1) Li *et al.* scheme (**CExp**), (2) Cai *et al.* scheme (**SgExp**) and (3) Zhou *et al.* scheme (**ExpSOS**).

3.1   Li *et al.* Scheme (CExp) and Its Weakness

First, we present Li *et al.* scheme briefly and then show its security weakness of not achieving full verifiability as claimed by the authors.

CExp Algorithm. We use the same notations followed by Li *et al.* [16] to describe their outsourcing algorithm. Let $N = pq$ for two large primes p and q. Let CExp be an algorithm which outputs $u^d \bmod N$ upon accepting $u \in \mathbb{Z}_N^*$ and $d \in \mathbb{Z}_{\phi(N)}^*$ as inputs. The assumption is that the inputs u and d are secret or (honest/adversarial) protected. Hence the inputs need to be computationally blinded (masked) by the delegator $\mathcal{C}$ before passing them to the untrusted server $\mathcal{U}$.

Masking the Inputs. To mask the inputs, the algorithm CExp uses preprocessing technique RandN for efficient generation of pairs of the form $(x, x^e \bmod N)$ for a fixed e. (For more details about the description, analysis and efficiency of these pre-processing techniques, please refer to [4, Sect. 2] and [20].)

The client $\mathcal{C}$ runs RandN to generate four pairs $(g_1, g_1^e), (g_2, g_2^e), (g_3, g_3^e),$ (g_4, g_4^e). Let $v_1 = g_1^e \bmod N, w_1 = g_2^e \bmod N, v_2 = g_3^e \bmod N$, and $w_2 = g_4^e \bmod N$. To logically split base u and exponent d into random looking pieces, the first logical divisions are done as follows:

$$u^d = v_1 c_1^{r_1} w_1 w^{\ell_1} (w^{k_1} g_2)^{t_1}, \tag{1}$$

where $c_1 = g_1/g_2, r_1 = d - e, w = u/g_1, t_1 = r_1 - e$ and $d = \ell_1 + k_1 t_1$. For second logical divisions, $\mathcal{C}$ computes an integer a such that $ad \equiv 1 \bmod \phi(N)$. Then $\mathcal{C}$ computes $r^a \bmod N$ for a randomly chosen $r \in \{2, 10\}$ and sets $u' = u \cdot r^a$.

$$(u')^d = v_2 c_2^{r_1} w_2 (w')^{\ell_1} ((w')^{k_1} g_4)^{t_1}, \tag{2}$$

where $c_2 = g_3/g_4, r_1 = d - e, w' = u'/g_3, t_1 = r_1 - e$ and $d = \ell_1 + k_1 t_1$. It is recommended to choose the random blinding factor t_1 such that $t_1 \geq 2^\lambda$, where λ being the security parameter should be at least 64 bits long.

Queries to $\mathcal{U}$. $\mathcal{C}$ queries $\mathcal{U}$ in random order as follows:

1. $(r_1, c_1) \to c_1^{r_1}$;
2. $(r_1, c_2) \to c_2^{r_1}$;
3. $(\ell_1, w) \to w^{\ell_1}$;
4. $(k_1, w) \to w^{k_1}$;
5. $(\ell_1, w') \to (w')^{\ell_1}$;
6. $(k_1, w') \to (w')^{k_1}$.

Verifying the correctness of $\mathcal{U}$'s outputs. The client $\mathcal{C}$ checks whether

$$r\left(v_1 c_1^{r_1} w_1 w^{\ell_1} (w^{k_1} g_2)^{t_1}\right) \stackrel{?}{=} v_2 c_2^{r_1} w_2 (w')^{\ell_1} ((w')^{k_1} g_4)^{t_1} \tag{3}$$

Recovering $u^d \bmod N$. If the above check passes, $\mathcal{C}$ computes the result as

$$u^d = v_1 c_1^{r_1} w_1 w^{\ell_1} (w^{k_1} g_2)^{t_1}. \tag{4}$$

Otherwise $\mathcal{C}$ outputs error message.

Attack on CExp Algorithm. Li *et al.* claim that their algorithm CExp is 1-verifiable; that is, it allows the client $\mathcal{C}$ to verify the outputs returned by $\mathcal{U}$ with probability 1. We show that with a minimal effort, $\mathcal{U}$ can cheat $\mathcal{C}$ with the malformed outputs and hence CExp offers unfortunately 0-verifiability.

The attacker's strategy is to identify and segregate just 2 out of 6 queries for which the exponent is same. For instance, let $\mathcal{U}$ choose (r_1, c_1) and (r_1, c_2). Note that this separation is easier since each base value c_i is distinct and does not appear twice. Now the malicious $\mathcal{U}$ manipulates the outputs corresponding to these two queries only by multiplying them with a random $\delta \in \{1, N\}$ and proceeds as follows.

1. $(r_1, c_1) \to \delta c_1^{r_1}$;
2. $(r_1, c_2) \to \delta c_2^{r_1}$;
3. $(\ell_1, w) \to w^{\ell_1}$;
4. $(k_1, w) \to w^{k_1}$;
5. $(\ell_1, w') \to (w')^{\ell_1}$;
6. $(k_1, w') \to (w')^{k_1}$.

After receiving the outputs, $\mathcal{C}$ checks if

$$r\left(v_1 \delta c_1^{r_1} w_1 w^{\ell_1} (w^{k_1} g_2)^{t_1}\right) \stackrel{?}{=} v_2 \delta c_2^{r_1} w_2 (w')^{\ell_1} ((w')^{k_1} g_4)^{t_1}.$$

Or equivalently, $r \delta u^d \stackrel{?}{=} \delta (u')^d$.

Since the check has been passed, $\mathcal{C}$ finally computes the unintended output:

$$u^d = v_1 \delta c_1^{r_1} w_1 w^{\ell_1} (w^{k_1} g_2)^{t_1}.$$

3.2   Cai *et al.* Scheme (SgExp) and Its Weakness

For the single untrusted server model, Cai *et al.* [5] proposed two algorithms to securely outsource single and simultaneous modular exponentiations with verifiability probability being close to 1. We show in this section that both their variants fail to detect wrong values output by the malicious server.

SgExp Algorithm. We use the same notations followed by Cai *et al.* to describe their outsourcing algorithm SgExp.

Masking the Inputs. To mask the inputs, the algorithm SgExp used the pre-processing techniques BPV^+ or SMBL to generate four pairs (α, g^α), (β, g^β), (ϵ, g^ϵ), (θ, g^θ) denoted by A, B, C and D respectively. Let $w = u/A \mod p$ and $v = u/C \mod p$. Then C represents u in the following two ways:
  - $u^a = (Aw)^a = g^{a\alpha}w^a = g^\beta g^\gamma w^a \mod p$, where $\gamma = (a\alpha - \beta) \mod q$;
  - $u^a = (Cv)^a = g^{a\epsilon}v^a = g^\theta g^\tau v^a \mod p$, where $\tau = (a\epsilon - \theta) \mod q$.

  To implicitly mask a in w^a and v^a, C randomly chooses i, j such that $2^i \neq 2^j < a$ and computes
  - $a_1 = a - 2^i$
  - $a_2 = a - 2^j$.

Queries to $\mathcal{U}$. C runs BPV^+ or SMBL to generate 8 pairs $(t_1, g^{t_1}), (t_2, g^{t_2})$ and $(s_1, g^{s_1}), (s_2, g^{s_2}) \cdots (s_6, g^{s_6})$. Then C query $\mathcal{U}$ in random order after choosing $m_1, \cdots m_{i-1}, m_{i+1}, \cdots m_{j-1}, m_{j+1}, \cdots m_n$ as follows:
  - $(g^{t_1}, \gamma/t_1, p) \to g^\gamma$;
  - $(wg^{s_1}, a_1, p) \to R_{11} = w^{a_1}g^{s_1a_1}$;
  - $(g^{s_3}, \frac{s_1a_1-s_2}{s_3}, p) \to R_{12} = g^{s_1a_1-s_2}$;
  - $(g^{t_2}, \tau/t_2, p) \to g^\tau$;
  - $(vg^{s_4}, a_2, p) \to R_{21} = v^{a_2}g^{s_4a_2}$;
  - $(g^{s_6}, \frac{s_4a_2-s_5}{s_6}, p) \to R_{22} = g^{s_4a_2-s_5}$;
  - $(m_1, 2) \to m[1] = m_1^2$;
  - $(m_2, 2^2) \to m[2] = m_2^4$;
  - $\cdots$
  - $(w, 2^i) \to m[i] = w^{2^i}$;
  - $\cdots$
  - $(v^{-1}, 2^j) \to m[j] = v^{-2^j}$;
  - $\cdots$
  - $(m_n, 2^n) \to m[n] = m_n^{2^n}$;

Verifying the correctness of $\mathcal{U}$'s outputs. The client C computes

$$w^{a_1} = R_{11}(R_{12}g^{s_2})^{-1}$$

$$v^{a_2} = R_{21}(R_{22}g^{s_5})^{-1}$$

and checks whether

$$Bg^\gamma w^{a_1}m[i]m[j] \mod p \stackrel{?}{\equiv} Dg^\tau v^{a_2} \mod p \tag{5}$$

Recovering u^a. If the above check passes, C computes

$$u^a \equiv Bg^\gamma w^{a_1}m[i] \mod p.$$

Otherwise C outputs the error symbol $\perp$.

Attack on SgExp Algorithm. Cai *et al.* claim that their algorithm SgExp preserves secrecy and the client $\mathcal{C}$ can verify the outputs returned by $\mathcal{U}$ with probability $1 - 1/n^2$. We show that even a minimal effort from $\mathcal{U}$ could lead to cheating the client $\mathcal{C}$ with the malformed outputs and hence SgExp is unfortunately 0-verifiable.

The attacker's strategy is to identify and segregate n out of $n+6$ queries for which the first argument is a power of 2. Then the attacker forms two bins; to fill n number of queries of the form 2^i in one bin and the remaining 6 queries in the other bin. Note that this kind of separation of $n+6$ queries is possible as exponent is a power of 2 in n queries. After segregation of n queries, the adversary manipulates the outputs of remaining 6 queries only by multiplying them with a random $\delta \in \mathbb{G}$ and proceeds as follows.

- $(g^{t_1}, \gamma/t_1, p) \rightarrow \delta g^\gamma$;
- $(wg^{s_1}, a_1, p) \rightarrow R_{11} = \delta w^{a_1} g^{s_1 a_1}$;
- $(g^{s_3}, \frac{s_1 a_1 - s_2}{s_3}, p) \rightarrow R_{12} = \delta g^{s_1 a_1 - s_2}$;
- $(g^{t_2}, \tau/t_2, p) \rightarrow \delta g^\tau$;
- $(vg^{s_4}, a_2, p) \rightarrow R_{21} = \delta v^{a_2} g^{s_4 a_2}$;
- $(g^{s_6}, \frac{s_4 a_2 - s_5}{s_6}, p) \rightarrow R_{22} = \delta g^{s_4 a_2 - s_5}$;
- $(m_1, 2) \rightarrow m[1] = m_1^2$;
- $(m_2, 2^2) \rightarrow m[2] = m_2^4$;
- $\ldots$
- $(w, 2^i) \rightarrow m[i] = w^{2^i}$;
- $\ldots$
- $(v^{-1}, 2^j) \rightarrow m[j] = v^{-2^j}$;
- $\ldots$
- $(m_n, 2^n) \rightarrow m[n] = m_n^{2^n}$;

After receiving the outputs, $\mathcal{C}$ computes

$$w^{a_1} = \delta R_{11}(\delta R_{12} g^{s_2})^{-1} = R_{11}(R_{12} g^{s_2})^{-1}$$
$$v^{a_2} = \delta R_{21}(\delta R_{22} g^{s_5})^{-1} = R_{21}(R_{22} g^{s_5})^{-1}$$

and checks

$$B\delta g^\gamma w^{a_1} m[i] m[j] \mod p \stackrel{?}{=} D\delta g^\tau v^{a_2} \mod p. \tag{6}$$

Since the check has been passed, $\mathcal{C}$ finally computes the undesired output:

$$u^a = B\delta g^\gamma \delta w^{a_1} m[i] \mod p \neq Bg^\gamma w^{a_1} m[i] \mod p.$$

3.3 Zhou *et al.* Scheme (ExpSOS) and Its Weakness

Zhou *et al.* [25] proposed several algorithms for outsourcing variable-exponent variable-base modular exponentiation using only a single untrusted server. In this section, we consider only the most generic algorithm, namely **ExpSOS** under malicious model [25, Sect. IV]. However our observations here are also applicable to other versions in the paper [25].

ExpSOS Algorithm. Let N be either a prime number or an RSA modulus and $u, a \in \mathbb{Z}_N$. The aim of the client $\mathcal{C}$ is to compute $u^a \bmod N$ keeping the variable values u, a and u^a secret. The client runs the oracle $\mathcal{U}$ whose task is to return $i^j \bmod k$ on input (i, j, k). In order to maintain the secrecy of u and a they are computationally masked before being given as input to $\mathcal{U}$.

Masking the Inputs. To mask the inputs, the client $\mathcal{C}$ generates a large prime p and calculates $L = pN$ keeping N and p secret from the server $\mathcal{U}$. By choosing the random integers k_1, k_2, t_1, t_2, r such that $t_1, t_2 \leq b$, (where b is a security parameter) $\mathcal{C}$ calculates the following:
1. $A_1 = a + k_1\phi(N)$
2. $A_2 = t_1 a + t_2 + k_2\phi(N)$
3. $U = u + rN \bmod L$.
Queries to $\mathcal{U}$. $\mathcal{C}$ queries $\mathcal{U}$ in random order as follows:
1. $(U, A_1, L) \rightarrow R_1$
2. $(U, A_2, L) \rightarrow R_2$.
Verifying the correctness of $\mathcal{U}$'s outputs. The client $\mathcal{C}$ checks whether

$$R_1^{t_1} \cdot u^{t_2} \stackrel{?}{\equiv} R_2 \quad \bmod N. \tag{7}$$

Recovering u^a. If the above check passes, $\mathcal{C}$ computes

$$u^a \equiv R_1 \quad \bmod N. \tag{8}$$

Otherwise $\mathcal{C}$ outputs error message.

Attack on ExpSOS Algorithm. The first generic algorithm for outsourcing variable-exponent variable-base modular exponentiation using only a single untrusted server was due to Wang et al. [23]. In [7] Chevalier et al. presented a lattice-based attack on Wang et al.'s scheme recovering the secret exponent when it appears again in another invocation. In this section, we follow the approach of Chevalier et al. and describe a similar attack on **ExpSOS** when the same secret exponent is used in two or more runs. In fact we will show that an exponent in **ExpSOS** can be recovered in polynomial time when two exponentiations having the same exponent are outsourced to the server $\mathcal{U}$. The assumed scenario is evident from the first application proposed in [25, Sect. VI.A] to securely offload Inner Product Encryption for Biometric Authentication [13].

The considered attack scenario is the following: The client wants to compute $u^a \bmod N$ first and $(u')^a \bmod N$ later. Let (U, A_1, A_2) and (U', A_1', A_2') be the queries to $\mathcal{U}$ corresponding to two exponentiations such that

$$A_1 = a + k_1\phi(N); A_2 = t_1 a + t_2 + k_2\phi(N)$$

and

$$A_1' = a + k_3\phi(N); A_2' = t_3 a + t_4 + k_4\phi(N).$$

Now, subtracting the first exponents in two exponentiations gives $A_1 - A_1' = (k_1 - k_3)\phi(N)$. Thus, given a multiple of $\phi(N)$, $\mathcal{U}$ can recover the secret exponent a in polynomial time using the well-known Miller's algorithm. (Miller in [19] showed that factoring of n is possible given any multiple of $\phi(n)$).

Remark 1. The above attack breaks the secrecy of other versions of **ExpSOS** in [25]. In fact it is applicable even for **ExpSOS** under honest-but-curious server model in [25, Sect. III.C]. The malicious server could act benignly in computing the required values but can learn silently any reused secret exponent. In the next section we attempt to thwart this attack under malicious server model since there is no efficiency gain to consider having the semi-honest server.

4   Our Algorithms for Single and Simultaneous Exponentiations

We first present the algorithm for single exponentiation by revising the ExpSOS algorithm and then extend the resulting algorithm for simultaneous exponentiations.

4.1   Improved **ExpSOS** Scheme for Single Exponentiation

In this section we present an improved ExpSOS algorithm which resists attack described in Sect. 3.3. We use the same notations from Sect. 3.3 used to describe the ExpSOS algorithm.

The MExpSOS Algorithm Let N be either a prime number or an RSA modulus and $u, a \in \mathbb{Z}_N$. The aim of the client $\mathcal{C}$ is to compute $u^a \bmod N$ keeping the variable values u, a and u^a secret.

Masking the Inputs. To mask the inputs, the client $\mathcal{C}$ generates a large prime p and calculates $L = pN$ to keep N and p secret from the server $\mathcal{U}$. Select a random r such that $N' = rN$ is fixed for all invocations. By choosing the random integers k_1, k_2, t_1, t_2 such that $t_1, t_2 \le b$ (where b is a security parameter), $\mathcal{C}$ calculates the following:
    1. $a_1 = a - t_1$
    2. $A_1 = a_1 + k_1 \phi(N)$
    3. $A_2 = t_2 a + k_2 \phi(N)$
    4. $U = u + N' \bmod L$

Queries to $\mathcal{U}$. $\mathcal{C}$ queries $\mathcal{U}$ in random order as follows:
    1. $(U, A_1, L) \rightarrow R_1$
    2. $(U, A_2, L) \rightarrow R_2$

Verifying the correctness of $\mathcal{U}$'s outputs. The client $\mathcal{C}$ checks whether

$$(R_1 u^{t_1})^{t_2} \bmod N \stackrel{?}{=} R_2 \bmod N \tag{9}$$

Recovering u^a. If the above check passes, $\mathcal{C}$ computes

$$u^a = R_1 u^{t_1} \bmod N \tag{10}$$

Otherwise $\mathcal{C}$ outputs error message.

Remark 2. The performance gain of the above algorithm is that instead of a *full modular exponentiation* (on the size of an RSA private key, for example), the client device needs to do 2 *smaller* exponentiations with size comparable to the security parameter. Moreover the communication cost and the overhead for the third-party server are not prohibitive compared to the previously-known algorithms. Hence the proposed solution shall directly produce speed-ups in practice.

Lemma 1 *(Correctness).* *In the malicious model, the algorithms* $(\mathcal{C}, \mathcal{U})$ *are correct implementation of* MExpSOS.

Proof. Whenever $\mathcal{U}$ returns R_1 and R_2, $\mathcal{C}$ computes u^{t_1} on its own and then raising the value $R_1 u^{t_1} \mod N$ to the power t_2. Then the resultant $(R_1 u^{t_1})^{t_2} \mod N$ is compared with R_2. If the equality holds, then $\mathcal{C}$ computes the desired result $u^a = R_1 u^{t_1} \mod N$. $\qquad\square$

In the following theorem, we show that $(\mathcal{C}, \mathcal{U})$ is an outsource-secure implementation of MExpSOS using Hohenberger-Lysyanskaya security model for a single malicious server [12].

Theorem 1 *(Privacy).* *In the one malicious program model, the algorithms* $(\mathcal{C}, \mathcal{U})$ *are an outsource-secure implementation of* MExpSOS.

Proof. Assume that $\mathcal{A} = (\mathcal{E}, \mathcal{U}')$ be a probabilistic polynomial time (PPT) adversary which interacts with the PPT algorithm $\mathcal{C}$ in the one malicious program model.

Pair One: $(\mathcal{E}$ learns nothing) $\text{EVIEW}_{real} \sim \text{EVIEW}_{ideal}$

If the input (u, a, N) is honest, protected or adversarial protected, the simulator $\mathcal{S}_1$ behaves the same way as in the real experiment. If the input is honest and secret, then $\mathcal{S}_1$ ignores the received input in the ith round. The goal of $\mathcal{S}_1$ in this ith round is to query $\mathcal{U}'$ with the inputs (U^*, A_1^*, A_2^*, L^*) such that the inputs U^*, A_1^*, A_2^* and L^* are chosen at random by $\mathcal{S}_1$. After receiving the outputs, $\mathcal{S}_1$ saves both the states of $\mathcal{S}_1$ and $\mathcal{U}'$.

In real process, all the inputs that occur in the queries are re-randomized to give computational indistinguishability. Whereas $\mathcal{S}_1$ always set the queries to $\mathcal{U}'$ with the random input values. Hence the input distributions to $\mathcal{U}'$ are computationally indistinguishable both in the real and ideal process.

Pair Two: $(\mathcal{U}'$ learns nothing): $\text{UVIEW}_{real} \sim \text{UVIEW}_{ideal}$:

Let $\mathcal{S}_2$ be a PPT simulator that behaves in the same manner regardless of whether the input is honest, secret or honest, protected or adversarial protected. That is, $\mathcal{S}_2$ ignores the actual input in the ith round and set the queries to $\mathcal{U}'$ with the random value. Then $\mathcal{S}_2$ saves not only its state but also $\mathcal{U}'$'s state.

Whenever the inputs to the experiment are honest, protected and adversarial protected, $\mathcal{E}$ can easily distinguish ith round of two experiments. But it is of no help as $\mathcal{E}$ cannot communicate to $\mathcal{U}'$ and the inputs are computationally blinded by $\mathcal{C}$ before being given as input to $\mathcal{U}'$ in the ideal experiment. In the ideal experiment, the simulator $\mathcal{S}_2$ always query the values selected uniform at random from the same distribution. Hence $\text{UVIEW}^i_{real} \sim \text{UVIEW}^i_{ideal}$ for each round i. By the hybrid argument, it is easy to see that $\text{UVIEW}_{real} \sim \text{UVIEW}_{ideal}$. □

Theorem 2 *(verifiability). In the one malicious program model, the above algorithms* $(\mathcal{C}, \mathcal{U})$ *are an* $(3 + 1.5(\log t_1 + \log t_2), 1 - 1/2b)$*-outsource-secure implementation of* MExpSOS.

Proof. The computation of modular exponentiation $u^a \bmod N$ without outsourcing requires roughly $1.5 \log a$ modular multiplications (MM) using square and multiply method. As discussed in [25], the computational overhead to calculate $\phi(N), L$ and N' becomes negligible when the client $\mathcal{C}$ runs MExpSOS multiple times. The following paragraph shows that with outsourcing, the modular exponentiation computation is reduced to $3 + 1.5(\log t_1 + \log t_2)$ modular multiplications: the computation of A_1 and A_2 during the masking step requires two modular multiplication altogether. Then the verification step requires one modular exponentiation (u^{t_1}), one modular multiplication $(R_1 u^{t_1})$ and one modular exponentiation $((R_1 u^{t_1})^{t_2})$. Thus our algorithm MExpSOS requires 3 modular multiplications and two $b-$bit modular exponentiations. Therefore our algorithm $(\mathcal{C}, \mathcal{U})$ is an $(\frac{1}{2} \log_b a)-$efficient implementation of MExpSOS.

On the other hand the two outputs sent by $\mathcal{U}$ are verified as in Eq. 11. The server $\mathcal{U}$ can trick the client $\mathcal{C}$ if it correctly guesses t_2 as in the following:

- Assume that the malicious $\mathcal{U}$ sets A_1 as $A_1 + \theta$ and A_2 as $A_2 + \theta$
- Then the Eq. 11 becomes

$$\begin{aligned}
(U^{A_1+\theta} u^{t_1})^{t_2} &\equiv U^{A_2+\theta} \quad \bmod N \\
u^{a_1 t_2 + \theta t_2} u^{t_1 t_2} &\equiv u^{t_2 a + \theta} \quad \bmod N \\
u^{a t_2} u^{\theta t_2} &\equiv u^{t_2 a + \theta} \quad \bmod N
\end{aligned}$$

If the value t_2 is correctly guessed then the adversary can compute $u^{\theta t_2}$ and set A_1 as $A_1 + \theta$ and A_2 as $A_2 + \theta t_2$ to pass the verification. If t_2 is guessed with probability $1/b$ and θt_2 is inserted accordingly in one out of two queries sent in random order, the malicious server can pass the verification step with false outputs with probability $\frac{1}{2b}$. Hence our algorithm is a $(1 - \frac{1}{2b})$-verifiable implementation of MExpSOS.

The proof of the theorem completes by combining the above arguments. □

4.2   New Algorithm for Simultaneous Exponentiation

In this section, we present a generic algorithm πExpSOS for simultaneous exponentiation whose complexity grows linearly in size of the number of exponentiations. Simultaneous modular exponentiations appear predominantly in cryptographic primitives such as provable data possession [2] and proofs of retrievability [3].

The πExpSOS Algorithm. Let us follow the notations used to describe the MExpSOS algorithm. Let N be either a prime number or an RSA modulus and $u_i, a_i \in \mathbb{Z}_N$ for $i = 1, \ldots, n$. In order to maintain the secrecy of u_i and $a_i, i = 1, \ldots, n$ they are computationally masked before being given as input to $\mathcal{U}$.

Masking the Inputs. To mask the inputs, the client $\mathcal{C}$ generates a large prime p and calculates $L = pN$ to keep N and p secret from the server $\mathcal{U}$. Select a random r such that $N' = rN$ is fixed for all invocations. By choosing the random integers $k_{1i}, k_{2i}(i = 1, \ldots, n)$ and t_1, t_2 such that $t_1, t_2 \leq b$, $\mathcal{C}$ calculates the following for $i = 1, \ldots, n$:

1. $a_{1i} = a_i - t_1$
2. $A_{1i} = a_{1i} + k_{1i}\phi(N)$
3. $A_{2i} = t_2 a_i + k_{2i}\phi(N)$
4. $U_i = u_i + N' \mod L$

Queries to $\mathcal{U}$. $\mathcal{C}$ issues $2n$ queries to $\mathcal{U}$ in random order as follows:

1. $(U_i, A_{1i}, L) \rightarrow R_{1i}$
2. $(U_i, A_{2i}, L) \rightarrow R_{2i}$

Verifying the correctness of $\mathcal{U}$'s outputs. The client $\mathcal{C}$ checks whether

$$\left[\prod_{i=1}^{n} R_{1i}\left(\prod_{i=1}^{n} u_i\right)^{t_1}\right]^{t_2} \mod N \stackrel{?}{=} \prod_{i=1}^{n} R_{2i} \mod N \tag{11}$$

Recovering u^a. If the above check passes, $\mathcal{C}$ computes

$$\prod_{i=1}^{n} u_i^{a_i} \equiv \prod_{i=1}^{n} R_{1i}\left(\prod_{i=1}^{n} u_i\right)^{t_1} \mod N \tag{12}$$

Otherwise $\mathcal{C}$ outputs error message.

Lemma 2 *(Correctness). In the malicious model, the algorithms $(\mathcal{C}, \mathcal{U})$ are correct implementation of πExpSOS.*

Proof. Whenever $\mathcal{U}$ returns R_{1i} and R_{2i} for $i = 1 \ldots n$, $\mathcal{C}$ computes $(\prod_{i=1}^{n} u_i)^{t_1}$ on its own and then raising the value $\prod_{i=1}^{n} R_{1i}(\prod_{i=1}^{n} u_i)^{t_1} \mod N$ to the power t_2. Then the resultant $(\prod_{i=1}^{n} R_{1i}(\prod_{i=1}^{n} u_i)^{t_1})^{t_2} \mod N$ is compared with $\prod_{i=1}^{n} R_{2i}$. If the equality holds, then $\mathcal{C}$ computes the desired result

$$\prod_{i=1}^{n} u_i^{a_i} \equiv \prod_{i=1}^{n} R_{1i}\left(\prod_{i=1}^{n} u_i\right)^{t_1} \mod N. \qquad \square$$

In the following theorem, we show that $(\mathcal{C}, \mathcal{U})$ is an outsource-secure implementation of πExpSOS using Hohenberger-Lysyanskaya security model for a single malicious server [12].

Theorem 3. *(Privacy). In the one malicious program model, the algorithms* $(\mathcal{C}, \mathcal{U})$ *are an outsource-secure implementation of* πExpSOS.

We omit the proof to this theorem as this can be easily written using Theorem 1

Theorem 4. *(verifiability). In the one malicious program model, the above algorithms* $(\mathcal{C}, \mathcal{U})$ *are an* $(5n - 2 + 1.5(\log t_1 + \log t_2), 1 - 1/2b)$-*outsource-secure implementation of* πExpSOS.

Proof. The computation of modular exponentiation $\prod_{i=1}^{n} u_i^{a_i} \bmod N$ without outsourcing requires roughly $1.5n \log a$ modular multiplications (MM) using square and multiply method. Outsourcing the modular exponentiation computation reduces the cost to $2n + 3(n - 1) + 1.5(\log t_1 + \log t_2) + 1$ modular multiplications as detailed below: the computation of A_{1i} and A_{2i} during the masking step requires $2n$ modular multiplications altogether. Then the verification step requires $n - 1$ modular multiplications to compute $\prod_{i=1}^{n} u_i$, one modular exponentiation to compute $\prod_{i=1}^{n} u_i^{t_1}$, $n - 1$ modular multiplications to compute $\prod_{i=1}^{n} R_{1i}$, one modular multiplication to compute $\prod_{i=1}^{n} R_{1i} \prod_{i=1}^{n} u_i^{t_1}$, one modular exponentiation to compute $(\prod_{i=1}^{n} R_{1i} \prod_{i=1}^{n} u_i^{t_1})^{t_2}$ and $n - 1$ modular multiplications to compute $\prod_{i=1}^{n} R_{2i}$. Thus our algorithm πExpSOS requires $2n + 1 + 3(n - 1) = 5n - 2$ modular multiplications and 2 b-bit modular exponentiations. Therefore our algorithm $(\mathcal{C}, \mathcal{U})$ is an $(\frac{1}{2} \log_b a)$-efficient implementation of πExpSOS.

On the other hand the $2n$ outputs sent by $\mathcal{U}$ are verified as in Eq. 11. The server $\mathcal{U}$ can trick the client $\mathcal{C}$ if it correctly guess t_2 as in explained in Theorem 2 with probability $1/b$. Thus an adversary can pass the verification step with false outputs with probability $\frac{1}{2b}$. Hence our algorithm is a $(1 - \frac{1}{2b})$-verifiable implementation of πExpSOS. The proof of the theorem completes by combining the above arguments. $\square$

5   Conclusion

The need for reducing cost of cryptographic computations is growing especially in the case of devices having resource scarcity. We reviewed several algorithms for offloading single and simultaneous modular exponentiations to a single untrusted helper. In **CExp** and **SgExp** algorithms, we demonstrated that the falsified values of a malicious server could go undetected by the client in the verification and hence the client outputs the unintended value. For **ExpSOS**, we presented a practical attack revealing the secret exponent challenging the claimed security guarantees. We then proposed modifications to the **ExpSOS** algorithm and proved that the resulting algorithm **MExpSOS** meets the fundamental security requirements of the Hohenberger-Lysyanskaya security model. We finally solved an intriguing problem underlined in [6] by proposing π**ExpSOS**, the most efficient to-date algorithm using single untrusted (cloud) server for securely outsourcing (multi-)modular exponentiation. Our proposal being near error-free and preprocessing-free is of both theoretical and practical interest.

References

1. Abadi, M., Feigenbaum, J., Kilian, J.: On hiding information from an oracle. In: Proceedings of the Nineteenth Annual ACM Symposium on Theory of Computing (STOC 1987), pp 195–203. ACM (1987). https://doi.org/10.1145/28395.28417
2. Ateniese, G., et al.: Provable data possession at untrusted stores. In: Proceedings of the 14th ACM Conference on Computer and Communications Security, CCS 2007, pp. 598–609. ACM, New York (2007). https://doi.org/10.1145/1315245.1315318
3. Bowers, K.D., Juels, A., Oprea, A.: Proofs of retrievability: theory and implementation. In: Proceedings of the 2009 ACM Workshop on Cloud Computing Security, CCSW 2009, pp. 43–54. ACM, New York (2009). https://doi.org/10.1145/1655008.1655015
4. Boyko, V., Peinado, M., Venkatesan, R.: Speeding up discrete log and factoring based schemes via precomputations. In: Nyberg, K. (ed.) EUROCRYPT 1998. LNCS, vol. 1403, pp. 221–235. Springer, Heidelberg (1998). https://doi.org/10.1007/BFb0054129
5. Cai, J., Ren, Y., Huang, C.: Verifiable outsourcing computation of modular exponentiations with single server. Int. J. Network Secur. **19**(3), 449–457 (2017). http://ijns.jalaxy.com.tw/download_paper.jsp?PaperID=IJNS-2015-12-05-1&PaperName=ijns-v19-n3/ijns-2017-v19-n3-p449-457.pdf
6. Chen, X., Li, J., Ma, J., Tang, Q., Lou, W.: New algorithms for secure outsourcing of modular exponentiations. In: Foresti, S., Yung, M., Martinelli, F. (eds.) ESORICS 2012. LNCS, vol. 7459, pp. 541–556. Springer, Heidelberg (2012). https://doi.org/10.1007/978-3-642-33167-1_31
7. Chevalier, C., Laguillaumie, F., Vergnaud, D.: Privately outsourcing exponentiation to a single server: cryptanalysis and optimal constructions. In: Askoxylakis, I., Ioannidis, S., Katsikas, S., Meadows, C. (eds.) ESORICS 2016. LNCS, vol. 9878, pp. 261–278. Springer, Cham (2016). https://doi.org/10.1007/978-3-319-45744-4_13
8. Gennaro, R., Gentry, C., Parno, B.: Non-interactive verifiable computing: outsourcing computation to untrusted workers. In: Rabin, T. (ed.) CRYPTO 2010. LNCS, vol. 6223, pp. 465–482. Springer, Heidelberg (2010). https://doi.org/10.1007/978-3-642-14623-7_25
9. Girault, M., Lefranc, D.: Server-aided verification: theory and practice. In: Roy, B. (ed.) ASIACRYPT 2005. LNCS, vol. 3788, pp. 605–623. Springer, Heidelberg (2005). https://doi.org/10.1007/11593447_33
10. Golle, P., Mironov, I.: Uncheatable distributed computations. In: Naccache, D. (ed.) CT-RSA 2001. LNCS, vol. 2020, pp. 425–440. Springer, Heidelberg (2001). https://doi.org/10.1007/3-540-45353-9_31
11. Green, M., Hohenberger, S., Waters, B.: Outsourcing the decryption of ABEciphertexts. In: USENIX Security Symposium 2011. USENIX Association (2011). https://www.usenix.org/publications/proceedings/?f[0]=im_group_audience%3A277
12. Hohenberger, S., Lysyanskaya, A.: How to securely outsource cryptographic computations. In: Kilian, J. (ed.) TCC 2005. LNCS, vol. 3378, pp. 264–282. Springer, Heidelberg (2005). https://doi.org/10.1007/978-3-540-30576-7_15
13. Kim, S., Lewi, K., Mandal, A., Montgomery, H., Roy, A., Wu, D.J.: Function-Hiding Inner Product Encryption is Practical. Cryptology ePrint Archive, Report 2016/440 (2016), accepted at SCN 2018. https://eprint.iacr.org/2016/440
14. Kiraz, M.S., Uzunkol, O.: Efficient and verifiable algorithms for secure outsourcing of cryptographic computations. Int. J. Inf. Secur. **15**(5), 519–537 (2016). https://doi.org/10.1007/s10207-015-0308-7

15. Kuppusamy, L., Rangasamy, J.: CRT-based outsourcing algorithms for modular exponentiations. In: Dunkelman, O., Sanadhya, S.K. (eds.) INDOCRYPT 2016. LNCS, vol. 10095, pp. 81–98. Springer, Cham (2016). https://doi.org/10.1007/978-3-319-49890-4_5

16. Li, S., Huang, L., Fu, A., Yearwood, J.: CExp: secure and verifiable outsourcing of composite modular exponentiation with single untrusted server. Digit. Commun. Netw. 3(4), 236–241 (2017). https://doi.org/10.1016/j.dcan.2017.05.001

17. Matsumoto, T., Kato, K., Imai, H.: Speeding up secret computations with insecure auxiliary devices. In: Goldwasser, S. (ed.) CRYPTO 1988. LNCS, vol. 403, pp. 497–506. Springer, New York (1990). https://doi.org/10.1007/0-387-34799-2_35

18. McLoone, M., Robshaw, M.J.B.: Public key cryptography and RFID tags. In: Abe, M. (ed.) CT-RSA 2007. LNCS, vol. 4377, pp. 372–384. Springer, Heidelberg (2006). https://doi.org/10.1007/11967668_24

19. Miller, G.L.: Riemann's hypothesis and tests for primality. In: Rounds, W.C., Martin, N., Carlyle, J.W., Harrison, M.A. (eds.) Proceedings of the ACM Symposium on Theory of Computing (STOC) 1975, pp. 234–239. ACM (1975). https://doi.org/10.1145/800116.803773

20. Nguyen, P.Q., Shparlinski, I.E., Stern, J.: Distribution of modular sums and the security of the server aided exponentiation. In: Proceedings of the Cryptography and Computational Number Theory Workshop, pp. 257–268. Birkhäuser (2001). https://doi.org/10.1007/978-3-0348-8295-8_24

21. Schnorr, C.P.: Efficient identification and signatures for smart cards. In: Brassard, G. (ed.) CRYPTO 1989. LNCS, vol. 435, pp. 239–252. Springer, New York (1990). https://doi.org/10.1007/0-387-34805-0_22

22. Schnorr, C.P.: Efficient signature generation by smart cards. J. Cryptology 4(3), 161–174 (1991). https://doi.org/10.1007/BF00196725

23. Wang, Y., et al.: Securely outsourcing exponentiations with single untrusted program for cloud storage. In: Kutyłowski, M., Vaidya, J. (eds.) ESORICS 2014. LNCS, vol. 8712, pp. 326–343. Springer, Cham (2014). https://doi.org/10.1007/978-3-319-11203-9_19

24. Wu, W., Mu, Y., Susilo, W., Huang, X.: Server-aided verification signatures: definitions and new constructions. In: Baek, J., Bao, F., Chen, K., Lai, X. (eds.) ProvSec 2008. LNCS, vol. 5324, pp. 141–155. Springer, Heidelberg (2008). https://doi.org/10.1007/978-3-540-88733-1_10

25. Zhou, K., Afifi, M.H., Ren, J.: ExpSOS: secure and verifiable outsourcing of exponentiation operations for mobile cloud computing. IEEE Trans. Inf. Forens. Secur. 12(11), 2518–2531 (2017). https://doi.org/10.1109/TIFS.2017.2710941

Keyword Search Meets Membership Testing: Adaptive Security from SXDH

Sanjit Chatterjee and Sayantan Mukherjee[✉]

Department of Computer Science and Automation, Indian Institute of Science,
Bangalore, India
{sanjit,sayantanm}@iisc.ac.in

Abstract. Searchable encryption (SE) allows users to securely store sensitive data in encrypted form on cloud and at the same time perform keyword search over the encrypted documents. In this work, we focus on variants of SE schemes that along with keyword search, also support membership testing. The problem can be formulated in two flavors depending on whether the *search policy* is encoded in the ciphertext or in the trapdoor. The ciphertext-policy variant is called Broadcast Encryption with Keyword Search (BEKS) and allows only privileged users to perform keyword search on an encrypted file. Available dedicated constructions could achieve selective security under parameterized assumption. The key-policy variant, called Key-Aggregate Searchable Encryption (KASE), restricts the keyword search within a particular set of documents. Naive application of existing SE schemes in this scenario leads to inefficient protocols with either variable length trapdoor or exponential blowup of storage requirement in terms of the document set size. This therefore calls for an efficient solution that allows such subset based restricted search with constant trapdoor size.

In this work, we have presented adaptively secure solutions for both the above problems. Our BEKS construction achieves constant-size ciphertext whereas the KASE construction achieves constant-size trapdoor. Both the constructions are instantiated in prime-order bilinear groups and are proven anonymous CPA-secure under SXDH assumption by extending Jutla-Roy technique. Our proposed solutions improve upon the only other adaptively secure schemes that can be obtained using the generic technique of Ambrona et al.

1 Introduction

The advent of cloud has opened up the possibility of achieving a lot of functionalities without virtually any restriction on storage space and/or computational power. However, storing data in cloud naturally invites concerns regarding confidentiality. While conventional encryption schemes do provide confidentiality, they effectively destroy any possibility of performing computation in the encrypted domain. Several flavors of searchable encryption have been suggested in the literature [5,10,13,21] to address this question. In order to compute simple predicate such as searching a keyword in public key settings, Boneh et al. [3]

© Springer Nature Switzerland AG 2018
D. Chakraborty and T. Iwata (Eds.): INDOCRYPT 2018, LNCS 11356, pp. 21–43, 2018.
https://doi.org/10.1007/978-3-030-05378-9_2

proposed Public-Key Encryption with Keyword Search (PEKS). Given a trapdoor corresponding to a keyword and some encrypted text, PEKS allows a third party cloud to run the so-called Test algorithm to find whether the keyword is present in the encrypted text or not without learning any other information.

Attrapadung et al. [2] introduced a generalization of PEKS involving Broadcast Encryption (BE) [4] along with (hierarchical) identity-based encryption ((H)IBE) [11] for universe $\mathcal{U}$ and identity space $\mathcal{ID}$. Every user in [2] will have two identifiers – one is an index $x \in \mathcal{U}$ for broadcast encryption while the other is a (hierarchical) identity id $\in \mathcal{ID}^\ell$ corresponding to (H)IBE. Each user gets a key associated with both the identifiers (x, id) and can decrypt a ciphertext associated with a privileged set $\ddot{\Omega}$ and a (hierarchical) identity id$'$ if $x \in \ddot{\Omega}$ and id $\prec$ id$'$ where $\ddot{\Omega} \subset \mathcal{U}$, id, id$' \in \mathcal{ID}^\ell$, $\prec$ denotes the prefix relation and $\ell \in \mathbb{N}$. Attrapadung et al. [2] termed this primitive as (Hierarchical) Identity-Coupled Broadcast Encryption ((H)ICBE) and kick-started the study of amalgamation of broadcast encryption with other primitives. Such an amalgamation is useful in solving some real world applications of the following type.

Suppose an encrypted file sharing system, where files are stored in the cloud encrypted for some privileged set of users. A user can access a file if s/he is among the privileged set of users for that file. Boneh et al. [4] casted their broadcast encryption construction to achieve such a mechanism. Now, consider a scenario where a privileged user (x) wants to employ the cloud server to search whether a certain keyword is present in the file stored in the cloud. In other words, user x wants to find files containing keyword (ω) gives trapdoor (SK_x) to cloud server where x $= (x, \omega)$. The cloud server then searches the encrypted files and returns those files that contains the keyword ω and include x as privileged user. The security requirement of such a scheme is that neither the cloud server nor any unprivileged user will get to learn anything new about the file content. Attrapadung et al. [2] called such a scheme as Broadcast Encryption with Keyword Search (BEKS) and presented a construction from anonymous Identity-Coupled Broadcast Encryption (ICBE).

Note that in the application scenario described above, the search policy is decided while generating the encrypted file. A somewhat complementary problem of interest is where the trapdoor-index of a user is x $= (\Omega, \omega)$ for $\Omega \subset \mathcal{U}$ and the search policy is encoded in the corresponding trapdoor. To appreciate the problem better, first recall that any conventional searchable encryption (SE), roughly speaking, provides an all-or-nothing capability of search. Hence, given a trapdoor, there is no way to restrict the trapdoor holder within a particular set of documents. Naive solutions to restrict search for a keyword (ω) within a (permitted) set (Ω), either require larger trapdoor (dependent on the cardinality of Ω) or suffers from exponential blowup in the storage space [9]. The reason for this sort of blow-up is that the trapdoor $(\mathsf{SK}_{\Omega,\omega})$ must contain information about the set Ω it is restricted to and Ω could be any of the subsets of all files stored in the cloud.

1.1   Related Works

Boneh et al. [4] suggested few applications and extensions of broadcast encryption. Attrapadung et al. [2] presented two constructions for (hierarchical) identity-coupled broadcast encryption ((H)ICBE) and constructed the first broadcast encryption with keyword search from anonymous ICBE. Kiayias et al. [15] recently proposed another construction for BEKS where some unnatural restrictions are placed on the adversary's capability in order to achieve some form of function privacy. Also note that both [2,15] achieved selective security under parameterized q-type assumptions.

Related to the other problem, Cui et al. [9] introduced the notion of Key-Aggregate Searchable Encryption (KASE) based on the compact-key cryptosystem by Chu et al. [8]. Recently, another construction of KASE was proposed [17] and was claimed to be secure in the random oracle model.

Ambrona et al. in a recent work [1], proposed generic construction for different predicate encodings. Earlier, Chen et al. [6] proposed a compiler that constructs predicate encryption from predicate encoding in the prime-order bilinear groups. Thus, one can employ [1] on top of the generic technique of [6] to construct a BEKS or KASE. However, like any other generic constructions, we pay the price in terms of bigger public parameter, ciphertext and/or trapdoor size.

Jutla and Roy [14] proposed quasi-adaptive non-interactive zero knowledge (QA-NIZK) for linear subspaces over pairing groups. Unlike usual NIZK, the common reference string (CRS) in [14] can depend on the language. This effectively is quite a powerful tool as Jutla and Roy constructed a number of primitives like signature, CCA2-PKE, IBE, commitment etc. Later Kiltz and Wee [16] presented more intuitive approach towards various forms of QA-NIZK. JR-IBE, the identity-based encryption construction of [14] actually spawns from a variation of quasi-adaptive non-interactive zero knowledge called split-CRS QA-NIZK for tag-based languages. The verification CRS here is split into two components that will be combined during the verification by means of public tag. A number of follow-up works by Ramanna et al. [18–20] explored the portability of JR-IBE for different functionalities such as HIBE, IBBE and IPE respectively.

Chen and Gong [7] recently generalized the predicate encryption constructed from JR-IBE [18–20] that achieves better efficiency than [6]. Generic merger of [1] with [7] can also be casted to get efficient constructions for specific functionalities like BEKS and KASE.

1.2   Our Contribution

In this work, we present two constructions that respectively deal with both flavors of keyword search with membership testing: Broadcast Encryption with Keyword Search (BEKS) and Key-Aggregate Searchable Encryption (KASE). Note that, the ciphertext of BEKS is generated on data-index $\mathsf{y} = (\ddot{\Omega}, \omega')$ and trapdoor is generated on trapdoor-index $\mathsf{x} = (\varkappa, \omega)$ where $\varkappa \in \mathcal{U}$, $\ddot{\Omega} \subset \mathcal{U}$, $\omega, \omega' \in \mathcal{W}$. In case of KASE, the ciphertext is generated on data-index $\mathsf{y} = (\varkappa, \omega')$

and trapdoor is generated on trapdoor-index $\mathsf{x} = (\Omega, \omega)$ where $\varkappa \in \mathcal{U}$, $\Omega \subset \mathcal{U}$, $\omega, \omega' \in \mathcal{W}$ for $\mathcal{W}$ being the keyword space.

Both the proposed constructions achieve adaptive IND-CPA security as well as anonymity in standard model under SXDH assumption. The BEKS construction achieves constant size ciphertext whereas the KASE construction achieves constant size trapdoor. The non-triviality of these constructions comes from the fact that the ciphertext in BEKS (resp. the trapdoor in KASE) encodes the *privileged set* information employing a (small) constant number of components without recourse to random oracle.

Our BEKS construction uses GW-Hash ($\sum_{i \in \ddot{\Omega}} w_i$) [12] to encode the privileged set $\ddot{\Omega}$ where $(w_i)_i$ is the randomness shared between public parameter, trapdoor and ciphertext. We then merge ω' with the original GW-Hash via an affine relation to construct the ciphertext. Precisely, we hash both $\ddot{\Omega}$ (via GW-Hash) and ω' together to form an affine equation which effectively constructs our constant-size ciphertext out of data-index $\mathsf{y} = (\ddot{\Omega}, \omega')$. The trapdoors here are constructed in such a way that during decryption, given the set $\ddot{\Omega}$, one can choose appropriate components of the trapdoor thereby causing the trapdoor to contain $\mathsf{K}_1, \ldots, \mathsf{K}_n$ elements. More precisely, $\mathsf{K}_\varkappa$ of the trapdoor for trapdoor-index $\mathsf{x} = (\varkappa, \omega)$, contains another affine equation on ω (i.e. $w_{n+1}\omega + w_\varkappa$). Rest of the trapdoor components are given by w_i for $i \in [n] \setminus \{\varkappa\}$.

In case of KASE construction, we do the opposite namely we create such an affine equation for the trapdoor generation for trapdoor-index $\mathsf{x} = (\Omega, \omega)$. We add the affine equation to master secret key to generate a trapdoor that is in a way a secret sharing of the master secret key α. The ciphertext here contains $\mathsf{C}_1, \ldots, \mathsf{C}_n$ similar to trapdoor in BEKS. The $\mathsf{C}_\varkappa$ here also stores an affine equation $(w_{n+1}\omega' + w_\varkappa)$ whereas the rest of the ciphertext components are given by w_i for $i \in [n] \setminus \{\varkappa\}$ for data-index $\mathsf{y} = (\varkappa, \omega')$.

The JR-IBE [14] and all its descendants [18–20] used tags to argue the reduction via dual system encryption [22]. As noted earlier, Chen and Gong [7] generalized the approach by presenting a compiler that constructs predicate encryption from predicate encoding in the prime-order bilinear settings. Moreover, given an *attribute-hiding* predicate encoding, this compiler [7] can construct corresponding anonymous predicate encryption. Thus, one can simply use predicate encoding construction technique of Ambrona et al. [1] on the top of compiler by Chen and Gong [7] to achieve various generic predicate encryptions (including BEKS and KASE). We however present two dedicated constructions for BEKS and KASE that further improves the complexity of both the protocols. For that, our starting point was Ramanna's inner-product encryption constructions [18]. Our BEKS construction is a constant-size ciphertext construction whereas the KASE construction achieves constant-size trapdoor. Both these constructions achieve anonymity as well with smaller parameter size and are proven to be adaptive secure under SXDH assumption.

We have casted the generic construction of Ambrona et al. [1] on [6] and on [7] in the setting of BEKS and KASE and compared their performance with our schemes in Table 1. As opposed to the merger of [1,6], our constructions achieve exciting parameter size. We also achieve some improvement over generic integration of [1,7] in this respect as can be seen next.

Table 1. Comparison of Efficiency with [1] in terms of size of public parameter $|mpk|$, size of trapdoor $|SK|$, size of ciphertext $|CT|$ and number of primitive operations required in Test. Here $n = |\mathcal{U}|$, [P] denotes number of pairing operations, [M] denotes number of group element multiplications and [E] denotes number of group element exponentiations.

	$|mpk|$	$|SK|$	$|CT|$	Test
BEKS [1,6]	$(2n+6)G_1$	$(2n+4)G_2$	$6G_1 + G_T$	$4[P] + (2|\Omega|+2)[M]$
BEKS [1,7]	$(n+6)G_1$	$(2n+5)G_2$	$4G_1 + G_T + 2\mathbb{Z}_p$	$3[P] + (|\Omega|+3)[M] + 2[E]$
BEKS	$(n+3)G_1$	$(2n+3)G_2$	$3G_1 + G_T + \mathbb{Z}_p$	$3[P] + (|\Omega|+2)[M] + 2[E]$
KASE [1,6]	$(2n+8)G_1$	$8G_2$	$(2n+4)G_1 + G_T$	$4[P] + (2|\Omega|+6)[M]$
KASE [1,7]	$(n+7)G_1$	$9G_2$	$(n+3)G_1 + G_T + (n+1)\mathbb{Z}_p$	$3[P] + (|\Omega|+4)[M] + 2[E]$
KASE	$(n+3)G_1$	$5G_2$	$(n+2)G_1 + G_T + n\mathbb{Z}_p$	$3[P] + (|\Omega|+2)[M] + 2[E]$

As a minor contribution, we have shown a simple attack on the recent KASE construction of [17].

1.3   Organization of the Paper

In Sect. 2, we discuss the preliminaries such as notations, searchable encryption definition. In Sect. 3, we present the construction of BEKS and its security proof. Sect. 4 discusses secure construction of KASE. Section 5 concludes the paper. In Appendix A we show that the KASE construction of [17] to be insecure.

2   Preliminaries

Notations. Here we denote $[a,b] = \{i \in \mathbb{N} : a \leq i \leq b\}$ and for any $n \in \mathbb{N}$, $[n] = [1,n]$. The security parameter is denoted by 1^λ where $\lambda \in \mathbb{N}$. By $s \hookleftarrow S$ we denote a uniformly random choice s from S. By $\mathfrak{P}(S)$ we denote the power set of set S. Here, $\mathcal{U}$ is the universe, $\mathcal{W}$ is set of all keywords and $\mathcal{W}$ is set of all identifiers. We use $\mathsf{Adv}_{\mathcal{A}}^i(\lambda)$ to denote the advantage adversary $\mathcal{A}$ has when deciding the b in security game Game_i and $\mathsf{Adv}_{\mathcal{A}}^{\mathsf{HP}}(\lambda)$ is used to denote the advantage of $\mathcal{A}$ to solve the hard problem HP. Here $\mathsf{neg}(\lambda)$ denotes a negligible function.

2.1   Searchable Encryption

We give a definition of Searchable Encryption (SE) that generalizes all searchable encryption in public key settings (e.g. PEKS [3], BEKS [2] etc.).

This definition resembles to the definition of predicate encryption that generalized identity-based encryption and its descendants. For a predicate function $R : \mathcal{X} \times \mathcal{Y} \rightarrow \{0,1\}$, the searchable encryption is defined as a collection of following four probabilistic polynomial-time algorithms.

- KeyGen: It takes 1^λ and outputs public parameter mpk and master secret key msk.
- Trapdoor: It takes mpk, msk and a trapdoor-index $x \in \mathcal{X}$ and outputs a trapdoor $SK \in \mathcal{SK}$ corresponding to x.
- SrchEnc: It takes mpk and a data-index $y \in \mathcal{Y}$ and outputs ciphertext $CT \in \mathcal{C}$ corresponding to y and encapsulation key $\mathfrak{K} \in \mathcal{K}$.
- Test: It takes mpk, SK and $(CT, \mathfrak{K})$ as input. Outputs $\mathfrak{b} \in \{0,1\}$.

Correctness. For all $(mpk, msk) \leftarrow KeyGen(1^\lambda)$, all key-indices $x \in \mathcal{X}$, $SK \leftarrow Trapdoor(mpk, msk, x)$, all data-indices $y \in \mathcal{Y}$, $(CT, \mathfrak{K}) \leftarrow SrchEnc(mpk, y)$,

$$\Pr[Test(mpk, SK, (CT, \mathfrak{K})) = \mathfrak{b}] = 1 \iff R(x, y) = \mathfrak{b}.$$

For simplicity of presentation, we will sometimes adhere to implicit use of mpk by different algorithms. We sometime use the notation CT_y (resp. SK_x) to denote the ciphertext (resp. trapdoor) corresponding to data-index y (resp. trapdoor-index x).

2.2 Security of Searchable Encryption

The security of a searchable encryption scheme SE can be modeled as a security game between challenger $\mathcal{C}$ and adversary $\mathcal{A}$.

- **Setup:** $\mathcal{C}$ gives out mpk and keeps msk as secret.
- **Query Phase-I:** Queries are performed to available oracles as follows.
  - **Trapdoor Queries:** Given trapdoor-index x, the trapdoor oracle $\mathcal{O}_K$ returns $SK \leftarrow Trapdoor(msk, x)$.
- **Challenge:** $\mathcal{A}$ provides challenge data-index y^* (such that $R(x, y^*) = 0$ for any trapdoor query on x). $\mathcal{C}$ generates $(CT_0^*, \mathfrak{K}_0) \leftarrow SrchEnc(mpk, y^*)$ and chooses $(CT_1^*, \mathfrak{K}_1) \hookleftarrow \mathcal{C} \times \mathcal{K}$. Then it returns $(CT_\mathfrak{b}^*, \mathfrak{K}_\mathfrak{b})$ as challenge for $\mathfrak{b} \hookleftarrow \{0,1\}$.
- **Query Phase-II:** Queries are performed to available oracles as follows.
  - **Trapdoor Queries:** Given a trapdoor-index x such that $R(x, y^*) = 0$, trapdoor oracle $\mathcal{O}_K$ returns $SK \leftarrow Trapdoor(msk, x)$.
- **Guess:** $\mathcal{A}$ outputs its guess $\mathfrak{b}' \in \{0,1\}$ and wins if $\mathfrak{b} = \mathfrak{b}'$.

For any adversary $\mathcal{A}$ the advantage is,

$$Adv_{\mathcal{A}}^{SE}(\lambda) = |\Pr[\mathfrak{b} = \mathfrak{b}'] - 1/2|.$$

A searchable encryption scheme is said to be Id-CPA secure if for any efficient adversary $\mathcal{A}$, $Adv_{\mathcal{A}}^{SE}(\lambda) \leq neg(\lambda)$. If there is an **Init** phase before the **Setup** where the adversary $\mathcal{A}$ commits to the challenge data-index y^*, we call such security model as sId-CPA security model.

Remark 1. Both BEKS and KASE deal with two different predicates, namely a set-membership testing and an equality testing. The BEKS can be used to perform keyword search in secure file system and the KASE can be used to perform keyword search within a restricted set of files. It is therefore essential that, in both BEKS and KASE, the ciphertext should not give out any new information about the associated keyword. In case of BEKS, as mentioned earlier, the challenge data-index is $\mathsf{y}^* = (\ddot{\Omega}^*, \omega'^*)$. Following the security notion of anonymous (H)ICBE [2], we call a BEKS anonymous, if the challenge ciphertext $(\mathsf{CT}^*_b, \mathfrak{K}_b)$ hides ω'^* (i.e. associated challenge keyword). The (H)ICBE construction of Attrapadung et al. [2] is only sId-CPA secure scheme in this sense. Notice that, the other part of the data-index i.e. $\ddot{\Omega}^*$ must accompany the ciphertext in plaintext form for a correct decryption. This is quite natural as BEKS is a descendant of broadcast encryption where it is customary that the privileged set information is given out in plain for decryption purpose. Recall that, the challenge data-index in case of KASE is $\mathsf{y}^* = (\varkappa^*, \omega'^*)$. Unlike the case of BEKS, we call a KASE anonymous [9], if the challenge ciphertext $(\mathsf{CT}^*_b, \mathfrak{K}_b)$ hides both $\varkappa^*$ and ω'^*.

2.3 Hardness Assumption

Let $(p, \mathsf{G}_1, \mathsf{G}_2, \mathsf{G}_T, e) \leftarrow \mathcal{G}_{\mathsf{abg}}(1^\lambda)$ be the output of asymmetric bilinear group generator where $\mathsf{G}_1, \mathsf{G}_2, \mathsf{G}_T$ are cyclic groups of order a large prime p.

Symmetric External Diffie-Hellman Assumption (SXDH). The SXDH representation that is used in our work was introduced in [19]. The SXDH assumption in group $(\mathsf{G}_1, \mathsf{G}_2)$ is: DDH in G_1 and DDH in G_2 is hard. We rewrite DDH in G_1 in the form of 1-Lin assumption below and call it $\mathsf{DDH}_{\mathsf{G}_1}$. The $\mathsf{DDH}_{\mathsf{G}_2}$ denotes the hardness of DDH in G_2.

- $\mathsf{DDH}_{\mathsf{G}_1}$: $\{D, T_0\} \approx_{\epsilon_{\mathsf{DDH}_{\mathsf{G}_1}}} \{D, T_1\}$ for $T_0 = g_1^s$ and $T_1 = g_1^{s+\hat{s}}$ given $D = (g_1, g_2, g_1^b, g_1^{bs})$ where $g_1 \hookleftarrow \mathsf{G}_1$, $g_2 \hookleftarrow \mathsf{G}_2$, $b \hookleftarrow \mathbb{Z}_p^\times$, $s, \hat{s} \hookleftarrow \mathbb{Z}_p$. In other words, the advantage of any adversary $\mathcal{A}$ to solve the $\mathsf{DDH}_{\mathsf{G}_1}$ is

$$\mathsf{Adv}_{\mathcal{A}}^{\mathsf{DDH}_{\mathsf{G}_1}}(\lambda) = |\Pr[\mathcal{A}(D, T_0) \to 1] - \Pr[\mathcal{A}(D, T_1) \to 1]| \leq \epsilon_{\mathsf{DDH}_{\mathsf{G}_1}}.$$

$\mathsf{DDH}_{\mathsf{G}_1}$ is hard if advantage of $\mathcal{A}$ is negligible i.e. $\epsilon_{\mathsf{DDH}_{\mathsf{G}_1}} \leq \mathsf{neg}(\lambda)$.
- $\mathsf{DDH}_{\mathsf{G}_2}$: $\{D, T_0\} \approx_{\epsilon_{\mathsf{DDH}_{\mathsf{G}_2}}} \{D, T_1\}$ for $T_0 = g_2^{cr}$ and $T_1 = g_2^{cr+\hat{r}}$ given $D = (g_1, g_2, g_2^c, g_2^r)$ where $g_1 \hookleftarrow \mathsf{G}_1$, $g_2 \hookleftarrow \mathsf{G}_2$, $c, r, \hat{r} \hookleftarrow \mathbb{Z}_p$. In other words, the advantage of any adversary $\mathcal{A}$ to solve the $\mathsf{DDH}_{\mathsf{G}_2}$ is

$$\mathsf{Adv}_{\mathcal{A}}^{\mathsf{DDH}_{\mathsf{G}_2}}(\lambda) = |\Pr[\mathcal{A}(D, T_0) \to 1] - \Pr[\mathcal{A}(D, T_1) \to 1]| \leq \epsilon_{\mathsf{DDH}_{\mathsf{G}_2}}.$$

$\mathsf{DDH}_{\mathsf{G}_2}$ is hard if advantage of $\mathcal{A}$ is negligible i.e. $\epsilon_{\mathsf{DDH}_{\mathsf{G}_2}} \leq \mathsf{neg}(\lambda)$.

3   BEKS: Broadcast Encryption with Keyword Search

We now present our first construction that is BEKS. It is instantiated in the prime order bilinear groups and achieves anonymous adaptive CPA (Id-CPA) security under the SXDH assumption. Recall that the Trapdoor takes trapdoor-index $x = (\varkappa, \omega)$ and SrchEnc takes data-index $y = (\ddot{\Omega}, \omega')$. BEKS is defined via predicate function $R_v : (\mathcal{U} \times \mathcal{W}) \times (\mathfrak{P}(\mathcal{U}) \times \mathcal{W}) \to \{0,1\}$. The index satisfies the predicate function R_v if $\varkappa \in \ddot{\Omega}$ and $\omega = \omega'$. On such occasion, the Test outputs 1.

3.1   Construction

BEKS for universe $\mathcal{U} = [n]$ where $n = \mathsf{poly}(\lambda)$ and identity space $\mathcal{W}$ of size $\mathsf{exp}(\lambda)$ is defined as following four algorithms.

- KeyGen($1^\lambda, \mathcal{U}, \mathcal{W}$): The asymmetric bilinear group generator outputs $(p, \mathbb{G}_1, \mathbb{G}_2, \mathbb{G}_T, e) \leftarrow \mathcal{G}_{\mathsf{abg}}(1^\lambda)$ where $\mathbb{G}_1, \mathbb{G}_2, \mathbb{G}_T$ are cyclic groups of order p. Choose generators $g_1 \hookleftarrow \mathbb{G}_1$ and $g_2 \hookleftarrow \mathbb{G}_2$ and define $g_T = e(g_1, g_2)$. Choose $\alpha_1, \alpha_2, c, d, (u_i, v_i)_{i \in [n+1]} \hookleftarrow \mathbb{Z}_p$, $b \hookleftarrow \mathbb{Z}_p^\times$ and define $\alpha = (\alpha_1 + b\alpha_2)$ to set $g_T^\alpha = e(g_1, g_2)^{(\alpha_1 + b\alpha_2)}$. For $i \in [n+1]$, define $g_1^{w_i} = g_1^{u_i + bv_i}$ and $g_1^w = g_1^{c+bd}$. Define the $\mathsf{msk} = (g_2, g_2^c, \alpha_1, \alpha_2, d, (u_i, v_i)_{i \in [n+1]})$ and the public parameter is defined as

$$\mathsf{mpk} = \left(g_1, g_1^b, (g_1^{w_i})_{i \in [n+1]}, g_1^w, g_T^\alpha \right).$$

- Trapdoor($\mathsf{msk}, x = (\varkappa, \omega)$): Given trapdoor-index x such that $\varkappa \in \mathcal{U}$ and $\omega \in \mathcal{W}$, choose $r \hookleftarrow \mathbb{Z}_p$ and compute the trapdoor $\mathsf{SK}_x = (\mathsf{K}_1, \mathsf{K}_2, (\mathsf{K}_{3,i})_{i \in [n]}, \mathsf{K}_4, (\mathsf{K}_{5,i})_{i \in [n]})$ where

$$\mathsf{K}_1 = g_2^r, \mathsf{K}_2 = g_2^{cr}, \mathsf{K}_{3,i} = \begin{cases} g_2^{\alpha_1 + r(u_{n+1}\omega + u_\varkappa)} & \text{if } i = \varkappa \\ g_2^{ru_i} & \text{otherwise} \end{cases},$$

$$\mathsf{K}_4 = g_2^{dr}, \mathsf{K}_{5,i} = \begin{cases} g_2^{\alpha_2 + r(v_{n+1}\omega + v_\varkappa)} & \text{if } i = \varkappa \\ g_2^{rv_i} & \text{otherwise} \end{cases}.$$

- SrchEnc($\mathsf{mpk}, y = (\ddot{\Omega}, \omega')$): Given data-index y such that $\ddot{\Omega} \subseteq \mathcal{U}$ and $\omega' \in \mathcal{W}$, choose $s, \ddot{t} \hookleftarrow \mathbb{Z}_p$. Compute $\mathfrak{K} = e(g_1, g_2)^{\alpha s}$ and $\mathsf{CT}_y = (\mathsf{C}_0, \mathsf{C}_1, \mathsf{C}_2, \ddot{t})$, where,

$$\mathsf{C}_0 = g_1^s, \mathsf{C}_1 = g_1^{bs}, \mathsf{C}_2 = g_1^{s\left(w_{n+1}\omega' + \sum_{l \in \ddot{\Omega}} w_l + w\ddot{t}\right)}.$$

- Test($\mathsf{SK}_x, (\mathsf{CT}_y, \mathfrak{K}, \ddot{\Omega})$): Let $x = (\varkappa, \omega)$ and $y = (\ddot{\Omega}, \omega')$; output 1 iff $\mathfrak{K} = B/A$ where,

$$A = e(\mathsf{C}_2, \mathsf{K}_1), \quad B = e\left(\mathsf{C}_0, \mathsf{K}_2^{\ddot{t}} \prod_{i \in \ddot{\Omega}} \mathsf{K}_{3,i}\right) e\left(\mathsf{C}_1, \mathsf{K}_4^{\ddot{t}} \prod_{i \in \ddot{\Omega}} \mathsf{K}_{5,i}\right).$$

It is easy to verify correctness of this protocol which is omitted due to space constraints.

3.2 Security Proof

Theorem 1. *For any adversary $\mathcal{A}$ of BEKS construction BEKS in the Id-CPA model (ID-CPA) that makes at most q many trapdoor queries, there exist PPT adversaries $\mathcal{B}_1$, $\mathcal{B}_2$ such that*

$$\mathsf{Adv}^{\mathsf{BEKS}}_{\mathcal{A},\mathsf{ID\text{-}CPA}}(\lambda) \leq \mathsf{Adv}^{\mathsf{DDH}_{\mathsf{G}_1}}_{\mathcal{B}_1}(\lambda) + q \cdot \mathsf{Adv}^{\mathsf{DDH}_{\mathsf{G}_2}}_{\mathcal{B}_2}(\lambda) + 3/p.$$

Proof Sketch. We propose a hybrid argument based proof that uses dual system proof technique [22] at its core. In case of dual system encryption, the protocol creates *normal* ciphertext and *normal* trapdoor. For the sake of security argument, another form of ciphertext and trapdoors are defined that are called *semi-functional*. By definition, a semi-functional trapdoor cannot decrypt a semi-functional ciphertext even if their associate indexes satisfy each other. The proof technique changes the normal ciphertext to semi-functional ciphertext first, followed by changing queried normal trapdoors into semi-functional trapdoors individually. As the challenge ciphertext is semi-functional and trapdoors are semi-functional, in the final game, simulating them becomes easy. The proof completes by showing that the challenge ciphertext is indistinguishable from a random ciphertext. The crux of any dual system-based argument is to show that normal and semi-functional forms are indistinguishable to the adversary. The simulator $\mathcal{B}$ uses hard problem to establish the such indistinguishability as we will show next.

As noted, in the first game Game_0 of this sequence of game based argument, both the challenge ciphertext and trapdoors are normal. The ciphertext is changed first to semi-functional in Game_1. Then the trapdoors are changed to semi-functional in a series of games ($\mathsf{Game}_{2,k})_k$ for $k \in [q]$. For all $k \in [q]$, in each $\mathsf{Game}_{2,k}$, all the jth queried trapdoor for $1 \leq j \leq k$ are semi-functional whereas all of the following (i.e. $k < j \leq q$) trapdoors are normal. We proceed till all the trapdoors become semi-functional. In the final game Game_3, the encapsulation key $\mathfrak{K}$ and ciphertext CT are replaced by a uniform random choice from $\mathcal{K}$ and $\mathcal{C}$ respectively. We show that the semi-functional components of challenge ciphertext and trapdoors in Game_3 supply enough entropy to hide the encapsulation key $\mathfrak{K}$ as well as the data-index in ciphertext CT. Hence $\mathfrak{K}$ and CT jointly are distributed identically to random choice from $\mathcal{K} \times \mathcal{C}$. We also denote Game_1 by $\mathsf{Game}_{2,0}$. Thus, the advantage an efficient adversary $\mathcal{A}$ has, to break BEKS in Id-CPA security model is not more than sum of advantage of games in the hybrid i.e.

$$\mathsf{Adv}^{\mathsf{BEKS}}_{\mathcal{A},\mathsf{ID\text{-}CPA}}(\lambda) \leq \left|\mathsf{Adv}^0_{\mathcal{A}}(\lambda) - \mathsf{Adv}^1_{\mathcal{A}}(\lambda)\right| + \left|\mathsf{Adv}^1_{\mathcal{A}}(\lambda) - \mathsf{Adv}^{2,0}_{\mathcal{A}}(\lambda)\right|$$
$$+ \sum_{k \in [1,q]} \left|\mathsf{Adv}^{2,k-1}_{\mathcal{A}}(\lambda) - \mathsf{Adv}^{2,k}_{\mathcal{A}}(\lambda)\right| + \left|\mathsf{Adv}^{2,q}_{\mathcal{A}}(\lambda) - \mathsf{Adv}^3_{\mathcal{A}}(\lambda)\right| + \mathsf{Adv}^3_{\mathcal{A}}(\lambda).$$

Thus it is sufficient to argue that all the difference of advantages in above equation are either negligible or zero. We show that Game_0 and Game_1 are negligibly close in Lemma 1. The indistinguishability of $\mathsf{Game}_{2,k-1}$ and $\mathsf{Game}_{2,k}$ for arbitrary $k \in [1,q]$ is proved in Lemma 2. We then show that $\mathsf{Game}_{2,q}$ and Game_3 are negligibly close in Lemma 4. Finally, we argue that no efficient adversary has any advantage in the Game_3 resulting in $\mathsf{Adv}_{\mathcal{A}}^3(\lambda) = 0$. Since, Game_1 and $\mathsf{Game}_{2,0}$ are same due to our notation, the advantage difference between the two is 0.

Broadly, our proof strategy resembles to that of [18–20] in constructing the semi-functional form of trapdoors and ciphertexts. Recall that, our BEKS construction despite being a constant-size ciphertext construction like IPE_1 [18], also achieves anonymity. Here, we note the structural difference between the two constructions. In BEKS there is a *single* component in the trapdoor that encodes both master secret key α and trapdoor-index $\mathsf{x} = (\Omega, \omega)$ in an affine equation whereas in IPE_1 they are encoded in separate components. Informally speaking, IPE_1 [18] decoupled α and x via new randomness and introduced another tag ktag to argue security. Appropriating this technique in our context however does not ensure hiding of ω'^* in the challenge ciphertext thereby failing anonymity. In fact encoding the α and x into a single trapdoor-component via affine equation allows us to achieve anonymous adaptive CPA-security following an argument that is in some sense closer to security argument of IPE_2 [18].

3.2.1 Semi-functional Algorithms

- $\mathsf{sfTrapdoor}(\mathsf{msk}, \mathsf{x} = (\varkappa, \omega))$: Let the normal trapdoor be $\mathsf{SK}'_\mathsf{x} = (\mathsf{K}'_1, \mathsf{K}'_2, (\mathsf{K}'_{3,i})_{i \in [n]}, \mathsf{K}'_4, (\mathsf{K}'_{5,i})_{i \in [n]}) \leftarrow \mathsf{Trapdoor}(\mathsf{msk}, \mathsf{x})$ where r is the corresponding randomness used in $\mathsf{Trapdoor}$. Choose $\hat{r}, (t_i)_{i \in [n]} \xleftarrow{} \mathbb{Z}_p$. Compute the semi-functional trapdoor $\mathsf{SK}_\mathsf{x} = (\mathsf{K}_1, \mathsf{K}_2, (\mathsf{K}_{3,i})_{i \in [n]}, \mathsf{K}_4, (\mathsf{K}_{5,i})_{i \in [n]})$ as follows:

$$\mathsf{K}_1 = \mathsf{K}'_1 = g_2^r, \mathsf{K}_2 = \mathsf{K}'_2 \cdot g_2^{\hat{r}} = g_2^{cr+\hat{r}},$$
$$\mathsf{K}_{3,i} = \mathsf{K}'_{3,i} \cdot g_2^{\hat{r}t_i},$$
$$= \begin{cases} g_2^{\alpha_1 + r(u_{n+1}\omega + u_\varkappa) + \hat{r}t_\varkappa} & \text{if } i = \varkappa \\ g_2^{ru_i + \hat{r}t_i} & \text{otherwise} \end{cases},$$
$$\mathsf{K}_4 = \mathsf{K}'_4 \cdot g_2^{-\hat{r}b^{-1}} = g_2^{dr - \hat{r}b^{-1}},$$
$$\mathsf{K}_{5,i} = \mathsf{K}'_{5,i} \cdot g_2^{-\hat{r}t_\varkappa b^{-1}}$$
$$= \begin{cases} g_2^{\alpha_2 + r(v_{n+1}\omega + v_\varkappa) - \hat{r}t_\varkappa b^{-1}} & \text{if } i = \varkappa \\ g_2^{rv_i - \hat{r}t_i b^{-1}} & \text{otherwise} \end{cases}.$$

- $\mathsf{sfSrchEnc}(\mathsf{mpk}, \mathsf{msk}, \mathsf{y} = (\ddot{\Omega}, \omega'))$: Let the normal encapsulation key and normal ciphertext be $(\mathcal{K}', \mathsf{CT}'_\mathsf{y}) \leftarrow \mathsf{SrchEnc}(\mathsf{mpk}, \mathsf{y})$ where s is the corresponding randomness and $\ddot{t}$ is the random tag used in $\mathsf{SrchEnc}$ and $\mathsf{CT}'_\mathsf{y} = (\mathsf{C}'_0, \mathsf{C}'_1, \mathsf{C}'_2, \ddot{t})$.

Choose $\hat{s} \hookleftarrow \mathbb{Z}_p$. Compute the semi-functional ciphertext $\mathcal{K}$ and $\mathsf{CT} = (\mathsf{C}_0, \mathsf{C}_1, \mathsf{C}_2, \ddot{t})$ where,

$$\mathcal{K} = \mathcal{K}' \cdot g_T^{\alpha_1 \hat{s}} = e(g_1, g_2)^{\alpha s + \alpha_1 \hat{s}}, \mathsf{C}_0 = \mathsf{C}_0' \cdot g_1^{\hat{s}} = g_1^{s+\hat{s}}, \mathsf{C}_1 = \mathsf{C}_1' = g_1^{bs},$$

$$\mathsf{C}_2 = \mathsf{C}_2' \cdot g_1^{\hat{s}(u_{n+1}\omega' + \sum\limits_{l \in \ddot{\Omega}} u_l + c\ddot{t})}$$

$$= g_1^{s(w_{n+1}\omega' + \sum\limits_{l \in \ddot{\Omega}} w_l + w\ddot{t}) + \hat{s}(u_{n+1}\omega' + \sum\limits_{l \in \ddot{\Omega}} u_l + c\ddot{t})}.$$

Note that, Testing a semi-functional ciphertext against a semi-functional trapdoor fails as the encapsulation key $\mathcal{K}$ is blinded by a random G_T component namely $g_T^{\hat{s}\hat{r}(\ddot{t} + \sum\limits_{l \in \ddot{\Omega}} t_l)}$. $\square$

3.2.2   Sequence of Games

The idea is to change each game only by a small margin and prove indistinguishability of two consecutive games.

Lemma 1 (Game$_0$ to Game$_1$). *For any efficient adversary $\mathcal{A}$ that makes at most q trapdoor queries, there exists a PPT algorithm $\mathcal{B}$ such that $\left| \mathsf{Adv}_{\mathcal{A}}^0(\lambda) - \mathsf{Adv}_{\mathcal{A}}^1(\lambda) \right| \leq \mathsf{Adv}_{\mathcal{B}}^{\mathsf{DDH}_{\mathsf{G}_1}}(\lambda).$*

Proof. The solver $\mathcal{B}$ is given the $\mathsf{DDH}_{\mathsf{G}_1}$ problem instance $D = (g_1, g_2, g_1^b, g_1^{bs})$ and the target $T = g_1^{s+\hat{s}}$ where $\hat{s} = 0$ or chosen uniformly random from $\mathbb{Z}_p^\times$.

Setup. $\mathcal{B}$ chooses $\alpha_1, \alpha_2, (u_i, v_i)_{i \in [n+1]}, c, d \hookleftarrow \mathbb{Z}_p$. As both α_1 and α_2 are available to $\mathcal{B}$, it can generate $g_T^\alpha = e(g_1^{\alpha_1} \cdot (g_1^b)^{\alpha_2}, g_2)$. Hence, $\mathcal{B}$ outputs the public parameter mpk. Notice that msk is available to $\mathcal{B}$.

Query Phase-I. Since $\mathcal{B}$ knows the msk, it can answer with normal trapdoors on any query of $\mathsf{x} = (\varkappa, \omega)$.

Challenge. Given the challenge y^*, $\mathcal{B}$ chooses $\ddot{t} \hookleftarrow \mathbb{Z}_p$ where $\mathsf{y}^* = (\ddot{\Omega}^*, \omega'^*)$. It then computes the encapsulation key as $\mathcal{K}_0$ and the challenge ciphertext $\mathsf{CT}_0^* = (\mathsf{C}_0, \mathsf{C}_1, \mathsf{C}_2, \ddot{t})$ using the problem instance where,

$$\mathcal{K}_0 = e(\mathsf{C}_0, g_2)^{\alpha_1} \cdot e(\mathsf{C}_1, g_2)^{\alpha_2} = e(g_1^{s(\alpha_1 + b\alpha_2) + \hat{s}\alpha_1}, g_2) = g_T^{\alpha s + \alpha_1 \hat{s}},$$

$$\mathsf{C}_0 = T = g_1^{s+\hat{s}}, \mathsf{C}_1 = g_1^{bs}, \mathsf{C}_2 = \mathsf{C}_0^{(u_{n+1}\omega' + \sum\limits_{l \in \ddot{\Omega}} u_l + c\ddot{t})} \mathsf{C}_1^{(v_{n+1}\omega' + \sum\limits_{l \in \ddot{\Omega}} v_l + d\ddot{t})}.$$

$\mathcal{B}$ then chooses $\mathcal{K}_1 \hookleftarrow \mathcal{K}$, $\mathsf{CT}_1^* \hookleftarrow \mathcal{C}$ and returns $\left(\mathcal{K}_\mathfrak{b}, \mathsf{CT}_\mathfrak{b}^* \right)$ as the challenge for $\mathfrak{b} \hookleftarrow \{0,1\}$.

Query Phase-II. Same as **Query Phase-I**.

Guess. $\mathcal{A}$ output $\mathfrak{b}' \in \{0,1\}$. $\mathcal{B}$ outputs 1 if $\mathfrak{b} = \mathfrak{b}'$ and 0 otherwise.

It is easy to see that the challenge ciphertext follows proper distribution. The randomness $\hat{s}$ that is due to the DDH instance, is injected into C_2 as semi-functional randomness $\hat{s}$ via C_0. Notice that, if $\hat{s}$ in DDH_{G_1} problem instance is 0, then the challenge ciphertext CT_0^* is normal. Otherwise the challenge ciphertext CT_0^* is semi-functional. Thus if any efficient adversary $\mathcal{A}$ can distinguish between the two, $\mathcal{B}$ can use such $\mathcal{A}$ to find out if $\hat{s} \overset{?}{=} 0$. □

Lemma 2. (Game$_{2,k-1}$ to Game$_{2,k}$). *For any efficient adversary $\mathcal{A}$ that makes at most q trapdoor queries, there exists a PPT algorithm $\mathcal{B}$ such that* $|\mathsf{Adv}_{\mathcal{A}}^{2,k-1}(\lambda) - \mathsf{Adv}_{\mathcal{A}}^{2,k}(\lambda)| \leq \mathsf{Adv}_{\mathcal{B}}^{DDH_{G_2}}(\lambda)$.

Proof. The solver $\mathcal{B}$ is given the DDH_{G_2} problem instance $D = (g_1, g_2, g_2^c, g_2^r)$ and the target $T = g_2^{cr+\hat{r}}$ where $\hat{r} = 0$ or chosen uniformly random from $\mathbb{Z}_p^{\times}$.

Setup. $\mathcal{B}$ chooses $b \hookleftarrow \mathbb{Z}_p^{\times}$, $\alpha, \alpha_1, w, (p_i, q_i, w_i)_{i \in [n+1]} \hookleftarrow \mathbb{Z}_p$. It sets $\alpha_2 = b^{-1}(\alpha - \alpha_1)$, $d = b^{-1}(w - c)$, $u_i = p_i + cq_i$, $v_i = b^{-1}(w_i - u_i)$. Note that, as c is not known to $\mathcal{B}$ explicitly, all but α_2 assignment has been done implicitly. The public parameters mpk are generated as $(g_1, g_1^b, (g_1^{w_i})_{i \in [n+1]}, g_1^w, g_T^\alpha)$ where $g_T = e(g_1, g_2)$. However, few components of msk, precisely $(d, (u_i, v_i)_{i \in [n+1]})$ are unavailable to $\mathcal{B}$.

Query Phase-I. Given the jth trapdoor query on $x_j = (x, \omega)$,

- If $j > k$: $\mathcal{B}$ has to return a normal trapdoor. However, $(d, (u_i, v_i)_{i \in [n+1]})$ of msk are unavailable to $\mathcal{B}$ as mentioned earlier. Still, $\mathcal{B}$ could simulate the normal trapdoors as it chose the b during **Setup**. $\mathcal{B}$ chooses $r_j \hookleftarrow \mathbb{Z}_p$. Computes the trapdoor is $SK_{x_j} = (K_1, K_2, (K_{3,i})_{i \in [n]}, K_4, (K_{5,i})_{i \in [n]})$ such that,

$$K_1 = g_2^{r_j}, K_2 = (g_2^c)^{r_j} = g_2^{cr_j},$$

$$K_{3,i} = \begin{cases} g_2^{\alpha_1} \cdot K_1^{(p_{n+1}\omega + p_x)} \cdot K_2^{(q_{n+1}\omega + q_x)} & \text{if } i = x \\ K_1^{p_i} \cdot K_2^{q_i} & \text{otherwise} \end{cases},$$

$$K_4 = K_1^{b^{-1}w} \cdot K_2^{-b^{-1}} = g_2^{wr_j b^{-1}} \cdot g_2^{-cr_j b^{-1}} = g_2^{dr_j},$$

$$K_{5,i} = \begin{cases} K_1^{b^{-1}(w_{n+1}\omega + w_x)} \cdot K_{3,x}^{-b^{-1}} & \text{if } i = x \\ K_1^{b^{-1}w_i} \cdot K_{3,i}^{-b^{-1}} & \text{otherwise} \end{cases},$$

$$= \begin{cases} g_2^{b^{-1}r_j(w_{n+1}\omega + w_x)} \cdot g_2^{-b^{-1}r_j(u_{n+1}\omega + u_x)} & \text{if } i = x \\ g_2^{b^{-1}r_j w_i} \cdot g_2^{-b^{-1}r_j u_i} & \text{otherwise} \end{cases},$$

$$= \begin{cases} g_2^{r_j(v_{n+1}\omega + v_x)} & \text{if } i = x \\ g_2^{r_j v_i} & \text{otherwise} \end{cases}.$$

Notice that SK_{x_j} is identically distributed to output of $\mathsf{Trapdoor(msk, x_j)}$. Thus $\mathcal{B}$ has managed to simulate the normal trapdoor without knowing the msk completely.

- If $j < k$: $\mathcal{B}$ has to return a semi-functional trapdoor. It first creates normal trapdoors as above and chooses $\hat{r}, (t_i)_{i\in[n]} \hookleftarrow \mathbb{Z}_p$ to create semi-functional trapdoors following sfTrapdoor.
- If $j = k$: $\mathcal{B}$ will use $\mathsf{DDH}_{\mathbb{G}_2}$ problem instance to simulate the trapdoor. It first sets $(t_i)_{i\in[n]}$ to define $\mathsf{SK}_\varkappa = (\mathsf{K}_1, \mathsf{K}_2, (\mathsf{K}_{3,i})_{i\in[n]}, \mathsf{K}_4, (\mathsf{K}_{5,i})_{i\in[n]})$ as follows.

$$t_i = \begin{cases} (q_{n+1}\omega + q_\varkappa) & \text{if } i = \varkappa \\ q_i & \text{otherwise} \end{cases}.$$

$$\mathsf{K}_1 = g_2^r = \mathsf{K}_1', \mathsf{K}_2 = T = g_2^{cr+\hat{r}} = \mathsf{K}_2' \cdot g_2^{\hat{r}},$$

$$\mathsf{K}_{3,i} = \begin{cases} g_2^{\alpha_1} \cdot \mathsf{K}_1^{(p_{n+1}\omega + p_\varkappa)} \cdot \mathsf{K}_2^{(q_{n+1}\omega + q_\varkappa)} & \text{if } i = \varkappa \\ \mathsf{K}_1^{p_i} \cdot \mathsf{K}_2^{q_i} & \text{otherwise} \end{cases},$$

$$= \begin{cases} g_2^{\alpha_1} \cdot \mathsf{K}_1'^{(p_{n+1}\omega + p_\varkappa)} \cdot \mathsf{K}_2'^{(q_{n+1}\omega + q_\varkappa)} \cdot g_2^{\hat{r}(q_{n+1}\omega + q_\varkappa)} & \text{if } i = \varkappa \\ \mathsf{K}_1'^{p_i} \cdot \mathsf{K}_2'^{q_i} \cdot g_2^{\hat{r}q_i} & \text{otherwise} \end{cases},$$

$$= \mathsf{K}_{3,i}' \cdot g_2^{\hat{r}t_i}.$$

$$\mathsf{K}_4 = \mathsf{K}_1^{b^{-1}w} \cdot \mathsf{K}_2^{-b^{-1}} = g_2^{wrb^{-1}} \cdot g_2^{-b^{-1}(cr+\hat{r})},$$

$$= g_2^{dr} \cdot g_2^{-\hat{r}b^{-1}} = \mathsf{K}_4' \cdot g_2^{-\hat{r}b^{-1}},$$

$$\mathsf{K}_{5,i} = \begin{cases} g_2^\alpha \cdot \mathsf{K}_1^{b^{-1}(w_{n+1}\omega + w_\varkappa)} \cdot \mathsf{K}_{3,\varkappa}^{-b^{-1}} & \text{if } i = \varkappa \\ \mathsf{K}_1^{b^{-1}w_i} \cdot \mathsf{K}_{3,i}^{-b^{-1}} & \text{otherwise} \end{cases},$$

$$= \begin{cases} g_2^\alpha \cdot \mathsf{K}_1'^{b^{-1}(w_{n+1}\omega + w_\varkappa)} \cdot \mathsf{K}_{3,\varkappa}'^{-b^{-1}} \cdot g_2^{-\hat{r}t_\varkappa b^{-1}} & \text{if } i = \varkappa \\ \mathsf{K}_1'^{b^{-1}w_i} \cdot \mathsf{K}_{3,i}'^{-b^{-1}} \cdot g_2^{-\hat{r}t_i b^{-1}} & \text{otherwise} \end{cases},$$

$$= \mathsf{K}_{5,i}' \cdot g_2^{-\hat{r}t_i b^{-1}}.$$

Notice that if $\hat{r} = 0$ then the trapdoor is normal (i.e. $\mathsf{K}_1', \mathsf{K}_2', (\mathsf{K}_{3,i}')_{i\in[n]}$, $\mathsf{K}_4', (\mathsf{K}_{5,i}')_{i\in[n]}$); otherwise it is semi-functional trapdoor.

Challenge. Given the challenge $\mathsf{y}^* = (\ddot{\Omega}^*, \omega'^*)$, $\mathcal{B}$ chooses $s, \hat{s} \hookleftarrow \mathbb{Z}_p$. It then defines the encapsulation key as $\mathcal{K}_0$ and the challenge ciphertext $\mathsf{CT}_0^* = (\mathsf{C}_0, \mathsf{C}_1, \mathsf{C}_2, \ddot{t})$ where,

$$\mathcal{K}_0 = e(g_1, g_2)^{\alpha s + \alpha_1 \hat{s}} = \mathcal{K}_0' \cdot g_T^{\alpha_1 \hat{s}}, \mathsf{C}_0 = g_1^{s+\hat{s}} = \mathsf{C}_0' \cdot g_1^{\hat{s}}, \mathsf{C}_1 = g_1^{bs},$$

$$\mathsf{C}_2 = \mathsf{C}_2' \cdot g_1^{\hat{s}(u_{n+1}\omega'^* + \sum_{l\in\ddot{\Omega}^*} u_l + c\ddot{t})},$$

$$= g_1^{s(w_{n+1}\omega'^* + \sum_{l\in\ddot{\Omega}^*} w_l + w\ddot{t}) + \hat{s}(u_{n+1}\omega'^* + \sum_{l\in\ddot{\Omega}^*} u_l + c\ddot{t})},$$

$$= g_1^{s(w_{n+1}\omega'^* + \sum_{l\in\ddot{\Omega}^*} w_l + w\ddot{t}) + \hat{s}(p_{n+1}\omega'^* + \sum_{l\in\ddot{\Omega}^*} p_l + c(q_{n+1}\omega'^* + \sum_{l\in\ddot{\Omega}^*} q_l) + c\ddot{t})}.$$

However, g_1^c is not available to $\mathcal{B}$. Therefore, we implicitly set the tag $\ddot{t} = -(q_{n+1}\omega'^* + \sum_{l\in\ddot{\Omega}^*} q_l)$. Then, $\mathsf{C}_2 = g_1^{s(w_{n+1}\omega'^* + \sum_{l\in\ddot{\Omega}^*} w_l + w\ddot{t}) + \hat{s}(p_{n+1}\omega'^* + \sum_{l\in\ddot{\Omega}^*} p_l)}.$

As $\mathcal{B}$ chose $(p_i)_{i \in [n+1]}$, $(w_i)_{i \in [n+1]}$, s and $\hat{s}$, it can compute C_2. Thus $\mathcal{B}$ can properly generate CT_0^*.

$\mathcal{B}$ then chooses $\mathfrak{K}_1 \hookleftarrow \mathcal{K}$, $\mathsf{CT}_1^* \hookleftarrow \mathcal{C}$ and returns $\left(\mathfrak{K}_\mathfrak{b}, \mathsf{CT}_\mathfrak{b}^* \right)$ as the challenge ciphertext for $\mathfrak{b} \hookleftarrow \{0, 1\}$.

Notice that, the challenge $(\mathfrak{K}_0, \mathsf{CT}_0^*)$ is identically distributed to the output of $\mathsf{sfSrchEnc}(\mathsf{mpk}, \mathsf{msk}, \mathsf{y}^*)$. Hence, the ciphertext is semi-functional.

Query Phase-II. Same as **Query Phase-I.**

Guess. $\mathcal{A}$ output $\mathfrak{b}' \in \{0, 1\}$. $\mathcal{B}$ outputs 1 if $\mathfrak{b} = \mathfrak{b}'$ and 0 otherwise.

As noted earlier, if $\hat{r}$ in $\mathsf{DDH}_{\mathsf{G}_2}$ problem instance is 0, then the kth trapdoor is normal. Otherwise the kth trapdoor is semi-functional. The challenge ciphertext is semi-functional.

As defined in Sect. 3.2.1, the tags that are used in the semi-functional trapdoor and semi-functional ciphertext, are independently chosen to be uniformly random. Thus, we need to argue that the tags (i.e. semi-functional ciphertext tag $\ddot{t}$ and semi-functional trapdoor tag $(t_i)_{i \in [n]}$) are jointly uniform random and independent. This will ensure that the challenge ciphertext and k^{th} trapdoor pair are jointly semi-functional where the challenge ciphertext is associated with data-index $\mathsf{y}^* \in \mathcal{Y}$ and the kth trapdoor is associated with trapdoor-index $\mathsf{x}_k \in \mathcal{X}$. Thus the semi-functional trapdoor and semi-functional ciphertext that are simulated are properly distributed. The following lemma ensures proper distribution of the tags and completes the proof of Lemma 2. We give the proof of Lemma 3 later.

Lemma 3. *Let* $(q_1, \ldots, q_{n+1}) \hookleftarrow \mathbb{Z}_p^{n+1}$. *Given* $\mathsf{x}_k = (\varkappa, \omega)$ *and* $\mathsf{y}^* = (\ddot{\Omega}^*, \omega'^*)$, *if* $\mathsf{R}_v(\mathsf{x}_k, \mathsf{y}^*) = 0$, *then* $(t_1, \ldots, t_n, \ddot{t})$ *is identically distributed to* $t \hookleftarrow \mathbb{Z}_p^{n+1}$ *where*

$$t_i = \begin{cases} (q_{n+1}\omega + q_\varkappa) & \text{if } i = \varkappa \\ q_i & \text{otherwise} \end{cases} \text{ and } \ddot{t} = -(q_{n+1}\omega'^* + \sum_{l \in \ddot{\Omega}^*} q_l). \qquad \square$$

Lemma 4 (Game$_{2,q}$ to Game$_3$). *For any efficient adversary* $\mathcal{A}$ *that makes at most* q *trapdoor queries,* $\left| \mathsf{Adv}_{\mathcal{A}}^{2,q}(\lambda) - \mathsf{Adv}_{\mathcal{A}}^3(\lambda) \right| \leq 3/p$.

Proof. In Game$_{2,q}$, all the queried trapdoors and the challenge ciphertext are transformed into semi-functional. To argue that the challenge $(\mathfrak{K}_0, \mathsf{CT}_0^*)$ hides the data-index $\mathsf{y}^* = (\ddot{\Omega}^*, \omega'^*)$ completely, we perform a conceptual change on the parameters of Game$_{2,q}$. Informally, we remove the effect of $(u_i)_{i \in [n]}$ from the semi-functional component of the trapdoors and use these *free variables* $(u_i)_{i \in [n]}$ to hide ω'^* in the challenge ciphertext.

Setup. Choose $b \hookleftarrow \mathbb{Z}_p^\times$, $\alpha_1, \alpha, c, w, (u_i, w_i)_{i \in [n+1]} \hookleftarrow \mathbb{Z}_p$. Set $\alpha_2 = b^{-1}(\alpha - \alpha_1)$, $d = b^{-1}(w - c)$, $v_i = b^{-1}(w_i - u_i)$. The public parameters are generated as $\mathsf{mpk} = (g_1, g_1^b, (g_1^{w_i})_{i \in [n+1]}, g_1^w, g_T^\alpha)$ where $g_T = e(g_1, g_2)$. Notice that g_T^α is independent of α_1 as α was chosen independently. Since w and $(w_i)_{i \in [n+1]}$ are chosen uniformly random, resulting mpk does not depend on neither c nor $(u_i)_{i \in [n+1]}$.

Query Phase-I. Given trapdoor query on $x = (x, \omega)$, choose $r, r', (t'_i)_{i \in [n]} \hookleftarrow \mathbb{Z}_p$. Compute the trapdoor $\mathsf{SK}_x = (\mathsf{K}_1, \mathsf{K}_2, (\mathsf{K}_{3,i})_{i \in [n]}, \mathsf{K}_4, (\mathsf{K}_{5,i})_{i \in [n]})$ as follows.

$$\mathsf{K}_1 = g_2^r, \mathsf{K}_2 = g_2^{r'}, \mathsf{K}_{3,i} = g_2^{t'_i},$$

$$\mathsf{K}_4 = \mathsf{K}_1^w \cdot \mathsf{K}_2^{b-1}, \mathsf{K}_{5,i} = \begin{cases} g_2^\alpha \cdot \mathsf{K}_1^{(w_{n+1}\omega + w_x)} \cdot \mathsf{K}_{3,x}^{-b-1} & \text{if } i = x \\ \mathsf{K}_1^{w_i} \cdot \mathsf{K}_{3,i}^{-b-1} & \text{otherwise} \end{cases}.$$

The reduction sets $t'_i = \begin{cases} \alpha_1 + r(u_{n+1}\omega + u_x) + \hat{r}t_x & \text{if } i = x \\ ru_i + \hat{r}t_i & \text{otherwise} \end{cases}$. Due to independent random choice of α_1 and t'_x, the t_x is uniformly random. Here the point of focus is that both $\mathsf{K}_{3,x}$ and $\mathsf{K}_{5,x}$ are generated using t'_x that is independent of α_1 which is absent in rest of the trapdoor components. The trapdoor SK_Ω is therefore independent of α_1 if $\hat{r} \neq 0$. This happens with probability $1 - 1/p$.

Challenge. On challenge $y^* = (\ddot{\Omega}^*, \omega'^*)$, choose $s, \hat{s}, \ddot{t} \hookleftarrow \mathbb{Z}_p$. Compute the encapsulation key $\mathfrak{K}_0$ and challenge ciphertext $\mathsf{CT}_0^* = (\mathsf{C}_0, \mathsf{C}_1, \mathsf{C}_2, \ddot{t})$ where,

$$\mathfrak{K}_0 = e(g_1, g_2)^{\alpha s + \alpha_1 \hat{s}} = g_T^{\alpha s} \cdot g_T^{\alpha_1 \hat{s}}, \mathsf{C}_0 = g_1^{s + \hat{s}}, \mathsf{C}_1 = g_1^{bs},$$

$$\mathsf{C}_2 = g_1^{s(w_{n+1}\omega'^* + \sum_{l \in \ddot{\Omega}^*} w_l + w\ddot{t}) + \hat{s}(u_{n+1}\omega'^* + \sum_{l \in \ddot{\Omega}^*} u_l + c\ddot{t})}.$$

Query Phase-II. Same as **Query Phase-I**.
Guess. $\mathcal{A}$ output $b' \in \{0, 1\}$. Output 1 if $b = b'$ and 0 otherwise.

We already have established that all the scalars used in mpk are independent of α_1, u_{n+1} and c. As the trapdoor does not contain any of α_1, u_{n+1} or c, the trapdoors $(\mathsf{SK}_{x_j})_{j \in [q]}$ are also independent of those scalars chosen during KeyGen. None of the ciphertext components but $\mathfrak{K}_0$ contain uniformly random α_1. This essentially allows the replacement of $\mathfrak{K}_0$ by a uniform random choice $\mathfrak{K}_1 \hookleftarrow \mathcal{K}$ provided $\hat{s} \neq 0$. Notice that, C_2 alone contains uniformly random u_{n+1} and c. Thus, if $\hat{s} \neq 0$ and $\ddot{t} \neq 0$, C_2 becomes uniformly random. For $\hat{s} \neq 0$, $\hat{s}$ is chosen uniformly at random from $\mathbb{Z}_p^\times$. It thus hides s in C_0 which in turn makes C_1 uniformly random as $b \hookleftarrow \mathbb{Z}_p^\times$. As a result, the challenge ciphertext CT_0^* is identically distributed to CT_1^* where $\mathsf{CT}_1^* \hookleftarrow \mathcal{C}$ provided $\hat{s} \neq 0$ and $\ddot{t} \neq 0$. Thus,

$$\left| \mathsf{Adv}_{\mathcal{A}}^{2,q}(\lambda) - \mathsf{Adv}_{\mathcal{A}}^3(\lambda) \right| \leq \Pr[\hat{r} = 0] + \Pr[\hat{s} = 0] + \Pr[\ddot{t} = 0] \leq 3/p.$$

As challenge $\mathfrak{K}_b$ and CT_b^* output in Game$_3$ completely hides b and ω'^*, for any adversary $\mathcal{A}$, $\mathsf{Adv}_{\mathcal{A}}^3(\lambda) = 0$. $\square$

Remark 2. Recall, in Remark 1, we mentioned that BEKS hides the identity ω'^* whereas the other part of challenge data-index that is the privileged set $\ddot{\Omega}^*$ has to be given in public for proper decryption. Our simulation of final game that hides ω'^* thus is in keeping with the security requirement of BEKS.

Proof of Lemma 3. Given $x = (x, \omega)$ and $y^* = (\ddot{\Omega}^*, \omega'^*)$, as $R_v(x, y^*) = 0$, there could be two mutually exclusive and exhaustive cases.

1. $x \notin \ddot{\Omega}^*$: The linear equations $(\ddot{t}, t_1, \ldots, t_n)$ can be expressed as system of linear equation $\mathbf{t} = \mathbf{Aq}$ as presented in Eq. (1) where $b_i = 1$ iff $i \in \ddot{\Omega}^* \setminus \{x\}$ and $b_x = 0$.

$$
\begin{bmatrix}
\ddot{t} \\ t_1 \\ t_2 \\ t_3 \\ \vdots \\ t_x \\ \vdots \\ t_n
\end{bmatrix}
=
\begin{bmatrix}
\omega'^* & b_1 & b_2 & b_3 & \cdots & b_x & \cdots & b_n \\
0 & 1 & 0 & 0 & \cdots & 0 & \cdots & 0 \\
0 & 0 & 1 & 0 & \cdots & 0 & \cdots & 0 \\
0 & 0 & 0 & 1 & \cdots & 0 & \cdots & 0 \\
\vdots & \vdots & \vdots & \vdots & \ddots & \vdots & \ddots & \vdots \\
\omega & 0 & 0 & 0 & \cdots & 1 & \cdots & 0 \\
\vdots & \vdots & \vdots & \vdots & \ddots & \vdots & \ddots & \vdots \\
0 & 0 & 0 & 0 & \cdots & 0 & \cdots & 1
\end{bmatrix}
\cdot
\begin{bmatrix}
q_{n+1} \\ q_1 \\ q_2 \\ q_3 \\ \vdots \\ q_x \\ \vdots \\ q_n
\end{bmatrix}
\tag{1}
$$

Since $\mathbf{A}$ above has $n+1$ many pivots, $\mathbf{A}$ is non-singular. As $(q_1, \ldots, q_{n+1}) \hookleftarrow \mathbb{Z}_p^{n+1}$ then $(\ddot{t}, t_1, t_2, t_3, \cdots, t_n)$ is identically distributed to $\mathbf{t} \hookleftarrow \mathbb{Z}_p^{n+1}$.

2. $x \in \ddot{\Omega}^*$ and $\omega \neq \omega'^*$: In this case we do a simple restructuring of $\mathbf{t} = \mathbf{Aq}$ presented above to $\mathbf{t}' = \mathbf{Bq}'$. Precisely, $\mathbf{t}'[i] = \begin{cases} \mathbf{t}[i] & \text{if } i = 1 \\ \mathbf{t}[x+1] & \text{if } i = 2 \\ \mathbf{t}[i-1] & \text{for all } i \in [3, n] \end{cases}$

and do corresponding changes in both $\mathbf{A}$ and $\mathbf{q}$ to get $\mathbf{B}$ and $\mathbf{q}'$ respectively. This makes the analysis of the system of linear equation easier, namely the determinant computation of $\mathbf{B}$ (in Eq. (3)) is now extremely straight forward.

$$
\begin{bmatrix}
\ddot{t} \\ t_x \\ t_1 \\ t_2 \\ t_3 \\ \vdots \\ t_{x-1} \\ t_{x+1} \\ \vdots \\ t_n
\end{bmatrix}
=
\begin{bmatrix}
\omega'^* & b_x & b_1 & b_2 & b_3 & \cdots & b_{x-1} & b_{x+1} & \cdots & b_n \\
\omega & 1 & 0 & 0 & 0 & \cdots & 0 & 0 & \cdots & 0 \\
0 & 0 & 1 & 0 & 0 & \cdots & 0 & 0 & \cdots & 0 \\
0 & 0 & 0 & 1 & 0 & \cdots & 0 & 0 & \cdots & 0 \\
0 & 0 & 0 & 0 & 1 & \cdots & 0 & 0 & \cdots & 0 \\
\vdots & \vdots & \vdots & \vdots & \vdots & \ddots & \vdots & \vdots & \ddots & \vdots \\
0 & 0 & 0 & 0 & 0 & \cdots & 1 & 0 & \cdots & 0 \\
0 & 0 & 0 & 0 & 0 & \cdots & 0 & 1 & \cdots & 0 \\
\vdots & \vdots & \vdots & \vdots & \vdots & \ddots & \vdots & \vdots & \ddots & \vdots \\
0 & 0 & 0 & 0 & 0 & \cdots & 0 & 0 & \cdots & 1
\end{bmatrix}
\cdot
\begin{bmatrix}
q_{n+1} \\ q_x \\ q_1 \\ q_2 \\ q_3 \\ \vdots \\ q_{x-1} \\ q_{x+1} \\ \vdots \\ q_n
\end{bmatrix}
\tag{2}
$$

The determinant of above matrix $\mathbf{B}$ is,

$$
\det(\mathbf{B}) = \omega'^* \cdot \det(\mathbf{B}') - \omega \cdot \det(\mathbf{B}'') \tag{3}
$$

Now, $\mathbf{B}' = \mathbf{I}_n$ and $\mathbf{B}'' = \begin{bmatrix} b_x & b_1 & b_2 & b_3 & \cdots & b_n \\ 0 & 1 & 0 & 0 & \cdots & 0 \\ 0 & 0 & 1 & 0 & \cdots & 0 \\ 0 & 0 & 0 & 1 & \cdots & 0 \\ \vdots & \vdots & \vdots & \vdots & \ddots & \vdots \\ 0 & 0 & 0 & 0 & \cdots & 1 \end{bmatrix}$ where $b_x = 1$.

It is easy to see that $\mathbf{B}''$ can be row-reduced into $\mathbf{I}_n$. Thus from Eq. (3), $\det(\mathbf{B}) = \omega'^* - \omega$ (as determinant of both $\mathbf{B}'$ and $\mathbf{B}''$ is 1). As $\omega'^* \neq \omega$, then $\mathbf{B}$ is non-singular. As $(q_1, \ldots, q_{n+1}) \hookleftarrow \mathbb{Z}_p^{n+1}$ then $(\ddot{t}, t_1, t_2, t_3, \cdots, t_n)$ is identically distributed to $t \hookleftarrow \mathbb{Z}_p^{n+1}$.

4 KASE: Key-Aggregate Searchable Encryption

We now present our construction for KASE. It is also instantiated in the prime order bilinear groups and achieves anonymous adaptive CPA security under SXDH assumption. Recall that KASE is the key-policy variant of the *keyword search with membership testing* and thus can be viewed as a dual of the BEKS. Precisely, the Trapdoor here takes trapdoor-index $x = (\Omega, \omega)$ and SrchEnc takes data-index $y = (x, \omega')$. This is defined via predicate function $R_{\ell} : (\mathfrak{P}(\mathcal{U}) \times \mathcal{W}) \times (\mathcal{U} \times \mathcal{W}) \to \{0, 1\}$. The indexes here satisfy the predicate function R_{ℓ} if $x \in \Omega$ and $\omega = \omega'$. On such occasion, the Test outputs 1.

4.1 Construction

KASE for universe $\mathcal{U} = [n]$ for $n = \mathsf{poly}(\lambda)$ and keyword space $\mathcal{W}$ of size $\exp(\lambda)$ is defined as following four algorithms.

- KeyGen$(1^\lambda, \mathcal{U}, \mathcal{W})$: The asymmetric bilinear group generator outputs $(p, \mathsf{G}_1, \mathsf{G}_2, \mathsf{G}_T, e) \leftarrow \mathcal{G}_{\mathsf{abg}}(1^\lambda)$ where $\mathsf{G}_1, \mathsf{G}_2, \mathsf{G}_T$ are cyclic groups of order p. Choose generators $g_1 \hookleftarrow \mathsf{G}_1$ and $g_2 \hookleftarrow \mathsf{G}_2$ and define $g_T = e(g_1, g_2)$. Choose $\alpha_1, \alpha_2, c, d, (u_i, v_i)_{i \in [n+1]} \hookleftarrow \mathbb{Z}_p$, $b \hookleftarrow \mathbb{Z}_p^\times$ and define $\alpha = (\alpha_1 + b\alpha_2)$ to set $g_T^\alpha = e(g_1, g_2)^{(\alpha_1 + b\alpha_2)}$. For $i \in [n+1]$, define $g_1^{w_i} = g_1^{u_i + bv_i}$ and $g_1^w = g_1^{c+bd}$. Define the msk $= (g_2, g_2^c, \alpha_1, \alpha_2, d, (u_i, v_i)_{i \in [n+1]})$ and the public parameter is defined as
$$\mathsf{mpk} = \left(g_1, g_1^b, (g_1^{w_i})_{i \in [n+1]}, g_1^w, g_T^\alpha\right).$$

- Trapdoor$(\mathsf{msk}, x = (\Omega, \omega))$: Given trapdoor-index x such that $\Omega \subseteq \mathcal{U}$ and $\omega \in \mathcal{W}$, choose $r \hookleftarrow \mathbb{Z}_p$. Compute the trapdoor $\mathsf{SK}_x = (\mathsf{K}_1, \mathsf{K}_2, \mathsf{K}_3, \mathsf{K}_4, \mathsf{K}_5)$ where
$$\mathsf{K}_1 = g_2^r, \mathsf{K}_2 = g_2^{cr}, \mathsf{K}_3 = g_2^{\alpha_1 + r(u_{n+1}\omega + \sum_{l \in \Omega} u_l)},$$
$$\mathsf{K}_4 = g_2^{dr}, \mathsf{K}_5 = g_2^{\alpha_2 + r(v_{n+1}\omega + \sum_{l \in \Omega} v_l)}.$$

- SrchEnc(mpk, y $= (\varkappa, \omega')$): Given data-index y such that $\varkappa \in \mathcal{U}$ and $\omega' \in \mathcal{W}$, choose $s, (t_i)_{i\in[n]} \hookleftarrow \mathbb{Z}_p$. Compute $\mathfrak{K} = e(g_1, g_2)^{\alpha s}$ and $\mathsf{CT}_y = (\mathsf{C}_0, \mathsf{C}_1, (\mathsf{C}_{2,i}, t_i)_{i\in[n]})$ where,

$$\mathsf{C}_0 = g_1^s, \mathsf{C}_1 = g_1^{bs}, \mathsf{C}_{2,i} = \begin{cases} g_1^{s(w_{n+1}\omega' + w_\varkappa + wt_\varkappa)} & \text{if } i = \varkappa \\ g_1^{s(w_i + wt_i)} & \text{otherwise} \end{cases}.$$

- Test$((\mathsf{SK}_x, \Omega), (\mathsf{CT}_y, \mathfrak{K}))$: Let $x = (\Omega, \omega)$ and $y = (\varkappa, \omega')$; output 1 iff $\mathfrak{K} = B/A$ where

$$A = e\left(\prod_{i\in\Omega} \mathsf{C}_{2,i}, \mathsf{K}_1\right), B = e\left(\mathsf{C}_0, \mathsf{K}_3 \prod_{i\in\Omega} \mathsf{K}_2^{t_i}\right) e\left(\mathsf{C}_1, \mathsf{K}_5 \prod_{i\in\Omega} \mathsf{K}_4^{t_i}\right).$$

It is easy to verify correctness of this protocol which is omitted due to space constraints.

4.2 Security Proof

Theorem 2. *For any adversary $\mathcal{A}$ of KASE construction KASE in the Id-CPA model (ID-CPA) that makes at most q many trapdoor queries, there exist PPT adversaries $\mathcal{B}_1, \mathcal{B}_2$ such that*

$$\mathsf{Adv}_{\mathcal{A},\mathsf{ID\text{-}CPA}}^{\mathsf{KASE}}(\lambda) \leq \mathsf{Adv}_{\mathcal{B}_1}^{\mathsf{DDH}_{G_1}}(\lambda) + q \cdot \mathsf{Adv}_{\mathcal{B}_2}^{\mathsf{DDH}_{G_2}}(\lambda) + 2/p.$$

Proof Sketch. The proof, though similar in nature to that of Theorem 1, has its own intricacies due to structural difference between the two constructions as well as the notion of anonymity. More precisely, as the trapdoor in BEKS now looks like ciphertext in KASE and vice-versa, the tags that are indispensable part of the proof behave little differently. So, we present an intuitive sketch of the proof.

Here again, we use dual system encryption [22] to prove Theorem 2. We first transform the normal ciphertext into semi-functional. Then, each trapdoor is transformed into semi-functional trapdoor individually. Finally, we replace the ciphertext component $\mathfrak{K}$ and $(\mathsf{C}_{2,i})_{i\in[n]}$ by uniformly random elements from $\mathcal{K} \times \mathcal{C}$. This proves that the data-index $y^* = (\varkappa^*, \omega'^*)$ is completely hidden in the ciphertext.

We now give a brief idea of the similarity and the difference between proof of Theorem 1 and this one primarily focusing on the tags that are used in semi-functional ciphertext and semi-functional trapdoors. The detailed proof is omitted due to space constraint. The first lemma that changes normal ciphertext into semi-functional, is proved similarly as was done in Lemma 1 with a small difference. The difference stems from the fact that here we define n tags $(t_i)_{i\in[n]}$ for $\mathcal{O}(n)$ ciphertext components as opposed to BEKS where we required only one ciphertext tag $(\ddot{t})$ to be defined. During the translation of normal trapdoor to semi-functional, the proof strategy shifts from the proof of Lemma 2 where k^{th} trapdoor set n tags $(t_i)_{i\in[n]}$ and challenge ciphertext defines one tag $\ddot{t}$. Whereas, here, in the game where we perform the translation of the normal trapdoors into semi-functional one-by-one, the target k^{th} trapdoor sets only one tag π and challenge ciphertext defines n tags $(t_i)_{i \in [n]}$. At this point, we define,

$$\pi = (q_{n+1}\omega + \sum_{l \in \Omega} q_l) \quad \text{and} \quad t_i = \begin{cases} -(q_{n+1}\omega'^* + q_{x^*}) & \text{if } i = x^* \\ -q_i & \text{otherwise} \end{cases}.$$

It is easy to see that the proof of Lemma 3 ensures independence of tags used here. This allows us to perform a conceptual change on all the $(C_{2,i})_{i \in [n]}$ in the final game. Note that, this strategy lets us hide whole data-index $\mathbf{y} = (x^*, \omega'^*)$ unlike Lemma 4 where we could hide only ω'^*. □

5   Conclusion

In this work, we have extended the Jutla-Roy's technique of secure IBE construction to search keywords along with membership testing. We have dealt with both the ciphertext-policy (BEKS) and key-policy (KASE) flavors of such functionality and have achieved adaptive security as well as anonymity in both the cases. As shown in Table 1, our dedicated solution compares favorably with the generic construction based on [1]. An interesting open problem would be to achieve sub-linear ciphertext size (resp. trapdoor size) in KASE (resp. BEKS).

Acknowledgement. We thank the anonymous reviewers of INDOCRYPT 2018 for their valuable suggestions.

A   Insecurity of KASE Construction of [17]

Recently, [17] presented a construction of key-aggregate searchable encryption that they named *controlled-access searchable encryption* (CASE). The primary emphasis of [17] is to propose an FPGA implementation of their construction. They also argued the security of their scheme.

We however demonstrate a simple mix-and-match attack on the construction in their security model. The security model, described in [17, Definition A.3], is a weaker version of IND-CKA1 [10]. Here we first present the security model followed by an attack on the construction of [17]. The description of their construction can be found in [17, Sect. 3.3]. To understand our attack, it is enough to take a look at the GenTrpdr function of their description. Essentially, our attack exploits the deterministic nature of the GenTrpdr of [17].

A.1   Security Model

We discuss the security game between the challenger $\mathcal{C}$ and adversary $\mathcal{A}$ below. Let $\mathbf{D}$ be a dataset and we perform search on key-indices $\{(S_j, \omega_j)\}_j$. Then, *trace* (τ) of a search history $(\mathbf{D}, \{(S_j, \omega_j)\}_j)$ is defined to be the list of *access pattern*. Informally, *access pattern* of $(\mathbf{D}, (\Omega, \omega))$ is the *result* denoted by $\delta(\mathbf{D}, (\Omega, \omega))$

where δ is a function that takes dataset $\mathbf{D}$ and the trapdoor-index $\mathsf{x} = (\Omega, \omega)$ and outputs document identifiers that satisfy this trapdoor-index.

Informally, the adversary $\mathcal{A}$ gives two datasets $(\mathbf{D}_0, \mathbf{D}_1)$ of its own choice. It is allowed to make queries that does not trivially distinguish the secure indexes $\mathbf{I}_0$ and $\mathbf{I}_1$ where $\mathbf{I}_i \leftarrow \mathsf{BuildIndex}(\mathsf{pk}, \mathbf{D}_i)$ for $i \in \{0, 1\}$. At the end, the adversary has to distinguish if $\mathbf{I}_0$ or $\mathbf{I}_1$ was given as the challenge secure index. We now formally define the model where non-trivially is ensured by the restriction due to *trace* τ.

- **Setup.** $\mathcal{C}$ generates msk, pk and gives pk to $\mathcal{A}$.
- **Trapdoor Queries.** Given j^{th} trapdoor query $\mathsf{x}_j = (S_j, \omega_j)$, $\mathcal{C}$ outputs $\Gamma \leftarrow \mathsf{GenTrpdr}(\mathsf{msk}, \mathsf{x}_j)$.
- **Challenge.** On receiving two file collections $\mathbf{D}_0$ and $\mathbf{D}_1$ as challenge with the restriction $\tau(\mathbf{D}_0, \{(S_j, \omega_j)\}_j) = \tau(\mathbf{D}_1, \{(S_j, \omega_j)\}_j)$, $\mathcal{C}$ picks $b \hookleftarrow \{0, 1\}$ and outputs $\mathbf{I}_b \leftarrow \mathsf{BuildIndex}(pk, \mathbf{D}_b)$.
- **Key Queries.** $\mathcal{A}$ continues querying with $\mathsf{x} = (\Omega, \omega)$ with the restriction $\tau(\mathbf{D}_0, \{(\Omega, \omega)\}) = \tau(\mathbf{D}_1, \{(\Omega, \omega)\})$.
- **Guess.** $\mathcal{A}$ outputs a guess b' and wins if $b = b'$.

Intuitively, for a secure searchable scheme, for every j^{th} query x_j, if $\delta(\mathbf{D}_0, \mathsf{x}_j) = \delta(\mathbf{D}_1, \mathsf{x}_j)$, $\mathcal{A}$ will not be able to guess b except with negligible probability.

A.2   Attack Details

We present a simple attack on the KASE construction of [17]. Informally, in this attack, we make few permitted queries to get corresponding trapdoors. Then we mix-and-match those trapdoors to create a new trapdoor. The new trapdoor will allow us to guess the challenge bit b with probability 1. Thereby, the adversary can trivially distinguish $\mathbf{I}_0$ and $\mathbf{I}_1$. Observe that, the $\mathsf{GenTrpdr}$ [17, Sect. 3.3] is deterministic. Thus each trapdoor does not have their own randomness. We exploit this property to mount a mix-n-match attack on the said construction that we present next. Note that, the natural restriction allows the adversary to query for trapdoor-index x only if $\tau(\mathbf{D}_0, \mathsf{x}) = \tau(\mathbf{D}_1, \mathsf{x})$.

Now, for $\mathsf{msk} = (\mathfrak{a}, \mathfrak{b})$, the trapdoor is $(H_1(\omega), (\mathfrak{a}F_S(\alpha) + \mathfrak{b}H_1(\omega))P_2) \leftarrow \mathsf{GenTrpdr}(\mathsf{msk}, (\Omega, \omega))$ where P_2 is group generator, H_1 is CRHF and for any set Ω, the *unique signature of* Ω is $F_S(\mathsf{x}) = \prod_{\mathsf{x} \in \Omega} (\mathsf{x} - i)$ (see [17, Sect. 3.3] for more details). We discuss the attack below as a game between challenger $\mathcal{C}$ and adversary $\mathcal{A}$.

- **Setup.** $\mathcal{C}$ gives pk to $\mathcal{A}$ and keeps msk. Let $\mathsf{msk} = (\mathfrak{a}, \mathfrak{b})$. $\mathcal{A}$ directly goes for challenge phase.
- **Challenge.** $\mathcal{A}$ defines document collection $\mathbf{D}_0 = \{i_0\}$ and $\mathbf{D}_1 = \{i_0, i_1\}$ where file i_0 contains keyword ω and file i_1 contains keyword ω^*. Lets assume $\Omega^* = \{i_0, i_1, i_2\}$ and $\Omega = \{i_2\}$. $\mathcal{A}$ sends $(\mathbf{D}_0, \mathbf{D}_1)$ to $\mathcal{C}$ who picks $b \hookleftarrow \{0, 1\}$ and returns $\mathbf{I}_b \leftarrow \mathsf{BuildIndex}(pk, \mathbf{D}_b)$.
- **Key Queries.** $\mathcal{A}$ makes following 3 queries.

1. On query $\mathsf{x}_1 = (\Omega, \omega)$: As both $i_0, i_1 \notin \Omega$, result is ϕ in both the cases. Therefore $\tau(\mathbf{D}_0, \mathsf{x}_1) = \tau(\mathbf{D}_1, \mathsf{x}_1)$. $\mathcal{C}$ runs $\mathsf{GenTrpdr}(msk, \mathsf{x}_1)$ to compute $\Gamma_1 = (H_1(\omega), (\mathfrak{a}F_S(\alpha) + \mathfrak{b}H_1(\omega))P_2)$.
2. On query $\mathsf{x}_2 = (\Omega, \omega^*)$: As both $i_0, i_1 \notin \Omega$, result is ϕ in both the cases. Therefore $\tau(\mathbf{D}_0, \mathsf{x}_2) = \tau(\mathbf{D}_1, \mathsf{x}_2)$. $\mathcal{C}$ runs $\mathsf{GenTrpdr}(msk, \mathsf{x}_2)$ to compute $\Gamma_2 = (H_1(\omega^*), (\mathfrak{a}F_S(\alpha) + \mathfrak{b}H_1(\omega^*))P_2)$.
3. On query $\mathsf{x}_3 = (\Omega^*, \omega)$: Here both $i_0, i_1 \in \Omega^*$. As only i_0 contain ω, result is $\{i_0\}$ in both the cases. Therefore $\tau(\mathbf{D}_0, \mathsf{x}_3) = \tau(\mathbf{D}_1, \mathsf{x}_3)$. $\mathcal{C}$ runs $\mathsf{GenTrpdr}(msk, \mathsf{x}_3)$ to compute $\Gamma_3 = (H_1(\omega^*), (\mathfrak{a}F_S(\alpha) + \mathfrak{b}H_1(\omega^*))P_2)$.

Now $\mathcal{A}$ computes, $Z = \frac{\Gamma_2[2]}{\Gamma_1[2]} = \mathfrak{b}(H_1(\omega^*) - H_1(\omega))P_2$.

Then it computes $\widehat{Z} = Z \times \Gamma_3[2]$
$$= \mathfrak{b}(H_1(\omega^*) - H_1(\omega))P_2 + (\mathfrak{a}F_{\Omega^*}(\alpha) + \mathfrak{b}H_1(\omega))P_2$$
$$= (\mathfrak{a}F_{\Omega^*}(\alpha) + \mathfrak{b}H_1(\omega^*))P_2.$$

Then it defines $\widehat{\Gamma}_3$ to be a valid trapdoor for (Ω^*, ω^*).

$$\widehat{\Gamma}_3 = (\Gamma_2[1], \widehat{Z}) = (H_1(\omega^*), (\mathfrak{a}F_{\Omega^*}(\alpha) + \mathfrak{b}H_1(\omega^*))P_2). \qquad (4)$$

- **Guess.** $\mathcal{A}$ outputs $b' = 0$ if $\mathsf{Search}(\mathbf{I}_b, \widehat{\Gamma}_3, \Omega^*) = \phi$, else outputs $b' = 1$.

We already have shown, $\mathcal{A}$ gets hold of a valid trapdoor $\widehat{\Gamma}_3$ on (Ω^*, ω^*) in Eq. (4). As,

- $\delta(\mathbf{D}_0, (\Omega^*, \omega^*)) = \phi$ as $\mathbf{D}_0 = \{i_0\}$ and i_0 doesn't contain keyword ω^*.
- $\delta(\mathbf{D}_1, (\Omega^*, \omega^*)) = \{i_1\}$ as $\mathbf{D}_1 = \{i_0, i_1\}$, $i_1 \in \Omega^*$ and i_1 contains keyword ω^*.

Therefore $\mathcal{A}$ wins the game with probability 1. This simple attack renders the KASE construction by [17] insecure.

References

1. Ambrona, M., Barthe, G., Schmidt, B.: Generic transformations of predicate encodings: constructions and applications. In: Katz, J., Shacham, H. (eds.) CRYPTO 2017. LNCS, vol. 10401, pp. 36–66. Springer, Cham (2017). https://doi.org/10.1007/978-3-319-63688-7_2
2. Attrapadung, N., Furukawa, J., Imai, H.: Forward-secure and searchable broadcast encryption with short ciphertexts and private keys. In: Lai, X., Chen, K. (eds.) ASIACRYPT 2006. LNCS, vol. 4284, pp. 161–177. Springer, Heidelberg (2006). https://doi.org/10.1007/11935230_11
3. Boneh, D., Di Crescenzo, G., Ostrovsky, R., Persiano, G.: Public key encryption with keyword search. In: Cachin, C., Camenisch, J.L. (eds.) EUROCRYPT 2004. LNCS, vol. 3027, pp. 506–522. Springer, Heidelberg (2004). https://doi.org/10.1007/978-3-540-24676-3_30
4. Boneh, D., Gentry, C., Waters, B.: Collusion resistant broadcast encryption with short ciphertexts and private keys. In: Shoup, V. (ed.) CRYPTO 2005. LNCS, vol. 3621, pp. 258–275. Springer, Heidelberg (2005). https://doi.org/10.1007/11535218_16

5. Chang, Y.C., Mitzenmacher, M.: Privacy preserving keyword searches on remote encrypted data. In: Ioannidis, J., Keromytis, A., Yung, M. (eds.) ACNS 2005. LNCS, vol. 3531, pp. 442–455. Springer, Heidelberg (2005). https://doi.org/10.1007/11496137_30

6. Chen, J., Gay, R., Wee, H.: Improved dual system ABE in prime-order groups via predicate encodings. In: Oswald, E., Fischlin, M. (eds.) EUROCRYPT 2015. LNCS, vol. 9057, pp. 595–624. Springer, Heidelberg (2015). https://doi.org/10.1007/978-3-662-46803-6_20

7. Chen, J., Gong, J.: ABE with tag made easy. In: Takagi, T., Peyrin, T. (eds.) ASIACRYPT 2017. LNCS, vol. 10625, pp. 35–65. Springer, Cham (2017). https://doi.org/10.1007/978-3-319-70697-9_2

8. Chu, C.K., Chow, S.S.M., Tzeng, W.G., Zhou, J., Deng, R.H.: Key-aggregate cryptosystem for scalable data sharing in cloud storage. IEEE Trans. Parallel Distrib. Syst. **25**(2), 468–477 (2014)

9. Cui, B., Liu, Z., Wang, L.: Key-aggregate searchable encryption (KASE) for group data sharing via cloud storage. IEEE Trans. Comput. **65**(8), 2374–2385 (2016)

10. Curtmola, R., Garay, J.A., Kamara, S., Ostrovsky, R.: Searchable symmetric encryption: improved definitions and efficient constructions. J. Comput. Secur. **19**(5), 895–934 (2011)

11. Gentry, C., Silverberg, A.: Hierarchical ID-based cryptography. In: Zheng, Y. (ed.) ASIACRYPT 2002. LNCS, vol. 2501, pp. 548–566. Springer, Heidelberg (2002). https://doi.org/10.1007/3-540-36178-2_34

12. Gentry, C., Waters, B.: Adaptive security in broadcast encryption systems (with short ciphertexts). In: Joux, A. (ed.) EUROCRYPT 2009. LNCS, vol. 5479, pp. 171–188. Springer, Heidelberg (2009). https://doi.org/10.1007/978-3-642-01001-9_10

13. Goh, E.: Secure indexes. IACR Cryptology ePrint Archive 2003, 216 (2003). http://eprint.iacr.org/2003/216

14. Jutla, C.S., Roy, A.: Shorter quasi-adaptive NIZK proofs for linear subspaces. J. Cryptol. **30**(4), 1116–1156 (2017)

15. Kiayias, A., Oksuz, O., Russell, A., Tang, Q., Wang, B.: Efficient encrypted keyword search for multi-user data sharing. In: Askoxylakis, I., Ioannidis, S., Katsikas, S., Meadows, C. (eds.) ESORICS 2016. LNCS, vol. 9878, pp. 173–195. Springer, Cham (2016). https://doi.org/10.1007/978-3-319-45744-4_9

16. Kiltz, E., Wee, H.: Quasi-adaptive NIZK for linear subspaces revisited. In: Oswald, E., Fischlin, M. (eds.) EUROCRYPT 2015. LNCS, vol. 9057, pp. 101–128. Springer, Heidelberg (2015). https://doi.org/10.1007/978-3-662-46803-6_4

17. Patranabis, S., Mukhopadhyay, D.: Spot the black hat in a dark room: parallelized controlled access searchable encryption on FPGAs. Cryptology ePrint Archive, Report 2017/668 (2017)

18. Ramanna, S.C.: More efficient constructions for inner-product encryption. In: Manulis, M., Sadeghi, A.-R., Schneider, S. (eds.) ACNS 2016. LNCS, vol. 9696, pp. 231–248. Springer, Cham (2016). https://doi.org/10.1007/978-3-319-39555-5_13

19. Ramanna, S.C., Sarkar, P.: Efficient (Anonymous) compact HIBE from standard assumptions. In: Chow, S.S.M., Liu, J.K., Hui, L.C.K., Yiu, S.M. (eds.) ProvSec 2014. LNCS, vol. 8782, pp. 243–258. Springer, Cham (2014). https://doi.org/10.1007/978-3-319-12475-9_17

20. Ramanna, S.C., Sarkar, P.: Efficient adaptively secure IBBE from the SXDH assumption. IEEE IT **62**(10), 5709–5726 (2016)

21. Song, D.X., Wagner, D., Perrig, A.: Practical techniques for searches on encrypted data. In: Symposium on Security and Privacy, pp. 44–55. IEEE (2000)
22. Waters, B.: Dual system encryption: realizing fully secure IBE and HIBE under simple assumptions. In: Halevi, S. (ed.) CRYPTO 2009. LNCS, vol. 5677, pp. 619–636. Springer, Heidelberg (2009). https://doi.org/10.1007/978-3-642-03356-8_36

Symmetric Key Cryptography and Format Preserving Encryption

Tweakable HCTR: A BBB Secure Tweakable Enciphering Scheme

Avijit Dutta$^{(\boxtimes)}$ and Mridul Nandi

Indian Statistical Institute, Kolkata, India
avirocks.dutta13@gmail.com, mridul.nandi@gmail.com

Abstract. HCTR, proposed by Wang et al., is one of the most efficient candidates of tweakable enciphering schemes that turns an n-bit block cipher into a variable input length tweakable block cipher. Wang et al. have shown that HCTR offers a cubic security bound against all adaptive chosen plaintext and chosen ciphertext adversaries. Later in FSE 2008, Chakraborty and Nandi have improved its bound to $O(\sigma^2/2^n)$, where σ is the total number of blocks queried and n is the block size of the block cipher. In this paper, we propose **tweakable HCTR** that turns an n-bit tweakable block cipher to a variable input length tweakable block cipher by replacing all the block cipher calls of HCTR with tweakable block cipher. We show that when there is no repetition of the tweak, tweakable HCTR enjoys the optimal security against all adaptive chosen plaintext and chosen ciphertext adversaries. However, if the repetition of the tweak is limited, then the security of the construction remains close to the security bound in no repetition of the tweak case. Hence, it gives a graceful security degradation with the maximum number of repetition of tweaks.

Keywords: Tweakable enciphering scheme · HCTR · TSPRP
H-Coefficient.

1 Introduction

TWEAKABLE ENCIPHERING SCHEME. A block cipher is a fundamental primitive in symmetric key cryptography that processes only fixed length messages. Examples of such block ciphers are DES [29], AES [10] etc. The general security notion for a block cipher is pseudorandom permutation (PRP) which says that any computationally bounded adversary should be unable to distinguish between a random permutation and a permutation picked at random from a keyed family of permutations over the input set. A stronger security notion for block cipher called strong pseudorandom permutation (SPRP) requires computationally bounded adversary should be unable to distinguish between a random permutation and its inverse from a permutation and its inverse, picked at random from the keyed family of permutations. A *mode of operation* of a block cipher specifies a particular way the block cipher should be used to process arbitrary and variable length messages; hence extending the domain of applicability

© Springer Nature Switzerland AG 2018
D. Chakraborty and T. Iwata (Eds.): INDOCRYPT 2018, LNCS 11356, pp. 47–69, 2018.
https://doi.org/10.1007/978-3-030-05378-9_3

from fixed length messages to long and variable length messages. As its security requirement, we require that it should be secure if the underlying block cipher is a secure PRP, then the extended domain mode of operation also satisfies an appropriate notion of security.

The two major goals of a mode of operation that it wants to achieve are confidentiality and integrity. For example, CBC [36] mode provides only confidentiality whereas CBC-MAC [1] is a mode of operation that guarantees only integrity. OCB [33] is a mode of operation which provides both confidentiality and integrity. A mode of operation that can encrypt arbitrary length messages and provides SPRP security is called a *length preserving transformation* for which no tag is produced. In that case, a change in the ciphertext remains undetected but the decryption of a tampered ciphertext results in a plaintext which is indistinguishable from a random string. The detection of tampering is possible by allowing additional redundancy in the message by higher level applications as discussed by Bellare and Rogaway [2].

A *Tweakable Enciphering Scheme* (TES) is a keyed family of length preserving transformations $\mathcal{E} : \mathcal{K} \times \mathcal{T} \times \mathcal{M} \rightarrow \mathcal{M}$ where $\mathcal{K}$ and $\mathcal{T}$ are the finite and nonempty set of keys and tweaks respectively and $\mathcal{M}$ is a message space such that for all $K \in \mathcal{K}$ and all $T \in \mathcal{T}$, $\mathcal{E}_K(T, \cdot)$ is a length preserving permutation[1] over $\mathcal{M}$ and there must be an inverse $\mathcal{D}_K(T, \cdot)$ to $\mathcal{E}_K(T, \cdot)$. Unlike the key K, tweak T is public whose sole purpose is to introduce the variability of the ciphertext, similar to that of the role of IV in the mode of encryption.

The general security notion of a TES is tweakable strong pseudorandom permutation (TSPRP) which is to say that it is computationally infeasible for an adversary to distinguish the oracle that maps (T, M) into $\mathcal{E}_K(T, M)$ and maps (T, C) into $\mathcal{D}_K(T, C)$ when the key K is random and secret from an oracle that realizes a T-indexed family of random permutations and their inverses. A TSPRP secure TES is a desirable tool for solving the disk encryption problem as pointed out in [14] where the sector address of the disk plays the role of the tweak in TES.

1.1   Different Paradigm of Designing TES

In the past few years there have been various proposals of designing TES. If we categorize all these proposals, then we see that all the proposals falls under one of the following three categories:

HASH-ENCRYPT-HASH. Naor and Reingold [28] designed a wide block SPRP using a invertible ECB mode of encryption sandwiched between two invertible pairwise independent hash functions. This paradigm of construction is known as Hash-Encrypt-Hash. However, as discussed in [14] that the description given in [28] is at a top level and also the latter work [27] does not fully specify a mode of operation. Moreover, the construction was not a tweakable SPRP. Later in 2006, Chakraborty and Sarkar [8] first instantiated Hash-Encrypt-Hash mode with

---

[1] A length preserving permutation over $\mathcal{M}$ is a permutation π such that for all $M \in \mathcal{M}$, $|\pi(M)| = |M|$.

PEP by sandwiching a ECB type encryption layer in between of two layers of polynomial hashing. TET, a more efficient version of PEP, was later proposed by Halevi [13]. HEH, an improvement upon TET, is also reported in [34].

ENCRYPT-MIX-ENCRYPT. CBC-Mix-CBC (CMC), proposed by Halevi and Rogaway [14], is the first TES construction in which a mixing layer is sandwiched between two CBC layers; hence the design is inherently sequential. Later, Halevi and Rogaway proposed a parallel construction, called EME [15] in which the encryption layers are of ECB type. Later EME was extended to EME* [12] for handling arbitrary length messages. All of these constructions follow the same design principle where a simple mixing layer is sandwiched between two invertible encryption layers. Recently, Bhaumik et al. [3] proposed FMix, a variant of CMC, that uses a single block cipher key (instead of two block cipher keys used in CMC) and lifted up the requirement of the block cipher invertibility.

HASH-COUNTER-HASH. This paradigm is similar to the Hash-Encrypt-Hash, but instead of a ECB layer, a counter mode encryption layer is sandwiched between two almost-xor universal hash function[2]. The advantage of using the counter mode encryption is to tackle the variable length messages easily. XCB [20] is the first Hash-Counter-Hash type construction that requires 5 block cipher keys and two block cipher calls (apart from block cipher calls in counter mode encryption). Later, Wang et al. proposed HCTR [35] with a single block cipher key and removed one extra invocation of block cipher call. FAST, a pseudorandom function (PRF) based TES construction following the Hash-Counter-Hash paradigm has recently been proposed by Chakraborty et al. [5].

Amongst the above mentioned constructions, only CMC and EME* are block cipher based constructions with a light weight masking layer in between of two encryption layers, whereas the other two paradigms require the field multiplication (as a part of the hash function evaluation) along with the block cipher evaluation. Thus, the only significant cost for Encrypt-Mix-Encrypt type constructions are the block cipher calls, whereas for the other two paradigms the cost involved in both evaluating the block cipher calls and the finite field multiplications. A detailed comparison of the performance and efficiency of different TES can be found in [7,13,34]. This comparison study along with [19] suggests that HCTR is one of the most efficient candidates amongst all proposed TES.

However, unlike other TES proposals which have the usual "**birthday bound**" type security, HCTR was initially shown to have the **cubic** security bound [35]. Later, the bound was improved to the birthday bound by Chakraborty and Nandi [6]. Chakraborty and Sarkar [7] proposed HCH, a simple variant of HCTR, in which they introduce one more block cipher call before initializing the counter and shown to have the birthday bound security.

---

[2] An almost-xor universal hash function is a keyed hash function such that for any two distinct messages, the probability, over the random draw of a hash key, the hash differential being equal to a specific output is small.

1.2 Our Contribution

In this paper, we propose **tweakable HCTR**, a variant of the HCTR construction, that yields a variable input length *tweakable block cipher* (TBC)[3] from a fixed input length tweakable block cipher, in which all the block ciphers of HCTR are replaced with TBC. In HCTR, the tweak is one of the inputs of the upper and lower layer hash function (i.e., H_{K_h} in Fig. 2), but in our construction, we process the tweak through another independent keyed $(n + m)$-bit hash function H'_L where the m-bit hash value becomes the tweak of the underlying tweakable block cipher and the remaining n-bit hash output is used to mask the input and the output of the leftmost TBC (see Fig. 1). We process tweak through an independent keyed hash function for allowing large sized tweaks.

We have shown that if there is no repetition of tweaks, or in other words, all the queried tweaks are distinct, then tweakable HCTR is secure upto 2^n many message blocks against any computationally unbounded chosen plaintext chosen ciphertext adaptive adversaries. Moreover, when the repetition of the tweak is limited, then the security we obtain is close to the optimal one. This is in contrast to the security of other nonce based constructions (e.g., Wegman-Carter MAC [4], AES-GCM [21] etc.) where a single time repetition of the nonce completely breaks the scheme. This property is called the **graceful degradation of security** when tweak repeats. Gracefully degrading secured construction based on tweakable block ciphers has been studied in [32] and the notion of tweak repetition has been studied in [22] by Mennink for proving $3n/4$-bit security of CLRW2. In [22], Mennink stated that:

"The condition on the occurrence of the tweak seems restrictive, but many modes of operation based on a tweakable block cipher query their primitives for tweaks that are constituted of a nonce or random number concatenated with a counter value: in a nonce-respecting setting, every nonce appears at most $1 + q_f$ times, where q_f is the amount of forgery attempts."

In practical settings like disk-encryption problem where the sector address plays the role of the tweak, tweak is not repeated arbitrarily and therefore the security of any tweakable scheme where the tweak repeats in a limited way, is worth to study.

1.3 Comparison with Minematsu-Iwata Proposal [24]

Hash-Sum-Expansion or (HSE) due to Minematsu and Matsushima [26] is a generic structure that underlies the construction of HCTR and HCH. HSE is instantiated with a TBC and a weak pseudorandom function (wPRF) [26] and its security proof shows that the expansion function of HCTR and HCH, which is achieved through the counter mode encryption, can be instantiated with any secure wPRF. However, HSE is shown to have the birthday bound security.

---

[3] A tweakable block cipher is basically a simple block cipher with an additional parameter called tweak.

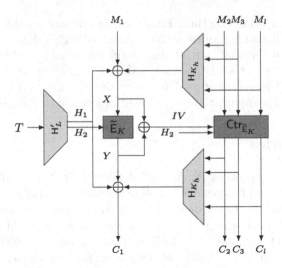

Fig. 1. Tweakable HCTR construction with tweak T and message $M_1\|M_2\|\ldots\|M_l$ and the corresponding ciphertext $C_1\|C_2\|\ldots\|C_l$. H_{K_h} is an n-bit almost-xor universal and almost regular hash function with hash key K_h. H'_L is an $(n+m)$-bit partial almost xor universal hash function with hash key L and $\mathsf{H}'_L(T) = (H_1, H_2)$, where H_1 is of size n bits and H_2 is of size m bits. $\widetilde{\mathsf{E}}_K$ is the tweakable block cipher and $\mathsf{Ctr}_{\widetilde{\mathsf{E}}_K}$ is the tweakable block cipher based counter mode encryption.

Later, Minematsu and Iwata [24] designed a block cipher for processing arbitrary length messages. For processing messages of shorter length than $2n$ bits, they proposed **Small-Block Cipher**, which is instantiated with two independent keyed TBCs with tweak size (m) < block size (n) and an n-bit PolyHash function Poly_{K_h} which eventually provides sprp security upto $(n+m)/2$ bits[4]. The construction is identical to a scheme of [23]. To process messages larger than $2n$ bits, they proposed **Large Block-Cipher, Method 1** and **Large Block-Cipher, Method-2**. The former one is structurally similar to HCTR and hence is of interest to us. LBC-1 (abbreviation for Large Block-Cipher, Method 1) uses (a) a $2n$-bit block cipher E_{2n}, (b) a $2n$-bit keyed hash function H_K in upper and lower layer and (c) a wPRF F. It has been shown [24] that LBC-1 provides the optimal (i.e., 2^n) sprp security, where block size and tweak size is of n bits.

Now, to instantiate each of the primitives, (a) E_{2n} is instantiated through **Small-Block Cipher** method and hence it requires two independent keyed TBCs with tweak size and block size n and an n-bit PolyHash function. (b) $2n$-bit keyed hash function H_K is instantiated through the concatenation of two independent keyed n-bit PolyHash functions and (c) the wPRF F is instantiated through a counter mode of encryption based on two independent invocations of TBCs with tweak size and block size n. Therefore, LBC-1 requires altogether two independent keyed TBCs with n-bits tweak and block along with three

---

[4] This security bound is beyond birthday in terms of the block size n, but with respect to the input size of TBC (i.e., $n+m$ bits), it is the birthday bound.

independent keyed n-bit PolyHash functions. In contrast to this, our proposal requires an n-bit almost xor universal hash function (e.g., polyhash) in upper and lower layers, an $(n + m)$-bit partial-almost xor universal hash function[5] [16,25] and a single instance of a TBC with tweak size m. Note that, in our case the tweak is provided as an additional input to the construction unlike to LBC-1 where the part of the input message is served as a tweak to the underlying TBC.

2   Preliminaries

BASIC NOTATIONS. For a set $\mathcal{X}$, $X \leftarrow_\$ \mathcal{X}$ denotes that X is sampled uniformly at random from $\mathcal{X}$ and independent of all other random variables defined so far. For two sets $\mathcal{X}$ and $\mathcal{Y}$, $\mathcal{X} \sqcup \mathcal{Y}$ denotes the disjoint union, i.e, when there is no common elements in $\mathcal{X}$ and $\mathcal{Y}$. $\{0, 1\}^n$ denotes the set of all binary strings of length n and $\{0, 1\}^*$ denotes the set of all binary strings of arbitrary length. 0^i denotes the string of length i with all bits zero. For any element $X \in \{0, 1\}^*$, $|X|$ denotes the number of bits of X. For any two elements $X, Y \in \{0, 1\}^*$, $X \| Y$ denotes the concatenation of X followed by Y. For $X, Y \in \{0, 1\}^n$, we write $X \oplus Y$ to denote the xor of X and Y. For any $X \in \{0, 1\}^*$, we parse X as $X = X_1 \| X_2 \| \ldots \| X_l$ where for each $i = 1, \ldots, l - 1$, X_i is an element of $\{0, 1\}^n$ and $1 \leq |X_l| \leq n$. We call each X_i a *block*. When there is a sequence of elements $X_1, X_2, \ldots, X_s \in \{0, 1\}^*$, we write X_a^i to denote the a-th block of the i-th element X_i. For any integer j, $\langle j \rangle$ denotes the n-bit binary representation of integer j. For integers $1 \leq b \leq a$, we write $(a)_b$ to denote $a(a - 1) \ldots (a - b + 1)$, where $(a)_0 = 1$ by convention. We write $[q]$ to refer to the set $\{1, \ldots, q\}$.

For a function $\Phi : \mathcal{X} \to \mathcal{Y}_1 \times \mathcal{Y}_2$, we write $\Phi(x) = (\phi_1, \phi_2)$ for all $x \in \mathcal{X}$. $\Phi[1]$ is the function from $\mathcal{X}$ to $\mathcal{Y}_1$ such that for all $x \in \mathcal{X}$, $\Phi[1](x) = \phi_1$. Similarly, $\Phi[2]$ is a function from $\mathcal{X}$ to $\mathcal{Y}_2$ such that $\Phi[2](x) = \phi_2$ for all $x \in \mathcal{X}$.

BLOCK CIPHERS. A *block cipher* (BC) with key space $\mathcal{K}$ and domain $\mathcal{X}$ is a mapping $\mathsf{E} : \mathcal{K} \times \mathcal{X} \to \mathcal{X}$ such that for all key $K \in \mathcal{K}$, $X \mapsto \mathsf{E}(K, X)$ is a permutation of $\mathcal{X}$. We denote $\mathsf{BC}(\mathcal{K}, \mathcal{X})$ the set of all block ciphers with key space $\mathcal{K}$ and domain $\mathcal{X}$. A *permutation* Π with domain $\mathcal{X}$ is a bijective mapping of $\mathcal{X}$ and $\mathsf{Perm}(\mathcal{X})$ denotes the set of all permutations over $\mathcal{X}$. $\mathsf{E} \in \mathsf{BC}(\mathcal{K}, \mathcal{X})$ is said to be a strong pseudorandom permutation or equivalently a strong block cipher if the sprp advantage of E against any chosen plaintext chosen ciphertext adaptive adversary A with oracle access to a permutation and its inverse with domain $\mathcal{X}$, defined as follows

$$\mathbf{Adv}_{\mathsf{E}}^{\mathrm{SPRP}}(\mathsf{A}) := |\Pr[K \leftarrow_\$ \mathcal{K} : \mathsf{A}^{\mathsf{E}_K, \mathsf{E}_K^{-1}} = 1] - \Pr[\Pi \leftarrow_\$ \mathsf{Perm}(\mathcal{X}) : \mathsf{A}^{\Pi, \Pi^{-1}} = 1]| \tag{1}$$

---

[5] Informally, a keyed hash function is said to be a partial-almost xor universal hash function, if for any two distinct inputs, the probability over the random draw of the hash key, that the first n-bit part of the sum of their hash output takes any value and the remaining m-bit part of the hash value collides, is very small.

that makes at most q queries with maximum running time t, is very small. When the adversary is given access only to the permutation and not its inverse, then we say the PRP advantage of A against E.

TWEAKABLE BLOCK CIPHERS. A *tweakable block cipher* (TBC) with key space $\mathcal{K}$, tweak space $\mathcal{T}$ and domain $\mathcal{X}$ is a mapping $\widetilde{\mathsf{E}} : \mathcal{K} \times \mathcal{T} \times \mathcal{X} \to \mathcal{X}$ such that for all key $k \in \mathcal{K}$ and all tweak $t \in \mathcal{T}$, $x \mapsto \widetilde{\mathsf{E}}(k, t, x)$ is a permutation of $\mathcal{X}$. We often write $\widetilde{\mathsf{E}}_k(t, x)$ or $\widetilde{\mathsf{E}}_k^t(x)$ for $\widetilde{\mathsf{E}}(k, t, x)$. We call a tweakable block cipher as (m, n) tweakable block cipher if $\mathcal{T} = \{0, 1\}^m$ and $\mathcal{X} = \{0, 1\}^n$. We denote $\mathsf{TBC}(\mathcal{K}, \mathcal{T}, \mathcal{X})$ the set of all such (m, n) tweakable block ciphers with key space $\mathcal{K}$, tweak space $\mathcal{T}$ and domain $\mathcal{X}$. A *tweakable permutation* with tweak space $\mathcal{T}$ and domain $\mathcal{X}$ is a mapping $\widetilde{\Pi} : \mathcal{T} \times \mathcal{X} \to \mathcal{X}$ such that for all tweak $T \in \mathcal{T}$, $X \mapsto \widetilde{\Pi}(T, X)$ is a permutation of $\mathcal{X}$. We often write $\widetilde{\Pi}^T(X)$ for $\widetilde{\Pi}(T, X)$. $\mathsf{TP}(\mathcal{T}, \mathcal{X})$ denotes the set of all (m, n) tweakable permutations with tweak space $\mathcal{T}(= \{0, 1\}^m)$ and domain $\mathcal{X}(= \{0, 1\}^n)$.

ADVERSARIAL MODEL FOR TBC. An adversary A for TBC has access to either of the pair of oracles $(\widetilde{\mathsf{E}}_K(\cdot, \cdot), \widetilde{\mathsf{E}}_K^{-1}(\cdot, \cdot))$ for some fixed key $K \in \mathcal{K}$ or access to the pair of oracles $(\widetilde{\Pi}(\cdot, \cdot), \widetilde{\Pi}^{-1}(\cdot, \cdot))$ oracles for some $\widetilde{\Pi} \in \mathsf{TP}(\mathcal{T}, \mathcal{X})$. Adversary A queries to the pair of oracles in an interleaved and adaptive way and after the interaction is over, it outputs a single bit b. We assume that A can query any tweak for at most μ times in all its encryption and decryption queries, which is called the *maximum tweak multiplicity*, i.e., if $\mu = 1$ then each queried tweak is distinct. Moreover, we assume that A does not repeat any query to the encryption or the decryption oracle. We also assume that A does not query the decryption oracle (resp. the encryption oracle) with the value that it obtained as a result of a previous encryption query (resp. decryption query). We call such an adversary A, a *non-trivial* (μ, q, t) chosen plaintext chosen ciphertext adaptive adversary, where A makes total q many encryption and decryption queries with running time at most t and maximum tweak multiplicity μ. Sometimes we write $(\mu, q, \ell, \sigma, t)$ chosen plaintext chosen ciphertext adaptive adversary A to emphasize that the maximum number of message blocks in a queried message of A is ℓ and the total number of message blocks that A can query is σ. When the parameters $\ell = \sigma = 0$, then we simply write (μ, q, t).

Definition 1 (TSPRP Security). *Let* $\widetilde{\mathsf{E}} \in \mathsf{TBC}(\mathcal{K}, \mathcal{T}, \mathcal{X})$ *be a tweakable block cipher and* A *be a non-trivial* (μ, q, t) *chosen plaintext chosen ciphertext adaptive adversary with oracle access to a tweakable permutation and its inverse with tweak space* $\mathcal{T}$ *and domain* $\mathcal{X}$. *The advantage of* A *in breaking the TSPRP security of* $\widetilde{\mathsf{E}}$ *is defined as*

$$\mathbf{Adv}_{\widetilde{\mathsf{E}}}^{\mathrm{TSPRP}}(\mathsf{A}) := |\Pr[K \leftarrow_{\$}\mathcal{K} : \mathsf{A}^{\widetilde{\mathsf{E}}_K, \widetilde{\mathsf{E}}_K^{-1}} = 1] - \Pr[\widetilde{\Pi} \leftarrow_{\$}\mathsf{TP}(\mathcal{T}, \mathcal{X}) : \mathsf{A}^{\widetilde{\Pi}, \widetilde{\Pi}^{-1}} = 1]|,$$
$$(2)$$

where the adversary queries with tweak $T \in \mathcal{T}$ *and input* $X \in \mathcal{X}$. *When the adversary is given access only to the tweakable permutation and not its inverse, then we say the tweakable pseudorandom permutation (TPRP) advantage of*

A against $\widetilde{\mathsf{E}}$. Informally, $\widetilde{\mathsf{E}}$ is said to be a tweakable strong pseudorandom permutation or equivalently a tweakable strong block cipher when the TSPRP advantage of $\widetilde{\mathsf{E}}$ against any adversary A that makes at most q queries with maximum running time t, as defined in Eq. (2), is very small.

ALMOST (XOR) UNIVERSAL AND ALMOST REGULAR HASH FUNCTION. Let $\mathcal{K}_h, \mathcal{X}$ be two non-empty finite sets and H be an n-bit keyed function $\mathsf{H} : \mathcal{K}_h \times \mathcal{X} \to \{0,1\}^n$. Then,

- H is said to be an ϵ-almost xor universal (AXU) hash function if for any distinct $X, X' \in \mathcal{X}$ and for any $Y \in \{0,1\}^n$,

$$\Pr[K_h \leftarrow_\$ \mathcal{K}_h : \mathsf{H}_{K_h}(X) \oplus \mathsf{H}_{K_h}(X') = Y] \leq \epsilon. \tag{3}$$

As a special case, when $Y = 0^n$, then H is said to be an ϵ-almost universal (AU) hash function.

- H is said to be an ϵ-almost regular hash function if for any $X \in \mathcal{X}$ and for any $Y \in \{0,1\}^n$,

$$\Pr[K_h \leftarrow_\$ \mathcal{K}_h : \mathsf{H}_{K_h}(X) = Y] \leq \epsilon. \tag{4}$$

It is easy to see that PolyHash with an n-bit key, as defined in [11,24], is an $\ell/2^n$-AXU and $\ell/2^n$-almsot regular hash function, where ℓ is the maximum number of message blocks. Proof of this result can be found in [11].

PARTIAL ALMOST (XOR) UNIVERSAL HASH FUNCTION. Let $\mathcal{K}_h, \mathcal{X}$ be two non-empty finite sets and H be an $(n + m)$-bit keyed function $\mathsf{H} : \mathcal{K}_h \times \mathcal{X} \to \{0,1\}^n \times \{0,1\}^m$. Then, H is said to be an (n, m, ϵ)-partial almost xor universal (pAXU) hash function if for any distinct $X, X' \in \mathcal{X}$ and for any $Y \in \{0,1\}^n$,

$$\Pr[K_h \leftarrow_\$ \mathcal{K}_h : \mathsf{H}_{K_h}(X) \oplus \mathsf{H}_{K_h}(X') = (Y, 0^m)] \leq \epsilon. \tag{5}$$

Note that, an ϵ-AXU $(n + m)$-bit keyed hash function is an (n, m, ϵ)-pAXU. We write $\mathsf{H}_{K_h}(X) = (H_1, H_2)$, where $H_1 \in \{0,1\}^n$ and $H_2 \in \{0,1\}^m$.

3   Specification and Security Result of Tweakable HCTR

HCTR, as proposed by Wang et al. [35], is a mode of operation which turns an n-bit strong prp into a tweakable strong prp that supports arbitrary and variable length input and tweak which is no less than n bits. For any message $M \in \{0,1\}^*$ and a tweak T, HCTR works as follows: it first parses the message M into l many blocks such that its first $l - 1$ message blocks are of length n-bits and the length of the last block is at most n. Then, it applies an n-bit PolyHash function on the string $M_2\| \ldots M_l\|T$ and xor its n-bit output value with the first message block M_1 to produce X. This X is then fed into an n-bit block cipher E whose output Y is xor-ed with X to produce an IV value which acts a counter in the counter mode encryption to produce the ciphertext blocks $C_2\| \ldots \|C_l$. Finally, the first ciphertext block C_1 is generated by applying the same PolyHash on

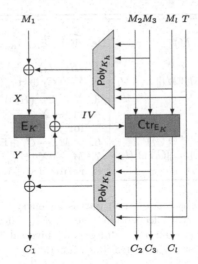

Fig. 2. HCTR construction with tweak T and message $M_1\|M_2\|\ldots\|M_l$ and the corresponding ciphertext $C_1\|C_2\|\ldots\|C_l$. Poly_{K_h} is the polynomial hash function with hash key K_h. Ctr_{E_K} is the block cipher based counter mode of encryption.

$C_2\|\ldots\|C_l\|T$ and xor its output with Y. Schematic diagram of HCTR is shown in Fig. 2.

Wang et al. [35] have shown HCTR to be a secure TES against all adaptive chosen plaintext and chosen ciphertext adversaries that make roughly $2^{n/3}$ encryption and decryption queries. Later in FSE 2008, Chakraborty and Nandi [6] have improved its security bound to $O(\sigma^2/2^n)$.

3.1 Specification of Tweakable HCTR

Our proposal Tweakable HCTR, which we denote as $\widetilde{\mathsf{HCTR}}$, closely resembles to the original HCTR with the exception that (i) the strong block cipher of HCTR is replaced by a (m, n) tweakable strong block cipher, where m is the size of the block cipher tweak and n is the block size of the TBC and (ii) the tweak used for the construction, which is processed through the upper and lower hash function in HCTR, is now processed through an independent keyed $(n + m)$-bit partial AXU hash function whose n-bit output is masked with the input and the output of the leftmost tweakable block cipher and the remaining m-bit output plays the role of the tweak of the underlying TBC. Moreover, all the block cipher calls of the counter mode encryption used in HCTR are replaced by TBCs where the same m-bit hash value of the tweak becomes the tweak of the underlying tweakable block cipher used in the tweakable counter mode of encryption. Schematic diagram of the construction is shown in Fig. 1 and its algorithmic description is shown in Fig. 3.

Enc.$\widehat{\text{HCTR}}[\widetilde{\mathsf{E}}_K, \mathsf{H}_{K_h}, \mathsf{H}'_L](T, M)$	Dec.$\widehat{\text{HCTR}}[\widetilde{\mathsf{E}}_K, \mathsf{H}_{K_h}, \mathsf{H}'_L](T, C)$
1. $(H_1, H_2) \leftarrow \mathsf{H}'_L(T)$	1. $(H_1, H_2) \leftarrow \mathsf{H}'_L(T)$
2. $X \leftarrow H_1 \oplus M_1 \oplus \mathsf{H}_{K_h}(M_2\|M_3\| \ldots \|M_l)$	2. $Y \leftarrow C_1 \oplus \mathsf{H}_{K_h}(C_2\|C_3\| \ldots \|C_l) \oplus H_1$
3. $Y \leftarrow \widetilde{\mathsf{E}}_K(H_2, X)$	3. $X \leftarrow \widetilde{\mathsf{E}}_K^{-1}(H_2, Y)$
4. $IV \leftarrow X \oplus Y$	4. $IV \leftarrow X \oplus Y$
5. **for** $j = 2$ to l:	5. **for** $j = 2$ to l:
6. $\quad C_j \leftarrow M_j \oplus \widetilde{\mathsf{E}}_K(H_2, IV \oplus \langle j \rangle)$	6. $\quad M_j \leftarrow C_j \oplus \widetilde{\mathsf{E}}_K(H_2, IV \oplus \langle j \rangle)$
7. $C_1 \leftarrow Y \oplus \mathsf{H}_{K_h}(C_2\|C_3\| \ldots \|C_l) \oplus H_1$	7. $M_1 \leftarrow X \oplus \mathsf{H}_{K_h}(M_2\|M_3\| \ldots \|M_l) \oplus H_1$
8. **return** $(C_1\|C_2\| \ldots \|C_l)$	8. **return** $(M_1\|M_2\| \ldots \|M_l)$

Fig. 3. Tweakable HCTR Construction. Left part is the encryption algorithm of tweakable HCTR and right part is its decryption algorithm. $\langle j \rangle$ denotes the n-bit binary representation of integer j. H is an n-bit almost xor universal hash function and H' is an $(n + m)$-bit partial almost xor universal hash function.

As can be seen from the algorithm there are three basic building blocks used in the construction of $\widehat{\text{HCTR}}$; an n-bit keyed AXU hash function H, an $(n+m)$-bit keyed pAXU hash function H' and a tweakable counter mode of encryption.

Given an n-bit string IV, we define a sequence $(IV_1, \ldots, IV_l)$, where each IV_i is some function of IV. Given such a sequence $(IV_1, \ldots, IV_l)$, a key K, a message $M = M_1\|M_2\| \ldots \|M_l$ (for simplicity we assume that $|M|$ is a multiple of n) and the hash value of an $(n + m)$-bit keyed pAXU hash function of the tweak T (i.e., $\mathsf{H}'_L(T)$), the tweakable counter mode is defined as follows:

$$\mathsf{Ctr}_{\widetilde{\mathsf{E}}_K^{H_2}, IV}(M_1, \ldots, M_l) = \left(M_1 \oplus \widetilde{\mathsf{E}}_K(H_2, IV_1), \ldots, M_l \oplus \widetilde{\mathsf{E}}_K(H_2, IV_l) \right),$$

where $IV_i = IV \oplus \langle i \rangle$ and $\mathsf{H}'_L(T) = (H_1, H_2)$. In case the last block M_l is incomplete then $M_l \oplus \widetilde{\mathsf{E}}_K(H_2, IV_l)$ is replaced by $M_l \oplus \mathsf{drop}_r(\widetilde{\mathsf{E}}_K(H_2, IV_l))$, where $r = n - |M_l|$ and $\mathsf{drop}_r(\widetilde{\mathsf{E}}_K(H_2, IV_l))$ is the first $(n - r)$ bits of $\widetilde{\mathsf{E}}_K(H_2, IV_l)$. If $l = 1$ (when we have one block message), we ignore line 4 and 5 of both the encryption and the decryption algorithm of $\widehat{\text{HCTR}}$ construction.

3.2 Security Result of Tweakable HCTR

In this section, we state the security result of $\widehat{\text{HCTR}}$. In specific, we state that if $\widetilde{\mathsf{E}}$ is a (m, n) tweakable strong block cipher, H is an ϵ-axu n-bit keyed hash function, H' is a δ-partial AXU $(n+m)$-bit keyed hash function, and H'[2] is a δ_{au} almost universal m-bit keyed hash function, then $\widehat{\text{HCTR}}$ is a secure TES against all $(\mu, q, \ell, \sigma, t)$ chosen plaintext and chosen ciphertext adaptive adversaries that make roughly $2^n/\mu\ell$ many encryption and decryption queries, where ℓ is the maximum number of message blocks among all q queries and σ is the total number of message blocks queried. Formally, the following result bounds the tsprp advantage of $\widehat{\text{HCTR}}$.

Theorem 1. *Let $\mathcal{M}, \mathcal{T}, \mathcal{K}, \mathcal{K}_h$ and $\mathcal{L}$ be finite and non-empty sets. Let $\widetilde{\mathsf{E}}$: $\mathcal{K} \times \{0,1\}^m \times \{0,1\}^n \to \{0,1\}^n$ be a (m,n) tweakable strong block cipher, H : $\mathcal{K}_h \times \mathcal{M} \to \{0,1\}^n$ be an ϵ-AXU and ϵ_1-almost regular n-bit keyed hash function and H' : $\mathcal{L} \times \mathcal{T} \to \{0,1\}^n \times \{0,1\}^m$ be an (n, m, δ)-partial AXU $(n + m)$-bit keyed hash function and $\mathsf{H}'[2]$ is a δ_{au}-almost universal m-bit keyed hash function. Then, for any $(\mu, q, \ell, \sigma, t)$ chosen plaintext chosen ciphertext adaptive adversary A against the tsprp security of $\widehat{\mathsf{HCTR}}[\widetilde{\mathsf{E}}, \mathsf{H}, \mathsf{H}']$, there exists a (μ, σ, t') chosen plaintext chosen ciphertext adaptive adversary A' against the tsprp security of $\widetilde{\mathsf{E}}$, where $t' = O(t + \sigma + q(2t_\mathsf{H} + t_{\mathsf{H}'}))$, σ is the total number of message blocks queried, t_H be the time for computing the hash function H, $t_{\mathsf{H}'}$ be the time for computing the hash function H' and $\mu \leq \min\{|\mathcal{T}|, q\}$, such that*

$$\mathbf{Adv}^{\mathrm{TSPRP}}_{\widehat{\mathsf{HCTR}}[\widetilde{\mathsf{E}}, \mathsf{H}, \mathsf{H}']}(\mathsf{A}) \leq \mathbf{Adv}^{\mathrm{TSPRP}}_{\widetilde{\mathsf{E}}}(\mathsf{A}') + 2(\mu - 1)(q\epsilon + \sigma/2^n) + 2q\sigma\delta_{\mathsf{au}}/2^n + q^2\delta$$

$$+ 2\max\{q\ell(\mu - 1)/2^n + q\sigma\delta_{\mathsf{au}}/2^n, \sigma\epsilon_1\}.$$

By assuming $\epsilon, \epsilon_1 \approx 2^{-n}$, $\delta_{\mathsf{au}} \approx 2^{-m}, \delta \approx 2^{-(n+m)}$ and $m > n$, $\widehat{\mathsf{HCTR}}$ is secured roughly upto $2^n/\mu\ell$ queries. Moreover, when all the tweaks are distinct, i.e., $\mu = 1$, then the tsprp security of $\widehat{\mathsf{HCTR}}$ becomes

$$\mathbf{Adv}^{\mathrm{TSPRP}}_{\widehat{\mathsf{HCTR}}[\widetilde{\mathsf{E}}, \mathsf{H}, \mathsf{H}']}(\mathsf{A}) \leq \mathbf{Adv}^{\mathrm{TSPRP}}_{\widetilde{\mathsf{E}}}(\mathsf{A}') + 2(\sigma\epsilon_1 + q\sigma\delta_{\mathsf{au}}/2^n) + q^2\delta.$$

Therefore, when all the tweaks in the encryption and decryption queries are distinct, then by assuming $\epsilon, \epsilon_1 \approx 2^{-n}$, $\delta_{\mathsf{au}} \approx 2^{-m}, \delta \approx 2^{(n+m)}$ and $m > n$, $\widehat{\mathsf{HCTR}}$ is secured roughly upto 2^n many message blocks.

4   Proof of Theorem 1

In this section, we prove Theorem 1. We would like to note that we will often refer to the construction $\widehat{\mathsf{HCTR}}[\widetilde{\mathsf{E}}, \mathsf{H}, \mathsf{H}']$ as simply $\widehat{\mathsf{HCTR}}$ when the underlying primitives are assumed to be understood.

As the first step of the proof, we replace $\widetilde{\mathsf{E}}_K$ with an (m, n)-bit tweakable uniform random permutation $\widetilde{\Pi}$ and denote the resulting construction as $\widehat{\mathsf{HCTR}}^*[\widetilde{\Pi}, \mathsf{H}, \mathsf{H}']$. It is easy to show that there exists an adversary against the tsprp security of $\widetilde{\mathsf{E}}$, making at most σ oracle queries and running in time at most $O(t + \sigma + q(2t_\mathsf{H} + t_{\mathsf{H}'}))$ with maximum tweak multiplicity μ, such that

$$\mathbf{Adv}^{\mathrm{TSPRP}}_{\widehat{\mathsf{HCTR}}[\widetilde{\mathsf{E}}, \mathsf{H}, \mathsf{H}']}(\mathsf{A}) \leq \mathbf{Adv}^{\mathrm{TSPRP}}_{\widetilde{\mathsf{E}}}(\mathsf{A}') + \underbrace{\mathbf{Adv}^{\mathrm{TSPRP}}_{\widehat{\mathsf{HCTR}}^*[\widetilde{\Pi}, \mathsf{H}, \mathsf{H}']}(\mathsf{A})}_{\delta^*}. \qquad (6)$$

Now, our goal is to upper bound δ^*. For doing this, we first describe how the ideal oracle works. Let us assume that $n\ell$ be the maximum size of any message M among all q many queries. Let $\mathcal{S}_i$ denotes the set of all binary strings of length i. Therefore, $\{0,1\}^{\leq n\ell}$, which denotes the set of all binary strings of length at

Ideal oracle ($) for Encryption On i^{th} input (T_i, M_i)	Ideal oracle ($^{-1}$) for Decryption On i^{th} input (T_i, C_i)

1. **if** $T_i = T_a$ **for some** $a \in [c]$
2.   **if** $M_i \in \mathcal{D}_a$, let $M_i = M_j$ for some
   $j < i$
3.   **then** $C_i \leftarrow C_j$
4.   **else** $C_i \leftarrow_\$ \mathcal{S}_{l_i} \setminus \mathcal{R}_a$
5. $\mathcal{D}_a = \mathcal{D}_a \cup \{M_i\}; \mathcal{R}_a = \mathcal{R}_a \cup \{C_i\}$
6. **else**
7. $c \leftarrow c + 1; T_c \leftarrow T_i$
8. $C_i \leftarrow_\$ \mathcal{S}_{l_i}$
9. $\mathcal{D}_c = \mathcal{D}_c \cup \{M_i\}; \mathcal{R}_c = \mathcal{R}_c \cup \{C_i\}$
10. **return** C_i
	1. **if** $T_i = T_a$ **for some** $a \in [c]$
	2. **if** $C_i \in \mathcal{R}_a$, let $C_i = C_j$ for
	   some $j < i$
	3. **then** $M_i \leftarrow M_j$
	4. **else** $M_i \leftarrow_\$ \mathcal{S}_{l_i} \setminus \mathcal{D}_a$
	5. $\mathcal{D}_a = \mathcal{D}_a \cup \{M_i\}; \mathcal{R}_a = \mathcal{R}_a \cup \{C_i\}$
	6. **else**
	7. $c \leftarrow c + 1; T_c \leftarrow T_i$
	8. $M_i \leftarrow_\$ \mathcal{S}_{l_i}$
	9. $\mathcal{D}_c = \mathcal{D}_c \cup \{M_i\}; \mathcal{R}_c = \mathcal{R}_c \cup \{C_i\}$
	10. **return** M_i

Fig. 4. Left part is the encryption algorithm of the ideal oracle and the right part is the decryption algorithm of the ideal oracle. c is the number of equivalent classes over the queried tweak space until the i-th query. $\mathcal{D}_a$ denotes the set of all already sampled output (for decryption) and queried input (for encryption) for a-th equivalent class and $\mathcal{R}_a$ denotes the set of all already sampled output (for encryption) and queried input (for decryption) for a-th equivalent class. l_i denotes the length of the i-th plaintext M_i, for encryption or the i-th ciphertext C_i for decryption.

most $n\ell$, can be written as $\mathcal{S}_1 \sqcup \mathcal{S}_2 \sqcup \ldots \sqcup \mathcal{S}_{n\ell}$. Now, for the i-th encryption or decryption query, the ideal oracle works as shown in Fig. 4.

In words, for the ith encryption query (T_i, M_i), the ideal oracle $ first checks if the tag T_i matches with some previous existing tags. If so, then it samples the ciphertext C_i without replacement from the set of all binary strings of length $|M_i|$; otherwise, it samples the C_i uniformly at random from $\mathcal{S}_{|M_i|}$. Decryption oracle also works in the similar way, except that the oracle samples the plaintext instead of ciphertext. Since, we have assumed the distinguisher is non-trivial, line 2–3 of both the algorithm will not be executed. Therefore, we write

$$\delta^* \leq \max_{D} \Pr[D^{\text{Enc}.\widetilde{\text{HCTR}}^*, \text{Dec}.\widetilde{\text{HCTR}}^*} = 1] - \Pr[D^{\$, \$^{-1}} = 1],$$

where the maximum is taken over all non-trivial distinguishers D that make total q many encryption and decryption queries with at most σ many blocks such that the maximum number of message blocks among all the queried messages is ℓ and the maximum tweak multiplicity μ. This formulation allows us to apply the H-Coefficient Technique [30,31], as we explain in more detail below, to prove

$$\delta^* \leq 2(\mu-1)(q\epsilon+\sigma/2^n)+2q\sigma\delta_{\text{au}}/2^n+q^2\delta+2\max\{q\ell(\mu-1)/2^n+q\sigma\delta_{\text{au}}/2^n, \sigma\epsilon_1\}. \tag{7}$$

H-Coefficient Technique. From now on, we fix a non-trivial distinguisher D that interacts with either (1) the real oracle $(\mathsf{Enc}.\widetilde{\mathsf{HCTR}}^*, \mathsf{Dec}.\widetilde{\mathsf{HCTR}}^*)$ for a (m, n)-bit tweakable random permutation $\widetilde{\Pi}$ and a pair of random hashing keys (K_h, L) or (2) the ideal oracle $(\$, \$^{-1})$, making q queries to its encryption and decryption oracle altogether with at most σ many blocks such that the maximum number of message blocks among all the queried messages is ℓ and the maximum tweak multiplicity is μ. When all the interactions between the oracle and D gets over, it outputs a single bit. We let,

$$\mathbf{Adv}(D) = \Pr[D^{\mathsf{Enc}.\widetilde{\mathsf{HCTR}}^*, \mathsf{Dec}.\widetilde{\mathsf{HCTR}}^*} = 1] - \Pr[D^{\$, \$^{-1}} = 1].$$

We assume that D is computationally unbounded and hence without loss of generality deterministic. Let

$$\tau := \{(T_1, M_1, C_1), (T_2, M_2, C_2), \ldots, (T_q, M_q, C_q)\}$$

be the list of all queries of D and its corresponding responses such that for all $i = 1, 2, \ldots, q$, $|C_i| = |M_i|$. Note that, as D is assumed to be non-trivial, there cannot be any repetition of triplet in τ. τ is called the *query transcript* of the attack. For convenience, we slightly modify the experiment where we reveal to the distinguisher (after it made all its queries and obtains the corresponding responses but before it output its decision) the hashing keys (K_h, L), if we are in the real world, or a pair of uniformly random dummy keys (K_h, L) if we are in the ideal world. All in all, the transcript of the attack is $\tau' = (\tau, K_h, L)$.

A transcript τ' is said to be an *attainable* (with respect to D) transcript if the probability to realize this transcript in the ideal world is non-zero. We denote $\mathcal{V}$ to be the set of all attainable transcripts and X_{re} and X_{id} denotes the probability distribution of transcript τ' induced by the real world and the ideal world respectively. We state in the following the main lemma of the H-Coefficient technique (see [9] for the proof of the lemma).

Lemma 1. *Let* D *be a fixed deterministic distinguisher and* $\mathcal{V} = \mathsf{GoodT} \sqcup \mathsf{BadT}$ *(disjoint union) be some partition of the set of all attainable transcripts. Suppose there exists* $\epsilon_{\mathrm{ratio}} \geq 0$ *such that for any* $\tau' \in \mathsf{GoodT}$,

$$\frac{\Pr[X_{\mathrm{re}} = \tau']}{\Pr[X_{\mathrm{id}} = \tau']} \geq 1 - \epsilon_{\mathrm{ratio}},$$

and there exists $\epsilon_{\mathrm{bad}} \geq 0$ *such that* $\Pr[X_{\mathrm{id}} \in \mathsf{BadT}] \leq \epsilon_{\mathrm{bad}}$. *Then,* $\mathbf{Adv}(D) \leq \epsilon_{\mathrm{ratio}} + \epsilon_{\mathrm{bad}}$.

The remaining of the proof of Theorem 1 is structured as follows: in Sect. 4.1 we define bad transcripts and upper bound their probability in the ideal world; in Sect. 4.2, we analyze good transcripts and prove that they are almost as likely in the real and the ideal world. Theorem 1 then follows easily by combining Lemma 1, Eqs. (6) and (7) above, and Lemmas 2 and 3 proven below.

4.1 Definition and Probability of Bad Transcripts

We begin with defining the bad transcripts and bound their probability in the ideal world. We denote $\widehat{M}_i$ as $M_2^i\|\ldots\|M_{l_i}^i$ and $\widehat{C}_i$ as $C_2^i\|\ldots\|C_{l_i}^i$. We recall that for a transcript $\tau' = (\tau, K_h, L)$, we denote $X_i = H_{K_h}(\widehat{M}_i) \oplus M_1^i \oplus H_{1,i}, Y_i = H_{K_h}(\widehat{C}_i) \oplus C_1^i \oplus H_{1,i}$ and $IV_a^i = X_i \oplus Y_i \oplus \langle a \rangle$, where $\mathsf{H}'_L(T_i) = (H_{1,i}, H_{2,i})$.

Definition 2. *An attainable transcript $\tau' = (\tau, K_h, L)$ is said to be a bad transcript if one of the following conditions are met*

    (B.1) *if there exists two queries $(T_i, M_i, C_i), (T_j, M_j, C_j)$ such that (a) $H_{2,i} = H_{2,j}$ and $X_i = X_j$ or (b) $H_{2,i} = H_{2,j}$ and $Y_i = Y_j$*

    (B.2) *if there exists two queries (T_i, M_i, C_i) and (T_j, M_j, C_j) such that $H_{2,i} = H_{2,j}$ and $IV_a^i = IV_b^j$ for $a \in [l_i]$ and $b \in [l_j]$.*

    (B.3) *if there exists distinct two queries (T_i, M_i, C_i) and (T_j, M_j, C_j) such that $H_{2,i} = H_{2,j}$ and $M_a^i \oplus C_a^i = M_b^j \oplus C_b^j$ for $a \in [l_i]$ and $b \in [l_j]$.*

    (B.4) *if there exists two queries (T_i, M_i, C_i) and (T_j, M_j, C_j) such that $H_{2,i} = H_{2,j}$ and $X_i = IV_a^j$ for $a \in [l_j]$.*

    (B.5) *if there exists two queries (T_i, M_i, C_i) and (T_j, M_j, C_j) such that $H_{2,i} = H_{2,j}$ and $Y_i = M_a^j \oplus C_a^j$ for $a \in [l_j]$.*

Note that in the ideal world, X_i and Y_i's are determined through the sampled random dummy hash key (K_h, L).

The underlying principle for identifying the bad events is that

if hash of two tweak value happens to collide in two different invocations of the cipher, then the block cipher input and output must not collide.

Let BadT denotes the set of all attainable transcripts τ' such that it satisfies either of the above conditions and the event B denotes B := B.1 $\vee$ B.2 $\vee$ B.3 $\vee$ B.4 $\vee$ B.5. We bound the probability of the event B in the ideal world as follows:

Lemma 2. *Let X_{id} and BadT be defined as above. Then we have,*

$$\Pr[X_{\mathrm{id}} \in \mathsf{BadT}] \le \epsilon_{\mathsf{bad}} = 2(\mu - 1)(q\epsilon + \sigma/2^n) + q^2\delta + 2q\sigma\delta_{\mathsf{au}}/2^n$$
$$+2\max\{q\ell(\mu - 1)/2^n + q\sigma\delta_{\mathsf{au}}/2^n, \sigma\epsilon_1\}.$$

Proof. We let Θ_i denote the set of attainable transcripts satisfying only (B.i) condition. Recall that, in the ideal world, the pair of hash keys (K_h, L) is drawn uniformly and independently from the query transcript. Moreover, K_h is drawn independent of L. We are going to consider every conditions in turn.

CONDITION B.1. We first fix two distinct queries (T_i, M_i, C_i) and (T_j, M_j, C_j). Now, we compute the following probability over the random draw of the hash keys L and K_h.

$$\Pr[H_{2,i} = H_{2,j}, X_i = X_j]. \tag{8}$$

We can write Eq. (8) as the joint probability of the following two events:

$$H_{2,i} = H_{2,j}, \quad \mathsf{H}_{K_h}(\widehat{M}_i) \oplus M_1^i \oplus H_{1,i} = \mathsf{H}_{K_h}(\widehat{M}_j) \oplus M_1^j \oplus H_{1,j}.$$

- **Case (a):** if $T_i = T_j$, then $(H_{1,i}, H_{2,i}) = (H_{1,j}, H_{2,j})$. Therefore, the above probability is bounded by ϵ, the AXU probability of H, as we assume the adversary is non-trivial. The number of choices of i is q and j is $\mu - 1$ and thus the overall probability becomes $q(\mu - 1)\epsilon$.
- **Case (b):** if $T_i \neq T_j$, then by conditioning the hash key K_h, the above probability is bounded by δ, the partial almost xor universal probability of the hash function H'. In this case, number of choices of (i, j) is $\binom{q}{2}$ and thus the overall probability becomes $\binom{q}{2}\delta$.

As a result, we have the following

$$\Pr[H_{2,i} = H_{2,j}, X_i = X_j] \leq q(\mu - 1)\epsilon + \binom{q}{2}\delta. \tag{9}$$

By doing the exact similar analysis, the probability over the random draw of the pair of hash keys (K_h, L),

$$\Pr[H_{2,i} = H_{2,j}, Y_i = Y_j] \leq q(\mu - 1)\epsilon + \binom{q}{2}\delta. \tag{10}$$

By summing Eqs. (9) and (10), the overall probability becomes

$$\Pr[X_{\mathrm{id}} \in \Theta_1] \leq 2(\mu - 1)q\epsilon + q^2\delta. \tag{11}$$

CONDITION B.2. We fix two distinct queries (T_i, M_i, C_i) and (T_j, M_j, C_j) and consider the joint probability of $H_{2,i} = H_{2,j}$ and $IV_a^i = IV_b^j$. Note that,

$$IV_a^i = \mathsf{H}_{K_h}(\widehat{M_i}) \oplus \mathsf{H}_{K_h}(\widehat{C_i}) \oplus M_1^i \oplus C_1^i \oplus \langle a \rangle. \tag{12}$$
$$IV_j^b = \mathsf{H}_{K_h}(\widehat{M_j}) \oplus \mathsf{H}_{K_h}(\widehat{C_j}) \oplus M_1^j \oplus C_1^j \oplus \langle b \rangle. \tag{13}$$

Without loss of generality we assume that $i < j$. Now, for a fixed choice of $a \in [l_i]$ and $b \in [l_j]$ and by fixing the hash key K_h, the probability over the random draw of C_1^j (if j-th query is an encryption query) or the random draw of M_i^j (if j-th query is a decryption query) that (12) = (13) is at most 2^{-n}. We have the following two cases:

- **Case (a):** if $T_i = T_j$, then the probability that $H_{2,i} = H_{2,j}$ is one. In this case, number of choices of (i, a) is at most σ and the number of choices of j is at most $\mu - 1$. Note that, the choices of b is only 1 as for fixed values of IV_a^i, IV_b^j and a that satisfies $IV_a^i \oplus IV_b^j = \langle a \rangle \oplus \langle b \rangle$, value of b is uniquely determined. Summing over every possible choices of (i, a, j, b), we get

$$\Pr[X_{\mathrm{id}} \in \Theta_2] \leq \sigma(\mu - 1)/2^n. \tag{14}$$

- **Case (b):** if $T_i \neq T_j$, then the probability that $H_{2,i} = H_{2,j}$ is at most δ_{au}, which follows from the almost universal property of H'[2]. As before, the number of choices of (i, a) is at most σ and the number of choices of j is at most q. Moreover, as argued before, there is a unique choice of b for a fixed values of IV_a^i, IV_b^j and a that satisfies $IV_a^i \oplus IV_b^j = \langle a \rangle \oplus \langle b \rangle$. Summing over every possible choices of (i, a, j, b), we get

$$\Pr[X_{\mathrm{id}} \in \Theta_2] \leq q\sigma\delta_{\mathrm{au}}/2^n. \tag{15}$$

By summing Eqs. (14) and (15), we have the following:

$$\Pr[X_{\mathrm{id}} \in \Theta_2] \leq \sigma(\mu-1)/2^n + q\sigma\delta_{\mathrm{au}}/2^n. \tag{16}$$

Note that, when $i = j$, then we cannot have $IV_a^i = IV_b^j$ for $a \neq b$ and hence in that case the probability will become 0.

CONDITION B.3. Analysis of this condition is exactly similar to that of condition B.2 and therefore, we have

$$\Pr[X_{\mathrm{id}} \in \Theta_3] \leq \sigma(\mu-1)/2^n + q\sigma\delta_{\mathrm{au}}/2^n. \tag{17}$$

CONDITION B.4. We first fix two distinct queries (T_i, M_i, C_i), (T_j, M_j, C_j) and compute the following:

$$\Pr[H_{2,i} = H_{2,j}, X_i = IV_a^j].$$

For a fixed index $a \in [l_j]$, we compute the probability of $X_i = IV_a^j$. Recall that, $X_i = H_{K_h}(\widehat{M_i}) \oplus M_1^i \oplus H_{1,i}$. Therefore, the probability of $X_i = IV_a^j$ is nothing but to calculate the probability of the event that

$$H_{K_h}(\widehat{M_i}) \oplus H_{K_h}(\widehat{M_j}) \oplus H_{K_h}(\widehat{C_j}) = M_1^i \oplus M_1^j \oplus C_1^j \oplus H_{1,i} \oplus \langle a \rangle. \tag{18}$$

Without loss of generality we assume that $i < j$. If the j-th query is an encryption query, then C_1^j is random and hence over the random draw of C_1^j, the probability of Eq. (18) is 2^{-n}. Similarly, if the j-th query is a decryption query, then M_1^j is random and hence over the random draw of M_1^j, the probability of Eq. (18) is 2^{-n}. We have the following two cases:

- **Case (a):** if $T_i = T_j$, then the probability that $H_{2,i} = H_{2,j}$ is one. In this case, the number of choices of i is q and (j, a) is at most $(\mu - 1)\ell$. Therefore, by summing over every possible choices of (i, j, a), we get

$$\Pr[X_{\mathrm{id}} \in \Theta_4] \leq q\ell(\mu-1)/2^n. \tag{19}$$

- **Case (b):** if $T_i \neq T_j$, then the probability that $H_{2,i} = H_{2,j}$ is at most δ_{au}, which follows from the almost universal property of H'[2]. Here, the number of choices of (j, a) is at most σ and the number of choices of i is at most q. Summing over every possible choices of (i, j, a), we get

$$\Pr[X_{\mathrm{id}} \in \Theta_4] \leq q\sigma\delta_{\mathrm{au}}/2^n. \tag{20}$$

By summing Eqs. (19) and (20), we obtain

$$\Pr[X_{id} \in \Theta_4] \leq q\ell(\mu - 1)/2^n + q\sigma\delta_{au}/2^n. \tag{21}$$

When $i = j$, then calculating the joint probability of $H_{2,i} = H_{2,j}, X_i = IV_a^j$ is nothing but to calculate the probability of the event that

$$\mathsf{H}_{K_h}(\widehat{C}_i) = C_1^i \oplus H_{1,i} \oplus \langle a \rangle. \tag{22}$$

Note that, when $i = j$, then the probability of $H_{2,i} = H_{2,j}$ is one. Now, for a fixed $i \in [q]$ and $a \in [l_i]$, over the random draw the hash key K_h, the probability of the above event is bounded by ϵ_1 due to the almost regular property of the hash function. Now, summing over all possible choices of (i, a) we get

$$\Pr[X_{id} \in \Theta_4] \leq \sigma\epsilon_1. \tag{23}$$

Therefore, from Eqs. (21) and (23) we have

$$\Pr[X_{id} \in \Theta_4] \leq \max\{q\ell(\mu - 1)/2^n + q\sigma\delta_{au}/2^n, \sigma\epsilon_1\}. \tag{24}$$

CONDITION B.5. Analysis of this condition is exactly similar to that of condition B.4. Therefore, we have

$$\Pr[X_{id} \in \Theta_5] \leq \max\{q\ell(\mu - 1)/2^n + q\sigma\delta_{au}/2^n, \sigma\epsilon_1\}. \tag{25}$$

The result follows by the union bound of these conditions in Eqs. (11), (16), (17), (24) and (25).

4.2   Analysis of Good Transcripts

In this section, we show that for a good transcript τ', realizing τ' is almost as likely in the real and the ideal world. Formally, we prove the following lemma.

Lemma 3. *Let $\tau' = (\tau, K_h, L)$ be a good transcript. Then*

$$\frac{\mathsf{p}_{re}(\tau')}{\mathsf{p}_{id}(\tau')} := \frac{\Pr[X_{re} = \tau']}{\Pr[X_{id} = \tau']} \geq 1.$$

Proof. Let $\tau' = (\tau, K_h, L) \in \mathsf{GoodT}$ and let $\tau = ((T_1, M_1, C_1), \ldots, (T_q, M_q, C_q))$. Now, we define an equivalence relation $\sim_\tau$ over τ such that two elements of τ are related through $\sim_\tau$, i.e., $(T_i, M_i, C_i) \sim_\tau (T_j, M_j, C_j)$, if and only if $\mathsf{H}'_L(T_i)[2] = \mathsf{H}'_L(T_j)[2]$. This equivalence relation induces a partition over τ and let $\mathcal{P}_1, \mathcal{P}_2, \ldots, \mathcal{P}_r$ be r many partitions of τ where $|\mathcal{P}_i| = \mathsf{q}_i$, called the multiplicity of the hash value of the tweak T_i. Therefore, we have $\mathsf{q}_1 + \mathsf{q}_2 + \ldots + \mathsf{q}_r = q$. Now, we consider any i-th partition $\mathcal{P}_i$ for $i = 1, \ldots, r$. Note that, $\mathcal{P}_i$ is of the form:

$$\mathcal{P}_i = ((T_{x_1}, M_{x_1}, C_{x_1}), \ldots, (T_{x_{q_i}}, M_{x_{q_i}}, C_{x_{q_i}})),$$

where $H'_L(T_{x_1})[2] = H'_L(T_{x_2})[2] = \ldots = H'_L(T_{x_{q_i}})[2]$. We say two elements (T_x, M_x, C_x) and (T_y, M_y, C_y) of $\mathcal{P}_i$ are related through an equivalence relation $\sim_\ell$ if and only if $|M_x| = |M_y|$ and hence $|C_x| = |C_y|$. Therefore, $\sim_\ell$ induces another v_i many *inner* partitions $\mathcal{C}_1, \mathcal{C}_2, \ldots, \mathcal{C}_{\mathsf{v}_i}$ of $\mathcal{P}_i$ such that

$$c_1 + c_2 + \ldots + c_{\mathsf{v}_i} = \mathsf{q}_i,$$

where $c_j = |\mathcal{C}_j|$ denotes the number of elements in the j-th partition $\mathcal{C}_j$. Moreover, for the simplicity of the analysis, we assume that the length of each queried message is a multiple of n.

IDEAL INTERPOLATION PROBABILITY. To compute the ideal interpolation probability for the fixed transcript $\tau' = (\tau, K_h, L)$, we first consider any partition $\mathcal{P}_i$ in which q_i many hash values of the tweaks attain the same value. Now, let us consider the j-th inner partition $\mathcal{C}_j$ of $\mathcal{P}_i$ for which we have c_j many (M, C) pairs having the same length nl_j. Therefore, for $\mathcal{C}_j$, the probability becomes $1/(2^{nl_j})_{c_j}$. Similarly, for other inner partition $\mathcal{C}_{j'}$ of $\mathcal{P}_i$ in which $c_{j'}$ many (M, C) pairs having the same length $nl_{j'}$, the probability becomes $1/(2^{nl_{j'}})_{c_{j'}}$. Thus, for a fixed partition $\mathcal{P}_i$, the probability becomes

$$\prod_{j=1}^{\mathsf{v}_i} \frac{1}{(2^{nl_j})_{c_j}}.$$

Since, we have r many such partitions, the overall probability becomes

$$\prod_{i=1}^{r} \prod_{j=1}^{\mathsf{v}_i} \frac{1}{(2^{nl_j})_{c_j}}.$$

By summarizing the above, we have

$$\Pr[X_{\mathrm{id}} = \tau'] = \frac{1}{|\mathcal{K}_h|} \frac{1}{|\mathcal{L}|} \cdot \prod_{i=1}^{r} \prod_{j=1}^{\mathsf{v}_i} \frac{1}{(2^{nl_j})_{c_j}}, \tag{26}$$

where nl_j is the length of every message in partition $\mathcal{C}_j$.

REAL INTERPOLATION PROBABILITY. To compute the real interpolation probability for the fixed good transcript $\tau' = (\tau, K_h, L)$, we first consider several lists created from τ:

$$\mathcal{L}_A = ((H_{2,1}, X_1, Y_1), (H_{2,2}, X_2, Y_2), \ldots, (H_{2,q}, X_q, Y_q)),$$

where $X_i = M_1^i \oplus H_{K_h}(\hat{M}_i) \oplus H_{1,i}$ and $Y_i = C_1^i \oplus H_{K_h}(\hat{C}_i) \oplus H_{1,i}$. Moreover, we also create q many different lists from τ as follows:

$$\mathcal{L}_1 = ((H_{2,1}, IV_1^1, Z_1^1), (H_{2,1}, IV_2^1, Z_2^1), \ldots, (H_{2,1}, IV_{l_1-1}^1, Z_{l_1-1}^1))$$
$$\mathcal{L}_2 = ((H_{2,2}, IV_1^2, Z_1^2), (H_{2,2}, IV_2^2, Z_2^2), \ldots, (H_{2,2}, IV_{l_2-1}^2, Z_{l_2-1}^2))$$
$$\vdots \qquad \vdots \qquad \vdots \qquad \vdots$$
$$\mathcal{L}_q = ((H_{2,q}, IV_1^q, Z_1^q), (H_{2,q}, IV_2^q, Z_2^q), \ldots, (H_{2,q}, IV_{l_q-1}^q, Z_{l_q-1}^q)),$$

where $IV_k^i = X_i \oplus Y_i \oplus \langle k \rangle$ and $Z_k^i = M_{k+1}^i \oplus C_{k+1}^i$. Now, we consider any partition $\mathcal{P}_i$, in which q_i many hash values of the tweaks (i.e., $H_{2,i}$) attain the same value. This implies that q_i many elements from the list $\mathcal{L}_A$ i.e.

$$((H_{2,k_1}, X_{k_1}, Y_{k_1}), \ldots, (H_{2,k_{q_i}}, X_{k_{q_i}}, Y_{k_{q_i}}))$$

will have the same tweak value, but all the $X_{k_1}, X_{k_2}, \ldots, X_{K_{q_i}}$ values are distinct. Similarly, all the $Y_{k_1}, Y_{k_2}, \ldots, Y_{K_{q_i}}$ values are distinct, otherwise condition B.1 would have been satisfied. Moreover, q_i many lists from $\mathcal{L}_1, \ldots, \mathcal{L}_q$ will also have the same tweak value i.e., $H_{2,k_1} = H_{2,k_2} = \ldots = H_{2,k_{q_i}}$ in

$$\mathcal{L}_{k_1} = ((H_{2,k_1}, IV_1^{k_1}, Z_1^{k_1}), (H_{2,k_1}, IV_2^{k_1}, Z_2^{k_1}), \ldots, (H_{2,k_1}, IV_{l_{k_1}-1}^{k_1}, Z_{l_{k_1}-1}^{k_1}))$$

$$\mathcal{L}_{k_2} = ((H_{2,k_2}, IV_1^{k_2}, Z_1^{k_2}), (H_{2,k_2}, IV_2^{k_2}, Z_2^{k_2}), \ldots, (H_{2,k_2}, IV_{l_{k_2}-1}^{k_2}, Z_{l_{k_2}-1}^{k_2}))$$

$$\vdots \qquad \vdots \qquad \vdots \qquad \vdots$$

$$\mathcal{L}_{k_{q_i}} = ((H_{2,k_{q_i}}, IV_1^{k_{q_i}}, Z_1^{k_{q_i}}), (H_{2,k_{q_i}}, IV_2^{k_{q_i}}, Z_2^{k_{q_i}}), \ldots, (H_{2,k_{q_i}}, IV_{l_{k_{q_i}}-1}^{k_{q_i}}, Z_{l_{k_{q_i}}-1}^{k_{q_i}}))$$

As τ' is a good transcript, it is evident that $IV_\beta^\alpha \neq IV_{\beta'}^{\alpha'}$ where $\alpha, \alpha' \in \{k_1, \ldots, k_{q_i}\}$ and $\beta \in [l_\alpha - 1], \beta' \in [l_{\alpha'} - 1]$ otherwise condition B.2 would have been satisfied. Similarly, as τ' is a good transcript, we have $Z_\beta^\alpha \neq Z_{\beta'}^{\alpha'}$ otherwise condition B.3 would have been satisfied. Moreover, due to condition B.4 and B.5, we also have $IV_\beta^\alpha \neq X_{\alpha'}$ and $Z_\beta^\alpha \neq Y_{\alpha'}$. This immediately gives us the probability for any such fixed partition $\mathcal{P}_i$ is

$$\frac{1}{(2^n)_{q_i+(l_{k_1}-1)+(l_{k_2}-1)+\ldots+(l_{k_{q_i}}-1)}} = \frac{1}{(2^n)_{l_{k_1}+l_{k_2}+\ldots+l_{k_{q_i}}}}.$$

Now, let us consider the j-th inner partition $\mathcal{C}_j$ of $\mathcal{P}_i$ for which we have c_j many (M, C) pairs having the same message length nl_j. Therefore, for the fixed partition $\mathcal{P}_i$, the eventual probability will be $1/(2^n)_{q_i+\theta}$, where $\theta = c_1(l_1 - 1) + c_2(l_2 - 1) + \ldots + c_{v_i}(l_{v_i} - 1)$. Summarizing above, we have

$$\Pr[X_{\mathrm{re}} = \tau'] = \frac{1}{|\mathcal{K}_h|} \cdot \frac{1}{|\mathcal{L}|} \cdot \prod_{i=1}^r \frac{1}{(2^n)_{q_i+\theta}} = \frac{1}{|\mathcal{K}_h|} \cdot \frac{1}{|\mathcal{L}|} \cdot \prod_{i=1}^r \frac{1}{(2^n)_{c_1 l_1 + c_2 l_2 + \ldots + c_{v_i} l_i}} \tag{27}$$

COMPUTE THE RATIO. Finally, by taking the ratio of Eqs. (27) to (26), we have

$$\frac{\Pr[X_{\mathrm{re}} = \tau']}{\Pr[X_{\mathrm{id}} = \tau']} = \prod_{i=1}^r \frac{\prod_{j=1}^{v_i} (2^{nl_j})_{c_j}}{(2^n)_{c_1 l_1 + c_2 l_2 + \ldots + c_{v_i} l_i}} = \prod_{i=1}^r \underbrace{\frac{(2^{nl_1})_{c_1} \cdot (2^{nl_2})_{c_2} \cdots (2^{nl_{v_i}})_{c_{v_i}}}{(2^n)_{c_1 l_1 + c_2 l_2 + \ldots + c_{v_i} l_i}}}_{(R)}$$

The following proposition shows that for any $i = 1, \ldots, r$, R ≥ 1 and hence the result follows. $\qquad\square$

Proposition 1. *For positive integers $c_1, \ldots, c_t$ and $l_1, \ldots, l_t$ such that $\sum\limits_{i=1}^{t} c_i l_i \leq$ 2^n, we have,*

$$(2^n)_{c_1 l_1 + c_2 l_2 + \ldots + c_t l_t} \leq \prod_{j=1}^{t} (2^{n l_j})_{c_j}.$$

Proof of the result is trivial and hence omitted.

COROLLARY OF THEOREM 1. When the input tweak size of the construction matches with the tweak size of the tweakable block cipher, then we can evade the hash function evaluation for processing tweaks. As a result, we directly feed the tweak of the construction to the tweakable block cipher and the security bound of the resulting construction is obtained as a simple corollary of Theorem 1. For an m-bit tweak T, we define the hash function $\mathsf{H}'_L(T)$ as $\mathsf{H}'_L(T) = (0^n, T)$. Note that, for this partial almost xor universal hash function, $\delta = 0$ and $\delta_{\mathsf{au}} = 0$. Therefore, following Theorem 1, the information theoretic security bound of tweakable HCTR^* for m-bit tweak becomes

$$\mathbf{Adv}^{\mathrm{TSPRP}}_{\widetilde{\mathsf{HCTR}}^*[\widetilde{\Pi}, \mathsf{H}, \mathsf{H}']}(\mathsf{A}) \leq 2(\mu - 1)(q\epsilon + \sigma/2^n) + 2\max\{q\ell(\mu - 1)/2^n, \sigma\epsilon_1\}.$$

When all the tweaks in the encryption and decryption queries are distinct (i.e., $\mu = 1$), then by assuming $\epsilon, \epsilon_1 \approx 2^{-n}$, $\widetilde{\mathsf{HCTR}}^*$ is secured roughly upto 2^n many message blocks.

5   Conclusion

HCTR is one of the most efficient TES candidates which turns an n-bit block cipher into a variable length TBC. In this paper, we have proposed tweakable HCTR, that turns an (m, n)-bit TBC into a variable length TBC, allowing to process arbitrary large tweaks, and proven its optimal security (in terms of the block size) for the case of distinct tweak. Moreover, we have shown that the construction gives a graceful security degradation with the maximum number of repetitions of tweak. It is evident that one can make the HCTR mode BBB secure by just doubling the size of all its primitives. Nevertheless, designing a double block sprp is not trivial. For example, 5 round Feistel construction [18] provides 2^n security against all adaptive chosen plaintext and chosen ciphertext adversaries. Thus, designing an efficient TES based on an n-bit block cipher with beyond the birthday bound security still remains an interesting open problem. However, following [17], analysis of multi-key security of HCTR will be similar to the analysis of ours.

Acknowledgements. Authors are supported by the WISEKEY project of R.C.Bose Centre for Cryptology and Security. The authors would like to thank all the anonymous reviewers of Indocrypt 2018 for their invaluable comments and suggestions that help to improve the overall quality of the paper.

References

1. Bellare, M., Kilian, J., Rogaway, P.: The security of cipher block chaining. In: Desmedt, Y.G. (ed.) CRYPTO 1994. LNCS, vol. 839, pp. 341–358. Springer, Heidelberg (1994). https://doi.org/10.1007/3-540-48658-5_32
2. Bellare, M., Rogaway, P.: Encode-then-encipher encryption: how to exploit nonces or redundancy in plaintexts for efficient cryptography. In: Okamoto, T. (ed.) ASIACRYPT 2000. LNCS, vol. 1976, pp. 317–330. Springer, Heidelberg (2000). https://doi.org/10.1007/3-540-44448-3_24
3. Bhaumik, R., Nandi, M.: An inverse-free single-keyed tweakable enciphering scheme. In: Iwata, T., Cheon, J.H. (eds.) ASIACRYPT 2015, Part II. LNCS, vol. 9453, pp. 159–180. Springer, Heidelberg (2015). https://doi.org/10.1007/978-3-662-48800-3_7
4. Carter, L., Wegman, M.N.: Universal classes of hash functions. J. Comput. Syst. Sci. 18(2), 143–154 (1979)
5. Chakraborty, D., Ghosh, S., Sarkar, P.: A fast single-key two-level universal hash function. IACR Trans. Symmetric Cryptol. 2017(1), 106–128 (2017)
6. Chakraborty, D., Nandi, M.: An improved security bound for HCTR. In: Nyberg, K. (ed.) FSE 2008. LNCS, vol. 5086, pp. 289–302. Springer, Heidelberg (2008). https://doi.org/10.1007/978-3-540-71039-4_18
7. Chakraborty, D., Sarkar, P.: HCH: a new tweakable enciphering scheme using the hash-encrypt-hash approach. In: Barua, R., Lange, T. (eds.) INDOCRYPT 2006. LNCS, vol. 4329, pp. 287–302. Springer, Heidelberg (2006). https://doi.org/10.1007/11941378_21
8. Chakraborty, D., Sarkar, P.: A new mode of encryption providing a tweakable strong pseudo-random permutation. In: Robshaw, M. (ed.) FSE 2006. LNCS, vol. 4047, pp. 293–309. Springer, Heidelberg (2006). https://doi.org/10.1007/11799313_19
9. Chen, S., Lampe, R., Lee, J., Seurin, Y., Steinberger, J.: Minimizing the two-round even-mansour cipher. In: Garay, J.A., Gennaro, R. (eds.) CRYPTO 2014, Part I. LNCS, vol. 8616, pp. 39–56. Springer, Heidelberg (2014). https://doi.org/10.1007/978-3-662-44371-2_3
10. Daemen, J., Rijmen, V.: Rijndael for AES. In: AES Candidate Conference, pp. 343–348 (2000)
11. Datta, N., Dutta, A., Nandi, M., Yasuda, K.: Encrypt or decrypt? to make a single-key beyond birthday secure nonce-based MAC. In: Shacham, H., Boldyreva, A. (eds.) CRYPTO 2018, Part I. LNCS, vol. 10991, pp. 631–661. Springer, Cham (2018). https://doi.org/10.1007/978-3-319-96884-1_21
12. Halevi, S.: EME*: extending EME to handle arbitrary-length messages with associated data. In: Canteaut, A., Viswanathan, K. (eds.) INDOCRYPT 2004. LNCS, vol. 3348, pp. 315–327. Springer, Heidelberg (2004). https://doi.org/10.1007/978-3-540-30556-9_25
13. Halevi, S.: Invertible universal hashing and the TET encryption mode. In: Menezes, A. (ed.) CRYPTO 2007. LNCS, vol. 4622, pp. 412–429. Springer, Heidelberg (2007). https://doi.org/10.1007/978-3-540-74143-5_23
14. Halevi, S., Rogaway, P.: A tweakable enciphering mode. In: Boneh, D. (ed.) CRYPTO 2003. LNCS, vol. 2729, pp. 482–499. Springer, Heidelberg (2003). https://doi.org/10.1007/978-3-540-45146-4_28
15. Halevi, S., Rogaway, P.: A parallelizable enciphering mode. In: Okamoto, T. (ed.) CT-RSA 2004. LNCS, vol. 2964, pp. 292–304. Springer, Heidelberg (2004). https://doi.org/10.1007/978-3-540-24660-2_23

16. Jha, A., List, E., Minematsu, K., Mishra, S., Nandi, M.: XHX - A framework for optimally secure tweakable block ciphers from classical block ciphers and universal hashing. IACR Cryptology ePrint Archive 2017, p. 1075 (2017)
17. Lee, J., Luykx, A., Mennink, B., Minematsu, K.: Connecting tweakable and multi-key blockcipher security. Des. Codes Crypt. **86**(3), 623–640 (2018)
18. Luby, M., Rackoff, C.: How to construct pseudorandom permutations from pseudorandom functions. SIAM J. Comput. **17**(2), 373–386 (1988)
19. Mancillas-López, C., Chakraborty, D., Rodríguez-Henríquez, F.: Efficient implementations of some tweakable enciphering schemes in reconfigurable hardware. In: Srinathan, K., Rangan, C.P., Yung, M. (eds.) INDOCRYPT 2007. LNCS, vol. 4859, pp. 414–424. Springer, Heidelberg (2007). https://doi.org/10.1007/978-3-540-77026-8_33
20. McGrew, D.A., Fluhrer, S.R.: The extended codebook (XCB) mode of operation. IACR Cryptology ePrint Archive 2004, p. 278 (2004)
21. McGrew, D.A., Viega, J.: The security and performance of the galois/counter mode (GCM) of operation. In: Canteaut, A., Viswanathan, K. (eds.) INDOCRYPT 2004. LNCS, vol. 3348, pp. 343–355. Springer, Heidelberg (2004). https://doi.org/10.1007/978-3-540-30556-9_27
22. Mennink, B.: Towards tight security of cascaded LRW2. IACR Cryptology ePrint Archive 2018, p. 434 (2018)
23. Minematsu, K.: Beyond-birthday-bound security based on tweakable block cipher. In: Dunkelman, O. (ed.) FSE 2009. LNCS, vol. 5665, pp. 308–326. Springer, Heidelberg (2009). https://doi.org/10.1007/978-3-642-03317-9_19
24. Minematsu, K., Iwata, T.: Building blockcipher from tweakable blockcipher: extending FSE 2009 Proposal. In: Chen, L. (ed.) IMACC 2011. LNCS, vol. 7089, pp. 391–412. Springer, Heidelberg (2011). https://doi.org/10.1007/978-3-642-25516-8_24
25. Minematsu, K., Iwata, T.: Tweak-length extension for tweakable blockciphers. In: Groth, J. (ed.) IMACC 2015. LNCS, vol. 9496, pp. 77–93. Springer, Cham (2015). https://doi.org/10.1007/978-3-319-27239-9_5
26. Minematsu, K., Matsushima, T.: Tweakable enciphering schemes from hash-sum-expansion. In: Srinathan, K., Rangan, C.P., Yung, M. (eds.) INDOCRYPT 2007. LNCS, vol. 4859, pp. 252–267. Springer, Heidelberg (2007). https://doi.org/10.1007/978-3-540-77026-8_19
27. Naor, M., Reingold, O.: A pseudo-random encryption mode. www.wisdom.weizmann.ac.il/naor
28. Naor, M., Reingold, O.: On the construction of pseudorandom permutations: Luby-rackoff revisited. J. Cryptol. **12**(1), 29–66 (1999)
29. National Bureau of Standards. Data encryption standard. Federal Information Processing Standard (1977)
30. Patarin, J.: A proof of security in $O(2^n)$ for the Xor of two random permutations. In: Safavi-Naini, R. (ed.) ICITS 2008. LNCS, vol. 5155, pp. 232–248. Springer, Heidelberg (2008). https://doi.org/10.1007/978-3-540-85093-9_22
31. Jacques, P.: The "Coefficients H" Technique. In: Selected Areas in Cryptography, SAC, , pp. 328–345 (2008)
32. Peyrin, T., Seurin, Y.: Counter-in-tweak: authenticated encryption modes for tweakable block ciphers. In: Robshaw, M., Katz, J. (eds.) CRYPTO 2016, Part I. LNCS, vol. 9814, pp. 33–63. Springer, Heidelberg (2016). https://doi.org/10.1007/978-3-662-53018-4_2

33. Phillip, R., Mihir, B., John, B., Krovetz, T.: O.C.B.: a block-cipher mode of operation for efficient authenticated encryption, In: Proceedings of the 8th ACM Conference on Computer and Communications Security, CCS 2001, Philadelphia, Pennsylvania, USA, 6-8 November 2001, pp. 196–205 (2001)
34. Sarkar, P.: Improving upon the TET mode of operation. In: Nam, K.-H., Rhee, G. (eds.) ICISC 2007. LNCS, vol. 4817, pp. 180–192. Springer, Heidelberg (2007). https://doi.org/10.1007/978-3-540-76788-6_15
35. Wang, P., Feng, D., Wu, W.: HCTR: a variable-input-length enciphering mode. In: Feng, D., Lin, D., Yung, M. (eds.) CISC 2005. LNCS, vol. 3822, pp. 175–188. Springer, Heidelberg (2005). https://doi.org/10.1007/11599548_15
36. Smith, J.L., Ehrsam, W.F., Meyer, C.H.W., Tuchman, W.L.: Message verification and transmission error detection by block chaining. US Patent 4074066 (1976)

Reconsidering Generic Composition:
The Tag-then-Encrypt Case

Francesco Berti[(⊠)], Olivier Pereira, and Thomas Peters

ICTEAM/ELEN/Crypto Group, Université catholique de Louvain,
1348 Louvain-la-Neuve, Belgium
{francesco.berti,olivier.pereira,thomas.peters}@uclouvain.be

Abstract. Authenticated Encryption (AE) achieves confidentiality and authenticity, the two most fundamental goals of cryptography, in a single scheme. A common strategy to obtain AE is to combine a Message Authentication Code (MAC) and an encryption scheme, either nonce-based or iv-based. Out of the 180 possible combinations, Namprempre et al. [20] proved that 12 were secure, 164 insecure and 4 were left unresolved: A10, A11 and A12 which use an iv-based encryption scheme and N4 which uses a nonce-based one. The question of the security of these composition modes is particularly intriguing as N4, A11, and A12 are more efficient than the 12 composition modes that are known to be provably secure.

We prove that: (i) N4 is not secure in general, (ii) A10, A11 and A12 have equivalent security, (iii) A10, A11, A12 and N4 are secure if the underlying encryption scheme is either misuse-resistant or "message malleable", a property that is satisfied by many classical encryption modes, (iv) A10, A11 and A12 are insecure if the underlying encryption scheme is stateful or untidy. All the results are quantitative.

1 Introduction

Authenticated Encryption and Generic Composition. From its start, the goal of cryptography is to prevent that anyone but the intended receiver can read a message (privacy) and that anyone can send a message impersonating someone else (authenticity). In order to answer this privacy (resp. authenticity) requirement, encryption schemes (resp. Message Authentication Codes (MACs)) were designed independently. When there is a need for both privacy and authenticity, Authenticated Encryption (AE) can be used [5,6,15,17]. Moreover, AE may be used to authenticate associated data (AD), which are data attached to a message which do not need to be private, but do need to be authenticated (e.g., message header [25]). We suppose that both the sender and the receiver share the same private key (symmetric scenario).

There are two possible ways to create an AE scheme: the first is to design it from scratch, using a single key, and the second is to combine an Encryption scheme with a MAC. Examples of the first path are AES-GCM [12], AES-CCM [19], CHACHA20_POLY305 [21] (used in TLS 1.3 [13]), SCT [24] and the

© Springer Nature Switzerland AG 2018
D. Chakraborty and T. Iwata (Eds.): INDOCRYPT 2018, LNCS 11356, pp. 70–90, 2018.
https://doi.org/10.1007/978-3-030-05378-9_4

CAESAR candidates [7]. When following the second path, the problem is to decide how to compose the ingredients. This problem is called *generic composition* and was introduced and studied first by Bellare and Namprempre [5]. They and Krawczyk proved the well-known result that *Encrypt-then*-MAC is secure [6,17]. Namprempre et al. have made a deeper analysis [20], which considered in detail the assumptions on the Encryption scheme, whether it is iv-based (ivE [with the iv randomly picked]) or nonce-based (nE [with the nonce n never repeated]) and assumed that the MACs are PRFs. Out of all the possible composition modes, 12 (9 with ivE, 3 with nE) were proved to be secure, 164 to be insecure and 4 were unresolved: N4 which uses a nE and A10, A11, A12 which use an ivE. These four modes, which are depicted in Fig. 1, are based on the Tag-then-Encrypt paradigm: given a nonce n, an associated data a and a message m, the resulting AEs simply output $c = \mathsf{Enc}_{k_E}^n(m\|\tau)$ or $c = \mathsf{Enc}_{k_E}^{iv}(m\|\tau)$ for some n/iv, where τ is the tag provided by the MAC, and is computed either as $\mathsf{Mac}_{k_M}(a, m)$ or as $\mathsf{Mac}_{k_M}(m)$ depending on the mode. When an ivE scheme is used, the iv is computed using a PRF MAC that takes as input either n or (n, a). Interestingly three of these modes (N4, A11 and A12) use the n, a and m only once in total during both the computation of iv ($\mathsf{Mac}_{k_M}^{\mathsf{IV}}$) and τ ($\mathsf{Mac}_{k_M}^{\mathsf{Tag}}$), which makes them the most efficient among all Tag-then-Encrypt schemes. In this paper, we investigate the security of these four composition modes, focusing on ciphertext integrity, as Namprempre et al. already established the expected confidentiality guarantees.

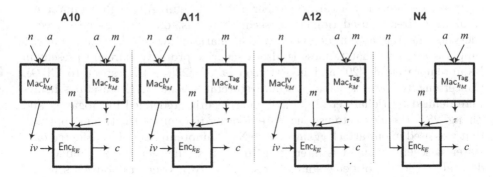

Fig. 1. The four modes A10, A11, A12, and N4.

Our Contribution. Our investigation gives several new results.

First, the mode N4 does not guarantee ciphertext integrity in general, and we offer a counterexample. The idea of this counterexample is to carefully inject a kind of Trojan in the nE encryption scheme, which can only be activated during the decryption queries using well-crafted ciphertext. The Trojan is triggered through the nonce and a block of the message.

Second, we show that A10, A11 and A12 have equivalent security, by offering security reductions between these three modes. Different techniques are used in these reductions, which are based on the uniqueness of the nonce and, in other cases, recrafting these nonces.

Third, we push our analysis further, by investigating the security of these 3 modes by making some additional hypothesis on the ivE scheme. We found that these modes are secure if the ivE scheme is either misuse-resistant (that is, repeated nonces can only lead to repeated ciphertexts without further security degradation) or "message-malleable" (that is, given a triple (iv, m, c) with $c = \mathsf{Enc}^{iv}(m)$, it is possible to compute correctly every other triple (iv, m', c') with $c' = \mathsf{Enc}^{iv}(m')$ [resp. $m' = \mathsf{Dec}^{iv}(c')$] for the same iv from m' [resp. c']). Many common schemes, like CTR and OFB [14], CHACHA20 [21] or any other stream ciphers, are "message-malleable", thus we have proved that the three composition modes are secure if implemented with these encryption schemes. This is another evidence of the "*generic composition's sensitivity to definitional and algorithmic adjustments*" [20]. While the proof for misuse-resistant ivE-schemes is relatively straightforward, the proof for "message-malleable" ivE is more interesting as it uses a reduction of a INT-CTXT (ciphertext integrity) adversary to a CPA (Chosen Plaintext Attack) adversary and not only to the properties of the MAC schemes. Interestingly, the N4 mode also becomes secure when the same extra requirements are made for the nE encryption scheme. With respect to the Namprempre et al. [20], we have still to use an additional hypothesis (they used Knowledge of Tag [KoT]), but ours are much easier to prove although they are less general.

Fourth, we find two insecure variants for all three modes, one if the ivE encryption scheme is not tidy, the other if it is stateful. Although Namprempre et al. [20] already used tidiness in security proofs, our ivE scheme correctly encrypts the tag and it decrypts in the "natural" way. Thus, our analysis supports the idea that tidiness is also a security property (already present in Namprempre et al. [20] and in Paterson et al. [23], with respect to CRD). Concerning the attack using a secure stateful scheme (AE stateful schemes were defined by Bellare et al. [3] and their security redefined by Rogaway and Zhang [28]), the idea is to use the state in order to emulate the trojan approach that was used in our attack against mode N4. Namprempre et al. considered only stateless schemes, but it is interesting to see how the security of a mode may depend on the fact of being stateful or stateless. Moreover, stateful AE schemes are an interesting subject of studies [4,11,16,23].

There is an extended version of this paper [9] containing a more detailed background and all the proofs.

Structure of the Paper. We give a section introducing all the notions we need (Sect. 2); after that we present the four modes N4, A10, A11 and A12 which we investigate (Sect. 3). Then, we show the proof that mode N4 is not secure (Sect. 4) and the security relations among modes A10, A11 and A12 (Sect. 5). After that, we prove that these modes are secure if we add some hypothesis on the ivE scheme (Sect. 6) and we end analyzing our insecure variants of modes A10, A11, and A12 (Sect. 7).

2 Background

2.1 Notations

We use finite binary strings. The length of the string x is denoted by $|x|$ and the concatenation of the strings x and y is denoted by $x\|y$. The set of all finite strings is denoted by $\{0,1\}^*$. We denote the set of all n-bit strings as $\{0,1\}^n$ and the set of strings of at most n bits as $\{0,1\}^{\leq n}$. Given a string $x = (x_1, x_2, ..., x_l)$ of l bits, we denote with $\pi_t(x)$ the string $(x_1, ..., x_T)$ where $T = min(|x|, t)$.

We reserve calligraphic notation for sets. In particular we denote with $\mathcal{K}, \mathcal{N}$, $\mathcal{IV}, \mathcal{A}, \mathcal{M}, \mathcal{TW}, \mathcal{T}, \mathcal{X}$ and $\mathcal{C}$ respectively the *key space, nonce space, iv-space, associated data space, message space, tweak space, tag space, input space of the* MAC and the *ciphertext space*. We suppose that $\mathcal{M} = \mathcal{A} = \{0,1\}^*$, that is, these spaces contain all the finite binary strings. We suppose that $\mathcal{C} \subseteq \{0,1\}^*$.

Given the set $\mathcal{Y}$, we write $y \leftarrow \mathcal{Y}$ to denote the uniformly random selection of y in $\mathcal{Y}$.

We reserve sans serif (Alg) notations for algorithms. If the algorithm Alg is probabilistic, we can think of its output as a distribution. We denote with $a \leftarrow$ Alg(b, c, d) the fact that we sample from the distribution induced by algorithm Alg on inputs (b, c, d), and we obtain a. We may write part of the arguments of the algorithm as subscripts or superscripts, that is, $\mathsf{Alg}_b^c(d) = \mathsf{Alg}_b(c, d) = \mathsf{Alg}(b, c, d)$.

A (q, t)-adversary A is a probabilistic algorithm which can make at most q queries to the oracle(s) he is granted access to, and runs in time bounded by t.

Let algorithm Alg be an algorithm whose inputs are in $\mathcal{S}^1 \times \cdots \times \mathcal{S}^n$ and whose output is in $\mathcal{Y}$. We say that algorithm Alg *does not reveal, via the length of its output, any information about its inputs apart from their lengths* if there exists a deterministic function $f : \mathbb{N}^n \longmapsto \mathbb{N}$ s.t. $|y| = f(|s_1|, ..., |s_n|)$ for all possible inputs $(s_1, ..., s_n)$. We assume that all the Enc and AEnc algorithms we use have this property.

Given a game, where the adversary A is allowed to query many oracles, we use a single counter for all the queries made by adversary A, during the game. The oracle $\perp(\cdot, \cdot)$ always answers $\perp$. When an adversary is playing a game where he has access either to an oracle implemented with algorithm $\mathsf{Alg}(\cdot, \cdot)$ or the oracle $\$(\cdot, \cdot)$ it means that the oracle $\$(\cdot, \cdot)$ answers a random bit string of length $|\mathsf{Alg}(\cdot, \cdot)|$. Moreover, it keeps in memory the answers it gives.

We denote with $c^i \leftarrow \mathsf{O}(a^i, b^i)$ the output c^i of the i-th query on oracle O with input (a^i, b^i). Usually we use only one counter for all the queries, that is the i query can be to oracle O_j and the $i+1$th to oracle O_l where oracle O_j and O_l are among the oracles the adversary is granted access. It can be that that the two oracle are the same, that is $\mathsf{O}_j = \mathsf{O}_l$.

We write $\Pr[B; A_1, A_2, ...]$ for the probability of event B after the experiment described by steps $A_1, A_2, ...$.

In the rest of this section we provide many standard definitions. The expert reader may skip all this section, except Definition 4, which is non-standard.

2.2   Pseudorandom Functions (PRF)

We now define the PRF-security notion, the base of many cryptographic primitives:

Definition 1. *A function* $F : \mathcal{K} \times \mathcal{M} \longmapsto \mathcal{T}$ *is a* (q, t, ϵ)*-pseudorandom function (PRF) if for every* (q, t) *adversary* A, *the advantage:*

$$\mathsf{Adv}_F^{\mathrm{PRF}}(A) := |\Pr[A^{F_k(\cdot)} \Rightarrow 1] - \Pr[A^{f(\cdot)} \Rightarrow 1]|$$

is upper bounded by ϵ *where* k *and* f *are chosen uniformly at random from their domains, namely* $\mathcal{K}$ *and the set of functions from* $\mathcal{M}$ *to* $\mathcal{T}$, $\mathcal{FUNC}(\mathcal{M}, \mathcal{T})$.

In a similar way, F is a pseudorandom permutation (PRP) if F_k is a permutation and the above advantage is ϵ bounded when f is selected as a random permutation.

We remind that a PRP is a PRF (see Proposition 3.27 [14]).

In some of our constructions, we will also use tweakable pseudorandom permutations [18]. They are PRPs with an additional input, the tweak: $E : \mathcal{K} \times \mathcal{TW} \times \mathcal{M} \longmapsto \mathcal{T}$, and their security advantage is then defined as $\mathsf{Adv}_E^{\mathrm{TPRP}}(A) := \mathsf{Adv}_F^{\mathrm{PRF}}(A)$ where $F(k, (tw, m)) := E(k, tw, m)$ and for any choice of k and tw $E_k(tw, \cdot)$ is a permutation.

2.3   Nonce-Based Authenticated Encryption (nAE) and Encryption (nE and ivE) Schemes

For the syntax of encryption schemes we follow the approach of Namprempre et al. [20] (taken by the work of Rogaway [26]) where the encryption algorithm is deterministic and an *"initialization vector"* (IV) iv is surfaced (and it may be seen as part of the AD [27]). Using this approach we classify encryption schemes according to the requirements of this extra input to provide CPA-security.

Definition 2 ([20]).   *A scheme for* nonce-based authenticated encryption (nAE) *is a triple* $\Pi := (\mathcal{K}, \mathsf{AEnc}, \mathsf{ADec})$, *where the keyspace* $\mathcal{K}$ *is a nonempty set, the encryption algorithm* AEnc *is a deterministic algorithm which takes as input the tuple* $(k, n, a, m) \in \mathcal{K} \times \mathcal{N} \times \mathcal{A} \times \mathcal{M}$ *and outputs a string* $c \leftarrow \mathsf{AEnc}_k^{n,a}(m)$ *called ciphertext.*

The decryption algorithm ADec *is a deterministic algorithm which takes as input the tuple* $(k, n, a, c) \in \mathcal{K} \times \mathcal{N} \times \mathcal{A} \times \mathcal{C}$ *and outputs* $m \leftarrow \mathsf{ADec}_k^{n,a}(c)$ *which is either a string* $m \in \mathcal{M}$ *or the symbol* $\perp$ *("invalid")*.

We require that the algorithms AEnc *and* ADec *are the inverse of each other, that is:*

- (Correctness) *if* $\mathsf{AEnc}_k^{n,a}(m) = c$ *then* $\mathsf{ADec}_k^{n,a}(c) = m$
- (Tidiness) *if* $\mathsf{ADec}_k^{n,a}(c) = m \neq \perp$ *then* $\mathsf{AEnc}_k^{n,a}(m) = c$

If $\mathsf{ADec}_k^{n,a}(c) = \perp$ *we say that the algorithm* rejects c, *otherwise it* accepts c.

A sloppy nAE *scheme satisfies everything but the tidiness condition.*

A nonce-based Encryption scheme (nE) *is a triple* $\Pi = (\mathcal{K}, \mathsf{Enc}, \mathsf{Dec})$, *where* Enc *and* Dec *do not take input the AD, that is,* $\mathsf{Enc} : \mathcal{K} \times \mathcal{N} \times \mathcal{M} \longmapsto \mathcal{C}$ *and* $\mathsf{Dec} : \mathcal{K} \times \mathcal{N} \times \mathcal{C} \longmapsto \mathcal{M}$.

An iv-*based encryption scheme* ivE *is syntactically equivalent to a* nE *scheme, with the only difference that the nonce space* $\mathcal{N}$ *is replaced with an IV space* $\mathcal{IV}$.

Tidiness, as correctness, is usually seen as a syntactic requirement (for example Namprempre et al. [20]). Instead, in this paper, we show an explicit case where this property is fundamental to provide security (see Sect. 7.1).

Paterson et al. [23] defined the "collision-resistant decryption" (CRD), which is a security property. Tidy schemes are inherently CRD-secure, since there is one and only valid ciphertext for each input, but the converse is not valid (because CRD-security is obtained when adversaries are able to break it with negligible probability, while tidiness always works).

The difference between nE schemes and ivE schemes lies in their security requirements. A complete survey about nAE, nE and ivE schemes can be found in the extended version.

2.4 Security for nAE, nE and ivE Schemes

The security definitions for nAE, nE and ivE schemes are inspired from those in [20,27].

Definition 3. *A nonce-based authenticated encryption scheme* (nAE) $\Pi := (\mathcal{K}, \mathsf{AEnc}, \mathsf{ADec})$ *is* (q, t, ϵ)-nAE-secure *if the advantage*

$$\mathsf{Adv}_{\Pi}^{\mathsf{nAE}}(A) := \left| \Pr\left[A^{\mathsf{AEnc}_k(\cdot,\cdot,\cdot), \mathsf{ADec}_k(\cdot,\cdot,\cdot)} \Rightarrow 1 \right] - \Pr\left[A^{\$(\cdot,\cdot,\cdot), \perp(\cdot,\cdot,\cdot)} \Rightarrow 1 \right] \right| \quad (1)$$

is bounded by ϵ *for every* (q, t)-*adversary* A *that respects the following two conditions: (i) If* A *queried the first (encryption) oracle on input* (n, a, m) *and was answered* c, *then he is not allowed to query the second (decryption) oracle on input* (n, a, c). *(ii)* A *is not allowed to repeat the first component (the nonce) on different left oracle queries.*

Π *is* (q, t, ϵ)-nAE-*E secure, if the advantage*

$$\mathsf{Adv}_{\Pi}^{\mathsf{nAE}\text{-}E}(A) := \left| \Pr\left[A^{\mathsf{AEnc}_k(\cdot,\cdot,\cdot)} \Rightarrow 1 \right] - \Pr\left[A^{\$(\cdot,\cdot,\cdot)} \Rightarrow 1 \right] \right| \quad (2)$$

is bounded by ϵ *for every* (q, t)-*adversary* A *that respects Condition (ii) above.*

A *nonce-based encryption scheme* (nE) $\Pi := (\mathcal{K}, \mathsf{Enc}, \mathsf{Dec})$ *is* (q, t, ϵ)-nE-secure *if the advantage,*

$$\mathsf{Adv}_{\Pi}^{\mathsf{nE}}(A) := \left| \Pr\left[A^{\mathsf{Enc}_k(\cdot,\cdot)} \Rightarrow 1 \right] - \Pr\left[A^{\$(\cdot,\cdot)} \Rightarrow 1 \right] \right| \quad (3)$$

is bounded by ϵ *for every* (q, t)-*adversary* A *that respects Condition (ii) above.*

An iv-*based encryption scheme* ivE $\Pi := (\mathcal{K}, \mathsf{Enc}, \mathsf{Dec})$ *is* (q, t, ϵ)-ivE-secure *if the advantage*

$$\mathsf{Adv}_{\Pi}^{\mathsf{ivE}}(A) := \left| \Pr\left[A^{\mathsf{Enc}_k^{\$}(\cdot)} \Rightarrow 1 \right] - \Pr\left[A^{\$(\cdot)} \Rightarrow 1 \right] \right| \quad (4)$$

is bounded by ϵ for every (q, t)-adversary. Here the oracle $\mathsf{Enc}^{\$}(m)$ picks a random iv $\leftarrow \mathcal{IV}$, then computes $c \leftarrow \mathsf{Enc}_k(iv, m)$ and returns (iv, c).

As a result of this definition, the only difference between ivE and nE security is the requirement on their auxiliary input: non-repeating nonces for nE and random *ivs* for ivE. We observe that ivE-security implies nE security when uniformly random *ivs* are expected to differ with overwhelming probability. The contrary does not hold: the CTR mode is well-known illustration (details are provided in the extended version).

In some cases, it is desirable to guarantee some security even if nonces are repeated: this is called resistance to nonce misuse, or simply misuse resistance.

Definition 4. *If we drop Condition (ii) on the non repetition of the nonces in the nE security definitions, then we augment the security notions with misuse resistance. Namely, we say that the nE scheme is (q, t, ϵ)-mrE secure.*

We point out that in the mrE definition the adversary has only access to an encryption oracle, differently from the standard *misuse resistance for authenticated encryption* mrAE [27]. An example of an nE scheme which is mrE and not mrAE is given in the extended version as well as many examples [8,24].

The mrE definition is trivially extended to ivE schemes, since the syntax of nE schemes and ivE schemes is identical.

2.5   Chosen-Plaintext Attack Security with Chosen Nonce

We define mCPA security for nE and ivE schemes, following the left-or-right definition of Bellare et al. [2], but adapted to the nonce-based setting. This definition is a multi-challenge definition, contrary to the common single-challenge variant [14]. The two versions of the definition are asymptotically equivalent, but come with a difference of a linear factor when quantitative bounds are used, as we do here.

Definition 5. *A nonce-based Encryption scheme* $\mathsf{nE}\,\Pi = (\mathcal{K}, \mathsf{Enc}, \mathsf{Dec})$ *is* (q, t, ϵ)*-mCPA secure, or* (q, t, ϵ)*-secure against chosen plaintext attacks for multiple encryptions, if:*

$$\mathsf{Adv}_{\Pi}^{\mathsf{mCPA}}(\mathsf{A}) := \left| \frac{1}{2} - \Pr\left[b' = b; b \leftarrow \{0, 1\}, \; b' \leftarrow \mathsf{A}^{\mathsf{Enc}_k^b(\cdot, \cdot, \cdot)} \right] \right|$$

is bounded by ϵ for any (q, t)-adversary. Here the oracle $\mathsf{Enc}_k^b(\cdot, \cdot, \cdot)$ is an oracle, which on input $(n, m_0, m_1) \in \mathcal{N} \times \mathcal{M}^2$ outputs $c \leftarrow \mathsf{Enc}_k(n, m_b)$ for a random secret bit $b \leftarrow \{0, 1\}$, which the oracle has picked at the start of the game. When the adversary A queries $\mathsf{Enc}_k^b(\cdot, \cdot, \cdot)$, he must choose two messages m_0 and m_1 s.t. $|m_0| = |m_1|$. Moreover he cannot repeat the first input (the nonce) in different queries.

There is a completely similar definition for ivE schemes. We only have to replace $\mathsf{Enc}_k^b(\cdot, \cdot, \cdot)$ with $\mathsf{Enc}_k^{b,\$}(\cdot, \cdot)$, and to adapt the $\$(\cdot, \cdot, \cdot)$ oracle accordingly. Similarly there is a similar notions for nAE schemes, obtained from the previous one by replacing $\mathsf{Enc}_k(\cdot, \cdot)$ with $\mathsf{AEnc}_k(\cdot, \cdot, \cdot)$ and adapting $\$(\cdot, \cdot, \cdot)$ accordingly.

2.6   Authenticity (INT-CTXT)

Following Bellare et al. [5], we focus on the notion of ciphertext integrity with a single decryption query.

Definition 6. *A nonce-based authenticated encryption scheme* nAE $\Pi = (\mathcal{K},$ AEnc, ADec) *is* (q, t, ϵ)-INT-CTXT1 (Ciphertext integrity with only 1 decryption query)-secure if*

$$\mathsf{Adv}_{\Pi}^{\mathsf{INT\text{-}CTXT1}}(\mathsf{A}) := \Pr\left[\bot \neq m^* \leftarrow \mathsf{ADec}_k(n^*, a^*, c^*); \ (n^*, a^*, c^*) \leftarrow \mathsf{A}^{\mathsf{AEnc}_k(\cdot, \cdot, \cdot)}\right]$$

is bounded by ϵ for every (q, t) adversary. The adversary A is not allowed to repeat the first component (the nonce) on different oracle queries. Moreover he is not allowed to output (n^*, a^*, c^*) if he received c^* as $c^* \leftarrow \mathsf{AEnc}_k(n^*, a^*, m^*)$ for a certain input (n^*, a^*, m^*) that he asked to the first oracle.

As we can expect, an nAE scheme that offers both mCPA and INT-CTXT1 security is an nAE scheme, (and we prove this the extended version).

2.7   Message Authentication Code (MAC)

Apart from an encryption scheme, all our composition modes are based on a deterministic notion of Message Authentication Code (MAC).

Definition 7. *A* Message Authentication Code MAC *is a triple* $\Pi = (\mathcal{K}, \mathsf{Mac}, \mathsf{Vrfy})$ *where the keyspace $\mathcal{K}$ is a non-empty set, the tag-generation algorithm* Mac *is a deterministic algorithm that takes as input the couple $(k, m) \in \mathcal{K} \times \mathcal{M}$ and outputs the tag $\tau \leftarrow \mathsf{Mac}_k(m)$ from the tag space $\mathcal{T}$. The verification algorithm* Vrfy *takes as input a triple (k, m, τ) in $\mathcal{K} \times \mathcal{M} \times \mathcal{T}$ and outputs $\top$ (accept) or $\bot$ (reject). We ask that* $\mathsf{Vrfy}(k, m, \mathsf{Mac}(k, m)) = \top$.

A string-input MAC strMAC *has as input space a set of strings, that is $\mathcal{M} \subseteq \{0, 1\}^*$.*

A vector-input MAC vecMAC *has as input space $\mathcal{M}$ which has one or more component and it can accept tuples of strings as input.*

Usually the security for MACs is expressed as unforgeability, but our composition modes rely on a Mac function that is a $(q, t, \epsilon_{\mathrm{PRF}}) - \mathrm{PRF}$.

Definition 8 ([20]). *A MAC $\Pi = (\mathcal{K}, \mathsf{Mac}, \mathsf{Vrfy})$ is $(q, t, \epsilon) - \mathrm{PRF}$-secure if*

$$\mathsf{Adv}_{\Pi}^{\mathrm{PRF}}(\mathsf{A}) := \mathsf{Adv}_{\mathsf{Mac}}^{\mathrm{PRF}}(\mathsf{B})$$

is bounded by ϵ for any (q, t) adversary B and if $\mathsf{Vrfy}(k, m, \mathsf{Mac}(k, m)) = \top$ *iff* $\tau = \mathsf{Mac}(k, m)$.

For completeness, the standard definitions are put in the extended version.

3   Problem

As discussed earlier, Namprempre et al. [20] left open the problem of the nAE security of 4 modes based on the Tag-then-Encrypt paradigm, which have been shown in Fig. 1.

Formally, the first three modes compose an ivE scheme $\Pi = (\mathcal{K}_E, \mathsf{Enc}, \mathsf{Dec})$ and two vecMAC schemes using the same key, $\mathsf{MAC}^{\mathsf{IV}} = (\mathcal{K}_M, \mathsf{Mac}^{\mathsf{IV}}, \mathsf{Vrfy}^{\mathsf{IV}})$ and $\mathsf{MAC}^{\mathsf{Tag}} = (\mathcal{K}_M, \mathsf{Mac}^{\mathsf{Tag}}, \mathsf{Vrfy}^{\mathsf{Tag}})$ in this way:

- A10: $\mathsf{AEnc}^{n,a}_{k_E, k_M}(m) := c$ with $iv = \mathsf{Mac}^{\mathsf{IV}}_{k_M}(n, a)$, $\tau = \mathsf{Mac}^{\mathsf{Tag}}_{k_M}(a, m)$ and $c = \mathsf{Enc}_{k_E}(iv, m\|\tau)$
- A11: $\mathsf{AEnc}^{n,a}_{k_E, k_M}(m) := c$ with $iv = \mathsf{Mac}^{\mathsf{IV}}_{k_M}(n, a)$, $\tau = \mathsf{Mac}^{\mathsf{Tag}}_{k_M}(m)$ and $c = \mathsf{Enc}_{k_E}(iv, m\|\tau)$
- A12: $\mathsf{AEnc}^{n,a}_{k_E, k_M}(m) := c$ with $iv = \mathsf{Mac}^{\mathsf{IV}}_{k_M}(n)$, $\tau = \mathsf{Mac}^{\mathsf{Tag}}_{k_M}(a, m)$ and $c = \mathsf{Enc}_{k_E}(iv, m\|\tau)$

The fourth mode composes a nE Encryption scheme $\Pi = (\mathcal{K}_E, \mathsf{Enc}, \mathsf{Dec})$ and a vecMAC = MAC = $(\mathcal{K}_M, \mathsf{Mac}, \mathsf{Vrfy})$:

- N4: $\mathsf{AEnc}^{n,a}_{k_E, k_M}(m) := c$ with $\tau = \mathsf{Mac}^{\mathsf{Tag}}_{k_M}(a, m)$ and $c = \mathsf{Enc}_{k_E}(n, m\|\tau)$

For clarity we reserve bold notations $\mathbf{m}$ for the messages inputs of the nAE scheme Π and normal notations m for the messages inputs to the underlying nE (or ivE)-scheme Π (so, we typically have that $m = \mathbf{m}\|\tau$).

If Π is tidy and the MAC is PRF-secure, then the AE scheme Π, obtained composing these components, is tidy. These modes also offer CPA security [20], which directly results from the underlying encryption schemes (a quantitative proof of this statement is available in the extended version [9]).

As a result, the open question lies in the INT-CTXT security of these modes.

4   Attack Against Mode N4

We provide here an attack against the mode N4, explicitly presenting an nAE-scheme Π, based on an nE Encryption scheme $\Pi = (\mathcal{K}_E, \mathsf{Enc}, \mathsf{Dec})$ and a vecMAC MAC = $(\mathcal{K}_M, \mathsf{Mac}, \mathsf{Vrfy})$ which is PRF-secure. For simplicity, we consider only the case when the message $\mathbf{m}$ of Π is λ-bit long and the tag is λ-bit long, leaving the general case to the extended version. The nE Encryption scheme, which encrypts 2λ-bit long message, is nE-secure and tidy, but the nAE-scheme Π obtained composing them according to mode N4, is not secure and, in particular, it is not INT-CTXT1-secure as we show a forgery.

The idea of the forgery is to force the tag τ of a couple $(a, \mathbf{m})$ to be encrypted identically for two different nonces, while keeping the nE-security.

4.1   Construction

Following the definition of mode $N4$, an authenticated ciphertext is computed as $c = \mathsf{Enc}_{k_E}(n, \mathbf{m} \| \mathsf{Mac}_{k_M}(a, \mathbf{m}))$, for which Mac is a PRF. We now define the nE scheme Π.

The keys produced in Π are made of two components (k, v^*): the key $k \in \mathcal{K}$ of a TPRP E and a random value v^*, which has the size of block of E, that is, λ bits. This value v^* will be leaked to Adv when asking for the encryption of a message with the nonce $n = 1$, and will then be used to trigger a kind of Trojan in the encryption scheme. That Trojan will have the following behavior: for nonces $n = 1, 2$, and if the first message block is v^*, then the last ciphertext block will be computed in a way that ignores the value of n.

This behavior is benign when considering the nE security of Π: the only way to observe it would be to make two encryption queries with nonces 1 and 2, and first message block v^*. But doing this would require guessing v^* before querying with nonce 1 (the nE adversary is nonce respecting), and this cannot be done but with probability $2^{-\lambda}$: it would require guessing v^*.

As we will see, it is not benign anymore when considering the ciphertext integrity property: there, Adv is free to use the nonces 1 and 2 in its decryption query, even if these nonces were used in encryption queries.

To make things concrete, we define the encryption process Enc of Π using a TPRP $\mathsf{E} : \mathcal{K} \times \mathcal{TW} \times \{0,1\}^\lambda \longmapsto \mathcal{T} = \{0,1\}^\lambda$ with tweak space $\mathcal{TW} = \{0,1,2\} \times \{0,1\}$. For a message $m = (m_1, m_2) \in \{0,1\}^{2\lambda}$, the ciphertext $\mathsf{Enc}_k^n(m)$ is made of three blocks (c_0, c_1, c_2) computed as follows:

- $c_0 = \mathsf{E}_k^{(0,0)}(n)$ unless $n = 1$, in which case $c_0 := v^*$.
- $c_1 = m_1 \oplus \mathsf{E}_k^{(1,0)}(n)$.
- $c_2 = m_2 \oplus \mathsf{E}_k^{(2,0)}(n)$, unless the condition $[[(n = 1 \vee n = 2) \wedge m_1 = v^*]$ is met, in which case $c_2 = m_2 \oplus \mathsf{E}_k^{(2,1)}(0)$.

With such a definition, the block c_0 looks random for any input, and its only purpose is to leak v^* when $n = 1$. The block c_1 is a traditional encryption of the message block m_1 using the TPRP E. The block c_2 is computed in the same way (just incrementing the tweak), except under a very specific condition: the nonce is either 1 or 2, and the first message block $m_1 = v^*$. Under that condition, the ciphertext block becomes independent of the nonce. As explained above, this condition is designed in such a way that it cannot lead to any observable event when Adv can only access an encryption oracle in a nonce respecting way: that would require querying Enc on a message starting with v^* on both $n = 1$ and $n = 2$, but v^* is only learned after a query with $n = 1$, and it is then not permitted to make a second query with $n = 1$ and v^* as message block.

The decryption of Π works in the natural way. In particular, in order to guarantee the tidiness of the nE encryption scheme, Dec must verify the correctness of the first ciphertext block c_0.

The proofs that Π is nE-secure can be found in the extended version.

4.2 Forgery

The composition of the previous nE scheme Π with a PRF-secure MAC according to mode N4 is not INT-CTXT1-secure. In fact, we provide a forgery where the adversary A asks the encryption of only two messages:

1. It first asks for an encryption of $(1, a, \mathbf{m})$, for arbitrary choices of a and $\mathbf{m}$. This returns a ciphertext whose first block is v^*, second block is $c_1 = \mathbf{m} \oplus E_k^{(1,0)}(1)$, and third block is ignored.
2. It then asks for an encryption of $(2, a, v^*)$. This returns a ciphertext whose last block is $c_2 = \mathsf{Mac}_{k_M}(a, v^*) \oplus E_k^{(2,1)}(0)$.

Eventually, Adv makes a decryption query on $(1, a, (v^*, c_1 \oplus \mathbf{m} \oplus v^*, c_2))$, which is different of the two previously obtained ciphertexts, and has a valid decryption to v^*, hence violating the ciphertext integrity property.

This shows that N4 is not a secure composition mode, in general.

Ciphertext Extension. We observe that the encryption scheme we use to break N4 uses the ciphertext expansion, that is the ciphertext is bigger than the plaintext (i.e. $|c| < |m|$ with $c = \mathsf{Enc}_k(m)$). It is left the question whether if the encryption scheme does not have the ciphertext expansion, scheme N4 is secure.

5   Security Relations Among A10, A11 and A12

While we are able to prove the generic insecurity of N4, we are not able to prove that modes A10, A11 and A12 are either secure or insecure in general. Still, in this section, we prove that these three modes are either all secure or all insecure.

To prove it we need to replace the two vecMACs vecMAC^{IV} and vecMAC^{Tag} with two $\overline{\text{vecMAC}}$s based on the random functions f^{IV} and f^{Tag}. Now the key of the new nAE scheme is $k := (k_E, f^{\mathsf{IV}}, f^{\mathsf{Tag}})$. To highlight these changes, we call the new modes $\overline{\text{A10}}, \overline{\text{A11}}$ and $\overline{\text{A12}}$ and the new nAE-schemes $\overline{\Pi}$. The security relations among modes $\overline{\text{A10}}, \overline{\text{A11}}$ and $\overline{\text{A12}}$ immediately lift to modes A10, A11 and A12. The standard details (replacing Mac^{IV} and Mac^{Tag} with two random functions) are discussed in the extended version, where we prove that if $\overline{A}i$ is $(q, t, \epsilon_{\mathsf{INT\text{-}CTXT}})$-INT-CTXT secure then Ai is $(q, t, \epsilon_{\mathsf{INT\text{-}CTXT}} + \epsilon_{\mathsf{PRF}})$-INT-CTXT secure provided that Mac^{IV} and Mac^{Tag} are $(q, t, \epsilon_{\mathsf{PRF}})$-PRF-secure.

We show the security equivalence of A10, A11 and A12 based on two events, B and C, that we define below. Consider a INT-CTXT1 adversary A against an nAE scheme $\overline{\Pi}$ (which is made according to any of $\overline{\text{A10}}, \overline{\text{A11}}$ or $\overline{\text{A12}}$). If the q-th decryption query (n^q, a^q, c^q) is valid, then $c^q = \overline{\mathsf{AEnc}}_k(n^q, a^q, \mathbf{m}^q)$ for a certain message $\mathbf{m}^q$, as a result of tidiness. Depending on the value of (n^q, a^q) (or only n^q for $\overline{\text{A12}}$), we distinguish between two possibilities, which define event B:

- (n^q, a^q) is fresh, that is, $(n^q, a^q) \neq (n^j, a^j) \; \forall j = 1, ..., q-1$ (we call this event B) [for mode $\overline{\text{A12}}$, we only demand that n^q is fresh, that is $n^q \neq n^j \; \forall j = 1, ..., q-1$].

– $(n^q, a^q) = (n^j, a^j)$ for a $j \in \{1, ..., q - 1\}$ (This j is unique since the nonce n cannot be repeated) [for mode $\overline{A12}$, we only demand that $n^q = n^j$ for a $j \in \{1, ..., q - 1\}$].

With regard to $(a^q, \mathbf{m}^q)$ (or only $\mathbf{m}^q$ for mode $\overline{A11}$), we again consider two possibilities, which define event C:

– $(a^q, \mathbf{m}^q)$ is fresh, that is $(a^q, \mathbf{m}^q) \neq (a^j, \mathbf{m}^j) \ \forall j = 1, ..., q - 1$ (we call this event C) [for mode $\overline{A11}$, we only demand that $\mathbf{m}^q$ is fresh, that is $\mathbf{m}^q \neq \mathbf{m}^j \ \forall j = 1, ..., q - 1$].
– $(a^q, \mathbf{m}^q) = (a^j, \mathbf{m}^j)$ for some $j \in \{1, ..., q - 1\}$ (there can be several such j's) [for mode $\overline{A11}$, we only demand $\mathbf{m}^q = \mathbf{m}^j$ for some $j \in \{1, ..., q - 1\}$].

Clearly by total law of probability

$$\Pr[\text{A wins}] = \Pr[\text{A wins } \cap C] + \Pr[\text{A wins } \cap B \cap \overline{C}] + \Pr[\text{A wins } \cap \overline{B} \cap \overline{C}]$$

With the following lemma we treat the first two addends of the previous equation:

Lemma 1. *Let* $f^{IV} : \mathcal{N} \times \mathcal{A} \longmapsto \mathcal{IV}$ *[for mode* $\overline{A12}$, $f^{IV} : \mathcal{N} \longmapsto \mathcal{IV}$*] and* $f^{Tag} : \mathcal{A} \times \mathcal{M} \longmapsto \mathcal{T}$ *[for mode* $\overline{A11}$, $f^{Tag} : \mathcal{M} \longmapsto \mathcal{T}$*] be two random functions and let* $\Pi = (\mathcal{K}_E, \mathsf{Enc}, \mathsf{Dec})$ *be a* (q, t, ϵ_{ivE})-*ivE-secure encryption scheme. Let* $\overline{\Pi}$ *be the* nAE *scheme obtained composing* f^{IV}, f^{Tag} *and* Π *according to mode* $\overline{A10}$ *or* $\overline{A11}$ *or* $\overline{A12}$. *Then we can bound*

$$\Pr[\text{A } wins \ \cap \overline{C}] + \Pr[\text{A } wins \ \cap B \cap \overline{C}] \leq q|\mathcal{T}|^{-1} + (q - 1)\epsilon_{ivE}$$

The proof is completely standard and can be found in the extended version (the ideas of this proof are already present in Namprempre et al. [20]. The proofs of the security implications between the 3 "A" modes then results from implications in the case A wins $\cap \overline{B} \cap \overline{C}$, which we examine in the rest of this section.

In order to make our notations more precise, if either f^{IV} or f^{Tag} have different signatures for two modes that we compare, we use a subscript to denote the mode that is used (e.g. $f^{IV_{10}}$ for mode A10).

In some proves we use hash function and their collision resistance, for more details about this see Katz and Lindell [14].

5.1 The INT-CTXT1-Security of $\overline{A12}$ Implies the INT-CTXT1-Security of $\overline{A10}$

Proposition 1. *Let* $f^{IV_{10}} : \mathcal{N} \times \mathcal{A} \longmapsto \mathcal{IV}$ *and* $f^{Tag} : \mathcal{A} \times \mathcal{M} \longmapsto \mathcal{T}$ *be two random functions and let* $\Pi = (\mathcal{K}_E, \mathsf{Enc}, \mathsf{Dec})$ *be a* (q, t, ϵ_{ivE})-*ivE-secure encryption scheme. Then, if mode* $\overline{A12}$ *implemented with the random function* $f^{IV_{12}} : \mathcal{N} \longmapsto \mathcal{IV}$ *is* $(q - 1, t, \epsilon_{INT\text{-}CTXT1})$-*INT-CTXT1-secure then mode* $\overline{A10}$ *is* $(q - 1, t, q|\mathcal{T}|^{-1} + (q - 1)\epsilon_{ivE} + \epsilon_{INT\text{-}CTXT1})$-*INT-CTXT1-secure.*

Let $f'^{IV_{12}} : \mathcal{N} \times \mathcal{A} \longmapsto \mathcal{IV}$ be defined $f'^{IV_{12}}(n,a) := f^{IV_{12}}(n) \; \forall n \in \mathcal{N}, a \in \mathcal{A}$ (it is an extension of $f^{IV_{12}}$). The proof is based on the fact that it is impossible using only encryption queries to mode $\overline{A10}$ to distinguish if it is used $f^{IV_{10}}$ or $f'^{IV_{12}}$ (as in mode $\overline{A12}$), since it is not possible for the adversary A to force the nAE algorithm to call $f^{IV_{10}}$ on inputs (n, a_1) and (n, a_2) (with $a_1 \neq a_2$) during encryption queries. Moreover, the couple (n^q, a^q) of the decryption query must not be fresh (due to event $\overline{B}$), thus, using $f'^{IV_{12}}$ is indistinguishable from using $f^{IV_{10}}$.

5.2   The INT-CTXT1-Security of $\overline{A11}$ Implies the INT-CTXT1-Security of $\overline{A10}$

Proposition 2. *Let* $f^{IV} : \mathcal{N} \times \mathcal{A} \longmapsto \mathcal{IV}$ *and* $f^{Tag_{10}} : \mathcal{A} \times \mathcal{M} \longmapsto \mathcal{T}$ *be two random functions and let* $\Pi = (\mathcal{K}_E, \mathsf{Enc}, \mathsf{Dec})$ *be a* (q, t, ϵ_{ivE})-ivE-secure encryption scheme. Let $H : \mathcal{A} \longmapsto \{0,1\}^N$ *be a* $(0, t, \epsilon_{cr})$ *collision resistant hash function. Then, if mode* $\overline{A11}$*, implemented with the random function* $f^{Tag_{11}} : \mathcal{M} \longmapsto \mathcal{T}$ *and with any* $(q, t, \epsilon_{ivE} + \frac{q^2}{2|\mathcal{IV}|})$-ivE-*secure Encryption scheme, is* $(q-1, t, \epsilon_{INT\text{-}CTXT1})$-*INT-CTXT1-secure then mode* $\overline{A10}$ *is* $(q-1, t, \epsilon)$-INT-CTXT1-*secure, where*

$$\epsilon = q|\mathcal{T}|^{-1} + (q-1)\epsilon_{ivE} + \epsilon_{cr} + \epsilon_{INT\text{-}CTXT1}.$$

The idea is to reduce the INT-CTXT1 adversary A against scheme $\overline{\Pi}$ (mode $\overline{A10}$), which uses the ivE scheme Π, to a INT-CTXT1 adversary C against scheme $\overline{\Pi'}$ (mode $\overline{A11}$), which uses the ivE scheme Π'. When the adversary A makes an encryption query $(n^i, a^i, \mathbf{m}^i)$ the adversary C makes an encryption query $(n^i, a^i, \mathbf{m}'^i)$ with $\mathbf{m}'^i = H(a^i) \| \mathbf{m}^i$. The ivE scheme Π' encrypts $m'^i = (H(a^i) \| m^i [= \mathbf{m}^i \| \tau^i])$ in this way: $\mathsf{Enc}'(m'^i) := H(a^i) \oplus f^{Enc}(iv^i) \| \mathsf{Enc}(iv^i, m^i)$, where f^{Enc} is a random function (and it is part of the key the scheme Π'). When the adversary A makes his decryption query (n^q, a^q, c^q) the adversary C simply asks the decryption of $(n^q, a^q, [f^{Enc}(iv^q) \oplus H(a^q)] \| c^q)$ (the iv^q must be not fresh due to event $\overline{B}$).

5.3   The INT-CTXT-Security of $\overline{A10}$ Implies the INT-CTXT-Security of $\overline{A12}$

Proposition 3. *Let* $f^{IV_{12}} : \mathcal{N} \longmapsto \mathcal{IV}$ *and* $f^{Tag} : \mathcal{A} \times \mathcal{M} \longmapsto \mathcal{T}$ *be two random functions and let* $\Pi = (\mathcal{K}_E, \mathsf{Enc}, \mathsf{Dec})$ *be a* (q, t, ϵ_{ivE})-ivE-secure encryption scheme. Let $\overline{\Pi}$ *be the* nAE-*scheme obtained composing these components according to mode* $\overline{A12}$*. Let* $H : \mathcal{A} \longmapsto \{0,1\}^N$ *be* $(0, t, \epsilon_{cr})$.

Then, if mode $\overline{A10}$*, implemented with the random function* $f^{IV_{10}} : \mathcal{N} \times \mathcal{A} \longmapsto \mathcal{IV}$ *and with any* $(q, t, \epsilon_{ivE} + \frac{q^2}{2|\mathcal{IV}|}) - $ ivE-*secure Encryption scheme, is* $(q, t, \epsilon_{INT\text{-}CTXT1})$-INT-CTXT1-*secure then mode* $\overline{A12}$ *is* $(q-1, t, \epsilon)$-INT-CTXT1-*secure with*

$$\epsilon = q|\mathcal{T}|^{-1} + (q-1)\epsilon_{ivE} + \epsilon_{cr} + \epsilon_{INT\text{-}CTXT1}.$$

The idea of the proof is similar to the previous one (Proposition 2), where we replace $\mathbf{m}^i$ with $\mathbf{m}'^i = (H(a^i) \| \mathbf{m}^i)$.

5.4   The INT-CTXT-Security of $\overline{A10}$ Implies the INT-CTXT-Security of $\overline{A11}$

Proposition 4. *Let* $f^{IV} : \mathcal{N} \times \mathcal{A} \longmapsto \mathcal{IV}$ *and* $f^{Tag_{11}} : \mathcal{M} \longmapsto \mathcal{T}$ *be two random functions and let* $\Pi = (\mathcal{K}_E, \mathsf{Enc}, \mathsf{Dec})$ *be a* (q, t, ϵ_{ivE})-ivE-*secure encryption scheme. Let* $\overline{\Pi}$ *be the* nAE *scheme obtained composing these components according to mode* $\overline{A11}$. *Let* $H : \mathcal{A} \longmapsto \{0,1\}^N$ *be a* $(0, t, \epsilon_{cr})$-*collision resistant hash function.*

Then, if mode $\overline{A10}$, *implemented with the random function* $f^{Tag_{10}} : \mathcal{A} \times \mathcal{M} \longmapsto \mathcal{T}$, *is* $(q, t, \epsilon_{INT\text{-}CTXT1})$-INT-CTXT1-*secure then mode* $\overline{A11}$ *is* $(q - 1, t, \epsilon')$-INT-CTXT1-*secure with*

$$\epsilon = q|\mathcal{T}|^{-1} + (q - 1)\epsilon_{ivE} + \epsilon_{cr} + \epsilon_{INT\text{-}CTXT1}.$$

The idea is to reduce the INT-CTXT1 adversary A against scheme $\overline{\Pi}$ (mode $\overline{A11}$) to a INT-CTXT1 adversary C against scheme $\overline{\Pi}_{10}$ (mode $\overline{A10}$). When the adversary A makes an encryption query $(n^i, a^i, \mathbf{m}^i)$, the adversary C makes an encryption query $(n^i \| H(a^i), a, \mathbf{m}^i)$. When the adversary A makes his decryption query (n^q, a^q, c^q) the adversary A' simply asks the decryption of $(n^q \| H(a^q), a, c^q)$.

6   Secure Variants of Modes N4, A10, A11 and A12

As a step towards the proof of the generic (in-)security of A10, A11 and A12, we consider two natural conditions on the ivE scheme that are sufficient to guarantee a secure composition. More precisely, we show that, if the ivE scheme is misuse resistant or if it is "message-malleable" (a condition that is satisfied by many standard modes, and that we formalize precisely below), then these modes are secure. Interestingly, these two properties are the two extreme of the range (clearly, it is impossible for a scheme to have both properties).

We prove everything only for mode A10, since the proofs can be straightforwardly extended to the other two modes. In this section we use the same replacement as in the previous one (we replace mode A10 with mode $\overline{A10}$). Surprisingly, we prove the same results for mode N4.

Then, we conclude this section, comparing our partial results about the (in-)security of modes A10, A11 and A12 with those of Namprempre et al. [20].

6.1   Misuse-Resistant ivE Scheme

The question is interesting since the misuse notion we consider (mrE, Definition 4) does not consider decryption queries.

Proposition 5. *Let the* ivE *scheme* Π *be a* (q, t, ϵ_{mrE})-*misuse resistant* mrE *and* (q, t, ϵ_{ivE}) − ivE *secure, let* $f^{IV} : \mathcal{N} \times \mathcal{A} \longmapsto \mathcal{IV}$ *and* $f^{Tag} : \mathcal{A} \times \mathcal{M} \longmapsto \mathcal{T}$ *be two random functions. Then, the scheme* $\overline{\Pi}$ *obtained composing these components according to mode* $\overline{A10}$, *is* $(q - 1, t, (q - 1)|\mathcal{T}|^{-1} + (q - 1)\epsilon_{ivE} + (q - 1)\epsilon_{mrE})$ − INT-CTXT1-*secure.*

As seen above, we only need to consider the case not studied in Lemma 1. The idea of the proof is to reduce the INT-CTXT1 adversary to a mrE-adversary. Since we are not in the cases studied in Lemma 1, the couples (n^q, a^q) and $(a^q, \mathbf{m}^q)$ are not fresh, and it is enough for the mrE adversary to ask one more encryption query guessing that the message encrypted $\mathbf{m}^q \| \tau^q$ is one of the message the INT-CTXT1 adversary has already asked to encrypt with the same AD a^q (that is, $\mathbf{m}^q \in \mathcal{M}_{a^q}$ where $\mathcal{M}_{a^q} := \{\mathbf{m}_i \ i = 1, ..., q-1 \ s.t. \ a^i = a^q\}$). If the ciphertext obtained is the ciphertext c^q that he is asked to decrypt, then he outputs 1 and, otherwise, 0. The mrE adversary wins only if he guesses correctly and he can guess correctly at most with probability $(q-1)^{-1}$.

Allowing the mrE adversary to ask $(2q-2)$ encryption queries the scheme $\overline{A10}$ would be $(q, t, \frac{2q-1}{|\mathcal{T}|} + 2\epsilon_{mrE}) -$ INT-CTXT1-secure, because the mrE adversary may try every possible message in $\mathcal{M}_{a^q}$.

We remember that for the misuse-resistance of Enc (Definition 3) the adversary has only access to encryption queries.

6.2 "Message-Malleable" nE Scheme

Definition 9. *A nonce-based encryption scheme* nE $\Pi = (\mathcal{K}, \mathsf{Enc}, \mathsf{Dec})$ *is message-malleable if, given an encryption c of a message m with nonce n, an adversary can efficiently decrypt all couples (n, c'), i.e., he is able to compute m' s.t. $m' \leftarrow \mathsf{Dec}_k(n, c')$ without having access to a decryption oracle.*

The same definition may be done for ivE schemes. Many schemes (as CTR and OFB [14]) have this "malleability" property when they are used for fixed length messages. We detail some examples the extended version [8,10]. Message-malleability is easy to prove in many cases, e.g., when the ciphertext $c = \mathsf{Enc}^{iv}(m)$ is computed as a pseudorandom bitstream r computed from the iv and it is XORed with the message m (that is, $c = r \oplus m$), then $\mathsf{Dec}_k^{iv}(c') = c \oplus c' \oplus m$.

Message-malleability allows us to prove the following proposition for $\overline{A10}$, which can be easily extended to modes $\overline{A11}$ and $\overline{A12}$).

Proposition 6. *Let the ivE scheme Π be (q, t, ϵ_{ivE})-ivE-secure, $(q-1, t, \epsilon_{mCPA})$-mCPA-secure and "message-malleable", let $\mathsf{f}^{IV} : \mathcal{N} \times \mathcal{A} \longmapsto \mathcal{IV}$ and $\mathsf{f}^{Tag} : \mathcal{A} \times \mathcal{M} \longmapsto \mathcal{T}$ be two random functions. Then, the scheme $\overline{\Pi}$ obtained composing these components according to mode $\overline{A10}$, is $(q, t, (q-1)\epsilon_{ivE} + q|\mathcal{T}|^{-1} + 8\epsilon_{mCPA}) -$ INT-CTXT1-secure.*

Again, we only need to consider the case that is not covered by Lemma 1. The idea of the proof is to reduce the INT-CTXT1 adversary to an mCPA-adversary. Since we are not in the cases studied in Lemma 1, the couples (n^q, a^q) (thus iv^q) and $(a^q, \mathbf{m}^q)$ are not fresh. The mCPA adversary, when he is asked to simulate the AEnc oracle on input (n^i, a^i, m^i) simply computes iv^i and τ^i using the appropriate functions and asks his Enc oracle on input $(iv^i, m^i \| \tau^i, m^i \| r^i)$ where r^i is a random value picked in $\mathcal{T}$, receiving c^i which he forwards to the INT-CTXT adversary. When this latter adversary outputs (n^q, a^q, c^q), the mCPA adversary computes iv^q, which, due to the fact that we are in the case not covered by Lemma 1, is iv^j for a $j \in \{1, ..., q - 1\}$. Now using the fact that Π is "nonce-message-malleable", he can decrypt c^q as if $c^i = \mathsf{Enc}_{k_E}^{iv^i}(m^i \| \tau^i)$. He outputs 0 if the decryption query is valid, 1 otherwise. We observe that if $c^i = \mathsf{Enc}_{k_E}^{iv^i}(m^i \| r^i)$ the decryption query may be valid with probability $|\mathcal{T}|^{-1}$ since the tags have never been used before the decryption query.

6.3   Extension to N4

Surprisingly, although mode N4 is not secure in general (see Sect. 4), if the nE scheme is either misuse-resistant or message-malleable, mode N4 is INT-CTXT1-secure and, thus, nAE secure. It is easy to prove easily adapting the proofs of Propositions 5 and 6 to the nE case.

This implies that for N4 it is capital that the adversary can efficiently decrypt *everything*. In fact, the nE scheme used in Sect. 4 is message-malleable except in the case if $n = 1$ or 2 when trying to decrypt or encrypt $(v^*, \cdot)$.

6.4   Comparison to Namprempre et al. [20]

Namprempre et al. [20] gave partial results using the Knowledge-of-Tag property (KoT) (introduced in the extended version). That is, adversaries must forge without any (extractable) knowledge of the tag used in the decryption query [20].

With respect to their work, although the main ideas of the proofs are very similar, it is much easier to prove that a scheme is mrE or message-malleable, than to prove that a scheme is KoT-secure (while it may be easy to prove that it is not KoT-secure). In fact, to prove that a scheme is message-malleable it is enough to provide an algorithm which efficiently computes the result. On the other hand to prove that a scheme is not message-malleable (a part from proving that it is mrE), it must be proved that *all* efficient adversaries are not able always to decrypt. Similarly to prove the KoT security it must be proved that for *all* possible efficient extractors the scheme has this property, while to prove that a scheme is not KoT secure, it is enough to provide a counterexample.

7   Insecure Variants of Modes A10, A11 and A12

While, in the previous section, we proved the security of A10, A11 and A12 by making some extra requirements on the ivE scheme, this section considers the relaxation of some of the requirements on ivE that makes these 3 modes to become insecure. More precisely, we show how to compute forgeries against the INT-CTXT property of mode A10 when the ivE scheme is non tidy or stateful. These attacks imply that the three modes are not nAE-secure, when implemented with such schemes.

7.1 Tidiness as a Security Property

Given an IV-based encryption $\Pi = (\mathcal{K}, \mathsf{Enc}, \mathsf{Dec})$, our idea is to turn Π into a sloppy IV-based scheme. This modification augments the ciphertext $c = \mathsf{Enc}_k(iv, m)$ with $c' = \mathsf{Enc}_{k'}(iv, m)$, leading to a double and independent encryption m with the same iv. It is easy to see that, for random iv, the new scheme has pseudorandom ciphertext $C = (c, c')$ as long as Π has pseudorandom ciphertext, (that is, if Π is ivE-secure). However, given iv, if we define the decryption of $C = (c, c')$ simply as $\mathsf{Dec}_k(iv, c)$ without any validity consideration on c', the new scheme is not tidy whether Π is tidy or not. Therefore, since the c' part of C is "out of control", any ciphertext $C' = (c, c'')$ decrypts to m and is deemed valid. Moreover, the A10 composition mode with two PRF-secure vecMACs does not rule out this malleability so that we can build a forgery with a single encryption query. Dropping the tidiness requirement of ivE, and then of nAE, is thus sufficient to leave a security breach in the resulting nAE.

More formally, we build $\Pi' = (\mathcal{K}, \mathsf{Enc}', \mathsf{Dec}')$ with keyspace $\mathcal{K}^2$, message space $\mathcal{M}$ and ciphertext space $\mathcal{C}^2$ as follows: $\mathsf{Enc}'_{(k,k')}(iv, m)$ outputs $C = (c, c')$ where $c = \mathsf{Enc}_k(iv, m)$ and $c' = \mathsf{Enc}_{k'}(iv, m)$; $\mathsf{Dec}'_{(k,k')}(iv, C)$ parses the ciphertext as $C = (c, c')$ and outputs $m = \mathsf{Dec}_k(iv, c)$. For any $c'' \neq c'$, we have $\mathsf{Enc}'(iv, \mathsf{Dec}'(iv, (c, c''))) = (c, c') \neq (c, c'')$ so that Π' is not tidy.

Let nAE be the authenticated encryption obtained from the A10 mode whose ciphertext has the form $C = (c, c')$ where $c = \mathsf{Enc}_k(iv, \mathbf{m} \| \tau)$ and $c' = \mathsf{Enc}_{k'}(iv, \mathbf{m} \| \tau)$ with $iv = \mathsf{Mac}^{\mathsf{IV}}_{k_M}(n, a)$ and $\tau = \mathsf{Mac}^{\mathsf{Tag}}_{k_M}(a, \mathbf{m})$. Now, we consider the forger A which makes a single encryption query on any triple $(n, a, \mathbf{m})$ and receives back $C = (c, c')$ as above. Then, A picks any (samplable) $c'' \in \mathcal{C}$ distinct of c' and outputs $C^\star = (c, c'')$. Following the description of the A10 mode we find that the decryption starts by running $iv = \mathsf{Mac}^{\mathsf{IV}}_{k_M}(n, a)$ and then $\mathsf{Dec}'_{(k,k')}(iv, C^\star) = \mathsf{Dec}_k(iv, c) = \mathbf{m} \| \tau$. Finally, since the check $\tau = \mathsf{Mac}^{\mathsf{Tag}}_{k_M}(a, \mathbf{m})$ passes $\mathbf{m} \neq \perp$ is returned although $C^\star \neq C$.

Message-Malleability. In order to further emphasize the crucial role of the tidiness in the insecurity of the authenticated encryption based on Π', we stress that if the underlying IV-based scheme Π is tidy and message-malleable (Definition 9), the A10 composition implemented with Π leads to an nAE-secure scheme (as shown in Sect. 6.2). However, even if Π' is not tidy, it is easy to see that Π' remains message-malleable while we proved that it never leads to a nAE-secure scheme. As a summary, (non) tidiness alone has an intrinsic propensity to degrade the nAE-security of the AEnc based on the Tag-then-Encrypt paradigm.

7.2 Forgery Against Stateful A10, A11 and A12

In stateful AE schemes the AEnc and ADec algorithms receive at the start of the game an additional input, the *state*, which is updated during every call and kept in memory to be reused in the following call. The scheme we use has a stateless ADec algorithm, that is, it does not use the state and every reordering and omission is tolerable (L_0 of Rogaway et al. [28]). With respect to their work we allow the adversary to choose the state at the start of the game.

The idea of this forgery is to use the state, which in our case is simply a counter of the encryption queries, as the nonce was used in the attack against mode N4 (Sect. 4). At the end of the section we discuss the meaning of tidiness for stateful schemes.

The ivE we present is an adaptation of the nE scheme used in Sect. 4. As there, we present it only for N-bit long message, leaving the general case to the extended version. The main changes are:

- We use a TPRP $E : \mathcal{K} \times \mathcal{TW} \times \{0,1\}^\lambda \longmapsto \{0,1\}^\lambda$.
- A new block c_{-1} is added to the ciphertext, in order to give the decryption algorithm the actual value of the counter ctr which is *an internal state only of the encryption device*, and $c_{-1} = E_k^{(0,1)}(ctr)$. The Dec algorithm inverts this to retrieve the correct ctr. The block c_{-1} is random since it is always obtained with different inputs (as long as the number of encryption queries is $\ll 2^n$).
- To compute this block, the TPRP E is called with a tweak $(0,1)$ that is never used else
- The boxed **if** is triggered by the value of the counter ctr (not of the nonce) and m
- The iv replaces the nonce n in the input of the TPRP E.

Note again that, due to mode A10, the messages which the nAE scheme Π can encrypt, are λ bits long, while those which Π can encrypt are 2λ bits long.

The forgery is an easy adaptation of that presented in Sect. 4 .

The scheme Π is clearly ivE-secure, the only important change with Sect. 4 is the fact that we have to consider also the block c_{-1}). Now we have to discuss what means for a stateful nAE (or nE or ivE) scheme to be *tidy*.

For stateless nAE schemes the definition was given in Definition 2 (similarly for nE and ivE): if $\mathsf{ADec}_k^{n,a}(c) = m \neq \perp$ then $\mathsf{AEnc}_k^{n,a}(m) = c$.

Now if the nAE scheme is stateful it means that $\mathsf{AEnc}_k^{n,a}(m)$ is no more defined, because the state s may influence the output of $\mathsf{AEnc}_k^{(\cdot,\cdot)}(\cdot)$. Thus, denoting with $\mathcal{S}$ the set of possible states, we redefine tidiness as:

Definition 10. *We say that an nAE scheme is* tidy *if* $\mathsf{ADec}_k^{n,a}(c) = m$ *then* $c \in \{\mathsf{AEnc}_{k,s}^{n,a}(m)\}_{s \in \mathcal{S}}$.

Similarly an nE (resp. an ivE) scheme is tidy *if* $\mathsf{Dec}_k^{n,a}(c) = m$ *(resp.* $\mathsf{Dec}_k^{iv,a}(c) = m$ *then* $c \in \{\mathsf{Enc}_{k,s}^{n,a}(m)\}_{s \in \mathcal{S}}$ *(resp.* $c \in \{\mathsf{Enc}_{k,s}^{iv,a}(m)\}_{s \in \mathcal{S}}$*).*

According with this new definition, the ivE scheme Π which we have just used, presented, is tidy, as it follows from a close inspection of the pseudocode provided, thus the nAE scheme Π is tidy.

We have also to redefine for stateful schemes all the notions presented in Sect. 2. We do it allowing the adversary *at the start of the game* to set the state of the scheme as he wishes.

8   Conclusion

In this paper we have studied four generic composition modes, N4, A10, A11 and A12, for building authenticated encryption for an encryption scheme and a PRF MAC. The security of these four modes was left open in previous works, and three of them are the most efficient among the 180 possible modes based on these building blocks.

We have proved that mode N4 is not secure in general, and that modes A10, A11 and A12 have equivalent security. Moreover we have proved that if these four modes are instantiated with many common schemes (like CTR, OFB) they are all secure. Finally, we have showed that tidiness (again) and being stateless can have a decisive impact on security, as the application of A10, A11 and A12 on untidy or stateful modes can lead to insecure solutions.

Having used an encryption scheme using ciphertext expansion to break N4, we leave as a question for future work the proof of the security of N4 if the encryption scheme does not expand the ciphertext.

Our analysis still leaves as an open problem to decide if modes A10, A11, and A12 are secure in general.

Acknowledgments. Thomas Peters is a postdoctoral researcher of the Belgian Fund for Scientific Research (F.R.S.-FNRS). This work has been funded in parts by the European Union (EU) and the Walloon Region through the FEDER project USER-Media (convention number 501907-379156) and the ERC project SWORD (convention number 724725).

References

1. Atluri, V. (ed.): Proceedings of the 9th ACM Conference on Computer and Communications Security, CCS 2002, Washington, DC, 18–22 November 2002. ACM (2002)
2. Bellare, M., Desai, A., Jokipii, E., Rogaway, P.: A concrete security treatment of symmetric encryption. In: 38th Annual Symposium on Foundations of Computer Science, FOCS 1997, Miami Beach, 19–22 October 1997, pp. 394–403. IEEE Computer Society (1997)
3. Bellare, M., Kohno, T., Namprempre, C.: Authenticated encryption in SSH: provably fixing the SSH binary packet protocol. In: Atluri [1], pp. 1–11
4. Bellare, M., Kohno, T., Namprempre, C.: Breaking and provably repairing the SSH authenticated encryption scheme: a case study of the encode-then-encrypt-and-MAC paradigm. ACM Trans. Inf. Syst. Secur. **7**(2), 206–241 (2004)
5. Bellare, M., Namprempre, C.: Authenticated encryption: relations among notions and analysis of the generic composition paradigm. In: Okamoto [22], pp. 531–545
6. Bellare, M., Rogaway, P.: Encode-then-encipher encryption: how to exploit nonces or redundancy in plaintexts for efficient cryptography. In: Okamoto [22], pp. 317–330
7. Bernstein, D.J.: Caesar call for submissions, final, 27 January 2014

8. Berti, F., Koeune, F., Pereira, O., Peters, T., Standaert, F.-X.: Ciphertext integrity with misuse and leakage: definition and efficient constructions with symmetric primitives. In: Kim, J., Ahn, G.-J., Kim, S., Kim, Y., López, J., Kim, T., (eds.) Proceedings of the 2018 on Asia Conference on Computer and Communications Security, AsiaCCS 2018, Incheon, Republic of Korea, 04–08 June 2018, pp. 37–50. ACM (2018)

9. Berti, F., Pereira, O., Peters, T.: Reconsidering generic composition: the tag-then-encrypt case. Cryptology ePrint Archive, Report 2018/991 (2018). https://eprint. iacr.org/2018/991

10. Berti, F., Pereira, O., Peters, T., Standaert, F.-X.: On leakage-resilient authenticated encryption with decryption leakages. IACR Trans. Symmetric Cryptol. **2017**(3), 271–293 (2017)

11. Boyd, C., Hale, B., Mjølsnes, S.F., Stebila, D.: From stateless to stateful: generic authentication and authenticated encryption constructions with application to TLS. In: Sako, K. (ed.) CT-RSA 2016. LNCS, vol. 9610, pp. 55–71. Springer, Cham (2016). https://doi.org/10.1007/978-3-319-29485-8_4

12. Dworkin, M.J.: Recommendation for block cipher modes of operation: Galois/counter mode (GCM) and GMAC. Technical report (2007)

13. IETF: The transport layer security (TLS) protocol version 1.3 draft-ietf-tls-tls13-28. Technical report (2018). https://tools.ietf.org/html/draft-ietf-tls-tls13-28

14. Katz, J., Lindell, Y.: Introduction to Modern Cryptography, 2nd edn. CRC Press, Boca Raton (2014)

15. Katz, J., Yung, M.: Unforgeable encryption and chosen ciphertext secure modes of operation. In: Goos, G., Hartmanis, J., van Leeuwen, J., Schneier, B. (eds.) FSE 2000. LNCS, vol. 1978, pp. 284–299. Springer, Heidelberg (2001). https://doi.org/10.1007/3-540-44706-7_20

16. Kohno, T., Palacio, A., Black, J.: Building secure cryptographic transforms, or how to encrypt and MAC. IACR Cryptology ePrint Archive, 2003:177 (2003)

17. Krawczyk, H.: The order of encryption and authentication for protecting communications (or: How secure is SSL?). In: Kilian, J. (ed.) CRYPTO 2001. LNCS, vol. 2139, pp. 310–331. Springer, Heidelberg (2001). https://doi.org/10.1007/3-540-44647-8_19

18. Liskov, M., Rivest, R.L., Wagner, D.: Tweakable block ciphers. In: Yung, M. (ed.) CRYPTO 2002. LNCS, vol. 2442, pp. 31–46. Springer, Heidelberg (2002). https://doi.org/10.1007/3-540-45708-9_3

19. McGrew, D.A.: An interface and algorithms for authenticated encryption. RFC **5116**, 1–22 (2008)

20. Namprempre, C., Rogaway, P., Shrimpton, T.: Reconsidering generic composition. In: Nguyen, P.Q., Oswald, E. (eds.) EUROCRYPT 2014. LNCS, vol. 8441, pp. 257–274. Springer, Heidelberg (2014). https://doi.org/10.1007/978-3-642-55220-5_15

21. Nir, Y., Langley, A.: Chacha20 and poly1305 for IETF protocols. RFC **7539**, 1–45 (2015)

22. Okamoto, T. (ed.): ASIACRYPT 2000. LNCS, vol. 1976. Springer, Heidelberg (2000). https://doi.org/10.1007/3-540-44448-3

23. Paterson, K.G., Ristenpart, T., Shrimpton, T.: Tag size *Does* matter: attacks and proofs for the TLS record protocol. In: Lee, D.H., Wang, X. (eds.) ASIACRYPT 2011. LNCS, vol. 7073, pp. 372–389. Springer, Heidelberg (2011). https://doi.org/10.1007/978-3-642-25385-0_20

24. Peyrin, T., Seurin, Y.: Counter-in-tweak: authenticated encryption modes for tweakable block ciphers. In: Robshaw, M., Katz, J. (eds.) CRYPTO 2016. LNCS, vol. 9814, pp. 33–63. Springer, Heidelberg (2016). https://doi.org/10.1007/978-3-662-53018-4_2

25. Rogaway, P.: Authenticated-encryption with associated-data. In: Atluri [1], pp. 98–107

26. Rogaway, P.: Nonce-based symmetric encryption. In: Roy, B., Meier, W. (eds.) FSE 2004. LNCS, vol. 3017, pp. 348–358. Springer, Heidelberg (2004). https://doi.org/10.1007/978-3-540-25937-4_22

27. Rogaway, P., Shrimpton, T.: A provable-security treatment of the key-wrap problem. In: Vaudenay, S. (ed.) EUROCRYPT 2006. LNCS, vol. 4004, pp. 373–390. Springer, Heidelberg (2006). https://doi.org/10.1007/11761679_23

28. Rogaway, P., Zhang, Y.: Simplifying game-based definitions. In: Shacham, H., Boldyreva, A. (eds.) CRYPTO 2018. LNCS, vol. 10992, pp. 3–32. Springer, Cham (2018). https://doi.org/10.1007/978-3-319-96881-0_1

On Diffusion Layers of SPN Based Format Preserving Encryption Schemes: Format Preserving Sets Revisited

Rana Barua[1], Kishan Chand Gupta[2], Sumit Kumar Pandey[3(✉)],
and Indranil Ghosh Ray[4]

[1] R.C. Bose Centre for Cryptology and Security,
Indian Statistical Institute, 203, B.T. Road, Kolkata 700108, India
rana@isical.ac.in
[2] Applied Statistics Unit, Indian Statistical Institute,
203, B.T. Road, Kolkata 700108, India
kishan@isical.ac.in
[3] Ashoka University, Sonepat, Haryana, India
emailpandey@gmail.com
[4] Department of Electrical and Electronic Engineering,
City University, London, London, UK
indranilgray@gmail.com

Abstract. In Inscrypt 2016, Chang et al. proposed a new family of substitution-permutation (SPN) based format preserving encryption algorithms in which a non-MDS (Maximum Distance Separable) matrix was used in its diffusion layer. In the same year in Indocrypt 2016 Gupta et al., in their attempt to provide a reason for choosing non-MDS over MDS matrices, introduced an algebraic structure called format preserving sets (FPS). They formalised the notion of this structure with respect to a matrix both of whose elements are coming from some finite field $\mathbb{F}_q$. Many interesting properties of format preserving sets $\mathbb{S} \subseteq \mathbb{F}_q$ with respect to a matrix $M(\mathbb{F}_q)$ were derived. Nevertheless, a complete characterisation of such sets could not be derived. In this paper, we fill that gap and give a complete characterisation of format preserving sets when the underlying algebraic structure is a finite field. Our results not only generalise and subsume those of Gupta et al., but also obtain some of these results over a more generic algebraic structure viz. ring $\mathcal{R}$. We obtain a complete characterisation of format preserving sets over rings when the sets are closed under addition. Finally, we provide examples of format preserving sets of cardinalities 10^3 and 26^3 with respect to 4×4 MDS matrices over some rings which are not possible over any finite field.

Keywords: Diffusion layer · Format preserving encryption
Format preserving set

© Springer Nature Switzerland AG 2018
D. Chakraborty and T. Iwata (Eds.): INDOCRYPT 2018, LNCS 11356, pp. 91–104, 2018.
https://doi.org/10.1007/978-3-030-05378-9_5

1 Introduction

In the current scenario, it seems easier to protect confidentiality of the data that resides in today's complex networks than to protect the networks themselves, and symmetric encryption is a good way to do that. Block ciphers like AES and DES are very popular and used all over the world to maintain confidentially of the message. However, in many applications, it is convenient to encrypt messages from an arbitrarily sized set onto the same set and here conventional block cipher modes such as ECB, CBC, or CTR are not suitable. For example, a valid Social Security Number (SSN) is encrypted into a valid SSN, a valid credit-card number (CCN) is encrypted into a valid credit-card number, etc. This has given rise to a new primitive called, *format preserving encryption* (FPE). With the advent of format preserving encryption, one can deterministically encrypt the data by preserving the format of the data, i.e., the ciphertext has the same format as the plaintext.

FPE gained popularity due to its usage in several financial sectors for encrypting credit-card numbers. Moreover, FPE allows to add encryption to legacy databases and applications without violating existing format constraints which makes it a natural choice for applications where encrypted data needs to be saved for future search.

The first known constructions for FPE were proposed in [3,4]. Later the approach was formalized by Bellare et al. [1]. Following these, several constructions on FPE were proposed in [2,3,5,9,10,12,15]. In their 2002 paper [3], the authors proposed a practical approach for constructing an FPE. There are three popular approaches for designing FPE based schemes, namely *prefix ciphering*, *cyclic walking* and a *Feistel* based construction. To the best of our knowledge, none of these schemes used substitution permutation network (SPN) as a core construction technique for FPE except the one by Chang et al. [5]. In an SPN, a diffusion layer may be modelled as a linear transformation. An SPN has three main components - confusion, diffusion and key mixing. Confusion is achieved by an S-box whereas diffusion is a linear transformation. FPE may be achieved in several ways in an SPN as follows:

1. Either format is preserved only by the plaintext and the ciphertext, or
2. Let r be the number of rounds and each round preserves the format or
3. Each component of SPN preserves the format.

In [5], the key idea is same as (3) where format-preserving transformations are used to ensure that the format of plaintext and ciphertext are always same. In [5], each round of SPN consists of these basic transformations: (1) Format-Preserving SubBytes (FPSB), (2) ShiftRows, (3) Format-Preserving MixColumns (FPMC), (4) Format-Preserving Key Addition (FPKA) and (5) Format-Preserving Tweak Addition (FPTA).

To understand the format preserving set in the context of diffusion layer, let us consider an $n \times n$ matrix, M, whose entries are from an algebraic structure $\mathbb{A}$ with $+$ and $\cdot$. Let X be any set and ϕ be an injective map from X to $\mathbb{A}$, i.e. $\phi : X \to \mathbb{A}$. We say that $\phi(X)$ is format preserving with respect to the matrix

M if $M\mathbf{v} \in \phi(X)^n$ for all $\mathbf{v} \in \phi(X)^n$. Building a perfect diffusion layer for FPE by constructing a format preserving set with respect to an MDS matrix is a nontrivial task. This is mainly because of the following reasons.

1. All standard cryptographic primitives operate on fixed-length binary domains, mostly the field $\mathbb{F}_{2^n}$. On the other hand, for FPE, the underlying domain is usually non-binary. For example, in the context of CCN, the underlying domain set is $\{0, 1, \ldots, 9\}$. Since the domain is of size 10 which is a product of two different primes, one possibility is to consider a format preserving set as a proper subset of a field or of a ring. However, not many systematic study and construction of MDS matrices for format a preserving set as a subset of a field was available until before [8], when Gupta et al. studied the problem under some assumptions. So a full characterization of format preserving sets was unavailable for subsets of a field.

2. In [8], a partial characterization of format preserving set was given when it is a subset of a finite field with additive identity. They proved that the format preserving set is a vector space and thus it should be of size $q = p^i$ where p is the *characteristic* of the field. They were unable to construct format preserving sets of cardinality of the form other than p^i. For example, a format preserving set for CCN is of size 10, which compels us to study format preserving sets over rings. In another instance, the $ANSI\ ASC\ X9.124$ standard adopted by the financial industry envisions applications with domains as small as two decimal digits. Using the techniques used in [8], no format preserving set can be designed to meet such requirements.

In [5], Chang et al. considered a 4×4 matrix

$$\begin{bmatrix} \bar{1} & \bar{1} & \bar{1} & \bar{0} \\ \bar{0} & \bar{1} & \bar{1} & \bar{1} \\ \bar{1} & \bar{0} & \bar{1} & \bar{1} \\ \bar{1} & \bar{1} & \bar{0} & \bar{1} \end{bmatrix}$$

whose entries are from the ring $\mathbb{Z}_{10}$. The entries $\bar{0}$ and $\bar{1}$ in the matrix are additive and multiplicative identities respectively. The choice of $\mathbb{Z}_{10}$ is natural if the underlying format is a string of decimal digits 0–9. But the choice of the matrix is not optimal as it is not MDS (maximum distance separable) and hence does not provide the optimal branch number. In fact, its branch number is one less than the optimal one. The question is - why was not an MDS matrix chosen over $\mathbb{Z}_{10}$? Or, why not a subset of any other algebraic structure, either a ring or a field, which could satisfy the requirements? Chang et al. could not provide satisfactory answers for these questions. This paper attempts to find answers for these questions.

Our Contribution: In [8], Gupta et al. could not provide a full characterisation of a format preserving set when the underlying structure is a field. We resolved this issue and give a full characterisation of a format preserving set over any finite field. Furthermore, we tried to find its structure over commutative rings with unity. Under some restrictive cases, we identified its structure over such rings also. Finally, we show how to construct format preserving sets of cardinalities 20 with respect to 3×3 MDS matrices and of cardinalities of 10^3 and 26^3 with respect to 4×4 MDS matrices. Though we could not find any format preserving set of cardinality 10 and 26 with respect to 4×4 MDS matrices yet, we feel, our results are some progress in this direction.

2   Notations and Preliminaries

In this paper we assume elementary knowledge of semigroups, groups, rings, fields, vector spaces and modules. For more details, see [6,11,13]. Here, we consider only abelian semigroups without mentioning it explicitly. Recall that a semigroup with an identity element is called a monoid.

Let S be a semigroup and $\mathcal{X} \subseteq S$. The sub-semigroup generated by the set $\mathcal{X}$ is the smallest semigroup which contains $\mathcal{X}$ and is denoted by $\langle \mathcal{X} \rangle_s$. The set $\langle \mathcal{X} \rangle_s$ can be written as

$$\langle \mathcal{X} \rangle_s = \{((\cdots(x_1 x_2)\cdots)x_n) \mid n \geq 1, x_i \in \mathcal{X}\}.$$

If $\mathcal{X}$ is finite, say $\mathcal{X} = \{x_1, x_2, \cdots, x_k\} \subseteq S$, then

$$\langle \mathcal{X} \rangle_s = \{(x_1^{r_1})(x_2^{r_2})\cdots(x_k^{r_k}) \mid r_i \geq 0 \text{ and } \sum_{i=1}^{k} r_i \geq 1\}.$$

If G is a (multiplicative) group and $S \subseteq G$ then the subgroup of G generated by S, denoted by $\langle S \rangle$, is the smallest subgroup of G which contains S and is given by

$$\langle S \rangle = \{s_1^{r_1} s_2^{r_2} \cdots s_n^{r_n} \mid n \in \mathbb{N} = \{0,1,2,3,\cdots\}, s_i \in S, r_i \in \{1,-1\}\}$$

with the convention that if $n = 0$, the product over the empty list is e (the identity element of G). If G is finite, $r_i \neq -1$ (-1 in the exponent not required). Let $S = \{g_1, g_2, \cdots, g_k\} \subseteq G$. If G is a finite abelian group, then

$$\langle S \rangle = \{g_1^{r_1} g_2^{r_2} \cdots g_k^{r_k} \mid r_i \in \mathbb{N}\}.$$

Let $h \in G$. By hS, we mean $\{hg_1, hg_2, \cdots, hg_k\}$.

Let $B = \{b_1, \cdots, b_k\} \subseteq \mathcal{R}$ and $a \in \mathcal{R}$, where $\mathcal{R}$ is a ring. Then $aB = \{ab_1, \cdots, ab_k\}$ and $a + B = \{a + b_1, \cdots, a + b_k\}$. We shall denote a field with q elements by $\mathbb{F}_q$, where $q = p^r$ for some prime p. We denote by $\mathbb{F}^*$ the multiplicative group of non-zero elements of $\mathbb{F}$.

In this paper, we assume that a module $\mathbb{M}$ is a unital $\mathcal{R}$-module, i.e. $\bar{1}\mathbf{m} = \mathbf{m}$ for all $\mathbf{m} \in \mathbb{M}$ where $\bar{1}$ is the unity in the ring $\mathcal{R}$. A unit in a ring $\mathcal{R}$ is an element of $\mathcal{R}$ which has its multiplicative inverse. A module $\mathbb{M}$ is a free $\mathcal{R}$-module if $\mathbb{M}$ has a free $\mathcal{R}$-module basis. A subset Γ of $\mathbb{M}$ is a free $\mathcal{R}$-module basis of $\mathbb{M}$ if Γ is an $\mathcal{R}$-module basis of $\mathbb{M}$ and Γ is linearly independent over $\mathcal{R}$. For more about rings and modules, see [16].

The quotient ring of integers modulo r is denoted as $\mathbb{Z}_r$. The set $\mathbb{Z}_r$ is commonly represented as $\{0, 1, 2, \cdots, r - 1\}$. Addition $(+)$ and multiplication $(*)$ operations on this set are defined as addition and multiplication on integers modulo r. The set of units of $\mathbb{Z}_r$, represented as $\mathbb{Z}_r^*$, are those elements from $\mathbb{Z}_r$ which are relatively prime to r over integers. For the notational convenience, $a * b$ will be written as ab.

An $m \times n$ matrix M is written $M(A)$ if the entries of M are from the set A. If entries are evident from the context, we simply denote $M(A)$ as M only. The $(i, j)^{\text{th}}$ entry of a matrix M where $1 \leq i \leq m$ and $1 \leq j \leq n$ is denoted as $m_{i,j}$ and the matrix M as $(m_{i,j})$. The transpose of a matrix $M = (m_{i,j})$ is denoted as $M^T = (m_{j,i})$ where $1 \leq i \leq m$ and $1 \leq j \leq n$. An $n \times n$ square matrix is said to be a matrix of order n. An identity matrix of order n is denoted as I_n.

A vector is a special kind of a matrix. If $m = 1$, then a $1 \times n$ matrix is called a row vector. If $n = 1$, then an $m \times 1$ matrix is called a column vector. A row vector is written as a horizontal array and column vector as a vertical array. If the vector $\mathbf{v}$ (either row or column) has n entries, then we say that the vector is n-dimensional. If the entries of the vector $\mathbf{v}$ are from the set A, then we write $\mathbf{v} \in A^n$. In this paper, by the term vector, we mean a column vector.

Given an $n \times n$ matrix M, we denote by Z the set

$$Z = \{m_{i,j} \mid m_{i,j} \neq \bar{0}, 1 \leq i, j \leq n\}$$

where $\bar{0}$ is the additive identity of the corresponding algebraic structure. We also set

$$\mathrm{m}_i = \Sigma_{j=1}^n m_{i,j} \text{ and } R = \{\mathrm{m}_i \mid \mathrm{m}_i \neq \bar{0}\}.$$

Definition 1. *Let $\mathcal{R}$ be a commutative ring with unity $(\bar{1})$. A set $\mathbb{S} \subseteq \mathcal{R}$ is said to be a format preserving set with respect to an $n \times n$ matrix $M(\mathcal{R})$ if $M\mathbf{v} \in \mathbb{S}^n$ for all $\mathbf{v} \in \mathbb{S}^n$.*

It may be noted that $\mathcal{R}$ may be a finite field $\mathbb{F}_q$. Other notations or any undefined terms in this paper have usual standard meanings.

3   Maximum Distance Separable (MDS) Codes

An $[n, k, d]$ linear code over any field satisfies $d \leq n - k + 1$ which is known as singleton bound. Maximum Distance Separable (MDS) codes are those codes which satisfy $d = n - k + 1$ [14]. Let a matrix $[I_m \mid M]$ be a generator matrix of an $[m + n, m, n + 1]$ MDS code. Then M is called an MDS matrix. There is an alternate way to define it which is as follows:

Definition 2 (Definition 4, [7]). *An $m \times n$ matrix M is called MDS if and only if all its square submatrices are non-singular.*

Equivalently,

Definition 3 (Fact 1, [7]). *An $m \times n$ matrix M is called MDS if and only if any $n \times n$ submatrix obtained after removing m rows from $\begin{bmatrix} I_n \\ M \end{bmatrix}$ is non-singular.*

And when $m = n$, an equivalent definition of MDS matrix can be obtained again which is as follows:

Definition 4 (Fact 1, [7]). *An $n \times n$ matrix M is called MDS if and only if any $n \times n$ submatrix obtained after removing n columns from $[I_n \mid M]$ is non-singular.*

The above three definitions are equivalent in characterising MDS matrix over a finite field. But it does not give the guarantee of the existence of an $[n, k, d]$ MDS code over $\mathbb{F}_q$. Unfortunately, we do not have any such result. However, there is a famous conjecture known as MDS conjecture which states the following.

Conjecture 1. For an $[n, k, d]$ linear MDS code over $\mathbb{F}_q$,

1. If $q = 2^h$ and $k = 3$, then $n \leq q + 2$, or
2. If $q = 2^h$ and $k = q - 1$, then $n \leq q + 2$, or
3. If $k \leq q$, then $n \leq q + 1$ otherwise.

3.1 MDS Matrices over a Finite Commutative Ring $\mathcal{R}$ with Unity

We assume that the ring $\mathcal{R}$ is a finite commutative ring with unity. An $n \times n$ matrix M over a finite commutative ring $\mathcal{R}$ with unity is said to be non-singular if the determinant of M is a unit in $\mathcal{R}$.

Definition 5. *An $n \times n$ matrix M over a finite commutative ring $\mathcal{R}$ with unity is called MDS if and only if all its square submatrices are non-singular.*

4   FPS Need Not Be an Invariant Subspace

A format preserving set $\mathbb{S}$ with respect to a matrix $M(\mathcal{R})$ need not be an invariant subspace. An invariant subspace W is a subspace of a vector space V that is preserved by a linear transformation $T : V \to V$, i.e. $T(W) \subseteq W$. The difference lies in the algebraic structure of $\mathbb{S}$ and W; $\mathbb{S}$ is merely a set whereas W is a vector space.

We emphasize that there exists a set $\mathbb{S}$ which is a format preserving set with respect to some matrix M but it is not a vector space. Consider $\mathbb{S} = \{\bar{1}, \alpha^3, \alpha^6, \alpha^9, \alpha^{12}\}$ where α is a primitive element of the field $\mathbb{F}_{2^4} = \mathbb{F}_2[x]/\langle x^4 + x + 1 \rangle$ and

$$
M = \begin{bmatrix} \alpha^3 & \bar{0} & \bar{0} & \bar{0} \\ \bar{0} & \alpha^6 & \bar{0} & \bar{0} \\ \bar{0} & \bar{0} & \alpha^9 & \bar{0} \\ \bar{0} & \bar{0} & \bar{0} & \alpha^{12} \end{bmatrix}
$$

In the example above, $\mathbb{S}$ is an FPS but not an invariant subspace. But, an invariant subspace W over a field $\mathbb{F}$ is a format preserving set with respect to any matrix $M(\mathbb{F})$.

5   Our Results

5.1   Over Fields

In [8], Gupta et al. showed the algebraic structure of a format preserving set $\mathbb{S}$ with respect to a matrix over a finite field when the additive identity $\bar{0}$ belongs to the set $\mathbb{S}$. It was not discussed what if the additive identity does not belong to the set. We fill this gap by showing that, in such case, an affine transformation of a format preserving set $\mathbb{S}$ results into a vector space over the smallest field containing entries of the matrix M if M has a row which contains at least two non-zero entries.

We assume that the set does not contain $\bar{0}$ and hence all elements are non-zero. Let $a \in \mathbb{S}$. Consider

$$\mathbb{S}' = a^{-1}\mathbb{S}$$

It is easy to check that $\bar{1} \in \mathbb{S}'$ and $\mathbb{S}'$ is a format preserving set if and only if $\mathbb{S}$ is a format preserving set with respect to the same matrix. Therefore, the algebraic structure of $\mathbb{S}'$ will determine the algebraic structure of $\mathbb{S}$ and hence we safely assume that $\bar{1} \in \mathbb{S}$.

Now consider,

$$\bar{\mathbb{S}} = \mathbb{S} - \bar{1}.$$

The set $\bar{\mathbb{S}}$ is closed under addition if and only if $s_1 + s_2 - \bar{1} \in \mathbb{S}$ for all $s_1, s_2 \in \mathbb{S}$. The reason behind considering $\bar{\mathbb{S}}$, instead of $\mathbb{S}$, is to show that $\bar{\mathbb{S}}$ is a vector space over the smallest field containing entries of the matrix M if M has a row which contains at least two non-zero entries. However, for the sake of completeness, we consider two cases:

1. If each row of M has at most one non-zero entry, and
2. If M has a row which contains at least two non-zero entries.

A similar result is obtained for the case (1) following the proof of the Theorem 1 discussed in [8]. The result is stated below. Recall that given an $n \times n$ matrix M, we denote by Z the set

$$Z = \{m_{i,j} \mid m_{i,j} \neq \bar{0}, 1 \leq i, j \leq n\}$$

where $\bar{0}$ is the additive identity of the corresponding algebraic structure. We also set

$$m_i = \Sigma_{j=1}^{n} m_{i,j} \text{ and } R = \{m_i \mid m_i \neq \bar{0}\}.$$

Theorem 1. *Let $\mathbb{S} \subseteq \mathbb{F}_q$. Suppose each row of $M(\mathbb{F}_q)$ contains at most one non-zero entry. Then, $\mathbb{S}$ is a format preserving set with respect to M if and only if there exists a set $H \subseteq \mathbb{S}$ such that $\mathbb{S} = \bigcup_{s \in H} s\langle Z \rangle = H\langle Z \rangle$.*

Proof. If each row of M contains at most one non-zero entry then $R = Z$ and hence $\langle R \rangle = \langle Z \rangle$. By Theorem 3 of [8], we have $\mathbb{S} = \bigcup_{s \in H} s \langle Z \rangle$ for some set $H \subseteq \mathbb{S}$.

Conversely, let $\mathbb{S} = \bigcup_{s \in H} s \langle Z \rangle$. Assume $s^{(i)} \in H$ and $\alpha_i \in \langle Z \rangle$ for all $1 \leq i \leq n$. Consider a vector $\mathbf{v} = [s^{(1)} \alpha_1 \quad s^{(2)} \alpha_2 \quad \cdots \quad s^{(n)} \alpha_n]^T$. It is easy to see that $\mathbf{v} \in \mathbb{S}^n$. Without loss of generality, we assume that each row of M has exactly one non-zero entry (the proof will be similar if each row has at most one non-zero entry). Suppose M has non-zero entries in columns $j_1, j_2, \cdots, j_n$ corresponding to rows $1, 2, \cdots, n$. Then $M\mathbf{v} = [s^{(j_1)} \alpha_{j_1} m_{1, j_1} \quad s^{(j_2)} \alpha_{j_2} m_{2, j_2} \quad \cdots \quad s^{(j_n)} \alpha_{j_n} m_{n, j_n}] \in \mathbb{S}^n$ since by our assumption, $s^{(j_i)} \alpha_{j_i} m_{i, j_i} \in \mathbb{S}$ for all $1 \leq i \leq n$. Therefore, $\mathbb{S}$ is a format preserving set with respect to M. Hence, the theorem. □

It is interesting to note that Theorem 1 does not assume that $\bar{0} \notin \mathbb{S}$ and hence it covers both cases when $\bar{0} \in \mathbb{S}$ and $\bar{0} \notin \mathbb{S}$.

Now, we consider the case (2) when M has a row which contains at least two non-zero entries. Let $\mathrm{m}_i = \sum_{j=1}^{n} m_{i,j}$.

Theorem 2. *Let $\bar{1} \in \mathbb{S} \subseteq \mathbb{F}_q$ and $\bar{0} \neq \mathrm{m}_i \in \mathbb{S}$ for $1 \leq i \leq n$. Suppose the $n \times n$ matrix $M(\mathbb{F}_q)$ has at least one row with at least two non-zero entries. Then $\mathbb{S}$ is a format preserving set with respect to M iff $\bar{\mathbb{S}}$ is a format preserving set with respect to M.*

Proof. Consider the map

$$\phi_{(i,j)} : \mathbb{S} \to \mathbb{F}_q$$

defined by

$$\phi_{(i,j)}(s) = m_{i,j}(s - \mathrm{m}_i^{-1}) + \bar{1}.$$

It is certain that m_i^{-1} exists because $\mathrm{m}_i \neq \bar{0}$. Under this map, the image $\phi_{(i,j)}(\mathbb{S}) \subseteq \mathbb{S}$. In-fact $\phi_{(i,j)}(\mathbb{S}) = \mathbb{S}$ because $\phi_{(i,j)}$ is a bijection from $\mathbb{S}$ to $\phi_{(i,j)}(\mathbb{S})$.

Suppose at row i, the j_1 and j_2^{th} columns of the matrix M contain non-zero entries. Choose $s_1, s_2 \in \mathbb{S}$. Since $\phi_{(i,j)}$ is a bijection from $\mathbb{S}$ to $\mathbb{S}$, there exist $s_1', s_2' \in \mathbb{S}$ such that $s_1 = \phi(s_1')$ and $\phi(s_2')$. Consider the vector $\mathbf{v} = [\mathrm{m}_i^{-1} \quad \mathrm{m}_i^{-1} \quad \cdots \quad s_1' \quad \cdots \quad s_2' \quad \cdots \quad \mathrm{m}_i^{-1}]^T \in \mathbb{S}^n$ where s_1' is at the j_1^{th} position, s_2' is at the j_2^{th} position of the vector $\mathbf{v}$ and the rest are m_i^{-1}. Take the vector $M\mathbf{v}$. The ith element of the vector $M\mathbf{v}$ will be $s_1 + s_2 - \bar{1} \in \mathbb{S}$ and thus $\bar{\mathbb{S}} = \mathbb{S} - \bar{1}$ is closed under addition. As $\bar{\mathbb{S}}$ contains $\bar{0}$ and is closed under addition, so is a format preserving set with respect to M if $\mathbb{S}$ is a format preserving set with respect to M too.

Conversely, if $\bar{\mathbb{S}}$ is a format preserving set with respect to M, then it is closed under addition because $\bar{0} \in \mathbb{S}$. Thus given any two elements $s_1, s_2 \in \mathbb{S}$, $s_1 + s_2 - \bar{1} \in \mathbb{S}$. Hence $\mathbb{S}$ too is a format preserving set with respect to M. □

Remark 1. Theorem 2 does not provide if and only if condition when the condition $\mathrm{m}_i \in \mathbb{S}$ is relaxed. We show by giving a simple example. Take any $M(\mathbb{F}_q)$ whose first row has exactly two non-zero entries such that their sum is neither $\bar{0}$ nor $\bar{1}$. For this, consider $q > 2$. Rest rows of the matrix M has exactly one non-zero entry. Take $\bar{\mathbb{S}} = \{\bar{0}\}$. Then $\mathbb{S} = \{\bar{1}\}$. It is easy to check that $\bar{\mathbb{S}}$ is a format preserving set with respect to M but not $\mathbb{S}$.

The paper [8] characterises format preserving sets $\mathbb{S}$ when $\bar{0} \in \mathbb{S}$. For format preserving sets $\mathbb{S}$ not containing $\bar{0}$, by invoking this theorem, we can obtain a complete characterisation of $\mathbb{S}$ via the format preserving sets $\bar{\mathbb{S}}$. If there is any such $\mathbb{S}$, its cardinality must be p^l for some prime p and some $l \geq 1$. And thus, it rules out the possibility of getting any format preserving set $\mathbb{S}$ over any finite field such that $|\mathbb{S}| = 10$ or 26.

5.2   Over Rings

The failure of obtaining format preserving sets of cardinality 10 or 26 with respect to cryptographically significant matrices over finite fields lead us to search over the next obvious algebraic structures which are rings. If $\mathbb{S} = \mathbb{Z}_{10}$ or $\mathbb{Z}_{26}$, it is a format preserving set with respect to any matrix $M(\mathbb{Z}_{10})$ or $M(\mathbb{Z}_{26})$ respectively. We observe the same in [5].

Though it provides a solution, the matrix M does not provide the optimal diffusion because any matrix $M(\mathbb{Z}_{2r})$ of order greater than or equal to 2 cannot be maximum distance separable. Mathematically, matrices which provide the optimal diffusion are known as maximum distance separable (MDS) matrices and in cryptography, such matrices play a significant role particularly in the diffusion layer of many cryptographic primitives.

Over Finite Commutative Rings with Unity: Our search will now mainly focus upon the format preserving sets over MDS matrices, however, we build our theory with respect to any arbitrary matrices. Instead of searching over any arbitrary rings, we focus upon finite commutative rings with unity. The result stated in Theorem 1 in [8] and in the previous subsection (Theorem 1) can be extended over the rings under appropriate restrictions. These results are summarised below.

Let $Z' = \{m_{i,j} \mid m_{i,j} \neq \bar{0}\}$. Let $Z = Z' \cup \{\bar{1}\}$ if $\bar{1} \notin Z'$ else $Z = Z'$. Recall that $\langle Z \rangle_s$ is a submonoid generated by Z of the ring $\mathcal{R}$ under multiplication. We first consider the case when each row of M has exactly one non-zero entry.

Theorem 3. *Let $\mathbb{S} \subseteq \mathcal{R}$. Suppose each row of $M(\mathcal{R})$ contains at most one non-zero entry. Then, $\mathbb{S}$ is a format preserving set with respect to M if and only if there exists a set $H \subseteq \mathbb{S}$ such that $\mathbb{S} = \bigcup_{s \in H} s\langle Z \rangle_s$.*

The theorem above assumes that the each row of the matrix M contains at most one non-zero entry. The next theorem covers the rest case but when the set $\mathbb{S}$ is closed under addition.

Define the following set

$$\mathbb{R} = \{k_1\alpha_1 + k_2\alpha_2 + \cdots + k_r\alpha_r \mid r \geq 0, k_i \geq 1, \alpha_i \in \langle Z \rangle_s\}$$

with the convention that if $r = 0$, the sum over the empty list is $\bar{0}$. It is not hard to check that the set $\mathbb{R}$ is in fact the smallest ring containing entries of Z. Note that Z contains $\bar{1}$ and therefore $\bar{1} \in \mathbb{R}$ too.

Theorem 4. *The following statements are equivalent.*

1. $\mathbb{S} \subseteq \mathcal{R}$ *is a format preserving set with respect to M and $\mathbb{S}$ is closed under $+$.*
2. $\mathbb{S}$ *is a module (unital) over the ring $\mathbb{R}$.*
3. $\mathbb{S}$ *is closed under $+$ and for all $s \in \mathbb{S}$ and $\alpha \in \mathbb{R}, s\alpha \in \mathbb{R}$.*

Proof. $(1) \rightarrow (2)$.

Assume (1) holds. We shall show that $\mathbb{S}$ is a module over $\mathbb{R}$. Since $\mathbb{S}$ is finite and closed under $+$, it follows that $\mathbb{S}$ is an additive subgroup of $\mathcal{R}$ and $\bar{0} \in \mathbb{S}$. Fix $s \in \mathbb{S}$ and $j, 1 \leq j \leq n$. Let $\mathbf{v}^{(j)} = [\bar{0}\,\bar{0} \,\cdots\, s \,\cdots\, \bar{0}\,\bar{0}]^T$ where s is at the jth position of the vector $\mathbf{v}^{(j)}$ and rest $\bar{0}$. Then $M\mathbf{v}^{(j)} = [sm_{0,j} \;\; sm_{1,j} \;\; \cdots \;\; sm_{n,j}]^T$ $\in \mathbb{S}^n$. Therefore, $sm_{i,j} \in \mathbb{S}$ for all $1 \leq i \leq n$. Thus $sm_{i,j} \in \mathbb{S}$ for $1 \leq i, j \leq n$. Repeated applications of this shows that $s\langle Z \rangle_s \subseteq \mathbb{S}$.

Take any $\alpha \in \mathbb{R}$. If $\alpha = \bar{0}$, then $\alpha s = \bar{0} \in \mathbb{S}$ for any $s \in \mathbb{S}$. Suppose $\alpha \neq \bar{0}$, say $\alpha = k_1\alpha_1 + k_2\alpha_2 + \cdots + k_r\alpha_r$ where $\alpha_i \in \langle Z \rangle_s$, $r \geq 1$ and $k_i \geq 1$ for all $1 \leq i \leq r$. Take $s \in \mathbb{S}$. Then $\alpha s = k_1(\alpha_1 s) + k_2(\alpha_2 s) + \cdots + k_r(\alpha_r s)$. From the preceding argument, $s\alpha_i \in \mathbb{S}$ for all $1 \leq i \leq r$. Since $\mathbb{S}$ is closed under $+$, we have $\alpha s = \sum_{i=1}^{r} k_i(\alpha_i s) \in \mathbb{S}$. All other axioms of a module hold since $\mathbb{S}$ and $\mathbb{R}$ are both subsets of $\mathcal{R}$.

$(2) \rightarrow (3)$ follows from the fact that $\mathbb{S}$ is a module over $\mathbb{R}$.

$(3) \rightarrow (1)$.

Consider the vector $\mathbf{v} = [s_1 \;\; s_2 \;\; \cdots \;\; s_n]^T \in \mathbb{S}^n$. Take $M\mathbf{v} = M[s_1 \;\; s_2 \;\; \cdots \;\; s_n]^T$. Then the ith element of the vector $M\mathbf{v}$ will be $\sum_{r=1}^{n} m_{i,r}s_r$. Now, each $m_{i,r} \in \mathbb{R}$ and so by assumption, each $m_{i,r}s_r \in \mathbb{S}$. Since $\mathbb{S}$ is closed under $+$, we have $\sum_{r=1}^{n} m_{i,r}s_r \in \mathbb{S}$. Thus, the ith element of the vector $M\mathbf{v}$ belongs to $\mathbb{S}$ and hence $M\mathbf{v} \in \mathbb{S}^n$. Therefore, $\mathbb{S}$ is a format preserving set with respect to M. This completes the proof. $\square$

When $\mathbb{S}$ is a free $\mathbb{R}$-module, then $|\mathbb{S}| = |\mathbb{R}|^l$ for some $l \geq 0$. For $|\mathbb{S}| = 10$ or 26, the requirement is $|\mathbb{R}| = 10$ or 26 respectively and $l = 1$. Since $\mathbb{R}$ is commutative (because $\mathbb{R} \subseteq \mathcal{R}$) and contains the unity, therefore $\mathbb{R} \cong \mathbb{Z}_{10}$ or $\mathbb{R} \cong \mathbb{Z}_{26}$ if $|\mathbb{R}| = 10$ or 26 respectively. And thus, in such cases, $M(\mathbb{R})$ cannot be MDS if the order of the matrix is greater than or equal to 2.

The Closure of $\mathbb{S}$: In general, $\mathbb{S}$ may not be closed under addition. For example, consider a 3×3 matrix $M(\mathbb{Z}_{10})$ whose each row contains three elements from the set $\{\bar{1}, \bar{3}, \bar{5}, \bar{7}, \bar{9}\}$. Then the set $\mathbb{S} = \{\bar{1}, \bar{3}, \bar{5}, \bar{7}, \bar{9}\}$ is a format preserving set with respect to M though not closed. Nevertheless, an interesting observation about this example is that the transformation $\mathbb{S} - \bar{1}$ is closed. It is merely an observation and we are not making any claim about those sets which are not closed.

The closure of $\mathbb{S}$ implies $\bar{0} \in \mathbb{S}$, but the converse need not be true. It can be easily shown using Theorem 3. But, Theorem 3 assumes that each row of the matrix has at most one non-zero entry. Now, we show it by an example that this is true even when the matrix has a row which contains at least two non-zero entries. Suppose $\mathcal{R} = \mathbb{F}_2[x]/\langle x^4 + x^3 \rangle$. Consider

$$
M = \begin{bmatrix} \bar{1} & \bar{0} & \bar{0} & \bar{0} \\ \bar{0} & \bar{x}^3 & \bar{x}^3 & \bar{0} \\ \bar{0} & \bar{0} & \bar{1} & \bar{0} \\ \bar{0} & \bar{0} & \bar{0} & \bar{1} \end{bmatrix}
$$

and $\mathbb{S} = \{\bar{0}, \bar{1}, \bar{x}^3\}$ where $\bar{x}$ is the residue class of polynomials which leave remainder $x \in \mathbb{F}_2[x]$ when divided by the polynomial $x^4 + x^3 \in \mathbb{F}_2[x]$. It is easy to check that $\mathbb{S}$ is a format preserving set with respect to $M(\mathcal{R})$, but not closed.

However, it is not hard to conclude that the set $\mathbb{S}$ is closed if $\bar{0} \in \mathbb{S}$ and the matrix M contains a row which has at least two units. The result is stated in the following lemma and the proof can be done in a similar manner as it was done as a part in the proof of Theorem 2.

Lemma 1. *Let $\mathbb{S} \subseteq \mathcal{R}$ be a format preserving set with respect to a matrix $M(\mathcal{R})$. Suppose there exists a row of M which contains at least two units. Then, $\mathbb{S}$ is closed under addition.*

Theorem 4 gives a nice characterisation of a format preserving set $\mathbb{S}$ over a commutative ring $\mathcal{R}$ with unity when $\mathbb{S}$ is closed under addition. Furthermore, it suggests that if $\mathbb{S}$ is a free module over $\mathbb{R}$, then it is impossible to construct a format preserving set of cardinality 10 or 26 with respect to an MDS matrix of order greater than or equal to 2. However, some cases are yet unexplored which are when (a) $\mathbb{S}$ is closed but not a free module over $\mathbb{R}$, (b) $\mathbb{S}$ is not closed and (c) $\mathcal{R}$ is any arbitrary ring. Thus, the possibility of getting $\mathbb{S}$ with respect to an MDS matrix such that $|\mathbb{S}| = 10$ or 26 is still open.

Though we could not provide $\mathbb{S}$ of cardinalities 10 or 26 with respect to an MDS matrix, the next subsection provides some constructions for the same when $|\mathbb{S}| = 10^3$ or 26^3.

5.3 Search for Format Preserving Sets with Respect to MDS Matrices over Rings

Let $n = p_1^{\alpha_1} p_2^{\alpha_2} \cdots p_r^{\alpha_r}$ where p_i's are prime numbers and $\alpha_i \geq 1$. A matrix $M(\mathbb{Z}_n)$ is MDS if and only if M modulo p_i is MDS for all $1 \leq i \leq r$. Suppose $p_1 < p_2 < \cdots < p_r$. Then, by the MDS conjecture, the maximum order of the matrix M so that it can be MDS over $\mathbb{Z}_n$ is $\lfloor (p_1 + 1)/2 \rfloor$. However, we can construct a ring $\mathcal{R}$ of n elements so that there exists an MDS matrix M over the ring $\mathcal{R}$ whose order is less than or equal to $\min\{3, \min\{\lfloor (p_i^{\alpha_i} + 1)/2 \rfloor\}_{i=2}^r\}$ if $p_1 = 2$ and $\alpha_1 = 2$, otherwise $\min\{\lfloor (p_i^{\alpha_i} + 1)/2 \rfloor\}_{i=1}^r$. Now we show how to construct $\mathcal{R}$.

Construction of $\mathcal{R}$: Suppose $n = p_1^{\alpha_1} p_2^{\alpha_2} \cdots p_r^{\alpha_r}$. Let $(\mathbb{F}_{p_i^{\alpha_i}}, +_i, \cdot_i)$ be fields for $1 \leq i \leq r$. For notational convenience, we simply denote $+_i$ and $\cdot_i$ by $+$ and $\cdot$ respectively if it is clear from the context. Then the ring $(\mathcal{R}, +, \cdot)$ is defined as

$$
\mathcal{R} = \mathbb{F}_{p_1^{\alpha_1}} \times \mathbb{F}_{p_2^{\alpha_2}} \times \cdots \times \mathbb{F}_{p_r^{\alpha_r}},
$$

where

$$(a_1, a_2, \cdots, a_r) + (b_1, b_2, \cdots, b_r) = (a_1 + b_1, a_2 + b_2, \cdots, a_r + b_r)$$

and

$$(a_1, a_2, \cdots, a_r) \cdot (b_1, b_2, \cdots, b_r) = (a_1 \cdot b_1, a_2 \cdot b_2, \cdots, a_r \cdot b_r)$$

It is easy to check that $\mathcal{R}$ is a commutative ring with unity $(\bar{1}, \bar{1}, \cdots, \bar{1})$. The additive identity is $(\bar{0}, \bar{0}, \cdots, \bar{0})$ and the number of elements in $\mathcal{R}$ is n. Now, we discuss how to construct an MDS matrix M over this ring $\mathcal{R}$.

Construction of an MDS Matrix M: For $1 \leq l \leq r$, let $M_l = (m_{ij}^{(l)})$ be MDS matrices over the field $\mathbb{F}_{p_l^{\alpha_l}}$ all of the same order k. Then, $M = ((m_{ij}^{(1)}, m_{ij}^{(2)}, \cdots, m_{ij}^{(r)}))$ will become an MDS matrix of order k over $\mathcal{R}$.

The construction above allows to have format preserving sets with respect to MDS matrices whose cardinalities can be other than p^r where p is a prime and $r > 0$. For example, it was not possible to have a format preserving set of cardinality 20 over any field. But, it is possible to have one over the ring $\mathbb{Z}_{20}$. But, the matrix cannot be MDS if the order is greater than one. But it is possible to have all when $\mathcal{R}' = \mathbb{F}_{2^2} \times \mathbb{F}_5$. Consider $M_1(\mathbb{F}_{2^2})$, $M_2(\mathbb{F}_5)$ and $M(\mathcal{R}')$ given by

$$M_1 = \begin{bmatrix} 1 & 1 & 1 \\ 1 & \alpha & \alpha^2 \\ 1 & \alpha^2 & \alpha \end{bmatrix} ; \quad M_2 = \begin{bmatrix} 1 & 1 & 1 \\ 1 & 2 & 3 \\ 1 & 3 & 4 \end{bmatrix} ; \quad M = \begin{bmatrix} (1,1) & (1,1) & (1,1) \\ (1,1) & (\alpha,2) & (\alpha^2,3) \\ (1,1) & (\alpha^2,3) & (\alpha,4) \end{bmatrix} ,$$

where α is a primitive element of the field $\mathbb{F}_{2^2}$. It is easy to check that M is MDS. Thus, $\mathbb{S} = \mathcal{R}'$ is a format preserving set with respect to MDS matrix M and $|\mathbb{S}| = 20$.

Remark 2. There does not exist any MDS matrix of order greater than or equal to 4 over $\mathcal{R}'$.

Construction of $\mathbb{S}$ with Respect to an MDS Matrix M and $|\mathbb{S}| = 10^3$: Though we are not successful to have $|\mathbb{S}| = 10$ yet, we can construct for $|\mathbb{S}| = 10^3$ with respect to a 4×4 MDS matrix. The construction is simple; take $\mathcal{R}' = \mathbb{F}_{2^3} \times \mathbb{F}_{5^3}$ and then construct 4×4 MDS matrices $M_1(\mathbb{F}_{2^3})$, $M_2(\mathbb{F}_{5^3})$ and $M(\mathcal{R}')$ similar as discussed previously.

The number 1000 is significant here because three digits can be taken at a time instead of one. It requires four bits to represent 9 in binary and hence altogether twelve bits for three digits each ranging from 0–9. Moreover, to represent $\mathcal{R}' = \mathbb{F}_{2^3} \times \mathbb{F}_{5^3}$, it needs $3 + 3 * 3 = 12$ bits. Thus, a suitable encoding is required before the diffusion layer which takes three digits as input and encodes into an element of $\mathcal{R}'$. Similarly, a decoder is required after the diffusion layer which decodes an element from $\mathcal{R}'$ to three digits. And $M(\mathcal{R}')$ can be used at the diffusion layer to provide the optimal diffusion. Of-course, it may not be profitable to use $M(\mathcal{R}')$ in the diffusion layer to achieve the optimal diffusion at the cost of extra hardware or software used for encoder and decoder before and after the diffusion layer. Nevertheless, it is significant because of the theoretical advancement towards achieving an optimal format preserving diffusion layer for digits 0–9 and alphabets A–Z.

Now, we construct a format preserving set having 26^3 elements with respect to a 4×4 MDS matrix M in a similar manner.

Construction of $\mathbb{S}$ with Respect to an MDS Matrix M and $|\mathbb{S}| = 26^3$:
Take $\mathcal{R}' = \mathbb{F}_{2^3} \times \mathbb{F}_{13^3}$ and then construct 4×4 MDS matrices $M_1(\mathbb{F}_{2^3})$, $M_2(\mathbb{F}_{13^3})$ and $M(\mathcal{R}')$ similar as discussed previously.

It requires five bits to represent 26 in binary and hence altogether fifteen bits for three alphabets each ranging from $A - Z$. Moreover, to represent $\mathcal{R}' = \mathbb{F}_{2^3} \times \mathbb{F}_{13^3}$, it needs $3 + 4 * 3 = 15$ bits. Thus again, a suitable encoding and decoding is required before and after the diffusion layer. The MDS matrix $M(\mathcal{R}')$ can be used at the diffusion layer.

6   Conclusion and Future Work

We have fully characterised the algebraic structure of a format preserving set over finite fields. Under some restrictive cases, we provided the structure over finite commutative rings with unity also. But the full characterisation over rings is still open. The whole problem had started with the question - can we have a format preserving set with respect to some MDS matrix and whose cardinality is 10 or 26? Though we could not provide a solution in this paper, we have given the same whose cardinalities are 10^3 or 26^3 which is a significant progress towards finding the solution.

References

1. Bellare, M., Ristenpart, T., Rogaway, P., Stegers, T.: Format-preserving encryption. In: Jacobson, M.J., Rijmen, V., Safavi-Naini, R. (eds.) SAC 2009. LNCS, vol. 5867, pp. 295–312. Springer, Heidelberg (2009). https://doi.org/10.1007/978-3-642-05445-7_19
2. Bellare, M., Rogaway, P.: On the construction of variable-input-length ciphers. In: Knudsen, L. (ed.) FSE 1999. LNCS, vol. 1636, pp. 231–244. Springer, Heidelberg (1999). https://doi.org/10.1007/3-540-48519-8_17
3. Black, J., Rogaway, P.: Ciphers with arbitrary finite domains. In: Preneel, B. (ed.) CT-RSA 2002. LNCS, vol. 2271, pp. 114–130. Springer, Heidelberg (2002). https://doi.org/10.1007/3-540-45760-7_9
4. Brightwell, M., Smith, H.: Using datatype-preserving encryption to enhance data warehouse security. In: 20th National Information Systems Security Conference Proceedings (NISSC), pp. 141–149 (1997)
5. Chang, D., et al.: SPF: a new family of efficient format-preserving encryption algorithms. In: Chen, K., Lin, D., Yung, M. (eds.) Inscrypt 2016. LNCS, vol. 10143, pp. 64–83. Springer, Cham (2017). https://doi.org/10.1007/978-3-319-54705-3_5
6. Grillet, P.A.: Semigroups: An Introduction to the Structure Theory. CRC Press, New York (1995)

7. Gupta, K.C., Pandey, S.K., Venkateswarlu, A.: Towards a general construction of recursive MDS diffusion layers. Des. Codes Cryptogr. **82**(1–2), 179–195 (2017)
8. Gupta, K.C., Pandey, S.K., Ray, I.G.: Format preserving sets: on diffusion layers of format preserving encryption schemes. In: Dunkelman, O., Sanadhya, S.K. (eds.) INDOCRYPT 2016. LNCS, vol. 10095, pp. 411–428. Springer, Cham (2016). https://doi.org/10.1007/978-3-319-49890-4_23
9. Halevi, S., Rogaway, P.: A tweakable enciphering mode. In: Boneh, D. (ed.) CRYPTO 2003. LNCS, vol. 2729, pp. 482–499. Springer, Heidelberg (2003). https://doi.org/10.1007/978-3-540-45146-4_28
10. Halevi, S., Rogaway, P.: A parallelizable enciphering mode. In: Okamoto, T. (ed.) CT-RSA 2004. LNCS, vol. 2964, pp. 292–304. Springer, Heidelberg (2004). https://doi.org/10.1007/978-3-540-24660-2_23
11. Herstein, I.N.: Topics in Algebra. Wiley, New York (1975)
12. Hoang, V.T., Rogaway, P.: On generalized Feistel networks. In: Rabin, T. (ed.) CRYPTO 2010. LNCS, vol. 6223, pp. 613–630. Springer, Heidelberg (2010). https://doi.org/10.1007/978-3-642-14623-7_33
13. Lidl, R., Niederreiter, H.: Finite Fields. Cambridge University Press, Cambridge (2008)
14. MacWilliams, F.J., Sloane, N.J.A.: The Theory of Error Correcting Codes, vol. 16. Elsevier, New York (1977)
15. Morris, B., Rogaway, P., Stegers, T.: How to encipher messages on a small domain. In: Halevi, S. (ed.) CRYPTO 2009. LNCS, vol. 5677, pp. 286–302. Springer, Heidelberg (2009). https://doi.org/10.1007/978-3-642-03356-8_17
16. Musili, C.: Introduction to Rings and Modules. Narosa Publishing House, New Delhi (1997)

Fault Attacks and Hash Functions

Differential Fault Attack on SIMON
with Very Few Faults

Ravi Anand[1]([✉]), Akhilesh Siddhanti[2], Subhamoy Maitra[3],
and Sourav Mukhopadhyay[1]

[1] Indian Institute of Technology Kharagpur, Kharagpur, India
`ravianandsps@gmail.com, msourav@gmail.com`
[2] Department of Computer Science and Mathematics,
BITS Pilani, Goa Campus, Goa, India
`akhileshsiddhanti@gmail.com`
[3] Applied Statistics Unit, Indian Statistical Institute, Kolkata, India
`subho@isical.ac.in`

Abstract. SIMON, a block cipher proposed by NSA (2013), has
received a lot of attention from the cryptology community. Several cryptanalytic results have been presented on its reduced-round variants. In
this work, we evaluate the cipher against Differential Fault Attack (DFA).
Our analysis shows that SIMON32/64, SIMON48/96 and SIMON64/128
can be attacked by injecting as little as 4, 6 and 9 faults respectively. We
first describe the process of identifying the fault locations after injecting
random faults. This exploits statistical correlations. Then we show how
one can recover the complete key using SAT solvers. To the best of our
knowledge, our results are much superior in terms of minimal number
of faults compared to the existing results. We also show our results are
superior in terms of injecting the faults in the earlier rounds compared
to the existing works.

Keywords: Block cipher · Correlation · Cryptanalysis
Differential fault attack · Simon

1 Introduction

There are several motivation towards designing lightweight cryptographic primitives to provide efficiency in resource-constrained environments – without com-
promising the confidentiality and integrity of the transmitted information. The
state size of such ciphers is kept as small as possible to minimize the area occupied by the cryptographic device. Typically, stream cipher designers keep the
state size equal to or slightly greater than twice the size of the secret key,
to provide security against the notorious Time-Memory-Data Tradeoff attacks.
Presently there are some efforts to obtain ciphers with state size lesser than
twice the key size, though the designs are still being actively evaluated. Such
secure stream ciphers are frequently exploited in designing RFID tags, smart

© Springer Nature Switzerland AG 2018
D. Chakraborty and T. Iwata (Eds.): INDOCRYPT 2018, LNCS 11356, pp. 107–119, 2018.
https://doi.org/10.1007/978-3-030-05378-9_6

cards, etc. In the domain of block ciphers, the scenario is not only related to the size of the states, but also on the design of the ciphers and the number of rounds. In 2013, the National Security Agency (NSA) proposed two families of light weight ciphers, SIMON and SPECK [3], which are flexible enough to provide very good performance in software as well as hardware environments. The motivation behind developing this family of block ciphers was to secure applications operating in resource-constrained environments, where general purpose ciphers such as AES might not be suitable. Since the structure of SIMON and SPECK are very simple, and more importantly it is proposed from NSA, the ciphers attracted serious attention from the cryptology community.

The different variants of SIMON are denoted by SIMON $2n/mn$, where $2n$ denotes the block size of the variant, and mn is the size of the secret key. Here n can take values from $16, 24, 32, 48$ or 64, and m from $2, 3$ or 4. For each combination of (m, n), the corresponding round number T is adopted. For SIMON32/64, $n = 16, m = 4$, and $T = 32$, for SIMON48/96 $n = 24, m = 4$, and $T = 36$ and for SIMON64/128 $n = 32, m = 4$, and $T = 44$.

It is natural for the cipher designers to ignore that the adversary can take advantage of the hardware implementation of the cipher and deduce critical information. One may consider fault resistance upto certain level in the design, but some assumptions must be considered and therefore, commercial ciphers cannot be made completely secure against different kinds of fault attacks. That is, under the assumptions of fault attack, the ciphers become inherently weaker. The idea here is to obtain secret information by injecting faults into the state of the cipher and then observing the differences between faulty and fault-free ciphertexts. This approach is called a Differential Fault Attack (also referred to as DFA). In the context of SIMON, there has already been some cryptanalysis using DFA that we discuss in the following section. However, our approach is different and it follows the overall scheme of Differential Fault Attack (DFA) that works against stream ciphers (see [8] and the references therein). We consider the L, R registers at certain round as states in comparison to stream cipher and inject random faults there in certain rounds. Then we compare the faulty and fault-free cipher-texts to identify the location of the faults. Once the fault locations are known, we can prepare certain equations and solve them using SAT solvers to obtain the secret key.

1.1 Existing DFAs on SIMON

SIMON was first cryptanalyzed using fault attack by Tupsamudre et al. in FDTC 2014 [10]. The authors proposed two models to retrieve the last-round key. The first one is a one-bit-flip model and the second one is a random one-byte model. In both models an adversary is able to inject a fault, flipping one bit, or one complete byte, into the left input of $(T - 2)^{th}$ round and thus recover the last n-bit round key. However, the complete secret key of SIMON, for mn-bit key size, is revealed only if all the last m round keys are known. This is because of the key scheduling algorithm. Thus, to recover the complete secret key in this model, more faults had to be injected in $(T - 3)^{th}, \cdots, (T - m - 1)^{th}$ rounds consecutively. Such a situation would increase the cost of the attack because the attacker has to control m round locations to inject the faults.

In ICISC 2014, Takashi et al. [9] presented a random fault attack in which the fault injection randomly changes the bits in left input of $(T-2)^{th}$ round. Their attack model required 3.05 faults on an average to recover the last round key of SIMON32/64. However, similar to [10], they had to inject faults in the $(T-3)^{th}, \cdots, (T-m-1)^{th}$ rounds respectively.

Further, Vasquez et al. presented an improved fault attack against SIMON in FDTC 2015 [11]. They considered the bit fault model and injected faults in the $(T-3)^{th}$ round instead of the $(T-2)^{th}$. When $m=2$, the last two round keys were enough to recover the complete secret key, but when $m=4$, more faults are required in the intermediate rounds.

Chen et al. presented an effective fault attack on SIMON under the random byte fault model in FDTC 2016 [6]. For SIMON, with key size mn, their attack needed to inject fault only once in the $(T-m-1)^{th}$ round to recover all the last m round keys without injecting faults in any other intermediate rounds. However for SIMON32/64, SIMON48/72, SIMON48/96 and SIMON64/128, their model needed to inject faults in more than one rounds to recover all the last m round keys. At the same time, how to attack these four SIMON instances by injecting fault only once in the intermediate round and how to reduce the number of faults required to attack SIMON were left as open problems by the authors.

1.2 Our Contribution

In this paper we demonstrate an improved fault attack on SIMON32/64, SIMON48/96 and SIMON64/128 using a transient single bit-flip model of attack. For SIMON, with the key word size m, we need to inject 1-bit faults into the state at the beginning $(T-m-1)^{th} = (T-5)^{th}$ round and recover the complete secret key. By observing the faulty and fault-free ciphertexts, we compute signatures to identify fault locations. Naturally, we would prefer injecting fault in *any* round of SIMON$2n/4n$. We note that the more earlier we inject a fault, the weaker is the signature, and more difficult it becomes to identify a fault location. For example, according to our methods used here, identifying the location of a fault injected in the 24th round or before of the 32-round SIMON32/64 is very difficult. This can be seen as a new way to analyze "the mixing strength" of SIMON.

We then explain how to recover the complete secret key. Comparisons of our effort with the existing works are presented in Tables 4 and 5. As we can see, our fault attack model requires much lesser faults to recover the complete key and injection is performed in only one specific round. We observe that faults can be injected upto 10 rounds and more before the final round of SIMON, however, the number of faults required increases required increases (upto 8 for SIMON32/64, 12 for SIMON48/96 and SIMON64/128) and the computation time required is higher. This is due to the rising non-linearity of equations. However, we show that injecting faults from the beginning and after of the 25th round of the 32-round SIMON32/64, the attack is successful with only 4 faults for deducing the key. Another important aspect of our model is that the faults can be injected in any location of register L or register R, since we are able to identify the fault locations in both cases, which the existing works tend to omit due to its complexity.

The rest of the article has been organized as follows. We illustrate the design of SIMON in Sect. 1.3 and explain our bit-flip transient model of attack in Sect. 2. In Sect. 2.1, we show how we identify the locations of injected faults, and in Sect. 2.2 we show how we obtain the complete secret key. Our experimental results for the complete DFA are presented in Sect. 2.3. Finally, we conclude our paper in Sect. 3.

1.3   Brief Description of Variants of SIMON

SIMON is a family of lightweight block ciphers released by the NSA in 2013. SIMON has ten variants supporting different block sizes and key sizes.

Table 1. SIMON parameters

Block Size ($2n$)	Key Size (mn)	word size (n)	keywords (m)	Rounds (T)
32	64	16	4	32
48	72,96	24	3,4	36,36
64	96,128	32	3,4	42,44
96	96, 144	48	2,3	52,54
128	128,192,256	64	2,3,4	68,69,72

SIMON is a two branch balanced Feistel network which consists of three operations: AND (&), XOR $\oplus$ and Rotation ($\lll$). The State Update Function of SIMON is shown in Fig. 1.

If (L_i, R_i) be the input to the ith round then the state is updated as follows:

$$F(x) = (x \lll 1)\&(x \lll 8) \bigoplus (x \lll 2)$$
$$L_{i+1} = R_i \bigoplus F(L_i) \bigoplus k_i$$
$$R_{i+1} = L_i$$

where k_i is the round key which is generated by the key scheduling algorithm described below.

The key scheduling algorithm of SIMON has three different procedures depending on the key size. We discuss the variants of SIMON only for the value of $m = 4$. Hence, from here, we will consider the value of m as 4. The first 4 round keys are initialized directly from the main key. The remaining $(T - 4)$ round keys are generated by the following procedure:

$$k_{i+4} = c \oplus (z_j)_i \oplus k_i \oplus (I \oplus S^{-1})(S^{-3}k_{i+3} \oplus k_{i+1})$$

For a detailed description of SIMON, the readers are referred to [3].

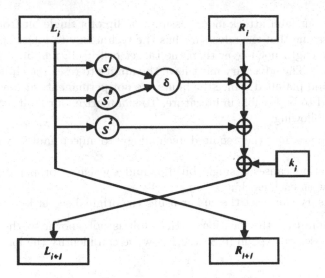

Fig. 1. SIMON round function

2   Proposed Differential Fault Attack

We follow a transient single bit-flip model of attack [4] here. It is *transient*, because the injected fault propagates to other locations with further encryption rounds. It is different from other fault attack models like *hard faults* [5], where a fault is *permanent* and sticks to a certain position. Moreover, hard faults can damage the device and prevent its re-usability.

As we discussed earlier, the attack model we employ here has been inspired from Differential Fault Attack on Stream Ciphers. In a stream cipher, knowledge of the plaintext is enough to know the generated keystream, using which fault attack is mounted. The fault can be injected in any particular state of the stream cipher and once injected, it is clocked ℓ times to generate an ℓ-length keystream sequence. Then the cipher is reset to the same state, a new fault is injected in a different register location of the state and is clocked again ℓ times to give out the faulty keystream. The process is repeated ρ many times, where ρ is the fault requirement of the cipher. There have been many works on fault attacks on stream ciphers, namely, [1,2,7,8]. In this work, we successfully adopt such models to a popular block cipher, SIMON$2n/4n$. However, for block ciphers, we need to assume that the plaintext remains fixed (contrary to stream ciphers where knowledge of the plaintext is enough) and the differences in faulty and fault-free ciphertexts are noted to deduce the secret key, hence $\ell = 2n$ for SIMON$2n/4n$. For the stream ciphers, the fault is injected in some initial rounds in PRGA. In a similar line, here the fault is injected in some unknown register location of L or R of SIMON$2n/4n$ at the beginning of some round, say $r = (T - 5)$. Then, in the analysis, all the 4 round keys are involved from round r to round T. Thus, the challenge is to mount the attack with fault injection at an earlier round.

Note that our fault attack model assumes a flip of a single bit from $1 \rightarrow 0$ or $0 \rightarrow 1$. We assume that the adversary has the technology to inject a single laser beam of wavelength not bigger than the target cell itself, i.e., only a single bit will be affected. The adversary must have the ability to reset the cipher with the original key and plaintext. Since the faults are not permanent, such assumptions are considered to be feasible in literature. To summarize, our fault attack model assumes the following.

1. The adversary has the required technology to inject faults, with precise timing.
2. Fault injection causes a single bit-flip, and the effect propagates to other locations with each clocking.
3. The adversary can reset the cipher using the original secret key.

We also assume that the location of the fault is not known to the adversary. For this purpose, we explain in Sect. 2.1 how we can identify the location of the injected fault.

2.1 Identifying the Fault Location Using Signatures

Consider an experiment where we encrypt a plaintext $P = \{p_0, p_1, \cdots, p_{(2n-1)}\}$ using key K to obtain the ciphertext $C = \{c_0, c_1, \cdots, c_{(2n-1)}\}$. Suppose we repeat the experiment, where the plaintext P is encrypted with the key K, but a 1-bit fault is injected in the rth round of SIMON, to obtain a faulty ciphertext $C^{(\gamma)} = \{c_0^{(\gamma)}, c_1^{(\gamma)}, \cdots, c_{(2n-1)}^{(\gamma)}\}$, where γ denotes the location of the injected fault. We take $r = T - 5$ here, that is, we inject fault in the 27th round of the 32 round SIMON32/64, 31st round of the 36 round SIMON48/96 and in the 39th round of the 44 round SIMON64/128. The objective of this section is to the determine the location of the injected fault, i.e., γ. The process of determining γ is same for all the three variants, and consists of two phases, the *offline phase* and the *online phase*.

The Offline Phase. In this phase, the adversary calculates *signatures* for each possible fault location of SIMON, and stores it in a tabular form for accessing in the online phase. The *signature vector* $S^{(j)}$ is calculated as:

$$S^{(j)} = (s_0^{(j)}, s_1^{(j)}, \cdots, s_{(2n-1)}^{(j)}), \tag{1}$$

where

$$s_i^{(j)} = \frac{1}{2} - Pr(c_i \neq c_i^{(j)}), \tag{2}$$

for $j = 0, 1, \cdots, (2n-1)$ and $i = 0, 1, \cdots, (2n-1)$.

The probability $(Pr(c_i \neq c_i^{(j)}))$ is calculated over a sufficient number of trials, where for each trial we consider a random key and a random plaintext. From experiments, we find that 2^{20} trials are sufficient for obtaining a reasonable accuracy. The signatures $S^{(0)}, S^{(1)}, \cdots, S^{(2n-1)}$ are stored for all possible $2n$ fault locations.

The Online Phase. For a plaintext P, the corresponding fault free ciphertext C is obtained using an unknown key K. Next, the adversary obtains a faulty ciphertext $C^{(\gamma)}$ from the same plaintext P and key K by injecting a fault at location γ, $0 \leq \gamma \leq 2n - 1$ in the internal state S_r. S_r is the internal state after r rounds of encryption.

Once the adversary has λ faulty ciphertexts, he calculates the *trail* of each faulty ciphertext $C^{(\gamma)}$ as:

$$\tau^{(\gamma)} = (\psi_0^{(\gamma)}, \psi_1^{(\gamma)}, \ldots, \psi_{(2n-1)}^{(\gamma)}), \tag{3}$$

where $\psi_i^{(\gamma)}$ is:

$$\psi_i^{(\gamma)} = \frac{1}{2} - (c_i \oplus c_i^{(\gamma)}). \tag{4}$$

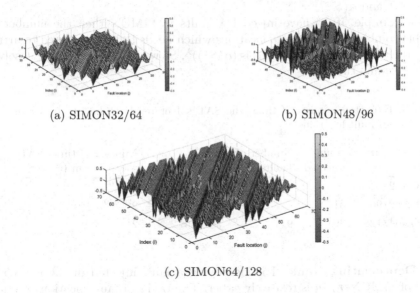

(a) SIMON32/64   (b) SIMON48/96

(c) SIMON64/128

Fig. 2. The plot of s_i^j on index (i) and fault location (j) for faults injected in $(T-5)^{th}$ round

Now the adversary has to identify γ for each faulty ciphertext $C^{(\gamma)}$, by matching the trail $\tau^{(\gamma)}$ to some suitable signature $S^{(j)}$. For this, a modified version of Pearson's correlation coefficient is used.

Using the modified Correlation Coefficient μ. Note that $-1 \leq \mu(S^{(j)}, \tau^{(\gamma)}) \leq 1$. A signature $S^{(j)} = (s_0^j, s_1^j, \cdots, s_{(2n-1)}^j)$ and a trail $\tau^{(\gamma)} = (\psi_0^{(\gamma)}, \psi_1^{(\gamma)}, \cdots, \psi_{(2n-1)}^{(\gamma)})$ are said to be a mismatch if there exists atleast one j, $0 \leq j \leq (2n-1)$ such that $(s_i^j = \frac{1}{2}, \psi_i^\gamma = -\frac{1}{2})$ or $(s_i^j = -\frac{1}{2}, \psi_i^\gamma = \frac{1}{2})$ holds true.

In case of a mismatch, we say $\mu(S^{(j)}, \tau^{(\gamma)}) = -1$, where $\mu(S^j, \tau^\gamma)$ is the correlation coefficient

$$\mu(x, y) = \frac{\sum_{i=1}^{n}(x_i - \bar{x})(y_i - \bar{y})}{\sqrt{\sum_{i=1}^{n}(x_i - \bar{x})^2}\sqrt{\sum_{i=1}^{n}(y_i - \bar{y})^2}} \tag{5}$$

For each trail $\tau^{(\gamma)}$, the adversary calculates the correlation $\mu(S^j, \tau^\gamma)$ and $\alpha(S^{(\gamma)}) = |\{j : (\mu(S^j, \tau^\gamma)) > \mu(S^\gamma, \tau^\gamma)\}|$.

For every unknown fault location γ, a table $T_{(\gamma)}$ is prepared, where each fault location j is arranged in the decreasing order of the correlation coefficient $\mu(S^{(j)}, \tau^{(\gamma)})$, taken between its signature S^j and the trail $\tau^{(\gamma)}$, i.e., $\mu(S^{(j)}, \tau^{(\gamma)})$.

Since the fault locations j are arranged in decreasing order of correlation coefficients $\mu(S^{(j)}, \tau^{(\gamma)})$, it is intuitive that we expect the first entry in the table to be the unknown fault location. However, our experiments show that it is not always the case. To arrive at the correct set of fault locations we need to consider all possible set $S^{(\gamma)}$ of fault locations, where $S^{(\gamma)} = \{j : (\mu(S^{(j)}, \tau^\gamma)) \geq \mu(S^\gamma, \tau^\gamma)\}$ and $\alpha(S^{(\gamma)}) = |S^{(\gamma)}|$.

For example, if we have injected λ faults in SIMON then the number of possible combinations of faults locations which needs to be considered to arrive at the correct set of fault locations is $(\alpha(S^\gamma))^\lambda$. So we have to run the SAT solver $(\alpha(S^\gamma))^\lambda$ many times.

Table 2. Expected number of times the SAT solver needs to be run to arrive at a correct set of fault locations.

SIMON$2n/4n$ Variant	Round injected	Number of Faults (λ)	$\alpha(S^\gamma)$	Number of times SAT solver is run $(=(\alpha(S^\gamma))^\lambda)$
SIMON32/64	27	4	$9.13 \approx 2^{3.191}$	$2^{12.764}$
SIMON48/96	31	6	$10.07 \approx 2^{3.345}$	$2^{20.070}$
SIMON64/128	39	9	$39.49 \approx 2^{5.311}$	$2^{47.799}$

The Depreciating Rank. For identifying a fault injected in the $(T - 5)^{th}$ round of SIMON$2n/4n$ is relatively easier. The ranks of fault locations average to 1, and very few combinations need to be checked for. However, while moving each subsequent round away (earlier) from the final round, the average rank falls pretty quickly due to further mixing. To be convinced, we have experimented and the trend can be seen in all the three variants of SIMON$2n/4n$. This means that it will become more and more difficult to identify location of the fault for rounds prior to $(T - 8)$ of SIMON$2n/4n$, which is round 24 for SIMON32/64, 28 for SIMON48/96 and 36 for SIMON64/128. Naturally, identifying faults injected in the beginning of first round of SIMON$2n/4n$ will mean a differential attack on SIMON$2n/4n$ itself. Further, the number of faults required to deduce the key also increases, due to the increasing non-linearity of the equations. This increases the overall number of combinations to check for (that is $(\alpha(S^{(\gamma)}))^\lambda$), since both α and λ increase.

While experimenting, we observed that the mixing is better in case of SIMON48/96 followed by SIMON32/64 and SIMON64/128. This needs further investigation, which is not in the scope of this paper. After the tables have been prepared for all the faulty ciphertexts, we deduce the key using a SAT solver. This is explained in the following section.

2.2   Recovering the Secret Key

Using our bit-flip model of attack mentioned in Sect. 2, we will show how we can recover the secret key of SIMON. Suppose the adversary is able to inject λ many faults in the rth round of the internal state register $\mathcal{S}_r$; resetting the cipher to its original state post every fault injection. Note that the fault can be injected in both L and R register locations of SIMON$2n/4n$. For simplicity, the fault-free ciphertext is denoted by C_0 and the λ faulty ciphertexts by $C_1, C_2, \ldots, C_\lambda$. We consider $r = T - 5$ for each variant of SIMON$2n/4n$ as mentioned before.

For every fault (out of λ many faults) injected in $\mathcal{S}_r$, the fault propagates as per the construction of SIMON to $\mathcal{S}_{r+1}$ using the same round key that was used to obtain C_0. A fault injected in register L_r propagates as follows:

$$L_{r+1} = F(L_r^*) \oplus R_i \oplus k_r \tag{6}$$
$$R_{r+1} = L_r^* \tag{7}$$

where L_r^* is the affected register. In case of injecting a fault in R_r, we have

$$L_{r+1} = F(L_r) \oplus R_i^* \oplus k_r \tag{8}$$
$$R_{r+1} = L_r \tag{9}$$

For every round r' from $r + 1$ to T, as illustrated in Sect. 1.3, we have two functions:

1. The State Update Function, and
2. The Round Key Function (or the key scheduling algorithm).

We initialize $2n$ variables $L_{r,0} \ldots, L_{r,(n-1)}$ and $R_{r,0} \ldots R_{r,(n-1)}$ for the state of SIMON at round r, where $L_{r,j}$ ($R_{r,j}$) is the jth bit of the left (right) block of the internal state $\mathcal{S}_r$. After every state update, the variables will be initialized as:

$$L_{r+1,j} = (L_{r,(j-1)\ mod(n)} \& L_{r,(j-8)\ mod(n)}) \oplus L_{r,(j-2)\ mod(n)} \tag{10}$$
$$\oplus R_{r,j} \oplus k_{r,j} \tag{11}$$
$$R_{r+1,j} = L_{r,j} \tag{12}$$

for $j = 0, 1, \ldots, (n - 1)$. However, every additional round of encryption will increase the degree of the polynomials, hence we introduce new state variables $L_{r'+1,0} \ldots, L_{r'+1,(n-1)}, R_{r'+1,0}, \ldots R_{r'+1,(n-1)}$ for each round r' and formulate equations accordingly. In a boolean polynomial system of equations, the same would be:

$$0 \equiv L_{r'+1,j} \oplus (L_{r',(j-1) \bmod(n)} \& L_{r',(j-8) \bmod(n)}) \oplus L_{r',(j-2) \bmod(n)}$$
$$\oplus R_{r',j} \oplus k_{r',j} \tag{13}$$
$$0 \equiv R_{r'+1,j} \oplus L_{r',j} \tag{14}$$

Hence, we obtain $5 \cdot 2n = 10n$ variables and $5 \cdot 2 \cdot 2n = 20n$ equations by repeating Eqs. (13) and (14). Now, we have $(\lambda + 1)$ cipher-texts, hence we have $10n \cdot (\lambda + 1)$ variables and $20n \cdot (\lambda + 1)$ equations, forming a system of Boolean equations. We consider using SAT solvers here, for example, Cryptominisat, which is available with SAGE. The SAT solver can return a solution set satisfying the equations. The key is then tested by encrypting a plaintext and verifying with the ciphertext.

As the number of rounds increases, the complexity of equations increases drastically. Thus, we guess a certain portion of the state of SIMON for the round in which the fault was injected. Here, we guess the entire register R. The guessed bits are directly substituted into the equations. We discuss our results in Sect. 2.3.

Because of the sharp rise in the non-linearity of the equations, we cannot formulate equations for faults injected in round 23 or before, even if we may be aware of the exact location of the injected faults. The computation time required for the same increases significantly (more than 6 hours of solving time), and also increasing the number of faults does not help – since this will only lead to more equations and variables. We are looking at possible optimizations to further improve this situation.

2.3   Experimental Results

In this section, we present the results of our fault attack.

Table 3. Fault requirements for DFA on SIMON.

	Round fault is injected in	Number of faults (λ)	Number of bits guessed in R	Time taken by SAT solver
SIMON32/64	27	4	16	191.230 s
SIMON48/96	31	6	24	290.997 s
SIMON64/128	39	9	32	403.035 s

These experiments were conducted as shown in Table 3 on a consumer grade laptop HP-15D103TX with CPU specifications Intel(R) Core(TM) i5-4200M CPU @ 2.50GHz running SageMath version 8.1 along with Cryptominisat package on Ubuntu Bionic Beaver (development branch). For the experiments, we have considered 2^{17} random plaintext and key pairs. For each pair λ faults are injected in the $(T-5)^{th}$ round. We find the location of the faults as described in Sect. 2.1. Then these fault locations are used to recover the key as in Sect. 2.2. Since we were successful in locating the faults and recovering the secret key correctly for all the 2^{17} pairs, we believe that our attack model will recover the key for any random plaintext and key. We claim a minimum requirement of 4, 6 and 9 faults for SIMON32/64, SIMON48/96 and SIMON64/128 respectively, to successfully recover the key.

2.4   Time Complexity Analysis

In all the previous papers, the discussions related to the time complexity for the fault attack has been limited to calculating the average number of fault injections required to deduce the round keys. The comparison of their attack models are based on the experimental number of fault injections required and the rounds in which the faults have been injected. However, we explain the fault requirements and time complexity clearly, as given in Tables 4 and 5 respectively.

Table 4. Comparison of the experimental number of the fault injections

SIMON2n/mn	Random n-bit model	Random byte model	Random bit model		
	[9]	[10]	[10]	[11]	Section 2.3
SIMON32/64	12.20	24	100	50.85	4
SIMON48/96	13.22	36	172	87.19	6
SIMON64/128	13.93	52	248	126.29	9

Table 5. Comparison of the rounds in which faults are injected in case of each attack model

SIMON2n/mn	Random n-bit model	Random byte model	Random bit model		
	[9]	[10]	[10]	[11]	Section 2.3
SIMON32/64	$L_{27}, L_{28}, L_{29}, L_{30}$	$L_{27}, L_{28}, L_{29}, L_{30}$	$L_{27}, L_{28}, L_{29}, L_{30}$	L_{27}, L_{29}	L_{27}, R_{27}
SIMON48/96	$L_{31}, L_{32}, L_{33}, L_{34}$	$L_{31}, L_{32}, L_{33}, L_{34}$	$L_{31}, L_{32}, L_{33}, L_{34}$	L_{31}, L_{33}	L_{31}, R_{31}
SIMON64/128	$L_{39}, L_{40}, L_{41}, L_{42}$	$L_{39}, L_{40}, L_{41}, L_{42}$	$L_{39}, L_{40}, L_{41}, L_{42}$	L_{39}, L_{41}	L_{39}, R_{39}

Since estimating the time complexity of the SAT solver compared to each evaluation of the cipher with a single key is not inside the scope of this paper, we give a rough estimate of the overall complexity of the attack. Our attack procedure consists of the following two steps:

1. locating the faults using correlation between faulty and fault-free keystreams,
2. deriving the secret key by formulating equations from keystreams.

Table 6. Time complexities of DFA on variants of SIMON.

	Fault Requirements	w	x	Time Complexity
SIMON32/64	4	16	12.76	$2^{28.76} \cdot c$
SIMON48/96	6	24	20.07	$2^{44.07} \cdot c$
SIMON64/128	9	32	47.80	$2^{79.80} \cdot c$

Consider that 2^x is the number of times the SAT solver needs to be run to arrive at a correct set of fault locations and w is the number of bits guessed by SAT solver to derive the key. Then the time complexity of the attack is $2^x * 2^w * c$, where c is the time complexity of each execution of the SAT solver.

While the time complexity of our attack cannot be immediately compared to the existing works, the fault requirement of our model is much lesser. Needless to mention that injecting too many faults can damage the device and thus we consider injecting as little faults as possible for a more practical attack. Further our attack also works for same or earlier rounds than the existing efforts.

3   Conclusion

In this work, we present a Differential Fault Attack on SIMON32. First, we show how one can identify the location of injected faults using signatures. Next, we describe that by injecting as few as 4, 6 and 9 faults in the $(T - m - 1)^{th}$ round of the SIMON32/64, SIMON48/96 and SIMON64/128, we can recover the complete key. Although our work does not compromise its security in normal mode, the attack is achievable under certain constrained environment. We will analyze the remaining seven SIMON instances using the same attack model and these results will be available in the journal version of our paper.

References

1. Banik, S., Maitra, S., Sarkar, S.: A differential fault attack on the grain family of stream ciphers. In: Prouff, E., Schaumont, P. (eds.) CHES 2012. LNCS, vol. 7428, pp. 122–139. Springer, Heidelberg (2012). https://doi.org/10.1007/978-3-642-33027-8_8
2. Banik, S., Maitra, S., Sarkar, S.: Improved differential fault attack on MICKEY 2.0. J. Cryptogr. Eng. **5**(1), 13–29 (2015). http://link.springer.com/article/10.1007%2Fs13389-014-0083-9
3. Beaulieu, R., Shors, D., Smith, J., Treatman-Clark, S., Weeks, B., Wingers, L.: The Simon and Speck families of lightweight block ciphers. Technical report, Cryptology ePrint Archive, Report 2013/404 (2013). http://eprint.iacr.org
4. Biham, E., Shamir, A.: A new cryptanalytic attack on DES, preprint, pp. 10–96 (1996)

5. Biham, E., Shamir, A.: Differential fault analysis of secret key cryptosystems. In: Kaliski, B.S. (ed.) CRYPTO 1997. LNCS, vol. 1294, pp. 513–525. Springer, Heidelberg (1997). https://doi.org/10.1007/BFb0052259

6. Chen, H., Feng, J., Rijmen, V., Liu, Y., Fan, L., Li, W.: Improved fault analysis on Simon block cipher family. In: 2016 Workshop on Fault Diagnosis and Tolerance in Cryptography (FDTC), pp. 16–24. IEEE (2016)

7. Hojsík, M., Rudolf, B.: Differential fault analysis of Trivium. In: Nyberg, K. (ed.) FSE 2008. LNCS, vol. 5086, pp. 158–172. Springer, Heidelberg (2008). https://doi.org/10.1007/978-3-540-71039-4_10

8. Maitra, S., Siddhanti, A., Sarkar, S.: A differential fault attack on plantlet. IEEE Trans. Comput. **66**(10), 1804–1808 (2017). https://doi.org/10.1109/TC.2017.2700469

9. Takahashi, J., Fukunaga, T.: Fault analysis on SIMON family of lightweight block ciphers. In: Lee, J., Kim, J. (eds.) ICISC 2014. LNCS, vol. 8949, pp. 175–189. Springer, Cham (2015). https://doi.org/10.1007/978-3-319-15943-0_11

10. Tupsamudre, H., Bisht, S., Mukhopadhyay, D.: Differential fault analysis on the families of Simon and Speck ciphers. In: Workshop on Fault Diagnosis and Tolerance in Cryptography (FDTC), pp. 40–48. IEEE (2014)

11. Vasquez, J.d.C.G., Borges, F., Portugal, R., Lara, P.: An efficient one-bit model for differential fault analysis on SIMON family. In: 2015 Workshop on Fault Diagnosis and Tolerance in Cryptography (FDTC), pp. 61–70. IEEE (2015)

Cryptanalysis of 2 Round Keccak-384

Rajendra Kumar[1(✉)], Nikhil Mittal[1], and Shashank Singh[2]

[1] Center for Cybersecurity, Indian Institute of Technology Kanpur, Kanpur, India
{rjndr,mnikhil}@iitk.ac.in
[2] Indian Institute of Science Education and Research Bhopal, Bhopal, India
shashank@iiserb.ac.in

Abstract. In this paper, we present a cryptanalysis of round reduced Keccak-384 for 2 rounds. The best known preimage attack for this variant of Keccak has the time complexity 2^{129}. In our analysis, we find a preimage in the time complexity of 2^{89} and almost same memory is required.

Keywords: Keccak · Sha-3 · Cryptanalysis · Hash functions
Preimage attack

1 Introduction

Cryptographic hash functions are the important component of modern cryptography. In 2008, U.S. National Institute of Standards and Technology (NIST) announced a competition for the Secure Hash Algorithm-3 (Sha-3). A total of 64 proposals were submitted to the competition. In the year 2012, NIST announced Keccak as the winner of the competition. The Keccak hash function was designed by Guido Bertoni, Joan Daemen, Michaël Peeters, and Gilles Van Assche [2]. Since 2015, Keccak has been standardized as Sha-3 by the NIST.

The Keccak hash function is based on sponge construction [3] which is different from previous Sha standards. Intensive cryptanalysis of Keccak is done since its inception [1,4–9,11,13–15]. In 2012, Dinur *et al.* gave a practical collision attack for 4 rounds of Keccak-224 and Keccak-256 using differential and algebraic techniques [5] and also provided attacks for 3 rounds for Keccak-384 and Keccak-512. They further gave collision attacks in 2013 for 5 rounds of Keccak-256 using internal differential techniques [6]. In 2016, using linear structures, Guo *et al.* proposed preimage attacks for 2 and 3 rounds of Keccak-224, Keccak-256, Keccak-384, Keccak-512 and for 4 rounds in case of smaller hash lengths [8]. Recently, in the year 2017, Kumar *et al.* gave efficient preimage and collision attacks for 1 round of Keccak [9]. There are hardly any attack for the full round Keccak, but there are many attacks for reduced round Keccak. Some of the important results are shown in the Tables 1 and 2.

© Springer Nature Switzerland AG 2018
D. Chakraborty and T. Iwata (Eds.): INDOCRYPT 2018, LNCS 11356, pp. 120–133, 2018.
https://doi.org/10.1007/978-3-030-05378-9_7

Our Contribution: We propose a preimage attack for 2 rounds of round-reduced KECCAK-384. The time complexity of attack is 2^{89} and the memory complexity is 2^{87}. The attack is not practical, but it outperforms the previous best-known attack [8], with a good gap. The proposed attack does not affect the security of full KECCAK.

Table 1. Preimage attack results.

No. of rounds	Hash length	Time complexity	Reference
1	KECCAK- 224/256/384/512	Practical	[9]
2	KECCAK- 224/256	2^{33}	[13]
2	KECCAK- 224/256	1	[8]
2	KECCAK- 384/512	$2^{129}/2^{384}$	[8]
3	KECCAK- 224/256/384/512	$2^{97}/2^{192}/2^{322}/2^{484}$	[8]
4	KECCAK- 224/256	$2^{213}/2^{251}$	[8]
4	KECCAK- 384/512	$2^{378}/2^{506}$	[10]

Table 2. Collision attack results.

No. of rounds	Hash length	Time complexity	Reference
1	KECCAK- 224/256/384/512	Practical	[9]
2	KECCAK- 224/256	2^{33}	[13]
3	KECCAK- 384/512	practical	[6]
4	KECCAK- 224/256	2^{24}	[5]
4	KECCAK- 224/256	2^{12}	[14]
4	KECCAK- 384	2^{147}	[6]
5	KECCAK- 224	2^{101}	[14]
5	KECCAK- 224	Practical	[15]
5	KECCAK- 256	2^{115}	[6]

2   Keccak Description and Notations

KECCAK is a family of sponge hash functions with arbitrary output length. A sponge construction consists of a permutation function, denoted by f, a parameter "rate", denoted by r, and a padding rule pad. The construction produces a sponge function which takes as input a bit string N and output length d. It is described below.

The bit string N is first padded based on the pad rule. The padded string is divided into blocks of length r. The function f maps a string of length b to another of same length. The capacity, denoted by c, is a positive integer such

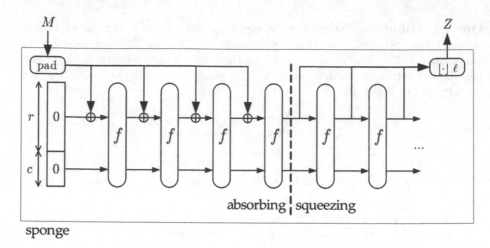

Fig. 1. The sponge construction [3]

that $r + c = b$. The initial state is a b-bit string which is set to all zeros. After a string N is padded, it undergoes two phases of sponge, namely absorbing and squeezing. In the absorbing phase, the padded string N' is split into r-bit blocks, say $N_1, N_2, N_3, \ldots, N_m$. The first r bits of initial state are XOR-ed with the first block and the remaining c bits are appended to the output of XOR. Then it is given as input to the function f as shown in the diagram given in the Fig. 1. The output of this f becomes the initial state for the next block and the process is repeated for all blocks of the message. After all the blocks are absorbed, let the resulting state be P.

In the squeezing phase, an string Z is initialized with the first r bits of the state P. The function f is applied on the state P and the first r bits of output, say P', is appended to Z. The P' is again passed to f and this process is repeated until $|Z| \geq d$. The output of sponge construction is given by the first d bits of Z.

The KECCAK family of hash functions is based on the sponge construction. The function f, in the sponge construction, is denoted by KECCAK-f $[b]$, where b is the length of input string. Internally KECCAK-f $[b]$ consists of a round function p which is recursively applied to a specified number of times, say n_r. More precisely KECCAK-f $[b]$ function is specialization of KECCAK-p $[b, n_r]$ family where $n_r = 12 + 2l$ and $l = \log_2(b/25)$ i.e.,

$$\text{KECCAK-}f\,[b] = \text{KECCAK-}p\,[b, 12 + 2l]\,.$$

The round function p in KECCAK consists of 5 steps, in each of which the state undergoes transformations specified by the step mapping. These step mappings are called θ, ρ, π, χ and ι. A state S, which is a b-bit string, in KECCAK is usually denoted by a 3-dimensional grid of size $(5 \times 5 \times w)$ as shown in the Fig. 2. The value of w depends on the parameters of KECCAK. For example in the case of KECCAK-f[1600], w is equal to 64. It is usual practice to represent a state in terms of rows, columns, lanes, and slices of the 3-dimensional grid. Given a bit location (x, y, z) in the grid, the corresponding row is given by $(S[x + i \pmod 5], y, z] : i \in [0, 4])$. Similarly the corresponding column is given by the bits $(S[x, y + i \pmod 5], z] : i \in [0, 4])$ and the corresponding lane is given by $(S[x, y, z + i \pmod w)] : i \in [0, w - 1])$. Further the slice corresponding to a location (x, y, z), consists of $(S[x + j \pmod 5], y + i \pmod 5], z] : i, j \in [0, 4])$ bits. It is pictorially shown in the Fig. 2.

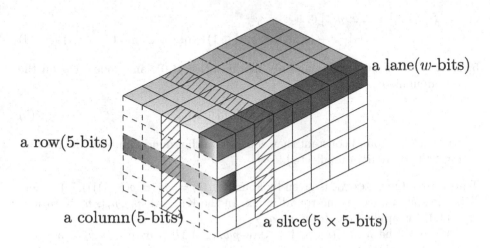

Fig. 2. A state in KECCAK

In the following, we provide a brief description of the step mappings. Let A and B respectively denote input and output states of a step mapping.

1. θ (**theta**): The theta step XORs each bit in the state with the parities of two neighboring columns. For a given bit position (x, y, z), one column is $((x - 1) \bmod 5, z)$ and the other is $((x + 1) \bmod 5, (z - 1) \bmod w)$. Thus, we have

$$B[x, y, z] = A[x, y, z] \oplus P[(x - 1) \bmod 5, z]$$
$$\oplus P[(x + 1) \bmod 5, (z - 1) \bmod w] \tag{1}$$

   where $P[x, z] = \oplus_{y=0}^{4} A[x, y, z]$.
2. ρ (**rho**): This step rotates each lane by a constant value towards the MSB i.e.,

$$B[x, y, z] = A[x, y, z + \rho(x, y) \bmod w], \tag{2}$$

   where $\rho(x, y)$ is the constant for lane (x, y). The constant value $\rho(x, y)$ is specified for each lane in the construction of KECCAK.

3. π (**pi**): It permutes the positions of lanes. The new position of a lane is determined by a matrix,

$$\begin{bmatrix} x' \\ y' \end{bmatrix} = \begin{bmatrix} 0 & 1 \\ 2 & 3 \end{bmatrix} \cdot \begin{bmatrix} x \\ y \end{bmatrix}, \tag{3}$$

where (x', y') is the position of lane (x, y) after π step.

4. χ (**chi**): This is a non-linear operation, where each bit in the original state is XOR-ed with a non-linear function of next two bits in the same row i.e.,

$$B[x, y, z] = A[x, y, z] \oplus$$
$$((A[(x + 1) \bmod 5, y, z] \oplus 1) \cdot A[(x + 2) \bmod 5, y, z])). \tag{4}$$

5. ι (**iota**): This step mapping only modifies the $(0, 0)$ lane depending on the round number i.e.,

$$B[0, 0] = A[0, 0] \oplus RC_i, \tag{5}$$

where RC_i is round constant that depends on the round number. The remaining 24 lanes remain unaffected.

Thus a round in KECCAK is given by $\text{Round}(A, i_r) = \iota(\chi(\pi(\rho(\theta(A))))), i_r)$, where A is the state and i_r is the round index. In the KECCAK-$p[b, n_r]$, n_r iterations of $\text{Round}(\cdot)$ is applied on the state A.

The SHA-3 hash function is KECCAK-$p[b, 12 + 2l]$, where $w = b/25$ and $l = \log_2(w)$. The value of b is 1600, so we have $l = 6$. Thus the f function in SHA-3 is KECCAK-$p[1600, 24]$.

The KECCAK team denotes the instances of KECCAK by KECCAK$[r, c]$, where $r = 1600 - c$ and the capacity c is chosen to be twice the size of hash output d, to avoid generic attacks with expected cost below 2^d. Thus the hash function with output length d is denoted by

$$\text{KECCAK-}d = \text{KECCAK}[r := 1600 - 2d, \ c := 2d], \tag{6}$$

truncated to d bits. The SHA-3 hash family supports minimum four different output length $d \in \{224, 256, 384, 512\}$. In the KECCAK-384, the size of $c = 2 \cdot d = 768$ and the rate $r = 1600 - c = 1600 - 768 = 832 = 13 \cdot 64$.

3   Preimage Attack for 2 Rounds of Round Reduced Keccak-384

In this section, we present a preimage attack for a round reduced KECCAK. We will show that the preimage can be found in 2^{88} time and 2^{87} memory for 2 rounds of round-reduced KECCAK-384. Although it is not a practical attack, but it is an improvement over the existing best attack, for 2 rounds of KECCAK-384, which takes 2^{129} time [8].

3.1 Notations and Observations

In the analysis, we will represent a state by the lanes. There are in total 5×5 lanes. Each lane in a state will be represented by a variable which is a 64-bit array. A variable with a number in round bracket "(.)" represents the shift of the bits in array towards MSB. A variable with a number in square bracket "[.]" represents the bit value of the variable at that index. If there are multiple numbers in the square bracket then it represents the corresponding bit values.

We are going to use the following observations in our analysis.

1. **Observation 1:** The χ is a row-dependent operation. Guo *et al.* in [8], observed that if we know all the bits of a row then we can invert χ for that row. It is depicted in the Fig. 3.

$$a_i' = a_i \oplus (a_{i+1} \oplus 1) \cdot (a_{i+2} \oplus (a_{i+3} \oplus 1) \cdot a_{i+4}) \tag{7}$$

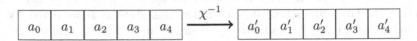

Fig. 3. Computation of χ^{-1}

2. **Observation 2:** When only one output bit is known after χ step, then the corresponding input bits have 2^4 possibilities. Kumar *et al.* [9] gave a way to fix the first output bit to be the same as input bit and the second bit as 1. It is shown in the Fig. 4.

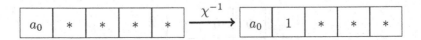

Fig. 4. Computation of χ^{-1}

3.2 Description of the Attack

The KECCAK-384 outputs 384 bits hash value, which is represented by the first 6 lanes in the state obtained in the start of the squeezing phase. The diagram in the Fig. 5 represents this state. The values of remaining lanes are represented by $\star$ and we do not care these values. We are interested in finding a preimage for which 6 lanes of corresponding state matches. We will call this state as *final state*. Furthermore, we can ignore the ι step without the loss of generality, as it does not affect the procedure of the attack. However it should be taken into account while implementing the attack.

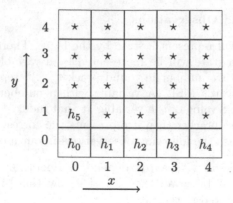

Fig. 5. Final state

We further note that the initial state, which is fed to KECCAK-f function, is the first message block which is represented by $25 - 2 \cdot 6$ i.e., 13 lanes. The remaining 12 lanes are set to 0. Pictorially, this state is represented by the diagram in the Fig. 6. We call this state *initial state*. Our aim is to find the values of $a_0, a_1, a_2, b_0, b_1, b_2, c_0, c_1, c_2, d_0, d_1$ and e_0, e_1 in the initial state which lead to a final state having first six lanes as h_0, h_1, h_2, h_3, h_4 and h_5.

We follow the basic idea of the attack, given in the paper [13]. We start the attack by setting variables in the initial state which ensures zero column parity. This is done by imposing the following restrictions.

$$a_2 = a_0 \oplus a_1, \quad b_2 = b_0 \oplus b_1, \quad c_2 = c_0 \oplus c_1$$
$$d_1 = 0, \quad d_0 = 0 \quad \text{and} \quad e_1 = e_0. \tag{8}$$

0	0	0	0	0
0	0	0	0	0
a_1	b_1	c_2	0	0
a_2	b_2	c_1	d_1	e_1
a_0	b_0	c_0	d_0	e_0

Fig. 6. Setting of initial state in the attack

This type of assignment to the initial state will make the θ step mapping, an identity mapping. Even though we have put some restrictions to the initial state, we still find the input space of KECCAK-384 (with 1 message block) large enough to ensure first 6 lanes of output state, the given hash value. We explain the details of the analysis below.

Note that the output of attack is an assignment to the variables $a_0, a_1, a_2,$ $b_0, b_1, b_2,$ $c_0, c_1, c_2,$ d_0, d_1 and e_0, e_1, which gives the target hash value. Recall that we are mounting an attack on the 2-Round KECCAK-384 (see the diagram in Fig. 7). The overall attack is summarized in the diagram given in the Fig. 8. The State 2, in the Fig. 8, represents the state after $\pi \circ \rho \circ \theta$ is applied to the State 1. The θ-mapping becomes identity due to the condition (Eq. 8) imposed on the initial state. The ρ and π mappings are, nevertheless, linear.

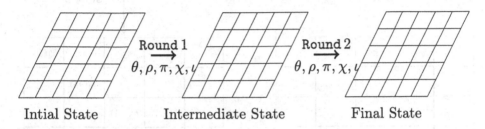

Fig. 7. Two round of Keccak-384

We are given with a hash value which is represented by first 6 lanes in the State 4 (Fig. 8). It represents the final state (**Round** 2) of KECCAK-384. The state can be inverted by applying $\chi^{-1} \circ \iota^{-1}$ mapping. The ι^{-1} is trivial and χ^{-1} can be computed using the Observations 3.1 and 3.1. The first 7 lanes of the output is $\{h'_0, h'_1, h'_2, h'_3, h'_4, h'_5, h'_6, 1\}$. We do not care the remaining lanes. Then the mappings π^{-1} and ρ^{-1} are applied, which are very easy to compute, to get the State 3 (Fig. 8). Note that, at this point, the blank lanes in the State 3, of the Fig. 8, could take any random value and this does not affect the target hash value. The number shown in round brackets along with the variable, in the State 2 and State 3 (Fig. 8), is due to ρ step mapping. On applying $\theta \circ \iota \circ \chi$, operation on the State 2, the output should match with the State 3 (Fig. 8). In the State 3, there are 7 lanes whose values are fixed. This will impose a total of 7×64 conditions on the variables we have set in the initial state. As mentioned earlier, we have also set 6 conditions (see the Eq. 8) on the initial state variable and this will further add 6×64 conditions. So there are in total 13×64 conditions. Since the number of variables and the number of conditions is equal, we can expect to find one solution and it is indeed the case. In the rest of this section, we provide an algorithm to get the unique solution. Our method is based on the technique proposed by Naya-Plasencia *et al.* in the paper [13].

We aim to find the assignment of bits to the initial state which leads to a target hash value. We proceed as follows. We start with all possible assignments in the groups successive 3 slices. Using the constraints (transformation from State 2 to State 3 (Fig. 8)), we discard some of the assignments, and store the remaining ones, out of which one would be a part of the solution. This is done

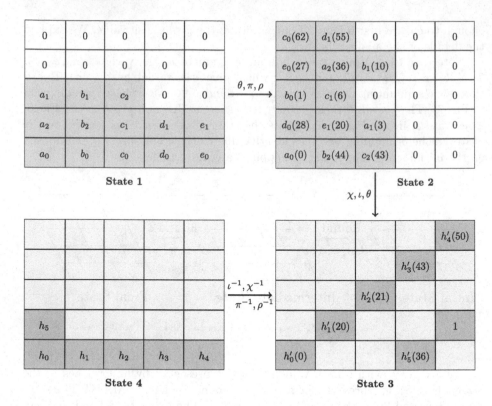

Fig. 8. Diagram for 2-round preimage attack on KECCAK-384

for every 3-slice. Next step is to merge the two successive 3-slices. Again we do discard certain choices of assignments and keep the remaining ones. This process is continued to fix a set of good assignments to the 6-slices, 12-slices, 16-slices, 24-slices and 48-slice. In the last, after combining all the assignments we are left with a unique assignment, which is the required preimage. We explain the details in the Sect. 3.3 below.

3.3   Finding Partial Solutions

We focus on the two intermediate states of the attack i.e., the State 2 and the State 3 (see the Fig. 9 below). Note that, since d_0 and d_1 are set to 0 in the beginning, we are now left with 11 lane variables $a_0, a_1, a_2, b_0, b_1, b_2, c_0, c_1, c_2, e_0$ and e_1 only. We can ignore the ι mapping in the transformation form State 2 to State 3, without the loss of generality. The χ-mapping depends only on the row, so it will not get affected by the bit values of the other slices. It is θ-mapping that depends on the values in the two slices; the slice on its original bit position and a slice just before it.

Possible Solutions for 3-slices. In a 3-slice there are $3 \cdot 11 = 33$ bit variables for which we have to find the possible assignments.

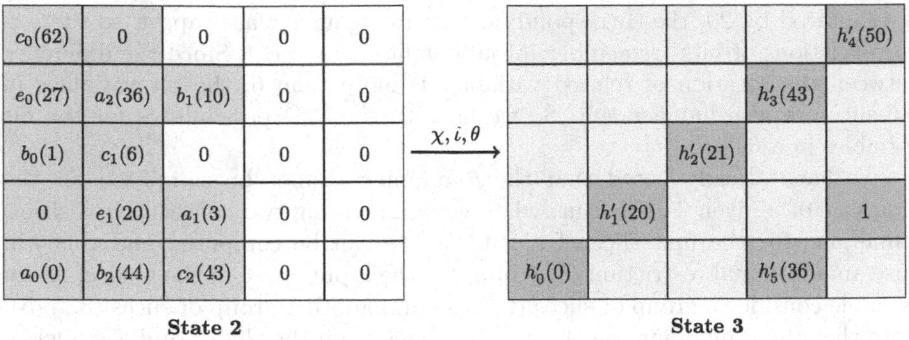

$c_0(62)$	0	0	0	0
$e_0(27)$	$a_2(36)$	$b_1(10)$	0	0
$b_0(1)$	$c_1(6)$	0	0	0
0	$e_1(20)$	$a_1(3)$	0	0
$a_0(0)$	$b_2(44)$	$c_2(43)$	0	0

State 2

$\xrightarrow{\chi,\iota,\theta}$

				$h'_4(50)$
			$h'_3(43)$	
		$h'_2(21)$		
	$h'_1(20)$			1
$h'_0(0)$			$h'_5(36)$	

State 3

Fig. 9. Intermediate states in 2-round preimage attack on KECCAK-384

Note that the bit variables, for example take $a_0[i]$, $a_1[i]$ and $a_2[i]$, are related $(a_2 = a_0 \oplus a_1)$, but due to rotation, they do not appear together when the successive 3 slices are considered. Similarly, the other variables are also independent when restricted to a 3-slice. This can be explained using the following example. If we take the first three slices then we get the following 33 independent variables, given in the Eq. 9.

$$
\begin{aligned}
&a_0[0,1,2], \quad a_1[3,4,5], \quad a_2[36,37,38], \\
&b_0[1,2,3], \quad b_1[10,11,12], \quad b_2[44,45,46], \\
&c_0[62,63,0], \quad c_1[6,7,8], \quad c_2[43,44,45], \\
&e_0[27,28,29], \quad e_1[20,21,22].
\end{aligned}
\tag{9}
$$

None of these variables have any dependency despite the initial restriction, given by Eq. 8. So we have an input space of 33 independent variables in a given 3-slice.

Given a 3-slice in the State 2, we need to apply $\theta \circ \iota \circ \chi$ mapping to get an output in the State 3. Since the θ mapping depends on the values of two slices; the current slice and one preceding it, we will only able to get the correct output for two slices. In the State 3, we have the values of 7 lanes available with us. So for the two slices, we have $7 \cdot 2$ fixed bit values. For each of 2^{33} assignments in a 3-slice of the State 2, we compute the output of $\theta \circ \iota \circ \chi$ mapping and match it to the 14 bit locations, the values of which are available in the State 3. Thus for each 3-slice, we get $2^{33-14} = 2^{19}$ solutions. This is repeated for 16 consecutive 3-slices, other than last 16 slices. We use the fact that the time complexity of building the list is given by the size of the list as stated in Sect. 6.4 of [13]. Thus the required time and memory complexity is of the order $16 \cdot 2^{19} = 2^{23}$.

Possible Solutions for 6-Slices. The possible solutions for a 6-slice are obtained by merging the possible solutions of its constituents two 3-slices. The variables restricted to the 6-slice is again independent. This can be explained in the following manner. Consider the rotated lanes $a_0(0)$, $a_1(3)$ and $a_2(36)$. Since the lane variable a_2 is rotated by 36 and a_1 is rotated by 3, the corresponding bits of original lanes are still 33 places apart. Similarly e_0 is rotated by 27 and

e_1 is rotated by 20, the corresponding bits are again 7 places apart, so there is no repetitions of bits (remember initial condition $e_0 = e_1$). Since the difference between the rotation of related variables is more than 6, the bit variables in a 6-slice are also independent. So we have $2^{19 \cdot 2} = 2^{38}$ possibilities for the bit variables in a 6-slice.

We have already noted that the θ-mapping cannot be computed for the first slice of a given 3-slice. But, when we are merging two consecutive 3-slices, θ-mapping for the first slice of second 3-slice can be computed and this will pose an additional restriction (of 7 bits) for the input space of the 6-slice. As an example consider a group of slices (0, 1, 2) and another group of slices (3, 4, 5). Note that the θ-mapping, on the slice 3, depends on the slice 3 and 2. So when we are merging these two 3-slices, we will have to satisfy the bits corresponding to slice 3, in the State 3.

So we get a total $2^{19 \cdot 2 - 7} = 2^{31}$ solutions. There are 8 number of 6-slices. The cost of this step is $8 \cdot 2^{31}$ in both time and memory. Note that the merging of two lists is done using the instant matching algorithm described in [12] by the method described in the Sect. 6.4 of the paper [13]. This method will be used in the following steps also, where the time complexity will be bounded by the number of solutions obtained. Thus this step has time and memory complexity of $8 \cdot 2^{31} = 2^{34}$.

Possible Solutions for 12-Slices. For computing the possible solutions for a 12-slice, we merge two of its constituents 6-slices, in a manner similar to what we did for a 6-slice. In this case, the number of repeated bits in merge is 5, because the corresponding bits in e_0 and e_1 are set 7 places apart by the rotation in the State 2. Thus total number of possible solutions for a 12-slice is $2^{31 \cdot 2 - 5 - 7} = 2^{50}$. There are 4 groups of 12 slices, so it has time and memory complexity of $4 \cdot 2^{50} = 2^{52}$.

Possible Solutions for 24-Slices. Similar to the previous cases, we merge each of its two consecutive 12-slices. In this case, the number of repeated bits is $24 - 7 = 17$, out of which $5 \cdot 2 = 10$ has already been considered, during the construction of possible solutions of 12-slices. So the number of new repeated bit variables are 7. Hence, the total number of possible solutions for this case is $2^{50 \cdot 2 - 7 - 7} = 2^{86}$. Note that the removal of addition seven bits is due to merging. There are 2 groups of 24 slices, so it has time and memory complexity of $2 \cdot 2^{86} = 2^{87}$.

Possible Solutions for 48-Slice. Finally, we merge the two groups of 24 slices. We have 2 sets of 24 slices as

1st group:

$$\left. \begin{array}{l} a_0 \rightarrow 0, 1, 2, \ldots, 23 \\ a_1 \rightarrow 3, 4, 5, \ldots, 26 \\ a_2 \rightarrow 36, 37, 38, \ldots, 59 \end{array} \right\} \tag{10}$$

2nd group:

$$\left.\begin{array}{l} a_0 \rightarrow 24,\ 25,\ 26,\ldots,\ 47 \\ a_1 \rightarrow 27,\ 28,\ 29,\ldots,\ 50 \\ a_2 \rightarrow 60,\ 61,\ 62,\ldots,\ 19 \end{array}\right\}. \qquad (11)$$

After Merging these two groups (Eqs. (10) and (11)) of 24 slices, we get

$$\left.\begin{array}{l} a_0 \rightarrow 0,\ 1,\ 2,\ldots,\ 47 \\ a_1 \rightarrow 3,\ 4,\ 5,\ldots,\ 50 \\ a_2 \rightarrow 36,\ 37,\ldots,\ 63,\ 0,\ 1,\ldots,\ 19 \end{array}\right\}. \qquad (12)$$

Here the common variables for $\langle a_0, a_1, a_2 \rangle$ are the bits with positions $36, 37, \ldots, 47$ and $3, 4, \ldots, 19$. They are total 29 in number. It will impose 29 conditions on the input space for the 48-slice. Similarly for the lanes $\langle b_0, b_1, b_2 \rangle$, we get 23 conditions and for $\langle c_0, c_1, c_2 \rangle$, we get 24 such conditions. On the other hand, there are 7 new repeated bits in the lanes e_0 and e_1. Thus the total number of solutions turns out to be $2^{86 \cdot 2 - (29 + 23 + 24 + 7) - 7} = 2^{82}$. Since, there is only one 48-slices, so it has time and memory complexity of 2^{82}.

Possible Solutions for Remaining 16 Slices. For finding solutions for the remaining 16 slices, we first find solutions for the 12 rightmost slices, the same way as before, and obtaining 2^{50} possible solutions. Next, we obtain the possible solutions for the remaining 4 slices, we have 44 variables and none of them are repeated. Since we can get the output of θ-mapping for the last 3 slices out of the 4. We have $2^{44 - 7 \cdot 3} = 2^{23}$ possible solutions for this 4-slice. Now, we can merge 12-slice and 4-slice to obtain possible solutions for the last 16 slices. Between 12-slice and 4-slice, there are 4 repetitions (due to e_0 and e_1) and there are additional 7 bits of restrictions due to merging. This gives us $2^{50 + 23 - 4 - 7} = 2^{62}$ possible solutions.

Final Solution(s) and Attack Complexity. Now, we have to merge the solutions for the group of first 48 slices and the group of last 16 slices. They have in common 35 bits from a_0, a_1 and a_2, 41 bits from b_0, b_1 and b_2, 40 bits from c_0, c_1 and c_2 and 14 bits from e_0 and e_1. Additionally, in merging, we can compute the θ mapping of the remaining two slices, in turn get the additional restriction of $2 \cdot 7$ bits. Thus the total number of possible solutions, we are left with, is $2^{82 + 62 - (35 + 41 + 40 + 14) - 2 \cdot 7} = 2^0 = 1$. This step has time complexity 2^{82}.

Total time complexity of the attack is given by $2^{33} + 2^{34} + 2^{52} + 2^{87} + 2^{63} + 2^{82}$, which is of the order 2^{88}. The total memory required is 2^{87}. This confirms that there exists a set of values for the variables such that the preimage can be obtained from the hash value for the KECCAK-384.

Remark: *In our attack, we have fixed d_0, d_1 lanes to be equal to 0 as shown in Eq. (8) because otherwise, these variables would have increased the number of solutions, due to shifting by ρ. And this would have increased the complexity of the attack. We chose to eliminate their effects by setting them to 0. For further implementation details, we refer to the Sect. 6.4 of the paper [13]. Also due to* *the padding rule on the message, the assignment to the $c_1[63]$ bit should be 1. This happens with probability $\frac{1}{2}$. On failure we can repeat the attack by setting any value to d_0, d_1 which satisfies $d_0[i] = d_1[i]$.*

In view of the above remark, the overall cost of the attack is $2 \cdot 2^{88}$ i.e., 2^{89}.

4   Conclusion and Future Works

In this paper, we have presented a preimage attack on the 2 rounds of round-reduced KECCAK-384. The attack is not yet practical but it is much better than the existing best-known attack in term of the time complexity. The basic idea of the attack can be used to mount a practical preimage attack on the KECCAK$[r := 400 - 192, c := 192]$ and KECCAK$[r := 800 - 384, c := 384]$. We are working on their implementations. We will make the source code public, once it is ready. Further, in future, we will try to explore a practical attack for the 2 or more rounds of round-reduced KECCAK-384.

Acknowledgement. We thank the reviewers of Indocrypt-2018 for providing comments which helped in improving the work. In particular, we thank an anonymous reviewer for suggesting us to implement the attack on the KECCAK$[r := 400 - 192, c := 192]$ and also providing insights to further improve the attack. We take it as the future work.

References

1. Bernstein, D.J.: Second preimages for 6 (7?(8??)) rounds of keccak. NIST mailing list (2010)
2. Bertoni, G., Daemen, J., Peeters, M., Van Assche, G.: Keccak specifications. Submission to NIST (Round 2) (2009)
3. Bertoni, G., Daemen, J., Peeters, M., Van Assche, G.: Cryptographic sponges (2011). http://sponge.noekeon.org
4. Chang, D., Kumar, A., Morawiecki, P., Sanadhya, S.K.: 1st and 2nd preimage attacks on 7, 8 and 9 rounds of Keccak-224,256,384,512. In: SHA-3 Workshop, August 2014
5. Dinur, I., Dunkelman, O., Shamir, A.: New attacks on Keccak-224 and Keccak-256. In: Canteaut, A. (ed.) FSE 2012. LNCS, vol. 7549, pp. 442–461. Springer, Heidelberg (2012). https://doi.org/10.1007/978-3-642-34047-5_25
6. Dinur, I., Dunkelman, O., Shamir, A.: Collision attacks on up to 5 rounds of SHA-3 using generalized internal differentials. In: Moriai, S. (ed.) FSE 2013. LNCS, vol. 8424, pp. 219–240. Springer, Heidelberg (2014). https://doi.org/10.1007/978-3-662-43933-3_12
7. Dinur, I., Dunkelman, O., Shamir, A.: Improved practical attacks on round-reduced Keccak. J. Cryptol. **27**(2), 183–209 (2014)
8. Guo, J., Liu, M., Song, L.: Linear structures: applications to cryptanalysis of round-reduced KECCAK. In: Cheon, J.H., Takagi, T. (eds.) ASIACRYPT 2016. LNCS, vol. 10031, pp. 249–274. Springer, Heidelberg (2016). https://doi.org/10.1007/978-3-662-53887-6_9

9. Kumar, R., Rajasree, M.S., AlKhzaimi, H.: Cryptanalysis of 1-round KECCAK. In: Joux, A., Nitaj, A., Rachidi, T. (eds.) AFRICACRYPT 2018. LNCS, vol. 10831, pp. 124–137. Springer, Cham (2018). https://doi.org/10.1007/978-3-319-89339-6_8

10. Morawiecki, P., Pieprzyk, J., Srebrny, M.: Rotational cryptanalysis of round-reduced KECCAK. In: Moriai, S. (ed.) FSE 2013. LNCS, vol. 8424, pp. 241–262. Springer, Heidelberg (2014). https://doi.org/10.1007/978-3-662-43933-3_13

11. Morawiecki, P., Srebrny, M.: A sat-based preimage analysis of reduced Keccak hash functions. Inf. Process. Lett. **113**(10–11), 392–397 (2013)

12. Naya-Plasencia, M.: How to improve rebound attacks. In: Rogaway, P. (ed.) CRYPTO 2011. LNCS, vol. 6841, pp. 188–205. Springer, Heidelberg (2011). https://doi.org/10.1007/978-3-642-22792-9_11

13. Naya-Plasencia, M., Röck, A., Meier, W.: Practical analysis of reduced-round KECCAK. In: Bernstein, D.J., Chatterjee, S. (eds.) INDOCRYPT 2011. LNCS, vol. 7107, pp. 236–254. Springer, Heidelberg (2011). https://doi.org/10.1007/978-3-642-25578-6_18

14. Qiao, K., Song, L., Liu, M., Guo, J.: New collision attacks on round-reduced Keccak. In: Coron, J.-S., Nielsen, J.B. (eds.) EUROCRYPT 2017. LNCS, vol. 10212, pp. 216–243. Springer, Cham (2017). https://doi.org/10.1007/978-3-319-56617-7_8

15. Song, L., Liao, G., Guo, J.: Non-full sbox linearization: applications to collision attacks on round-reduced KECCAK. In: Katz, J., Shacham, H. (eds.) CRYPTO 2017. LNCS, vol. 10402, pp. 428–451. Springer, Cham (2017). https://doi.org/10.1007/978-3-319-63715-0_15

Post Quantum Cryptography

A Faster Way to the CSIDH

Michael Meyer[1,2]([✉]) and Steffen Reith[1]

[1] Department of Computer Science, University of Applied Sciences Wiesbaden, Wiesbaden, Germany
[2] Department of Mathematics, University of Würzburg, Würzburg, Germany
{Michael.Meyer,Steffen.Reith}@hs-rm.de

Abstract. Recently Castryck, Lange, Martindale, Panny, and Renes published CSIDH, a new key exchange scheme using supersingular elliptic curve isogenies. Due to its small key sizes and the possibility of a non-interactive and a static-static key exchange, CSIDH seems very interesting for practical applications. However, the performance is rather slow. Therefore, we employ some techniques to speed up the algorithms, mainly by restructuring the elliptic curve point multiplications and by using twisted Edwards curves in the isogeny image curve computations, yielding a speed-up factor of 1.33 in comparison to the implementation of Castryck et al. Furthermore, we suggest techniques for constant-time implementations.

Keywords: CSIDH · Post-quantum cryptography
Supersingular elliptic curve isogenies

1 Introduction

Isogeny-Based Cryptography. Isogeny-based cryptography is one of the current proposals for post-quantum cryptography. Already proposed in a talk (but not published) by Couveignes in 1997 [12] and independently rediscovered by Rostovtsev and Stolbunov in 2004 [22], a Diffie-Hellman-style key exchange based on isogenies between ordinary elliptic curves was designed (called CRS in the following). In 2010, Childs, Jao and Soukharev [9] showed, that this scheme can be attacked by solving an abelian hidden shift problem, for which subexponential quantum algorithms are known to exist.

Due to this, Jao and De Feo [16] considered the use of supersingular elliptic curves, and designed the new key exchange scheme SIDH (supersingular isogeny Diffie-Hellman), based on random walks in isogeny graphs for supersingular elliptic curves defined over fields $\mathbb{F}_{p^2}$. The performance of their scheme was improved by Costello, Longa, and Naehrig [11], yielding an important step towards practical deployment of SIDH, and also causing an increase of attention and research for isogeny-based cryptography. This led to the development of SIKE [1], an isogeny-based key encapsulation scheme, as entry for the NIST post-quantum cryptography competition [23], that aims for the standardization

© Springer Nature Switzerland AG 2018
D. Chakraborty and T. Iwata (Eds.): INDOCRYPT 2018, LNCS 11356, pp. 137–152, 2018.
https://doi.org/10.1007/978-3-030-05378-9_8

of post-quantum schemes in order to start the transition to the practical use of quantum-resistant primitives. The main advantage of SIKE comes from its key sizes. Among all the submitted key encapsulation schemes, it provides the smallest public keys. However, the price for this is a rather bad performance. In comparison to the other competition entries, the running time of SIKE is slow[1].

Recently, De Feo, Kieffer, and Smith [15] published some new ideas for the optimization of the CRS scheme. Due to its commutative and non-interactive structure, it is still an interesting alternative to SIDH and SIKE. However, the performance is far from being practical. Therefore, Castryck, Lange, Martindale, Panny, and Renes [8] found that the optimizations, that De Feo, Kieffer, and Smith wanted to employ, work even better when adapting CRS to supersingular elliptic curves, i.e. working with supersingular elliptic curves over $\mathbb{F}_p$ rather than $\mathbb{F}_{p^2}$ like in SIDH. They obtain a non-interactive key exchange scheme with even smaller key sizes than in SIDH, called CSIDH (commutative SIDH, pronounced like "seaside"), that also allows static keys, since public keys can be validated, to detect active attacks. The performance is rather slow in comparison to SIDH and SIKE[2], which explains why it is an interesting and important task to optimize the running time of the scheme. However, we note that the security of the scheme and hence also the choice of parameters is still an open problem, which we will only briefly address in the next section.

Organization. In the following section, we give an introduction to CSIDH, mainly focusing on the implementer's point of view, and recall some aspects about Montgomery and twisted Edwards curves. We then introduce a way to restructure elliptic curve point multiplications in CSIDH, that allows a reduction of the computational effort. Thereafter, we review some methods to compute isogenies, i.e. point evaluations and computations of the image curves. In the first case, we employ an observation of Costello and Hisil [10] for a speed-up to the implementation of [8], whereas in the latter case, we exploit the well-known correspondence between Montgomery and twisted Edwards curves, to compute the image curves more efficiently. We give some implementation results according to our contributions, and give some remarks about constant-time implementations and bounds for their running time.

---

[1] In [1] it is stated that the best performance for SIKEp503 and SIKEp751 on a 3.4 GHz processor is 10.1 ms and 30.5 ms, respectively.

[2] The implementation results in [8] suggest that SIKE is about 10x faster than CSIDH at NIST security level 1. Note that SIKE uses a protected constant-time implementation, while the numbers for CSIDH are obtained from an unprotected non-constant-time implementation.

2   Preliminaries

2.1   CSIDH

Since our aim is to focus on implementations of CSIDH, we only give a very brief description of the mathematical background based on [8]. We recommend the lecture of [8] for a more detailed overview. We refer to [13] for additional information about isogenies and their cryptographic applications.

Mathematical Background. First consider a quadratic number field k and an order $\mathcal{O} \subset k$. The ideal class group of $\mathcal{O}$ is defined as

$$\text{cl}(\mathcal{O}) = I(\mathcal{O})/P(\mathcal{O}),$$

where $I(\mathcal{O})$ denotes the set of invertible fractional ideals and $P(\mathcal{O})$ the set of principal fractional ideals.

In the context of CRS and CSIDH, we use the group action of the ideal class group of an imaginary quadratic order $\mathcal{O}$ on ordinary and supersingular elliptic curves, respectively. In both cases the ideal class group $\text{cl}(\mathcal{O})$ acts on $\mathcal{E}\ell\ell_p(\mathcal{O})$ via isogenies, where $\mathcal{E}\ell\ell_p(\mathcal{O})$ is the set of elliptic curves E defined over $\mathbb{F}_p$ with $\text{End}_p(E) \cong \mathcal{O}$. By $\text{End}_p(E)$ we denote the subring of the endomorphism ring $\text{End}(E)$, that consists of endomorphisms defined over $\mathbb{F}_p$.

In CSIDH, a prime $p = 4 \cdot \ell_1 \cdot ... \cdot \ell_n - 1$ is chosen, where the ℓ_i are small distinct odd primes, and the elliptic curve $E_0 : y^2 = x^3 + x$ over $\mathbb{F}_p$, which is supersingular because $p \equiv 3 \pmod 4$. The supersingularity of E_0 (and of all curves that are isogenous to E_0) now guarantees the existence of elliptic curve points of order ℓ_i for all $i \in \{1, ..., n\}$.

The ideals $\ell_i \mathcal{O}$ split as $\ell_i \mathcal{O} = \mathfrak{l}_i \overline{\mathfrak{l}}_i$, where $\mathfrak{l}_i = (\ell_i, \pi - 1)$ and $\overline{\mathfrak{l}}_i = (\ell_i, \pi + 1)$ with the Frobenius endomorphism π. The kernel of the isogeny $\varphi_{\mathfrak{l}_i}$ then is the intersection of the kernels of the point multiplication $[\ell_i]$ and the endomorphism $\pi - 1$, i.e. a subgroup generated by a point P of order ℓ_i defined over $\mathbb{F}_p$. Analogously, the kernel of the isogeny $\varphi_{\overline{\mathfrak{l}}_i}$ is a subgroup generated by an order-ℓ_i point P defined over $\mathbb{F}_{p^2} \backslash \mathbb{F}_p$. Hence, the computation of the action of an ideal class $\prod \mathfrak{l}_i^{e_i}$ by computing the action of the $\mathfrak{l}_i$ resp. $\overline{\mathfrak{l}}_i$ can be done by efficient isogeny formulae: We have to find order-ℓ_i points defined over $\mathbb{F}_p$ resp. $\mathbb{F}_{p^2} \backslash \mathbb{F}_p$ and can apply efficient formulae such as Vélu-isogenies [24] or isogeny formulae for Montgomery curves like in [10] or [21]. By construction such points always exist. Ideal classes $\prod \mathfrak{l}_i^{e_i}$ can simply be represented by vectors $(e_1, ..., e_n)$.

Key Exchange. As already observed by Couveignes in [12], the commutativity of the class group action allows for a Diffie-Hellman-style key exchange in the following way:

Alice chooses a secret ideal class $[\mathfrak{a}]$, represented by a vector $(e_1, ..., e_n)$, computes $E_A = [\mathfrak{a}] \cdot E_0$ via isogenies, and sends the result to Bob as her public key in terms of a curve parameter. Bob proceeds in the same way, chooses a secret $[\mathfrak{b}]$ and computes his public key $E_B = [\mathfrak{b}] \cdot E_0$. Then, because of the commutativity, both parties can compute the shared secret $[\mathfrak{a}] \cdot [\mathfrak{b}] \cdot E_0 = [\mathfrak{a}] \cdot E_B = [\mathfrak{b}] \cdot E_A$.

Security of the Scheme. As for the CRS scheme, it is clear that the subexponential quantum attack from [9] also applies to CSIDH. However, Castryck et al. give some estimations for parameter sets for different security levels [8]. More recently, shortly after CSIDH was published, more analysis of this attack has been done [4,5]. Since we are only focusing on efficient implementations throughout this work, we will not discuss these attacks here, and we only note that the appropriate choice of parameters is still an open problem, that requires further analysis. However, our improvements don't rely on a special choice of parameters, and are thus independent of the selected parameters.

2.2   Implementation

We follow the implementation accompanying [8] here[3].

Algorithm 1. Evaluating the class group action.

Input : $A \in \mathbb{F}_p$ and a list of integers $(e_1, ..., e_n)$.
Output: A' such that $[\mathfrak{l}_1^{e_1} \cdots \mathfrak{l}_n^{e_n}]E_A = E_{A'}$.

1 **while** some $e_i \neq 0$ **do**
2 | Sample a random $x \in \mathbb{F}_p$.
3 | Set $s \leftarrow +1$ if $x^3 + Ax^2 + x$ is a square in $\mathbb{F}_p$, else $s \leftarrow -1$.
4 | Let $S = \{i \mid sign(e_i) = s\}$.
5 | **if** $S = \emptyset$ **then**
6 | | Go to line 2.
7 | $P = (x : 1)$, $k \leftarrow \prod_{i \in S} \ell_i$, $P \leftarrow [(p+1)/k]P$.
8 | **foreach** $i \in S$ **do**
9 | | $K \leftarrow [k/\ell_i]P$.
10 | | **if** $K \neq \infty$ **then**
11 | | | Compute a degree-ℓ_i isogeny $\varphi : E_A \rightarrow E_{A'}$ with $ker(\varphi) = \langle K \rangle$.
12 | | | $A \leftarrow A'$, $P \leftarrow \varphi(P)$, $k \leftarrow k/\ell_i$, $e_i \leftarrow e_i - s$.

First, we define a prime number $p = 4 \cdot \ell_1 \cdot ... \cdot \ell_n - 1$ as above, where $\ell_1, ..., \ell_n$ are small distinct odd primes. Then we choose a supersingular curve E_0 over $\mathbb{F}_p$. Therefore we have $\#E_0 = p + 1$, which means that there are points of order ℓ_i for $i = 1, ..., n$ on E_0. Note that the factor 4 is needed to ensure that we can use Montgomery curves.

The private key contains n integers sampled from an interval $[-m, m]$, i.e. has the form $(e_1, ..., e_n)$. For each i the absolute value $|e_i|$ determines how many isogenies of degree ℓ_i are to be computed, while the sign of e_i states if we have to use points defined over $\mathbb{F}_p$ or $\mathbb{F}_{p^2} \backslash \mathbb{F}_p$ to generate their kernels.

For the computation of isogenies, we choose a random point P by sampling a random $x \in \mathbb{F}_p$, and check in which of the cases above this leads us by checking

---

[3] We refer to the version from 27.04.2018 throughout this work.

the minimal field of definition of the corresponding y-coordinate by a square root check. We then eliminate the possible unwanted factors in the order of P by multiplying it by $4 \cdot \prod_{j \notin S} \ell_j$, where S is defined as in algorithm 1.

After that, we iterate over the ℓ_i for $i \in S$, removing all remaining possible factors of the order except for ℓ_i of our point by multiplications, and check whether the resulting point K can be used as kernel generator for computing an ℓ_i-isogeny, i.e. if $K \neq \infty$. If so, we compute the isogeny and push P through. Then we go to the next prime and proceed in the same way. However, we don't have to consider the previous ℓ_i in the multiplication, since the isogeny evaluations of P already eliminate the respective factors from its order, or in the other case, the order of P did not contain the previous ℓ_i as factors in the first place.

We proceed in the same way, and sample new random points, until all of the required isogenies are computed. The resulting curve then forms the public key, or the shared secret, respectively. Note that the computational effort in algorithm 1 highly depends on the private key. Therefore, for the practical usage of CSIDH, it is important to transform this into a constant-time scheme without adding too much computational overhead.

Public keys can also be validated by checking for supersingularity: We can simply sample a random point P on the curve corresponding to the received public key. For each ℓ_i we compute $Q_i = [(p+1)/\ell_i]P$. For all i with $Q_i \neq \infty$, we compute $[\ell_i]Q_i$ and $d = \prod \ell_i$. If any of these $[\ell_i]Q_i \neq \infty$, the curve cannot be supersingular, since $\#E(\mathbb{F}_p) \nmid p+1$. If this is not the case, and $d > 4\sqrt{p}$, the curve must be supersingular, as can be seen from the Hasse interval and Lagrange's theorem (see [8]). Otherwise, the procedure can be repeated with a different point P. Following this approach, it is not possible to wrongly classify an ordinary curve as supersingular. Therefore, we can check if a public key has been honestly generated, and thus can prevent certain kinds of active attacks.

Choice of Parameters. The following discussions and implementation results refer to the parameter set proposed in [8] for NIST's post-quantum security category I. They choose $p = 4 \cdot \ell_1 \cdot \ldots \cdot \ell_{74} - 1$, where $\ell_1, \ldots \ell_{73}$ are the 73 smallest distinct odd primes and $\ell_{74} = 587$. The elements of the private keys $(e_1, \ldots, e_{74})$ are chosen from the interval $[-5, 5]$. This parameter set leads to public key lengths of 64 bytes. As mentioned before, the appropriate choice of parameters is still an open problem, so the analysis of the actual security level of this parameter set is left for future work.

2.3 Montgomery Curves

Montgomery curves are given by an equation over a field k with $\text{char}(k) > 2$ of the form

$$E_{a,b} : by^2 = x^3 + ax^2 + x,$$

where $a \in k \backslash \{-2, 2\}$ and $b \in k \backslash \{0\}$. To avoid inversions during point additions and doublings, projective coordinates can be used. Furthermore, the efficient arithmetic given by Montgomery in [19] allows for dropping the Y-coordinate and still performing XZ-only point doublings and differential additions, that require the knowledge of the XZ-coordinates of P, Q, and $P - Q$ in order to compute $P + Q$.

In [11], Costello et al. propose to projectivize not only the point coordinates, but also the curve parameters. Instead of a Montgomery curve of the form given above, we work with an equation of the form

$$E_{(A:B:C)} : By^2 = Cx^3 + Ax^2 + Cx,$$

where $(A : B : C) \in \mathbb{P}^2(k)$, such that $a = A/C$ and $b = B/C$ for the corresponding curve $E_{a,b}$. However, in isogeny-based schemes it suffices to work with $(A : C) \in \mathbb{P}^1(k)$ in the projective model, since neither the Montgomery curve arithmetic, nor the isogeny computations require the coefficients b or B, respectively. In general a doubling then costs $4\mathbf{M} + 2\mathbf{S} + 8\mathbf{a}$, while a differential addition costs $4\mathbf{M} + 2\mathbf{S} + 6\mathbf{a}$. As usual, we denote field multiplications by $\mathbf{M}$, field squarings by $\mathbf{S}$, and field additions or subtractions by $\mathbf{a}$.

2.4   Twisted Edwards Curves

Introduced by Bernstein et al. in [2], twisted Edwards curves over k with $\mathrm{char}(k) > 2$ are given by equations of the form

$$E_{a,d} : aX^2 + Y^2 = 1 + dX^2Y^2,$$

with $ad \neq 0$, $d \neq 1$, and $a \neq d$. For $a = 1$ the twisted Edwards curve $E_{1,d} = E_d$ is called Edwards curve, originally proposed by Edwards in [14]. As in the Montgomery case, projective coordinates can be used in order to avoid inversions during additions and doublings. Note that in the Edwards case there are different models for doing this, as described in [3].

Similar to the XZ-only Montgomery curve arithmetic, Castryck, Galbraith, and Farashahi introduced a YZ-only doubling formula for twisted Edwards curves in [7] with a cost of $4\mathbf{M} + 5\mathbf{S}$. A formula for YZ-only differential addition of twisted Edwards curve points of odd order is derived in [17], using $6\mathbf{M} + 3\mathbf{S}$ in the projective case. Due to the fact that these operations are in general more expensive than the respective operations on the Montgomery curve, isogeny-based schemes usually use Montgomery curves (see [6,18] for a comparison to twisted Edwards curve point arithmetic in SIDH). However, in the following twisted Edwards curves are shown to be advantageous for the computation of isogenies.

3   Elliptic Curve Point Multiplications

Define $\alpha = \frac{p+1}{4} = \ell_1 \cdot \ell_2 \cdot \ldots \cdot \ell_n$. For the sake of simplicity, we consider a private key $(e_1, ..., e_n)$, where all $e_i > 0$, or all $e_i < 0$. We will return to the general case later on. The algorithm used by Castryck et al. then samples a random point P on the current curve E_0, checks if its y-coordinate is defined over the corresponding field $\mathbb{F}_p$ or $\mathbb{F}_{p^2} \setminus \mathbb{F}_p$, and if so, sets $P_0 = [4]P$ in order to remove the possible factor 4 from its order. Then they compute $K_0 = [\frac{\alpha}{\ell_1}]P_0$. If $K_0 = \infty$, the order of P does not contain the factor ℓ_1. We cannot use it to compute an isogeny of degree ℓ_1 and set $P_1 = P_0$ and $E_1 = E_0$. If however $K_0 \neq \infty$, then K_0 must have order ℓ_1 and can be used as generator of the kernel of an isogeny of degree ℓ_1, mapping to a curve E_1. In this case, we pull P_0 through the isogeny and obtain a point $P_1 \in E_1$. Note that this implies, that the order of P_1 does not contain the factor ℓ_1. Therefore, for checking if we can use P_1 to compute an isogeny of degree ℓ_2, it suffices to compute $K_1 = [\frac{\alpha}{\ell_1 \cdot \ell_2}]P_1$ and proceed as before. Following this approach, the required factor for the scalar multiplication of P_j reduces at each step, until only the factor ℓ_n remains at the last step of the loop.

Castryck et al. go through the primes in ascending order in their implementation, starting with small degree isogenies. However, we found it advantageous to change the direction of the loop, i.e. go through the primes in descending order. By doing this, we can eliminate the larger factors of $p+1$ first, and therefore end up with multiplications by significantly smaller factors as we proceed through the loop. Note that as soon as one isogeny degree is done, i.e. $|e_i|$ isogenies of degree ℓ_i were already computed, we include this factor in the first multiplication to compute P_0, making sure that the order of P_0 is not divided by ℓ_i. We can then ignore the factor ℓ_i in the loop, which slightly reduces the advantage of our approach every time this occurs. However, we note that our approach is still faster, as long as at least two factors are left in the loop.

Figure 1 shows the effect of our approach, compared to the implementation of [8]. Note that per bit of the factor of an elliptic curve point multiplication one step in the Montgomery ladder is carried out, i.e. one combined doubling and

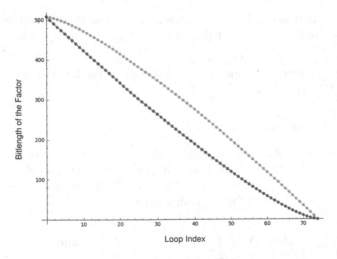

Fig. 1. Bitlengths of factors during the first loop, when all e_i have the same sign. The red line follows the algorithm of [8], the blue line follows our described approach. (Color figure online)

addition. Therefore, in the first loop, at each multiplication the computational effort is reduced by δ_i times the cost of a ladder step, where δ_i is the difference between the two plots for a given $i < n$, and hence $\delta_i \cdot (8\mathbf{M} + 4\mathbf{S} + 8\mathbf{a})$. As discussed before, the number of saved operations reduces in the following loops.

In the general case, our assumption that all elements of the private key share the same sign obviously does not hold. However, the described effect will translate at a lower scale to both of the somewhat distinct computations for the sets $S_+ = \{\ell_i \mid e_i > 0\}$ and $S_- = \{\ell_i \mid e_i < 0\}$ corresponding to the private key $(e_1, ..., e_n)$. Indeed, when plotting the bitlengths of the factors in the respective first loops in such cases, this leads to a similar result as in Fig. 1, only at a lower scale.

4   Isogeny Computations

The algorithm of [8] uses isogeny formulae for Montgomery curves by Costello and Hisil [10] and Renes [21]. We will treat point evaluations and computations of coefficients of image curves separately. First, we will state the isogeny formulae of [10], which can be used for the computation of isogenies in CSIDH.

Let K be a point of order $\ell = 2d + 1$ on a Montgomery curve $E : y^2 = x^3 + ax^2 + x$. Then we can compute the coordinate map of the unique (up to compositions by isomorphisms) ℓ-isogeny $\varphi : E \to E'$ with $ker(\varphi) = \langle K \rangle$ by

$$\varphi : (x, y) \mapsto (f(x), y \cdot f'(x)),$$

where

$$f(x) = x \cdot \prod_{i=1}^{d} \left(\frac{x \cdot x_{[i]K} - 1}{x - x_{[i]K}} \right)^2,$$

and $f'(x)$ is its derivative. The curve parameters a' and b' of E' can be computed by $a' = (6\sigma - 6\tilde{\sigma} + a) \cdot \pi^2$ and $b' = b \cdot \pi^2$, where we define $\sigma = \sum_{i=1}^{d} x_{[i]K}$, $\tilde{\sigma} = \sum_{i=1}^{d} 1/x_{[i]K}$, and $\pi = \prod_{i=1}^{d} x_{[i]K}$.

Note that the representation of $f(x)$ makes use of the fact that $x_{[i]K} = x_{[\ell-i]K}$ for all $k \in \{1, ..., (\ell - 1)/2\}$.

4.1   Point Evaluations

Since we work with XZ-only projective Montgomery coordinates, we have to represent $f(x)$ projectively. This is done in [10] by writing $(X_i : Z_i) = (x_{[i]K} : 1)$ for $i = 1, ..., d$, $(X : Z) = (x_P : 1)$ for the point P, at which the isogeny should be evaluated, and $(X' : Z')$ for the result. Then

$$X' = X \cdot \left(\prod_{i=1}^{d} (X \cdot X_i - Z_i \cdot Z) \right)^2, \quad \text{and}$$

$$Z' = Z \cdot \left(\prod_{i=1}^{d} (X \cdot Z_i - X_i \cdot Z) \right)^2.$$

In the implementation of [8], this is used directly by going through the $(X_i : Z_i)$ for $i = 1, ..., d$ and computing the pairs $(X \cdot X_i - Z_i \cdot Z)$ and $(X \cdot Z_i - X_i \cdot Z)$ at a cost of $4\mathbf{M} + 2\mathbf{a}$ per step. However, we can also use the observation by Costello and Hisil in [10] to reduce the cost to $2\mathbf{M} + 4\mathbf{a}$ per step by

$$X' = X \cdot \left(\prod_{i=1}^{d} \left[(X - Z)(X_i + Z_i) + (X + Z)(X_i - Z_i) \right] \right)^2, \quad \text{and}$$

$$Z' = Z \cdot \left(\prod_{i=1}^{d} \left[(X - Z)(X_i + Z_i) - (X + Z)(X_i - Z_i) \right] \right)^2,$$

assuming that $X + Z$ and $X - Z$ are precomputed, and hence save $d \cdot (2\mathbf{M} - 2\mathbf{a})$ per isogeny evaluation.

4.2   Computing the Image Curve

An efficient computation of the image curve parameters is not as straightforward as for the point evaluations. This is due to the fact that the required parameters σ and $\tilde{\sigma}$ consist of sums of fractions. Therefore, Costello and Hisil give two different approaches to compute the isogenous curve [10].

The first approach uses the fact that the projective parameters $(a' : 1) = (A' : C')$ of the isogenous curve E' can be recovered from the knowledge of the three 2-torsion points of E'. Therefore, it is possible to recover the required curve parameters of E' by computing the 2-torsion points of E and pushing one of these points through the odd-degree isogeny, which preserves its order on the image curve. However, in contrast to SIDH, we only work over the field $\mathbb{F}_p$ instead of $\mathbb{F}_{p^2}$, while the required points of order 2 are not defined over $\mathbb{F}_p$ in the CSIDH setting.

Their second approach uses the fact that the curve parameters can be recovered from the knowledge of the x-coordinates of two points on the curve, and their difference. While these points are typically available in SIDH during the key generation phase, this is not the case for CSIDH, where we only want to compute the isogenous curve and evaluate one point.

In [8], Castryck et al. compute the image curve by defining $c_j \in \mathbb{F}_p$ such that

$$\prod_{i=1}^{\ell-1}(Z_i w + X_i) = \sum_{j=0}^{\ell-1} c_j w^j$$

as polynomials in w. Then they observe that

$$(A' : C') = (\hat{\pi}(a - 3\hat{\sigma}) : 1) = (a c_0 c_{\ell-1} - 3(c_0 c_{\ell-2} - c_1 c_{\ell-1}) : c_{\ell-1}^2),$$

following the formulae and notation from Renes [21], where

$$\hat{\pi} = \prod_{i=0}^{\ell-1} x_{[i]K}, \quad \text{and} \quad \hat{\sigma} = \sum_{i=0}^{\ell-1} \left(x_{[i]K} - \frac{1}{x_{[i]K}} \right).$$

In their implementation, this is computed iteratively, going through the $(X_i : Z_i)$ for $i = 2, ..., d$, updating the required values at a cost of $6\mathbf{M} + 2\mathbf{a}$ per step. The final computations after that take further $8\mathbf{M} + 3\mathbf{S} + 6\mathbf{a}$ to compute the curve parameters $(A' : C')$.

Using Twisted Edwards Curves for the Image Curve Computation. Our idea to speed up this computation exploits the known correspondence between Montgomery and twisted Edwards curves. Given a Montgomery curve $E_{A,B} : Bv^2 = u^3 + Au^2 + u$, we can switch to a birationally equivalent twisted Edwards curve $E_{a,d} : ax^2 + y^2 = 1 + dx^2y^2$, where

$$A = \frac{2(a+d)}{a-d} \quad \text{and} \quad B = \frac{4}{a-d},$$

by the coordinate map

$$(u, v) \mapsto (x, y) = \left(\frac{u}{v}, \frac{u-1}{u+1} \right).$$

and back by its inverse

$$(x, y) \mapsto (u, v) = \left(\frac{1+y}{1-y}, \frac{1+y}{(1-y)x} \right)$$

In [18] it is shown how to switch to and from twisted Edwards curves in the SIDH setting, which also applies to CSIDH, where Montgomery XZ-only coordinates and projective curve parameters $(A : C)$ are used, ignoring the Montgomery parameter b. Following this and [7], a Montgomery point $(X^M : Z^M)$ can be transformed to the corresponding Edwards YZ-coordinates $(Y^E : Z^E)$ by the map

$$(X^M : Z^M) \mapsto (Y^E : Z^E) = (X^M - Z^M : X^M + Z^M),$$

and the Montgomery parameters $(A : C)$ to the corresponding twisted Edwards parameters (a_E, d_E) by

$$a_E = A + 2C \quad \text{and} \quad d_E = A - 2C.$$

As this allows us to switch efficiently between Montgomery and twisted Edwards curves in CSIDH at a cost of $3\mathbf{a}$ for the curve parameters and $2\mathbf{a}$ for point coordinates, we may as well use isogeny formulae for twisted Edwards curves. Therefore, we state the formulae given by Moody and Shumow in [20].

Let K be a point of order $\ell = 2d+1$ on a twisted Edwards curve $E : a_E x^2 + y^2 = 1 + d_E x^2 y^2$. Then we can compute the coordinate map of the unique (up to compositions by isomorphisms) ℓ-isogeny $\varphi : E \rightarrow E'$ with $\ker(\varphi) = \langle K \rangle$ by

$$\varphi(P) = \left(\prod_{Q \in \langle K \rangle} \frac{x_{P+Q}}{y_Q}, \prod_{Q \in \langle K \rangle} \frac{y_{P+Q}}{y_Q} \right).$$

The curve E' is defined by the parameters $a'_E = a^\ell_E$ and $d'_E = \pi^8_y d^\ell_E$, where $\pi_y = \prod_{i=1}^d y_{[i]K}$.

Since the coordinate map is not as simple to compute as for Montgomery curves, we are only interested in the computation of the image curve parameters. Writing $(Y^E_i : Z^E_i)$ for the projective coordinates of $[i]K$ for $i = 1, ..., d$, we can transform the formulae from above to the projective case by

$$a'_E = a^\ell_E \cdot \pi^8_Z, \quad \text{and} \quad d'_E = d^\ell_E \cdot \pi^8_Y,$$

where $\pi_Y = \prod_{i=1}^d Y^E_i$, and $\pi_Z = \prod_{i=1}^d Z^E_i$.

We can therefore use these formulae to compute the curve parameters of the image curves in CSIDH, by switching to twisted Edwards coordinates and points, and switch back after the computations by $(A' : C') = (2(a'_E + d'_E) : a'_E - d'_E)$, again at a cost of 3a.

Note that the parameters a'_E and d'_E can be computed efficiently: While going through the $(X_i : Z_i)$ on the Montgomery curve for $i = 1, ..., d$, we can compute the corresponding Edwards coordinates $(Y^E_i : Z^E_i)$ at a cost of 2a. However, the required sums and differences already occur at the point evaluation part, and hence do not add any computational cost at all. We can then compute π_Y and π_Z iteratively by 1M each per step. Compared to the algorithm of [8], this saves $4M + 2a$ per step. Furthermore, we have to compute π^8_Y and π^8_Z by three squarings each, and a^ℓ_E and d^ℓ_E, which can be done efficiently, e.g. by a square-and-multiply approach. We further note that the latter computation does not require any values generated during the loop through the $(X_i : Z_i)$. This means that especially hardware architectures that allow for parallel computations would benefit from this, since the computation of a^ℓ_E and d^ℓ_E can be done in parallel to the loop through the $(X_i : Z_i)$.

Figure 2 compares the costs of a combined image curve computation and point evaluation for different prime degrees, where the red line arises from using the Montgomery isogenies from [8], including the optimizations from [10], and the blue line from using our approach utilizing twisted Edwards curves to compute the isogenous curve. The cost is measured in field multiplications, assuming that $\mathbf{S} = 0.8\mathbf{M}$ and $20\mathbf{a} = \mathbf{M}$. In this case, we computationally derive a reduction of the costs by approximately 25% for the largest primes in the current parameter set. We note that different choices for the field operation ratios don't make a big difference, since the main difference between the approaches lies in the number of multiplications. To obtain the cost of the computations of a^ℓ_E and d^ℓ_E, we used a square-and-multiply approach. Since the exponents ℓ_i are small fixed numbers, it is also possible to precompute the optimal addition chains, and therefore save some computational effort compared to square-and-multiply. However, we found that even for the biggest ℓ_i from the current parameter set, this saves at most four multiplications. Hence, the benefit of this is rather small compared to the increased length of the code and the more complicated implementation.

Note that for $\ell \leq 5$, our approach is slightly more expensive than the Montgomery approach. Therefore, in this case, the Montgomery approach can be used. However, the benefit of this is rather small compared to the total computational

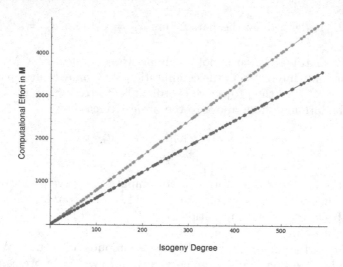

Fig. 2. Cost of different prime-degree isogeny computations. The red line uses Montgomery point evaluations from [10] and image curve computations from [8], while the blue line uses the same Montgomery point evaluations and twisted Edwards image curve computations.

effort, so it might be better to stick with our approach for all ℓ_i in order to keep the implementation simple.

It is further noted in [8], that for a fixed prime ℓ one could reduce the computational effort by finding an appropriate representative of the isogeny modulo (a factor of) the ℓ-division polynomial ψ_ℓ, as done in [11] for 3- and 4-isogenies. However, every required isogeny degree would have to be implemented separately, resulting in a much longer code.

5   Implementation Results

As a proof of concept and for measuring the efficiency of our work, we took the mentioned implementation of Castryck et al. accompanying [8] as reference, and added our optimizations. The implementation is written in C and uses $\mathbb{F}_p$-arithmetic in assembly. The parameters in use are the ones described in Sect. 2.2. The validation of keys is not included in the following discussion.

The first optimization is the precomputation of the curve parameters $(A + 2C : 4C)$ each time before entering the Montgomery ladder, as also done in SIDH [11]. This only saves a few additions per ladder step, but in total leads to a reduction of the computational effort by approximately 2%.

The other optimizations are as described above: One comes from rearranging the factors in the class group action evaluation algorithm, and the other one from more efficient isogeny computations by using the point evaluation from [10] and our twisted Edwards approach for the image curve computations.

Table 1. Performance comparison of the class group action evaluation in CSIDH with different optimizations applied. All timings are given in 10^6 clock cycles and were measured on an Intel Core i7-6500 Skylake processor running Ubuntu 16.04 LTS, averaged over 10 000 runs.

	Clock cycles $\times 10^6$	Acceleration factor
Castryck et al. [8][6]	138.6	-
Precomputation of $(A + 2C : 4C)$	135.7	1.021
Rearranging factors	126.5	1.096
Isogeny optimization	118.2	1.173
Combination of all optimizations	103.9	1.334

Table 1 lists the influence of the different optimizations on the overall performance. In the respective implementations, only the mentioned optimization was used, leaving the rest as in the reference implementation from [8]. For the last line, we combined all the described optimizations and therefore reduced the total computational effort by 25%, yielding a speed-up factor of 1.33. The latter implementation is available at https://zenon.cs.hs-rm.de/pqcrypto/faster-csidh.

6   On Constant-Time Implementations

As mentioned in the sections before, the discussed implementations do not include any protection from side-channel attacks. In particular, the running time depends on the private key, which corresponds to the number of isogenies, that have to be computed. Therefore, the first step to prevent simple timing attacks is a constant-time implementation.

One possibility to reach this is to fix the number of isogenies to the maximum for each degree, and use only as many of them as specified in the private key. However, in addition to lots of useless computational effort, this means that after each dummy isogeny, another multiplication by its degree ℓ_i is necessary for the point P, since the algorithm uses the fact, that by pushing P through the ℓ_i-isogeny, the order of the resulting point will not include the factor ℓ_i. The additionally required multiplications would then allow for a new timing attack, since many extra multiplications mean that many dummy isogenies were computed. This could be prevented by using constant-time ladders for the preparation of each kernel point. This however is undesirable, since it would further blow up the running time.

A possible tool for the design of a more optimized constant-time implementation could be specially tailored dummy isogenies, that, instead of computing an ℓ_i-isogeny and pushing the point P through, simply compute $[\ell_i]P$ and leaves the curve parameters unchanged. This is especially easy, since the ℓ_i-isogeny algorithm requires to compute all $[j]K$ for $j \in \{2, ..., (\ell_i - 1)/2\}$. Therefore, by replacing K with P and by two further differential additions, we can compute $[\ell_i]P$, and hence don't have to perform more multiplications to compensate for the dummy isogenies. Furthermore, the dummy isogenies can be designed to have the exact same number of field operations as the real isogeny computations.

Running Time. First, we note that this discussion refers to the case, that an implementation that follows algorithm 1 shall be transformed to a constant running-time. If the structure of algorithm 1 is changed, the results may vary accordingly.

It is obvious that the running time of a constant-time implementation must be at least as high as the highest possible running time of the fastest non-constant implementation. At first glance, this seems to be twice the average running time of the non-constant implementation, when we fix all elements of the private key to have the maximum absolute value. For a closer investigation, we consider the parameter set from above, and the private keys $e = (5, -5, 5, -5, ..., 5, -5)$ and $e' = (5, 5, ..., 5)$.

When comparing the performance for the private key e to the average case, our experiments suggest that indeed the running time roughly doubles. In fact, the computational cost for isogenies doubles, while the factor for point multiplications is slightly higher. One reason for this is that if we have to compute more isogenies, on average there will be more cases in which our randomly chosen points cannot be used to compute isogenies of certain degrees. Therefore, more points have to be chosen and their order checked by multiplications.

Now we want to compare the performance for the keys e and e'. For the sake of simplicity, assume that we can choose full order points, that allow for the computation of isogenies of all required degrees. Consider that we first want to compute one isogeny of each degree. For e', the bitlengths of the factors for the point multiplications are the ones shown in Fig. 1. After one loop, we have computed one isogeny of each required degree. For the key e, we have to perform one loop each for positive and negative key elements. When doing this and counting the bitlengths of the factors for the point multiplications, after computing one isogeny of each degree, we end up with only 0.54 times the sum of the bitlengths for e', already including the additional required multiplications from line 7 of algorithm 1. Therefore, since we simply perform five such rounds for e and e', the total computational effort for point multiplications for e' is 1.86 times as high as for e. When considering also isogeny computations, our experiments suggest that the total running time for the key e' is 1.49 times the running time for e. Therefore, we conclude that the running time of a constant-time implementation must be at least 2.98 times the running time of average-case measurements of non-constant implementations such as in Sect. 5. In practice, our experiments again suggest a higher factor for the running time for the key e', namely 3.07, for the same reasons as explained above[7].

However, we note that more careful analysis is required for an optimized constant-time implementation, and our proposal of dummy isogenies is merely

---

[7] We measured 318.9×10^6 clock cycles for the running time for the key e' in the setting from Table 1, using the optimized implementation.

a tool, that could possibly be used to design such an implementation, which we leave for future work.

7   Conclusion and Future Work

Although we gained a speed-up factor of 1.33 in our CSIDH implementation, it is still considerably slower than e.g. SIDH. Therefore, further research in that direction is necessary, to make the practical deployment of the scheme more attractive. In particular, side-channel protection, such as constant-time implementations, is required for that aim.

As mentioned before, also the security of CSIDH still requires some more detailed analysis on the implication of new attacks for specific parameter choices.

Acknowledgements. This work was partially supported by Elektrobit Automotive, Erlangen, Germany. We thank Fabio Campos, Marc Stöttinger, and the anonymous reviewers for their helpful and valuable comments.

References

1. Azarderakhsh, R., et al.: Supersingular isogeny key encapsulation, Round 1 submission, NIST Post-Quantum Cryptography Standardization (2017). https://sike.org/files/SIDH-spec.pdf
2. Bernstein, D.J., Birkner, P., Joye, M., Lange, T., Peters, C.: Twisted Edwards curves. In: Vaudenay, S. (ed.) AFRICACRYPT 2008. LNCS, vol. 5023, pp. 389–405. Springer, Heidelberg (2008). https://doi.org/10.1007/978-3-540-68164-9_26
3. Bernstein, D.J., Birkner, P., Lange, T., Peters, C.: ECM using Edwards curves. Math. Comput. **82**(282), 1139–1179 (2013)
4. Biasse, J.F., Jacobson Jr., M.J., Iezzi, A.: A note on the security of CSIDH. arXiv preprint arXiv:1806.03656 (2018)
5. Bonnetain, X., Schrottenloher, A.: Quantum security analysis of CSIDH and ordinary isogeny-based schemes. Cryptology ePrint Archive, Report 2018/537 (2018). https://eprint.iacr.org/2018/537
6. Bos, J.W., Friedberger, S.: Arithmetic considerations for isogeny based cryptography. Cryptology ePrint Archive, Report 2018/376 (2018). https://eprint.iacr.org/2018/376
7. Castryck, W., Galbraith, S., Farashahi, R.R.: Efficient arithmetic on elliptic curves using a mixed Edwards-Montgomery representation. Cryptology ePrint Archive, Report 2008/218 (2008), http://eprint.iacr.org/2008/218
8. Castryck, W., Lange, T., Martindale, C., Panny, L., Renes, J.: CSIDH: an efficient post-quantum commutative group action (2018). https://eprint.iacr.org/2018/383.pdf
9. Childs, A., Jao, D., Soukharev, V.: Constructing elliptic curve isogenies in quantum subexponential time. J. Math. Cryptology **8**(1), 1–29 (2014)
10. Costello, C., Hisil, H.: A simple and compact algorithm for SIDH with arbitrary degree isogenies. In: Takagi, T., Peyrin, T. (eds.) ASIACRYPT 2017. LNCS, vol. 10625, pp. 303–329. Springer, Cham (2017). https://doi.org/10.1007/978-3-319-70697-9_11

11. Costello, C., Longa, P., Naehrig, M.: Efficient algorithms for supersingular isogeny Diffie-Hellman. In: Robshaw, M., Katz, J. (eds.) CRYPTO 2016. LNCS, vol. 9814, pp. 572–601. Springer, Heidelberg (2016). https://doi.org/10.1007/978-3-662-53018-4_21

12. Couveignes, J.M.: Hard homogeneous spaces. Cryptology ePrint Archive, Report 2006/291 (2006). https://eprint.iacr.org/2006/291

13. De Feo, L.: Mathematics of isogeny based cryptography. Notes from a summer school on Mathematics for Post-quantum cryptography (2017). http://defeo.lu/ema2017/poly.pdf

14. Edwards, H.M.: A normal form for elliptic curves. Bull. Am. Math. Soc. **44**, 393–422 (2007)

15. Feo, L.D., Kieffer, J., Smith, B.: Towards practical key exchange from ordinary isogeny graphs. Cryptology ePrint Archive, Report 2018/485 (2018). https://eprint.iacr.org/2018/485

16. Jao, D., De Feo, L., Plût, J.: Towards quantum-resistant cryptosystems from supersingular elliptic curve isogenies. J. Math. Cryptology **8**(3), 209–247 (2014)

17. Marin, L.: Differential elliptic point addition in Twisted Edwards curves. In: 2013 27th International Conference on Advanced Information Networking and Applications Workshops, pp. 1337–1342 (2013)

18. Meyer, M., Reith, S., Campos, F.: On hybrid SIDH schemes using Edwards and Montgomery curve arithmetic. Cryptology ePrint Archive, Report 2017/1213 (2017). https://eprint.iacr.org/2017/1213

19. Montgomery, P.L.: Speeding the Pollard and Elliptic Curve methods of factorization. Math. Comput. **48**(177), 243–264 (1987)

20. Moody, D., Shumow, D.: Analogues of Vélu's formulas for isogenies on alternate models of elliptic curves. Math. Comput. **85**(300), 1929–1951 (2016)

21. Renes, J.: Computing Isogenies between montgomery curves using the action of (0, 0). In: Lange, T., Steinwandt, R. (eds.) PQCrypto 2018. LNCS, vol. 10786, pp. 229–247. Springer, Cham (2018). https://doi.org/10.1007/978-3-319-79063-3_11

22. Rostovtsev, A., Stolbunov, A.: Public-key cryptosystem based on isogenies. Cryptology ePrint Archive, Report 2006/145 (2006). http://eprint.iacr.org/2006/145

23. The National Institute of Standards and Technology (NIST): Submission requirements and evaluation criteria for the post-quantum cryptography standardization process (2016)

24. Vélu, J.: Isogénies entre courbes elliptiques. C.R. Acad. Sci. Paris, Série A **271**, 238–241 (1971)

A Note on the Security of CSIDH

Jean-François Biasse[1]([✉]), Annamaria Iezzi[1], and Michael J. Jacobson Jr.[2]

[1] Department of Mathematics and Statistics,
University of South Florida, Tampa, USA
{biasse,aiezzi}@usf.edu
[2] Department of Computer Science, University of Calgary, Calgary, Canada
jacobs@ucalgary.ca

Abstract. We propose a quantum algorithm for computing an isogeny between two elliptic curves E_1, E_2 defined over a finite field such that there is an imaginary quadratic order $\mathcal{O}$ satisfying $\mathcal{O} \simeq \mathrm{End}(E_i)$ for $i = 1, 2$. This concerns ordinary curves and supersingular curves defined over $\mathbb{F}_p$ (the latter used in the recent CSIDH proposal). Our algorithm has heuristic asymptotic run time $e^{O\left(\sqrt{\log(|\Delta|)}\right)}$ and requires polynomial quantum memory and $e^{O\left(\sqrt{\log(|\Delta|)}\right)}$ quantumly accessible classical memory, where Δ is the discriminant of $\mathcal{O}$. This asymptotic complexity outperforms all other available methods for computing isogenies.

We also show that a variant of our method has asymptotic run time $e^{\tilde{O}\left(\sqrt{\log(|\Delta|)}\right)}$ while requesting only polynomial memory (both quantum and classical).

1 Introduction

Given two elliptic curves E_1, E_2 defined over a finite field $\mathbb{F}_q$, the isogeny problem consists in computing an isogeny $\phi : E_1 \to E_2$, i.e. a non-constant morphism that maps the identity point on E_1 to the identity point on E_2. There are two different types of elliptic curves: ordinary and supersingular. The latter have very particular properties that impact the resolution of the isogeny problem. The first instance of a cryptosystem based on the hardness of computing isogenies was due to Couveignes [13], and its concept was independently rediscovered by Stolbunov [34]. Both proposals used ordinary curves.

Childs, Jao and Soukharev observed in [11] that the problem of finding an isogeny between two ordinary curves E_1 and E_2 defined over $\mathbb{F}_q$ and having the same endomorphism ring could be reduced to the problem of solving the Hidden Subgroup Problem (HSP) for a generalized dihedral group. More specifically, let $K = \mathbb{Q}(\sqrt{t^2 - 4q})$ where t is the trace of the Frobenius endomorphism of

---

Author list in alphabetical order; see https://www.ams.org/profession/leaders/culture/CultureStatement04.pdf. This work was supported by the U.S. National Science Foundation under grant 1839805, by NIST under grant 60NANB17D184, and by the Simons Foundation under grant 430128.

© Springer Nature Switzerland AG 2018
D. Chakraborty and T. Iwata (Eds.): INDOCRYPT 2018, LNCS 11356, pp. 153–168, 2018.
https://doi.org/10.1007/978-3-030-05378-9_9

the curves, and let $\mathcal{O} \subseteq K$ be the quadratic order isomorphic to the ring of endomorphisms of E_1 and E_2. Let $\mathrm{Cl}(\mathcal{O})$ be the ideal class group of $\mathcal{O}$. Classes of ideals act on isomorphism classes of curves with endomorphism ring isomorphic to $\mathcal{O}$. The problem of finding an isogeny between E_1 and E_2 can be then reduced to the problem of finding a "nicely" represented ideal $\mathfrak{a} \subseteq \mathcal{O}$ such that $[\mathfrak{a}] * \overline{E}_1 = \overline{E}_2$ where $*$ is the action of $\mathrm{Cl}(\mathcal{O})$, $[\mathfrak{a}]$ is the class of $\mathfrak{a}$ in $\mathrm{Cl}(\mathcal{O})$ and $\overline{E}_i$ is the isomorphism class of the curve E_i. Childs, Jao and Soukharev showed that this could be done by solving the HSP for $\mathbb{Z}_2 \ltimes \mathrm{Cl}(\mathcal{O})$. Let $N := |\mathrm{Cl}(\mathcal{O})| \sim \sqrt{|t^2 - 4q|}$. Using Kuperberg's sieve [27], this task requires $2^{O\left(\sqrt{\log(N)}\right)}$ queries to an oracle that computes the action of the class of an element in $\mathrm{Cl}(\mathcal{O})$. Childs et al. used a method with complexity in $2^{\tilde{O}\left(\sqrt{\log(N)}\right)}$ to evaluate this oracle, meaning that the total cost is $2^{\tilde{O}\left(\sqrt{\log(N)}\right)}$.

To avoid this subexponential attack, Jao and De Feo [23] described an analogue of these isogeny-based systems that works with supersingular curves. The endomorphism ring of such curves is a maximal order in a quaternion algebra. The non-commutativity of the (left)-ideals corresponding to isogenies between isomorphism classes of curves thwarts the attack mentioned above, but it also restricts the possibilities offered by supersingular isogenies, which are typically used for a Diffie-Hellman type of key exchange (known as SIDH) and for digital signatures. Most recently, two works revisited isogeny-based cryptosystems by restricting themselves to cases where the subexponential attacks based on the action of $\mathrm{Cl}(\mathcal{O})$ was applicable. The scheme known as CSIDH by Castryck et al. [10] uses supersingular curves and isogenies defined over $\mathbb{F}_p$, while the scheme of Feo, Kieffer and Smith [15] uses ordinary curves with many practical optimizations. In both cases, the appeal of using commutative structures is to allow more functionalities, such as static-static key exchange protocols that are not possible with SIDH without an expensive Fujisaki-Okamoto transform [2].

Contributions. Let E_1, E_2 be two elliptic curves defined over a finite field such that there is an imaginary quadratic order $\mathcal{O}$ satisfying $\mathcal{O} \simeq \mathrm{End}(E_i)$ for $i = 1, 2$. Let $\Delta = \mathrm{disc}(\mathcal{O})$. In this note, we provide new insight into the security of CSIDH as follows:

1. We describe a quantum algorithm for computing an isogeny between E_1 and E_2 with heuristic asymptotic run time in $e^{O\left(\sqrt{\log(|\Delta|)}\right)}$ and with quantum memory in $\mathrm{Poly}\left(\log(|\Delta|)\right)$ and quantumly accessible classical memory in $e^{O\left(\sqrt{\log(|\Delta|)}\right)}$.

2. We show that we can use a variant of this method to compute an isogeny between E_1 and E_2 in time $e^{\left(\frac{1}{\sqrt{2}} + o(1)\right)\sqrt{\ln(|\Delta|)\ln\ln(|\Delta|)}}$ with polynomial memory (both classical and quantum).

Our contributions bear similarities to the recent independent work of Bonnetain and Schrottenloher [7]. The main differences are that they rely on a generating set $\mathfrak{l}_1, \ldots, \mathfrak{l}_u$ of the class group, where $u \in \Theta(\log(|\Delta|))$, provided with the CSIDH protocol, and that they primarily focused on practical improvements and concrete security levels. Their method inherits Kuperberg's asymptotic complexity which is in $e^{\tilde{O}\left(\sqrt{\log(|\Delta|)}\right)}$. Section 4.2 elaborates on the differences between our algorithm and that of [7]. The run time of the variant described in Contribution 2 is asymptotically comparable to that of the algorithm of Childs, Jao and Soukharev [11], and to that of Bonnetain and Schrottenloher [7] (if its exact time complexity was to be worked out). The main appeal of our variant is the fact that it uses a polynomial amount of memory, which is likely to impact the performances in practice.

Our work is also connected to a recent and independent contribution of Jao, LeGrow, Leonardi and Ruiz-Lopez. The main claim of their work (slides are available online [24]) is an algorithm with heuristic time complexity in $e^{\tilde{O}\left(\sqrt{\log(|\Delta|)}\right)}$ that uses quantum polynomial memory and classical memory in $e^{O\left(\sqrt{\log(|\Delta|)}\right)}$. Compared to our Contribution 2, the difference in terms of performances is that our method requires only polynomial classical memory. The main technical difference between our work and that of Jao et al. is an alternative approach to lattice reduction. Both works rely on unproven heuristics: ours pertains to the connectivity of the Caley Graph of the ideal class group of the ring of endomorphisms, while Jao et al. make the assumption that this class group is cyclic, leaving the question of non-cyclic class groups open. Note that by design, the class group is very likely to be cyclic in instances of this problem pertaining to the cryptanalysis of CSIDH. Therefore, the generalization of their method would mostly be of fundamental interest.

2   Mathematical Background

An elliptic curve E defined over a finite field $\mathbb{F}_q$ of characteristic $p \neq 2, 3$ is a projective algebraic curve with an affine plane model given by an equation of the form $y^2 = x^3 + ax + b$, where a, $b \in \mathbb{F}_q$ and $4a^3 + 27b^2 \neq 0$. The set of points of an elliptic curve is equipped with an additive group law. Details about the arithmetic of elliptic curves can be found in many references, such as [33, Chap. 3].

Let E_1, E_2 be two elliptic curves defined over $\mathbb{F}_q$. An isogeny $\phi \colon E_1 \to E_2$ over $\mathbb{F}_q$ (resp. over $\overline{\mathbb{F}}_q$) is a non-constant rational map defined over $\mathbb{F}_q$ (resp. over $\overline{\mathbb{F}}_q$) which sends the identity point on E_1 to the identity point on E_2. The degree of an isogeny is its degree as a rational map, and an isogeny of degree ℓ is called an ℓ-isogeny. Two curves are isogenous over $\mathbb{F}_q$ if and only if they have the same number of points over $\mathbb{F}_q$ (see [36]). Moreover, E_1, E_2 are said to be isomorphic over $\mathbb{F}_q$, or $\mathbb{F}_q$-isomorphic, if there exist isogenies $\phi_1 \colon E_1 \to E_2$ and $\phi_2 \colon E_2 \to E_1$ over $\mathbb{F}_q$ whose composition is the identity. Two $\overline{\mathbb{F}}_q$-isomorphic elliptic curves have the same j-invariant given by $j := 1728\frac{4a^3}{4a^3 + 27b^2}$.

An order $\mathcal{O}$ in a number field K such that $[K : \mathbb{Q}] = n$ is a subring of K which is a $\mathbb{Z}$-module of rank n. The notion of ideal of $\mathcal{O}$ can be generalized to fractional ideals, which are sets of the form $\mathfrak{a} = \frac{1}{d}I$ where I is an ideal of $\mathcal{O}$ and

$d \in \mathbb{Z}_{>0}$. A fractional ideal I is said to be invertible if there exists a fractional ideal J such that $IJ = \mathcal{O}$. The invertible fractional ideals form a multiplicative group $\mathcal{I}$, having a subgroup consisting of the invertible principal ideals $\mathcal{P}$. The ideal class group $\mathrm{Cl}(\mathcal{O})$ is by definition $\mathrm{Cl}(\mathcal{O}) := \mathcal{I}/\mathcal{P}$. In $\mathrm{Cl}(\mathcal{O})$, we identify two fractional ideals $\mathfrak{a}, \mathfrak{b}$ if there is $\alpha \in K^*$ such that $\mathfrak{b} = (\alpha)\mathfrak{a}$, where $(\alpha) := \alpha\mathcal{O}$. We denote by $[\mathfrak{a}]$ the class of the fractional ideal $\mathfrak{a}$ in $\mathrm{Cl}(\mathcal{O})$. The ideal class group is finite and its cardinality is called the class number $h_{\mathcal{O}}$ of $\mathcal{O}$. For a quadratic order $\mathcal{O}$, the class number satisfies $h_{\mathcal{O}} \leq \sqrt{|\Delta|}\ln(|\Delta|)$ (see [12, Sect. 5.10.1]), where Δ is the discriminant of $\mathcal{O}$.

Let E be an elliptic curve defined over $\mathbb{F}_q$. An endomorphism of E is either an isogeny defined over $\overline{\mathbb{F}}_q$ between E and itself, or the zero morphism. The set of endomorphisms of E forms a ring that is denoted by $\mathrm{End}(E)$. For each integer m, the multiplication-by-m map $[m]$ on E is an endomorphism. Therefore, we always have $\mathbb{Z} \subseteq \mathrm{End}(E)$. Moreover, to each isogeny $\phi \colon E_1 \to E_2$ corresponds an isogeny $\hat{\phi} \colon E_2 \to E_1$ called its dual isogeny. It satisfies $\phi \circ \hat{\phi} = [m]$ where $m = \deg(\phi)$. For elliptic curves defined over a finite field, we know that $\mathbb{Z} \subsetneq \mathrm{End}(E)$. In this particular case, $\mathrm{End}(E)$ is either an order in an imaginary quadratic field (and has $\mathbb{Z}$-rank 2) or a maximal order in a quaternion algebra ramified at p (the characteristic of the base field) and ∞ (and has $\mathbb{Z}$-rank 4). In the former case, E is said to be ordinary while in the latter it is called supersingular. When a supersingular curve is defined over $\mathbb{F}_p$, then the ring of its $\mathbb{F}_p$-endomorphisms, denoted by $\mathrm{End}_{\mathbb{F}_p}(E)$, is isomorphic to an imaginary quadratic order, much like in the ordinary case.

The endomorphism ring of an elliptic curve plays a crucial role in most algorithms for computing isogenies between curves. Indeed, if E is ordinary (resp. supersingular over $\mathbb{F}_p$), the class group of $\mathrm{End}(E)$ (resp. $\mathrm{End}_{\mathbb{F}_p}(E)$) acts transitively on isomorphism classes of elliptic curves having the same endomorphism ring. More precisely, the class of an ideal $\mathfrak{a} \subseteq \mathcal{O}$ acts on the isomorphism class of a curve E with $\mathrm{End}(E) \simeq \mathcal{O}$ via an isogeny of degree $\mathcal{N}(\mathfrak{a})$ (the algebraic norm of $\mathfrak{a}$). Likewise, each isogeny $\varphi \colon E \to E'$ where $\mathrm{End}(E) \simeq \mathrm{End}(E') \simeq \mathcal{O}$ corresponds (up to isomorphism) to the class of an ideal in $\mathcal{O}$. From an ideal $\mathfrak{a}$ and the ℓ-torsion (where $\ell = \mathcal{N}(\mathfrak{a})$), one can recover the kernel of φ, and then using Vélu's formulae [37], one can derive the corresponding isogeny. We denote by $[\mathfrak{a}] * \overline{E}$ the action of the ideal class of $\mathfrak{a}$ on the isomorphism class of the curve E. The typical strategy to evaluate the action of $[\mathfrak{a}]$ is to decompose it as a product of classes of prime ideals of small norm ℓ, and evaluate the action of each prime ideal as an ℓ-isogeny. This strategy was described by Couveignes [13], Galbraith-Hess-Smart [16], and later by Bröker-Charles-Lauter [9] and reused in many subsequent works.

Notation: In this paper, log denotes the base 2 logarithm while ln denotes the natural logarithm.

3   The CSIDH Non-interactive Key Exchange

As pointed out in [17], the original SIDH key agreement protocol is not secure when using the same secret key over multiple instances of the protocol. This can be fixed by a Fujisaki–Okamoto transform [2] at the cost of a drastic loss of performance, requiring additional points in the protocol. These issues motivated the description of CSIDH [10] which uses supersingular curves defined over $\mathbb{F}_p$.

When Alice and Bob wish to create a shared secret, they rely on their secret keys $[\mathfrak{a}]$ and $[\mathfrak{b}]$ which are classes of ideals in the ideal class group of $\mathcal{O}$, where $\mathcal{O}$ is isomorphic to the $\mathbb{F}_p$-endomorphism ring of a supersingular curve E defined over $\mathbb{F}_p$. This key exchange procedure resembles the original Diffie–Hellman protocol [14]. Alice and Bob proceed as follow:

– Alice sends $[\mathfrak{a}] * \overline{E}$ to Bob.
– Bob sends $[\mathfrak{b}] * \overline{E}$ to Alice.

Then Alice and Bob can separately recover their shared secret

$$[\mathfrak{ab}] * \overline{E} = [\mathfrak{b}] * [\mathfrak{a}] * \overline{E} = [\mathfrak{a}] * [\mathfrak{b}] * \overline{E}.$$

The existence of a quantum subexponential attack forces the users to update the size of keys at a faster pace (or by larger increments) than in the regular SIDH protocol against which we only know quantum exponential attacks. This is partly compensated by the fact that elements are represented in $\mathbb{F}_p$, and are thus more compact than elements of $\mathbb{F}_{p^2}$ needed in SIDH (because the corresponding curves are defined over $\mathbb{F}_{p^2}$). Recommended parameter sizes and attack costs from [10] for 80, 128, and 256 bit security are listed in Table 1. In Table 1, the cost is in number of operations. These values do not account for the memory costs (the security estimates are therefore more conservative than if memory costs were accounted for). The NIST security levels are defined in the call for proposals for the Post Quantum Cryptography project [29]. Note that subsequent works such as that of Bonnetain and Schrottenloher [7] have suggested different values.

Table 1. Claimed security of CSIDH [10, Table 1].

NIST	$\log(p)$	Cost quantum attack	Cost classical attack
1	512	2^{62}	2^{128}
3	1024	2^{94}	2^{256}
5	1792	2^{129}	2^{448}

4   Asymptotic Complexity of Isogeny Computation

In this section, we show how to combine the general framework for computing isogenies between curves whose endomorphism ring is isomorphic to a quadratic order (due to Childs, Jao and Soukharev [11] in the ordinary case and to Biasse,

Jao and Sankar in the supersingular case [5]) with the efficient algorithm of Biasse, Fieker and Jacobson [4] for evaluating the class group action to produce a quantum algorithm that finds an isogeny between E_1 and E_2. We give two variants of our method:

- Heuristic time complexity $2^{O(\log(|\Delta|))}$, polynomial quantum memory and quantumly accessible classical memory in $2^{O(\log(|\Delta|))}$.
- Heuristic time complexity $e^{\left(\frac{1}{\sqrt{2}}+o(1)\right)\sqrt{\ln(|\Delta|)\ln\ln(|\Delta|)}}$ with polynomial memory (both classical and quantum).

4.1 Isogenies from Solutions to the Hidden Subgroup Problem

As shown in [5,11], the computation of an isogeny between E_1 and E_2 such that there is an imaginary quadratic order with $\mathcal{O} \simeq \text{End}(E_i)$ for $i = 1, 2$ can be done by exploiting the action of the ideal class group of $\mathcal{O}$ on isomorphism classes of curves with endomorphism ring isomorphic to $\mathcal{O}$. In particular, this concerns the cases of

- ordinary curves, and
- supersingular curves defined over $\mathbb{F}_p$.

Assume we are looking for $\mathfrak{a}$ such that $[\mathfrak{a}] * \overline{E}_1 = \overline{E}_2$. Let $A = \mathbb{Z}/d_1\mathbb{Z} \times \cdots \times \mathbb{Z}/d_k\mathbb{Z} \simeq \text{Cl}(\mathcal{O})$ be the elementary decomposition of $\text{Cl}(\mathcal{O})$. Then we define a quantum oracle $f : \mathbb{Z}/2\mathbb{Z} \ltimes A \to \{\text{quantum states}\}$ by

$$f(x, \boldsymbol{y}) := \begin{cases} |[\mathfrak{a}_{\boldsymbol{y}}] * \overline{E}_1\rangle & \text{if } x = 0, \\ |[\mathfrak{a}_{-\boldsymbol{y}}] * \overline{E}_2\rangle & \text{if } x = 1, \end{cases} \tag{1}$$

where $[\mathfrak{a}_{\boldsymbol{y}}]$ is the element of $\text{Cl}(\mathcal{O})$ corresponding to $\boldsymbol{y} \in A$ via the isomorphism $\text{Cl}(\mathcal{O}) \simeq A$. Let H be the subgroup of $\mathbb{Z}/2\mathbb{Z} \ltimes A$ of the periods of f. This means that $f(x, \boldsymbol{y}) = f(x', \boldsymbol{y}')$ if and only if $(x, \boldsymbol{y}) - (x', \boldsymbol{y}') \in H$. Then $H = \{(0,0), (1, \boldsymbol{s})\}$ where $\boldsymbol{s} \in A$ such that $[\mathfrak{a}_{\boldsymbol{s}}] * \overline{E}_1 = \overline{E}_2$. The computation of $\boldsymbol{s}$ can thus be done through the resolution of the Hidden Subgroup Problem in $\mathbb{Z}/2\mathbb{Z} \ltimes A$. In [11, Sect. 5], Childs, Jao and Soukharev generalized the subexponential-time polynomial space dihedral HSP algorithm of Regev [30] to the case of an arbitrary Abelian group A. Its run time is in $e^{(\sqrt{2}+o(1))\sqrt{\ln(|A|)\ln\ln(|A|)}}$ with a polynomial memory requirement. Kuperberg [27] describes a family of algorithms, one of which has running time in $e^{O(\sqrt{\log(|A|)})}$ while requiring polynomial quantum memory and $e^{O(\sqrt{\log(|A|)})}$ quantumly accessible classical memory. The high-level approach for finding an isogeny from the dihedral HSP is sketched in Algorithm 1.

Proposition 1. *Let $N = \# \text{Cl}(\mathcal{O}) \sim \sqrt{|\Delta|}$. Algorithm 1 is correct and requires:*

- *$e^{O(\sqrt{\log(N)})}$ queries to the oracle defined by (1) while requiring a $\text{Poly}(\log(N))$ quantum memory and $e^{O(\sqrt{\log(N)})}$ quantumly accessible classical memory overhead when using Kuperberg's second dihedral HSP algorithm [27] in Step 2.*

Algorithm 1. Quantum algorithm for evaluating the action in $\mathrm{Cl}(\mathcal{O})$

Input: Elliptic curves E_1, E_2, imaginary quadratic order $\mathcal{O}$ such that $\mathrm{End}(E_i) \simeq \mathcal{O}$
   for $i = 1, 2$ such that there is $[\mathfrak{a}] \in \mathrm{Cl}(\mathcal{O})$ satisfying $[\mathfrak{a}] * \overline{E}_1 = \overline{E}_2$.
Output: $[\mathfrak{a}]$
   1: Compute $A = \mathbb{Z}/d_1\mathbb{Z} \times \cdots \times \mathbb{Z}/d_k\mathbb{Z}$ such that $A \simeq \mathrm{Cl}(\mathcal{O})$.
   2: Find $H = \{(0,0), (1,s)\}$ by solving the HSP in $\mathbb{Z}/2\mathbb{Z} \ltimes A$ with oracle (1).
   3: **return** $[\mathfrak{a}_s]$

– $e^{(\sqrt{2}+o(1))\sqrt{\ln(N)\ln\ln(N)}}$ *queries to the oracle defined by* (1) *while requiring only polynomial memory overhead when using the dihedral HSP method of [11, Sect. 5] in Step 2.*

Remark 1. The cost of Algorithm 1 is dominated by Step 2. Indeed, Step 1 can be done by using an algorithm for solving the HSP in a commutative group. Even when the dimension grows to infinity, this step is known to run in polynomial time [6].

Remark 2. Algorithm 1 only returns the ideal class $[\mathfrak{a}]$ whose action on $\overline{E}_1$ gives us $\overline{E}_2$. This is all we are interested in as far as the analysis of isogeny-based cryptosystems goes. However, this is not an isogeny between E_1 and E_2. We can use this ideal to derive an actual isogeny by evaluating the action of $[\mathfrak{a}]$ using the oracle of Sect. 4.2 together with the method of [9, Algorithm 4.1]. This returns an isogeny $\phi : E_1 \rightarrow E_2$ as a composition of isogenies of small degree $\phi = \prod_i \phi_i^{e_i}$ with the same time complexity as Algorithm 1. Also note that the output fits in polynomial space if the product is not evaluated, otherwise, it needs $2^{\tilde{O}\left(\sqrt[3]{\log(N)}\right)}$ memory.

4.2   The Quantum Oracle

To compute the oracle defined in (1), Childs, Jao and Soukharev [11] used a purely classical subexponential method derived from the general subexponential class group computation algorithm of Hafner and McCurley [19]. This approach, mentioned in [10], was first suggested by Couveignes [13]. In a recent independent work [7], Bonnetain and Schrottenloher used a method that bears similarities with our oracle described in this section. They combined a quantum algorithm for computing the class group with classical methods from Biasse, Fieker and Jacobson [4, Algorithm 7] for evaluating the action of $[\mathfrak{a}]$ with a precomputation of $\mathrm{Cl}(\mathcal{O})$. More specifically, let $\mathfrak{l}_1, \ldots, \mathfrak{l}_u$ be prime ideals used to create the secret ideal $\mathfrak{a}$ of Alice. This means that there are (small) $(e_1, \ldots, e_u) \in \mathbb{Z}^u$ such that $\mathfrak{a} = \prod_i \mathfrak{l}_i^{e_i}$. Let $\mathcal{L}$ be the lattice of relations between $\mathfrak{l}_1, \ldots, \mathfrak{l}_u$, i.e. the lattice of all the vectors $(f_1, \ldots, f_u) \in \mathbb{Z}^u$ such that $\prod_i \mathfrak{l}_i^{f_i}$ is principal. In other words, the ideal class $\left[\prod_i \mathfrak{l}_i^{f_i}\right]$ is the neutral element of $\mathrm{Cl}(\mathcal{O})$. The high-level approach used in [7] deriving from [4, Algorithm 7] is the following:

1. Compute a basis B for $\mathcal{L}$.
2. Find a BKZ-reduced basis B' of $\mathcal{L}$.

3. Find $(h_1, \ldots, h_u) \in \mathbb{Z}^u$ such that $[\mathfrak{a}] = \left[\prod_i \mathfrak{l}_i^{h_i} \right]$.

4. Use Babai's nearest plane method on B' to find short $(h_1', \ldots, h_u') \in \mathbb{Z}^u$ such that $[\mathfrak{a}] = \left[\prod_i \mathfrak{l}_i^{h_i'} \right]$.

5. Evaluate the action of $\left[\prod_i \mathfrak{l}_i^{h_i'} \right]$ on $\overline{E}_1$ by applying repeatedly the action of the $\mathfrak{l}_i$ for $i = 1, \ldots, u$.

Steps 1 and 2 can be performed as a precomputation. Step 1 takes quantum polynomial time by using standard techniques for solving an instance of the Abelian Hidden Subgroup Problem in $\mathbb{Z}^u$ where $p = 4l_1 \cdots l_u - 1$ for small primes $l_1, \ldots, l_u$.

The oracle of Childs, Jao and Soukharev [11] has asymptotic time complexity in $2^{\tilde{O}\left(\sqrt{\log(|\Delta|)}\right)}$ and requires subexponential space due to the need for the storage of the ℓ-th modular polynomial $\Phi_\ell(X, Y)$ for ℓ up to $e^{\tilde{O}\left(\sqrt{\log(|\Delta|)}\right)}$. Indeed, the size of $\Phi_\ell(X, Y)$ is proportional to ℓ. The oracle of Bonnetain and Schrottenloher [7] relies on BKZ [31] lattice reduction in a lattice in $\mathbb{Z}^u$. Typically, $u \in \Theta(\log(p)) = \Theta(\log(|\Delta|))$, since $\sum_{q < l} \log(q) \in \Theta(l)$. In addition to not having a proven space complexity bound, the complexity of BKZ cannot be in $e^{\tilde{O}\left(\sqrt{\log(|\Delta|)}\right)}$ unless the block size is at least in $\Theta\left(\sqrt{\log(|\Delta|)}\right)$, which forces the overall complexity to be at best in $e^{\tilde{O}\left(\sqrt{\log(|\Delta|)}\right)}$.

Our strategy differs from that of Bonnetain and Schrottenloher on the following points:

- Our algorithm does not require the basis $\mathfrak{l}_1, \ldots, \mathfrak{l}_u$ provided with CSIDH.
- The complexity of our oracle is in $e^{\tilde{O}\left(\sqrt[3]{\log(|\Delta|)}\right)}$ (instead of $e^{\tilde{O}\left(\sqrt{\log(|\Delta|)}\right)}$ for the method of [7]), thus leading to an overall complexity of $e^{O\left(\sqrt{\log(|\Delta|)}\right)}$ (instead of $e^{\tilde{O}\left(\sqrt{\log(|\Delta|)}\right)}$ for the method of [7]).
- We specify the use of a variant of BKZ with a proven poly-space complexity.

To avoid the dependence on the parameter u, we need to rely on the heuristics stated by Biasse, Fieker and Jacobson [4] on the connectivity of the Caley graph of the ideal class group when a set of edges is $S \subseteq \{\mathfrak{p} : \mathcal{N}(\mathfrak{p}) \in \mathrm{Poly}(\log(|\Delta|))\}$ with $\#S \leq \log(|\Delta|)^{2/3}$ where Δ is the discriminant of $\mathcal{O}$. By assuming [4, Heuristic 2], we state that each class of $\mathrm{Cl}(\mathcal{O})$ has a representation over the class of ideals in S with exponents less than $e^{\log^{1/3}(|\Delta|)}$. A quick calculation shows that there are asymptotically many more such products than ideal classes, but their distribution is not well enough understood to conclude that all classes decompose over S with a small enough exponent vector. Numerical experiments reported in [4, Table 2] showed that decompositions of random ideal classes over the first $\log^{2/3}(|\Delta|)$ split primes always had exponents significantly less than $e^{\log^{1/3}(|\Delta|)}$.

Heuristic 1 (With parameter $c > 1$). *Let $c > 1$ and $\mathcal{O}$ be an imaginary quadratic order of discriminant Δ. Then there are $(\mathfrak{p}_i)_{i \leq k}$ for $k = \log^{2/3}(|\Delta|)$*

Table 2. Maximal exponent occurring in short decompositions (over 1000 random elements of the class group). Table 2 of [4].

$\log_{10}(|\Delta|)$	$\log^{2/3}(|\Delta|)$	Maximal coefficient	$e^{\log^{1/3}(|\Delta|)}$
20	13	6	36
25	15	8	48
30	17	7	61
35	19	9	75
40	20	10	91
45	22	14	110
50	24	13	130

split prime ideals of norm less than $\log^c(|\Delta|)$ whose classes generate $\mathrm{Cl}(\mathcal{O})$. Furthermore, each class of $\mathrm{Cl}(\mathcal{O})$ has a representative of the form $\prod_i \mathfrak{p}_i^{n_i}$ for $|n_i| \leq e^{\log^{1/3}|\Delta|}$.

A default choice for our set S could be the first $\log^{2/3}(|\Delta|)$ split primes of $\mathcal{O}$ (as in Table 2). We can derive our results under the weaker assumption that the $\log^{2/3}(|\Delta|)$ primes generating the ideal class group do not have to be the first consecutive primes. Assume we know that $\mathrm{Cl}(\mathcal{O})$ is generated by at most $\log^{2/3}(|\Delta|)$ distinct classes of the split prime ideals of norm up to $\log^c(|\Delta|)$ for some constant $c > 0$. Our algorithm needs to first identify these prime ideals as they might not be the first consecutive primes. Let $\mathfrak{p}_1, \ldots, \mathfrak{p}_k$ be the prime ideals of norm up to $\log^c(|\Delta|)$. We first compute a basis for the lattice $\mathcal{L}$ of vectors $(e_1, \ldots, e_k)$ such that $\prod_i \mathfrak{p}_i^{e_i}$ is principal (in other words, the ideal class $[\prod_i \mathfrak{p}_i^{e_i}]$ is trivial). Let M be the matrix whose rows are the vectors of a basis of $\mathcal{L}$. There is a polynomial time (and space) algorithm that finds a unimodular matrix U such that

$$UM = H = \begin{bmatrix} h_{1,1} & 0 & \cdots & 0 \\ \vdots & h_{2,2} & \ddots & \vdots \\ \vdots & \vdots & \ddots & 0 \\ * & * & \cdots & h_{k,k} \end{bmatrix},$$

where H is in Hermite Normal Form [35]. The matrix H represents the unique upper triangular basis of $\mathcal{L}$ such that $h_{i,i} > 0$, and $h_{j,j} > h_{i,j}$ for $i > j$. Every time $h_{i,i} = 1$, this means that we have a relation of the form $[\mathfrak{p}_i] = \left[\prod_{j<i} \mathfrak{p}_j^{-h_{i,j}}\right]$. In other words, $[\mathfrak{p}_i] \in \langle [\mathfrak{p}_1], \ldots, [\mathfrak{p}_{i-1}] \rangle$. On the other hand, if $h_{i,i} \neq 1$, then $[\mathfrak{p}_i] \notin \langle [\mathfrak{p}_1], \ldots, [\mathfrak{p}_{i-1}] \rangle$. Our algorithm proceeds by computing the HNF of M, and every time $h_{i,i} \neq 1$, it moves $\mathfrak{p}_i$ to the beginning of the list of primes, and moves the column i to the first column, recomputes the HNF and iterates the process. In the end, the first $\log^{2/3}(|\Delta|)$ primes in the list generate $\mathrm{Cl}(\mathcal{O})$.

Algorithm 2. Computation of $\log^{2/3}(|\Delta|)$ primes that generate $\mathrm{Cl}(\mathcal{O})$

Input: Order $\mathcal{O}$ of discriminant Δ and $c > 0$.
Output: $\log^{2/3}(|\Delta|)$ split primes whose classes generate $\mathrm{Cl}(\mathcal{O})$.
1: $S \leftarrow \{$Split primes $\mathfrak{p}_1, \ldots, \mathfrak{p}_k$ of norm less than $\log^c(|\Delta|)\}$.
2: $\mathcal{L} \leftarrow$ lattice of vectors $(e_1, \ldots, e_k)$ such that $\prod_i \mathfrak{p}_i^{e_i}$ is principal using [6].
3: Compute the matrix $H \in \mathbb{Z}^{k \times k}$ of a basis of $\mathcal{L}$ in HNF using [35, Ch. 6].
4: **for** $j = k$ down to $\log^{2/3}(|\Delta|) + 1$ **do**
5:      **while** $h_{j,j} \neq 1$ **do**
6:          Insert $\mathfrak{p}_j$ at the beginning of S.
7:          Insert the j-th column at the beginning of the list of columns of H.
8:          $H \leftarrow \mathrm{HNF}(H)$.
9:      **end while**
10: **end for**
11: **return** $\{\mathfrak{p}_1, \ldots, \mathfrak{p}_s\}$ for $s = \log^{2/3}(|\Delta|)$.

Proposition 2. *Assuming Heuristic 1 for the parameter c, Algorithm 2 is correct and runs in polynomial time in $\log(|\Delta|)$.*

Proof. Step 2 can be done in quantum polynomial time with the S-unit algorithm of Biasse and Song [6]. Assuming that $\log^{2/3}(|\Delta|)$ primes of norm less than $\log^c(|\Delta|)$ generate $\mathrm{Cl}(\mathcal{O})$, the loop of Steps 5 to 9 is entered at most j times as one of $[\mathfrak{p}_1], \ldots, [\mathfrak{p}_j]$ must be in the subgroup generated by the other $j - 1$ ideal classes. The HNF computation runs in polynomial time, therefore the whole procedure runs in polynomial time. $\qquad\square$

Once we have $\mathfrak{p}_1, \ldots, \mathfrak{p}_s$, we compute with Algorithm 3 a reduced basis B' of the lattice $\mathcal{L} \subseteq \mathbb{Z}^s$ of the vectors $(e_1, \ldots, e_s)$ such that $[\prod_i \mathfrak{p}_i^{e_i}]$ is trivial, and we compute the generators $\mathfrak{g}_1, \ldots, \mathfrak{g}_l$ such that $\mathrm{Cl}(\mathcal{O}) = \langle \mathfrak{g}_1 \rangle \times \cdots \times \langle \mathfrak{g}_l \rangle$ together with vectors $\boldsymbol{v}_i$ such that $\mathfrak{g}_i = \prod_j \mathfrak{p}_j^{v_{i,j}}$.

Lemma 1. *Let $\mathcal{L}$ be an n-dimensional lattice with input basis $B \in \mathbb{Z}^{n \times n}$, and let $\beta < n$ be a block size. Then the BKZ variant of [21] used with Kannan's enumeration technique [26] returns a basis $\boldsymbol{b}'_1, \ldots, \boldsymbol{b}'_n$ such that*

$$\|\boldsymbol{b}'_1\| \leq e^{\frac{n}{\beta} \ln(\beta)(1+o(1))} \lambda_1(\mathcal{L}),$$

using time $\mathrm{Poly}(n, \mathrm{Size}(B))\beta^{\beta\left(\frac{1}{2e} + o(1)\right)}$ and polynomial space.

Proof. According to [21, Theorem 1], $\|\boldsymbol{b}'_1\| \leq 4(\gamma_\beta)^{\frac{n-1}{\beta-1}+3} \lambda_1(\mathcal{L})$ where γ_β is the Hermite constant in dimension β. As asymptotically $\gamma_\beta \leq \frac{1.744\beta}{2\pi e}(1 + o(1))$ (see [25]), we get that $4(\gamma_\beta)^{\frac{n-1}{\beta-1}+3} \leq e^{\frac{n}{\beta} \ln(\beta)(1+o(1))}$. Moreover, this reduction is obtained with a number of calls to Kannan's algorithm that is bounded by $\mathrm{Poly}(n, \mathrm{Size}(B))$. According to [22, Theorem 2], each of these calls takes time $\mathrm{Poly}(n, \mathrm{Size}(B))\beta^{\beta\left(\frac{1}{2e} + o(1)\right)}$ and polynomial space, which terminates the proof. $\qquad\square$

Algorithm 3. Precomputation for the oracle

Input: Order $\mathcal{O}$ of discriminant Δ and $c > 0$.

Output: Split prime ideals $\mathfrak{p}_1, \ldots, \mathfrak{p}_s$ whose classes generate $\mathrm{Cl}(\mathcal{O})$ where $s = \log^{2/3}(|\Delta|)$, reduced basis B' of the lattice $\mathcal{L}$ of vectors $(e_1, \ldots, e_s)$ such that $\left[\prod_i \mathfrak{p}_i^{e_i}\right]$ is trivial, generators $\mathfrak{g}_1, \ldots, \mathfrak{g}_l$ such that $\mathrm{Cl}(\mathcal{O}) = \langle \mathfrak{g}_1 \rangle \times \cdots \times \langle \mathfrak{g}_l \rangle$ and vectors v_i such that $\mathfrak{g}_i = \prod_j \mathfrak{p}_j^{v_{i,j}}$.

1: $\mathfrak{p}_1, \ldots, \mathfrak{p}_s \leftarrow$ output of Algorithm 2.
2: $\mathcal{L} \leftarrow$ lattice of vectors $(e_1, \ldots, e_s)$ such that $\prod_i \mathfrak{p}_i^{e_i}$ is principal.
3: Compute a BKZ-reduced matrix $B' \in \mathbb{Z}^{s \times s}$ of a basis of $\mathcal{L}$ with block size $\log^{1/3}(|\Delta|)$.
4: Compute $U, V \in \mathrm{GL}_s(\mathbb{Z})$ such that $UB'V = \mathrm{diag}(d_1, \ldots, d_s)$ is the Smith Normal Form of B'.
5: $l \leftarrow \min_{i \leq s}\{i \mid d_i \neq 1\}$. For $i \leq l$, $v_i \leftarrow i$-th column of V.
6: $V' \leftarrow V^{-1}$. For $i \leq l$, $\mathfrak{g}_i \leftarrow \prod_{j \leq s} \mathfrak{p}_j^{v'_{i,j}}$.
7: **return** $\{\mathfrak{p}_1, \ldots, \mathfrak{p}_s\}$, B', $\{\mathfrak{g}_1, \ldots, \mathfrak{g}_l\}$, $\{v_1, \ldots, v_l\}$.

Proposition 3. *Assuming Heuristic 1 for c, Algorithm 3 is correct, runs in time $e^{\tilde{O}\left(\sqrt[3]{\log(|\Delta|)}\right)}$ and has polynomial space complexity.*

The precomputation of Algorithm 3 allows us to design the quantum circuit that implements the function described in (1). Generic techniques due to Bennett [3] convert any algorithm taking time T and space S into a reversible algorithm taking time $T^{1+\epsilon}$, for an arbitrary small $\epsilon > 0$, and space $O(S \log T)$. From a high-level point of view, this is simply the adaptation of the method of Biasse–Fieker–Jacobson [4, Algorithm 7] to the quantum setting.

Algorithm 4. Quantum oracle for implementing f defined in (1)

Input: Curves E_1, E_2. Order $\mathcal{O}$ of discriminant Δ such that $\mathrm{End}(E_i) \simeq \mathcal{O}$ for $i = 1, 2$. Split prime ideals $\mathfrak{p}_1, \ldots, \mathfrak{p}_s$ whose classes generate $\mathrm{Cl}(\mathcal{O})$ where $s = \log^{2/3}(|\Delta|)$, reduced basis B' of the lattice $\mathcal{L}$ of vectors $(e_1, \ldots, e_s)$ such that $\left[\prod_i \mathfrak{p}_i^{e_i}\right]$ is trivial, generators $\mathfrak{g}_1, \ldots, \mathfrak{g}_l$ such that $\mathrm{Cl}(\mathcal{O}) = \langle \mathfrak{g}_1 \rangle \times \cdots \times \langle \mathfrak{g}_l \rangle$ and vectors v_i such that $\mathfrak{g}_i = \prod_j \mathfrak{p}_j^{v_{i,j}}$. Ideal class $[\mathfrak{a}_y] \in \mathrm{Cl}(\mathcal{O})$ represented by the vector $y = (y_1, \ldots, y_l) \in \mathbb{Z}/d_1\mathbb{Z} \times \cdots \times \mathbb{Z}/d_l\mathbb{Z} \simeq \mathrm{Cl}(\mathcal{O})$, and $x \in \mathbb{Z}/2\mathbb{Z}$.

Output: $f(x, y)$.

1: $y \leftarrow \sum_{i \leq l} y_i v_i \in \mathbb{Z}^s$ (now $[\mathfrak{a}_y] = [\prod_i \mathfrak{p}_i^{y_i}]$).
2: Use Babai's nearest plane method with the basis B' to find $u \in \mathcal{L}$ close to y.
3: $y \leftarrow y - u$.
4: **If** $x = 0$ **then** $\overline{E} \leftarrow \overline{E}_1$ **else** $\overline{E} \leftarrow \overline{E}_2$.
5: **for** $i \leq s$ **do**
6:     **for** $j \leq y_i$ **do**
7:         $\overline{E} \leftarrow [\mathfrak{p}_i] * \overline{E}$.
8:     **end for**
9: **end for**
10: **return** $|\overline{E}\rangle$.

To bound the run time of Algorithm 4, we need to assume that the BKZ-reduced basis computed in Algorithm 3 has good geometric properties. We assume the following standard heuristic.

Heuristic 2 (Geometric Series Assumption). *The basis B' computed in Algorithm 3 satisfies the Geometric Series Assumption (GSA): there is $0 < q < 1$ such that $\|\widehat{b_i'}\| = q^{i-1}\|b_1\|$ where $\left(\widehat{b_i'}\right)_{i \leq n}$ is the Gram-Schmidt basis corresponding to B'.*

Proposition 4. *Assuming Heuristic 1 for some $c > 1$ and Heuristic 2, Algorithm 4 is correct and runs in quantum time $e^{\tilde{O}\left(\sqrt[3]{\log(|\Delta|)}\right)}$ and has polynomial space complexity.*

Proof. Each group action of Step 7 is polynomial in $\log(p)$ and in $\mathcal{N}(\mathfrak{p}_i)$. Moreover, Babai's algorithm runs in polynomial time and returns u such that

$$\|y - u\| \leq \frac{1}{2}\sqrt{\sum_i \|\widehat{b_i'}\|^2} \leq \frac{1}{2}\sqrt{n}\|b'_1\| \in e^{\tilde{O}\left(\sqrt[3]{\log(|\Delta|)}\right)}.$$

Therefore, the y_i are in $e^{\tilde{O}\left(\sqrt[3]{\log(|\Delta|)}\right)}$, which is the cost of Steps 5 to 9. The main observation allowing us to reduce the search to a close vector to the computation of a BKZ-reduced basis is that Heuristic 1 gives us the promise that there is $u \in \mathcal{L}$ at distance less than $e^{\sqrt[3]{\log(|\Delta|)}(1+o(1))}$ from y. □

Corollary 1. *Let E_1, E_2 be two elliptic curves and $\mathcal{O}$ be an imaginary quadratic order of discriminant Δ such that $\text{End}(E_i) \simeq \mathcal{O}$ for $i = 1, 2$. Then assuming Heuristic 1 for some constant $c > 0$, there is a quantum algorithm for computing $[\mathfrak{a}]$ such that $[\mathfrak{a}] * \overline{E}_1 = \overline{E}_2$ with:*

- *heuristic time complexity $e^{O\left(\sqrt{\log(|\Delta|)}\right)}$, polynomial quantum memory and $e^{O\left(\sqrt{\log(|\Delta|)}\right)}$ quantumly accessible classical memory,*
- *heuristic time complexity $e^{\left(\frac{1}{\sqrt{2}}+o(1)\right)\sqrt{\ln(|\Delta|)\ln\ln(|\Delta|)}}$ with polynomial memory (both classical and quantum).*

Remark 3. We referred to Heuristic 1 as Biasse, Fieker and Jacobson [4] provided numerical data supporting it. Heuristic 1 may be relaxed in the proof of the $e^{O\left(\sqrt{\log(|\Delta|)}\right)}$ asymptotic run time. As long as a number k in $\tilde{O}\left(\log^{1-\varepsilon}(|\Delta|)\right)$ of prime ideals of polynomial norm generate the ideal class group and that each class has at least one decomposition involving exponents less than $e^{\tilde{O}(\log^{1/2-\varepsilon}(|\Delta|))}$, the result still holds by BKZ-reducing with block size $\beta = \sqrt{k}$.

For the poly-space variant, these conditions can be relaxed even further. It is known under GRH that a number k in $\tilde{O}(\log(|\Delta|))$ of prime ideals of norm less than $12\log^2(|\Delta|)$ generate the ideal class group. We only need to argue that each class can be decomposed with exponents bounded by $e^{\tilde{O}\left(\sqrt{\log(|\Delta|)}\right)}$. Then by using the oracle of Algorithm 4 with block size $\beta = \sqrt{k}$, we get a run time of $e^{\tilde{O}\left(\sqrt{\log(|\Delta|)}\right)}$ with a poly-space requirement.

5   A Remark on Subgroups

It is well-known that the cost of quantum and classical attacks on isogeny based cryptosystems is more accurately measured by the size of the subgroup generated by the ideal classes used in the cryptosystem. As stated in [10, Sect. 7.1], in order to ensure that this is sufficiently large with high probability, the class group must have a large cyclic subgroup of order M, where M is not much smaller than the class number N. Assuming the Cohen-Lenstra heuristics this will be the case with high probability and, according to Hamdy and Saidak [20], one even expects a large prime-order subgroup.

It is an open problem as to whether the knowledge of smaller subgroups of the class group can be exploited to reduce the security of CSIDH; the current belief (see [10, p. 20]) is that there is no way to do this. There are nevertheless minor considerations that can easily be taken into account when selecting CSIDH parameters to minimize risk in this regard, stemming from the practical difficulties in constructing quadratic fields whose class numbers have a given divisor.

Constructing system parameters for which the class number has a known divisor could be done by a quantum adversary using the polynomial-time algorithm to compute the class group and trial-and-error. Using classical computation, this is in most cases infeasible because the recommended discriminant sizes are too large to compute the class number. Known methods to construct discriminants for which the class number has a given divisor M use a classical result of Nagell [28] relating the problem to finding discriminants $\Delta = c^2 D$ that satisfy $c^2 D = a^2 - 4b^M$ for integers a, b, c. These methods thus produce discriminants that are exponential in M, too large for practical purposes.

The one exception where classical computation can be used to find class numbers with a known divisor is when the divisor $M = 2^k$. Bosma and Stevenhagen [8] give an algorithm, formalizing methods described by Gauss [18, Sect. 286] and Shanks [32], to compute the 2-Sylow subgroup of the class group of a quadratic field. In addition to describing an algorithm that works in full generality, they prove that the algorithm runs in expected time polynomial in $\log(|\Delta|)$. Using this algorithm would enable an adversary to use trial-and-error efficiently to generate random primes p until the desired power of 2 divides the class number.

The primes p recommended for use with CSIDH are not amenable to this method, because they are congruent to $3 \bmod 4$, guaranteeing that the class number of the non-maximal order of discriminant $-4p$ is odd. However, in Sect. 5 of [10], the authors write that they pick $p \equiv 3 \pmod 4$ because it makes it easy to write down a supersingular curve, but that "in principle, this constraint is not necessary for the theory to work". We suggest that restricting to primes $p \equiv 3 \pmod 4$ is also desirable in order to avoid unnecessary potential vulnerabilities via the existence of even order subgroups.

6   Conclusion

We described two variants of a quantum algorithm for computing an isogeny between two elliptic curves E_1, E_2 defined over a finite field such that there is an imaginary quadratic order $\mathcal{O}$ satisfying $\mathcal{O} \simeq \mathrm{End}(E_i)$ for $i = 1, 2$ with $\Delta = \mathrm{disc}(\mathcal{O})$. Our first variant runs in heuristic asymptotic run time $2^{O\left(\sqrt{\log(|\Delta|)}\right)}$ and requires polynomial quantum memory and $2^{O\left(\sqrt{\log(|\Delta|)}\right)}$ quantumly accessible classical memory. The second variant of our algorithm relying on Regev's dihedral HSP solver [30] runs in time $e^{\left(\frac{1}{\sqrt{2}}+o(1)\right)\sqrt{\ln(|\Delta|)\ln\ln(|\Delta|)}}$ while relying only on polynomial (classical and quantum) memory. These variants of the HSP-based algorithms for computing isogenies have the best asymptotic complexity, but we left the assessment of their actual cost on specific instances such as the proposed CSIDH parameters [10] for future work. Some of the constants involved in lattice reduction were not calculated, and more importantly, the role of the memory requirement should be addressed in light of the recent results on the topic [1].

Acknowledgments. The authors thank Léo Ducas for useful comments on the memory requirements of the BKZ algorithm. The authors thank Noah Stephens-Davidowitz for information on the resolution of the approximate CVP. The authors also thank Tanja Lange and Benjamin Smith for useful comments on an earlier version of this draft.

References

1. Adj, G., Cervantes-Vázquez, D., Chi-Domínguez, J.-J., Menezes, A., Rodríguez-Henríquez, F.: The cost of computing isogenies between supersingular elliptic curves. Cryptology ePrint Archive, Report 2018/313 (2018). https://eprint.iacr.org/2018/313
2. Azarderakhsh, R., Jao, D., Leonardi, C.: Post-quantum static-static key agreement using multiple protocol instances. In: Adams, C., Camenisch, J. (eds.) SAC 2017. LNCS, vol. 10719, pp. 45–63. Springer, Cham (2018). https://doi.org/10.1007/978-3-319-72565-9_3
3. Bennett, C.H.: Time/space trade-offs for reversible computation. SIAM J. Comput. **18**(4), 766–776 (1989)
4. Biasse, J.-F., Fieker, C., Jacobson Jr., M.J.: Fast heuristic algorithms for computing relations in the class group of a quadratic order, with applications to isogeny evaluation. LMS J. Comput. Math. **19**(A), 371–390 (2016)
5. Biasse, J.-F., Jao, D., Sankar, A.: A quantum algorithm for computing isogenies between supersingular elliptic curves. In: Meier, W., Mukhopadhyay, D. (eds.) INDOCRYPT 2014. LNCS, vol. 8885, pp. 428–442. Springer, Cham (2014). https://doi.org/10.1007/978-3-319-13039-2_25
6. Biasse, J.-F., Song, F.: Efficient quantum algorithms for computing class groups and solving the principal ideal problem in arbitrary degree number fields. In: Krauthgamer, R. (ed.) Proceedings of the Twenty-Seventh Annual ACM-SIAM Symposium on Discrete Algorithms, SODA 2016, Arlington, VA, USA, 10–12 January 2016, pp. 893–902. SIAM (2016)

7. Bonnetain, X., Schrottenloher, A.: Quantum security analysis of CSIDH and ordinary isogeny-based schemes. Cryptology ePrint Archive, Report 2018/537 (2018). https://eprint.iacr.org/2018/537

8. Bosma, W., Stevenhagen, P.: On the computation of quadratic 2-class groups. Journal de Théorie des Nombres de Bordeaux **8**(2), 283–313 (1996)

9. Bröker, R., Charles, D., Lauter, K.: Evaluating large degree isogenies and applications to pairing based cryptography. In: Galbraith, S.D., Paterson, K.G. (eds.) Pairing 2008. LNCS, vol. 5209, pp. 100–112. Springer, Heidelberg (2008). https://doi.org/10.1007/978-3-540-85538-5_7

10. Castryck, W., Lange, T., Martindale, C., Panny, L., Renes, J.: CSIDH: an efficient post-quantum commutative group action. Cryptology ePrint Archive, Report 2018/383 (2018). https://eprint.iacr.org/2018/383. to appear in Asiacrypt 2018

11. Childs, A., Jao, D., Soukharev, V.: Constructing elliptic curve isogenies in quantum subexponential time. J. Math. Cryptol. **8**(1), 1–29 (2013)

12. Cohen, H.: A Course in Computational Algebraic Number Theory. Graduate Texts in Mathematics, vol. 138, p. xii+534. Springer, Berlin (1993). https://doi.org/10.1007/978-3-662-02945-9

13. Couveignes, J.-M.: Hard homogeneous spaces. http://eprint.iacr.org/2006/291

14. Diffie, W., Helman, M.: New directions in cryptography. IEEE Trans. Inf. Soc. **22**(6), 644–654 (1976)

15. Feo, L.D., Kieffer, J., Smith, B.: Towards practical key exchange from ordinary isogeny graphs. Cryptology ePrint Archive, Report 2018/485 (2018). https://eprint.iacr.org/2018/485. to appear in Asiacrypt 2018

16. Galbraith, S.D., Hess, F., Smart, N.P.: Extending the GHS weil descent attack. In: Knudsen, L.R. (ed.) EUROCRYPT 2002. LNCS, vol. 2332, pp. 29–44. Springer, Heidelberg (2002). https://doi.org/10.1007/3-540-46035-7_3

17. Galbraith, S.D., Petit, C., Shani, B., Ti, Y.B.: On the security of supersingular isogeny cryptosystems. In: Cheon, J.H., Takagi, T. (eds.) ASIACRYPT 2016. LNCS, vol. 10031, pp. 63–91. Springer, Heidelberg (2016). https://doi.org/10.1007/978-3-662-53887-6_3

18. Gauß, C.F., Waterhouse, W.C.: Disquisitiones Arithmeticae. Springer, New York (1986). https://doi.org/10.1007/978-1-4939-7560-0. translated by A.A. Clark

19. Hafner, J., McCurley, K.: A rigorous subexponential algorithm for computation of class groups. J. Am. Math. Soc. **2**, 839–850 (1989)

20. Hamdy, S., Saidak, F.: Arithmetic properties of class numbers of imaginary quadratic fields. JP J. Algebra Number Theory Appl. **6**(1), 129–148 (2006)

21. Hanrot, G., Pujol, X., Stehlé, D.: Terminating BKZ. IACR Cryptology ePrint Archive 2011, 198 (2011)

22. Hanrot, G., Stehlé, D.: Improved analysis of kannan's shortest lattice vector algorithm. In: Menezes, A. (ed.) CRYPTO 2007. LNCS, vol. 4622, pp. 170–186. Springer, Heidelberg (2007). https://doi.org/10.1007/978-3-540-74143-5_10

23. Jao, D., De Feo, L.: Towards quantum-resistant cryptosystems from supersingular elliptic curve isogenies. In: Yang, B.-Y. (ed.) PQCrypto 2011. LNCS, vol. 7071, pp. 19–34. Springer, Heidelberg (2011). https://doi.org/10.1007/978-3-642-25405-5_2

24. Jao, D., LeGrow, J., Leonardi, C., Ruiz-Lopez, L.: A subexponential-time, polynomial quantum space algorithm for inverting the cm action. In: Slides of Presentation at the MathCrypt Conference (2018). https://drive.google.com/file/d/15nkb9j0GKyLujYfAb8Sfz3TjBY5PWOCT/view

25. Kabatyanskii, A., Levenshtein, V.: Bounds for packings. On a sphere and in space. Proulcmy Peredacha informatsü **14**, 1–17 (1978)

26. Kannan, R.: Improved algorithms for integer programming and related lattice problems. In: Johnson, D., et al. (eds.) Proceedings of the 15th Annual ACM Symposium on Theory of Computing, 25–27 April, 1983, Boston, Massachusetts, USA, pp. 193–206. ACM (1983)
27. Kuperberg, G.: Another subexponential-time quantum algorithm for the dihedral hidden subgroup problem. In: Severini, S., Brandão, F. (eds.) 8th Conference on the Theory of Quantum Computation, Communication and Cryptography, TQC 2013, May 21–23, 2013, Guelph, Canada, vol. 22 of LIPIcs, pp. 20–34. Schloss Dagstuhl - Leibniz-Zentrum fuer Informatik (2013)
28. Nagell, T.: Über die Klassenzahl imaginär-quadratischer Zahlkörper. Abh. Math. Sem. Univ. Hamburg 1, 140–150 (1922)
29. National Institute of Standards and Technology. Post quantum cryptography project (2018). https://csrc.nist.gov/projects/post-quantum-cryptography
30. Regev, O.: A subexponential time algorithm for the dihedral hidden subgroup problem with polynomial space. arXiv:quant-ph/0406151
31. Schnorr, C.P., Euchner, M.: Lattice basis reduction: improved practical algorithms and solving subset sum problems. Math. Program. 66(2), 181–199 (1994)
32. Shanks, D.: Gauss's ternary form reduction and the 2-sylow subgroup. Math. Comput. 25(116), 837–853 (1971)
33. Silverman, J.H.: The Arithmetic of Elliptic Curves. Graduate Texts in Mathematics, vol. 106, p. xii+400. Springer, New York (1992). https://doi.org/10.1007/978-1-4757-1920-8
34. Stolbunov, A.: Constructing public-key cryptographic schemes based on class group action on a set of isogenous elliptic curves. Adv. Math. Commun. 4(2), 215–235 (2010)
35. Storjohann, A.: Algorithms for Matrix Canonical Forms. Ph.D. thesis, Department of Computer Science, Swiss Federal Institute of Technology - ETH (2000)
36. Tate, J.: Endomoprhisms of abelian varieties over finite fields. Inventiones Mathematica 2, 134–144 (1966)
37. Vélu, J.: Isogénies entre courbes elliptiques. C. R. Acad. Sci. Paris Sér. A-B 273, A238–A241 (1971)

Constructing Canonical Strategies for Parallel Implementation of Isogeny Based Cryptography

Aaron Hutchinson$^{(\boxtimes)}$ and Koray Karabina

Florida Atlantic University, Boca Raton, USA
hutchinsona2013@fau.edu

Abstract. Isogeny based cryptographic systems are one of the very competitive systems that are potentially secure against quantum attacks. The run time of isogeny based systems are dominated by a sequence of point multiplications and isogeny computations performed over supersingular elliptic curves in a specific order. The order of the sequence play an important role in the run time of the algorithms, and an optimal strategy can be efficiently determined yielding the minimum cost among all possible choices when a single processor is in use. In this paper, we generalize this idea and propose new algorithms that determine strategies for K processors under two different parallelization models: Per-Curve Parallelization (PCP) and Consecutive-Curve Parallelization (CCP). We present several recursive formulation of canonical strategies and their cost under the PCP model. As a result, we show how to construct the best (optimal) strategies under the PCP model. For some cryptographically interesting parameters, we obtain up to 24% (for $K = 2$), 40% (for $K = 4$), and 51% (for $K = 8$) theoretical speed ups over the optimal strategies with one processor. The more general CCP model offers a refinement of PCP, and yields up to 30% (for $K = 2$), 47% (for $K = 4$), and 55% (for $K = 8$) theoretical speed ups over the optimal strategies with one processor.

Keywords: SIDH · Isogeny-based cryptography · Parallelization

1 Introduction

Let E be a supersingular elliptic curve defined over a finite field $\mathbb{F}_q$ with q elements. Furthermore, assume that $q = p^2$ for some prime of the form $p = f\ell_A{}^{e_A}\ell_B{}^{e_B} \pm 1$. Here, ℓ_A and ℓ_B should be thought of as small primes, and f is called a cofactor. We choose E so that $|E(\mathbb{F}_q)| = (f\ell_A{}^{e_A}\ell_B{}^{e_B})^2$, and let $\{P_A, Q_A\}$ and $\{P_B, Q_B\}$ generate the $\ell_A{}^{e_A}$-torsion group $E[\ell_A{}^{e_A}] \cong \mathbb{Z}_{\ell_A{}^{e_A}} \oplus \mathbb{Z}_{\ell_A{}^{e_A}}$ and the $\ell_B{}^{e_B}$-torsion group $E[\ell_B{}^{e_B}] \cong \mathbb{Z}_{\ell_B{}^{e_B}} \oplus \mathbb{Z}_{\ell_B{}^{e_B}}$ of $E(\mathbb{F}_q)$, respectively. Under this setting, the supersingular isogeny-based Diffie-Hellman (SIDH) key exchange

© Springer Nature Switzerland AG 2018
D. Chakraborty and T. Iwata (Eds.): INDOCRYPT 2018, LNCS 11356, pp. 169–189, 2018.
https://doi.org/10.1007/978-3-030-05378-9_10

protocol between two parties A and B can be summarized at a high level as follows (see [3] for more details and the correctness):

1. $\mathbb{F}_q$, E, $\{P_A, Q_A\}$, $\{P_B, Q_B\}$ are published as the domain parameters of the protocol.
2. A chooses two random integers $m_A, n_A \in \mathbb{Z}_{\ell_A^{e_A}}$, and computes an elliptic curve E_A, where $\phi_A : E \to E_A$ is an isogeny with kernel $K_A = \langle m_A P_A + n_A Q_A \rangle$. A also computes the points $\phi_A(P_B)$ and $\phi_A(Q_B)$.
3. B chooses two random integers $m_B, n_B \in \mathbb{Z}_{\ell_B^{e_B}}$, and computes an elliptic curve E_B, where $\phi_B : E \to E_B$ is an isogney with kernel $K_B = \langle m_B P_B + n_B Q_B \rangle$. B also computes the points $\phi_B(P_A)$ and $\phi_B(Q_A)$.
4. A sends E_A, $\phi_A(P_B)$, and $\phi_A(Q_B)$ to B.
5. B sends E_B, $\phi_B(P_A)$, and $\phi_B(Q_A)$ to A.
6. A computes an elliptic curve E_{AB}, where $\phi'_A : E_B \to E_{AB}$ is an isogeny with kernel $K_{AB} = \langle m_A \phi_B(P_A) + n_A \phi_B(Q_A) \rangle$. A computes the j-invariant j_{AB} of E_{AB}.
7. B computes an elliptic curve E_{BA}, where $\phi'_B : E_A \to E_{BA}$ is an isogeny with kernel $K_{BA} = \langle m_B \phi_A(P_B) + n_B \phi_A(Q_B) \rangle$. B computes the j-invariant j_{BA} of E_{BA}.
8. A and B can now derive a shared key from their j-invariants because $E_{AB} \cong E_{BA}$ and $j_{AB} = j_{BA}$.

In [3], other isogeny based public key cryptosystems were proposed including zero knowledge proof of identity and public key encryption schemes. More recently, isogeny based public key signature schemes were proposed in [5]. The security of these schemes rely on the conjectural hardness of the computational problems in supersingular elliptic curve isogenies; see [6] for an extensive list of these problems. Currently, the best known classical and quantum algorithms to solve these problems run in exponential time, and therefore, supersingular isogeny based cryptosystems are believed to be quantum resistant.

The run time of SIDH (as well as the run time of other supersingular isogeny based cryptosystems) is dominated by point multiplications and isogeny computations performed over supersingular elliptic curves in a specific order. More specifically, given an elliptic curve $E/\mathbb{F}_q$ and a point $R \in E(\mathbb{F}_q)$ of order ℓ^n, one needs to compute a curve E_n and an isogeny $\phi : E \to E_n$ having kernel $\langle R \rangle$. Moreover, one needs to compute $\phi(P)$ and $\phi(Q)$ for some points P and Q on E. For example, see the steps 2, 3, 6, and 7 in the above description of SIDH. De Feo and Jao [3] describes a method to perform these computations efficiently. The main idea is to set $E_0 = E$, $R_0 = R$, and to factor an isogeny ϕ of degeree-ℓ^n as a composition of n degree-ℓ isogenies ϕ_i, $i = 0, ..., n-1$, satisfying:

$$\phi_i : E_i \to E_{i+1}, \ \text{Kernel}(\phi_i) = \ell^{n-i-1} R_i, \ R_{i+1} = \phi_i(R_i) \tag{1}$$

The decomposition yields

$$\phi = \phi_{n-1} \circ \phi_{n-2} \circ \cdots \circ \phi_1 \circ \phi_0, \ \phi : E \to E_n, \ \text{Kernel}(\phi) = R, \tag{2}$$

as required.

Given $E_0 = E$ and $R_0 = R$, there exist different strategies to compute E_n (and to evaluate ϕ at some given points on E). In [3], each strategy is associated with a subgraph S of a graph T_n. Here, T_n is a weighted directed graph whose vertices are lattice points of the unit triangular equilateral lattice between the x-axis, the line $y = \sqrt{3}x$, and the line $y = -\sqrt{3}(x - n + 1)$. For a given pair of vertices v and w in T_n, $\{v, w\}$ is an edge of T_n if and only if v and w are a unit away from each other, and the line connecting v and w is not horizontal. Each edge $\{v, w\}$ of T_n is directed in a top-down fashion from v to w. The top most vertex of T_n is labeled with the point $R_0 = R$, and each vertex in T_n corresponds to a point on an elliptic curve. An edge $\{P, Q\}$ with a positive slope in T_n correspond to multiplication by ℓ, that is, P and Q lie on the same curve E_i, and $Q = \ell P$. An edge $\{P, Q\}$ with a negative slope correspond to an isogeny computation, that is, P is on E_i, Q is on F_{i+1}, and $Q = \phi_i(P)$. The weights p (the cost of an ℓ-multiplication) and q (the cost of degree-ℓ isogeny computation) are assigned to the edges of T_n with positive and negative slopes, respectively. As an example, T_4 is shown in Fig. 1.

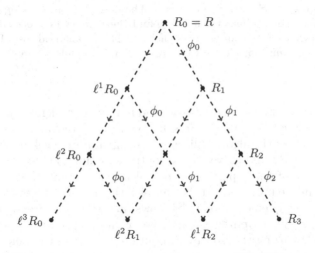

Fig. 1. Decomposing an isogeny for $n = 4$: $\phi = \phi_3 \circ \phi_2 \circ \phi_1 \circ \phi_0$.

In [3], a correspondence is established between *well-formed full strategies* S and binary trees with n leaves, which contain the strategies with minimal cost, and the cost of a strategy is computed as the sum of the weights of the edges of S. Furthermore, all well-formed full strategies can be partitioned such that binary trees of all strategies in the same class share the same *tree topology*. It is shown in [3] that a strategy with minimal cost must correspond to a strategy that is *canonically* constructed from the representative of its class, independent of the weights p and q. This correspondence yields the following recursive formula to determine the cost $C_{p,q}(n)$ of the optimal strategies with n leaves:

$$C_{p,q}(n) = \min_{i=1,\ldots,n-1} (C_{p,q}(i) + C_{p,q}(n - i) + (n - i)p + iq). \qquad (3)$$

Note that one can recover all optimal strategies by keeping track of the partitionings in the recursion. For example, there are two canonical strategies S_1 and S_2 for $n = 3$, as shown in Fig. 2. The cost of S_1 is $3p + 2q$, and the cost of S_2 is $2p + 3q$. Therefore, S_1 is optimal when $q \geq p$, and S_2 is optimal when $p \geq q$.

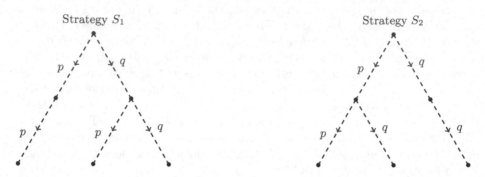

Fig. 2. Two strategies S_1 and S_2 for $n = 3$. The cost of S_1 is $3p + 2q$, and the cost of S_2 is $2p + 3q$. S_1 is optimal when $q \geq p$ and the number of processors is $K = 1$. Isogeny computations in S_1 cannot be parallelized. The parallelized cost of S_2 is $2p + 2q$. Therefore, S_2 is optimal for all p, q when $K = 2$, even though it is not optimal when $K = 1$ and $q > p$.

Recently, several high-speed software and hardware implementations of SIDH have been presented [1,2,4,7,8]. Speed ups are achieved as a combination of several techniques at the algorithmic and implementation level, including careful choices of parameters and curve models, optimizing finite field and elliptic curve arithmetic. Most of these papers do not utilize the parallelization of isogeny computations in the protocol and they all pick the optimal strategy as described in [3]. In fact, the implementation in [8] shows that SIDH can greatly benefit from parallelizing isogeny computations and they claim a speed-up by over a factor of 1.5 after parallelizing the isogeny computations. It is our understanding that the parallization technique deployed in [8] is based on a rather naive approach, which first determines an optimal strategy, and then parallelizes it. We observe though that this approach may not necessarily yield the best parallel strategy in general. For example, consider the canonical strategies S_1 and S_2, with costs $3p+2q$ and $2p+3q$, as in Fig. 2. The strategy S_2 cannot be optimal for $q > p$ but its cost can be reduced from $2p + 3q$ to $2p + 2q$ if the two isogeny computations ϕ_0 are performed in parallel. This makes S_2 an optimal parallel strategy for all p and q because no parallelization can be applied to S_1.

Contributions: We summarize our contributions as follows:

1. In Sect. 3 we detail the parallelization models considered in this work. We propose two models: a Per-Curve Parallelization (PCP) model, and a Consecutive-Curve Parallelization (CCP) model. In Sect. 4.1 we derive a recursive formula for $C_{p,q}^{K}(n)$, the cost of an optimal strategy with n leaves

using K processors, under the PCP model that determines the cost of an optimal parallelized strategy. Our formula can also be used to extract such a strategy explicitly. In Sect. 4.2, we refine our PCP model, and generalize it to the CCP model. We detail an algorithm for computing the cost of evaluating a canonical strategy S under the CCP model.

2. In Sect. 5 for the cryptographically interesting parameters $(n, p, q) = (239, 27.8, 17)$ and $(n, p, q) = (186, 25.8, 22.8)$ at the 124-bit quantum security level, our experiments with the PCP model found strategies which yield up to 24% (for $K = 2$), 40% (for $K = 4$), and 51% (for $K = 8$) theoretical speed ups over the optimal serial strategies for $n = 186$, and up to 23% (for $K = 2$), 39% (for $K = 4$), and 50% (for $K = 8$) theoretical speed ups over the optimal serial strategies for $n = 239$.

3. We observe that strategies constructed under the PCP model serves as a good basis to be evaluated under the more generalized CCP model. More specifically, for the CCP model, we were able to find strategies which yield up to 30% (for $K = 2$), 47% (for $K = 4$), and 55% (for $K = 8$) theoretical speed ups over the optimal serial strategies for $n = 186$, and up to 28% (for $K = 2$), 44% (for $K = 4$), and 52% (for $K = 8$) theoretical speed ups over the optimal serial strategies for $n = 239$.

2   Motivation

In this section we give motivation for both of the parallelization models to be defined in Sect. 3 and for our choice to restrict to looking only at canonical strategies. For this section we treat the models themselves as black boxes, and the interested reader may later verify the claims made in this section after reading Sects. 3 and 4.

In Sect. 3 we will define the PCP and CCP parallelization models. It will later be clear that the CCP model never performs worse than the PCP model. So why bother studying the PCP model? As it turns out, we are able to constructively characterize optimal canonical strategies with minimal costs under the PCP model. This can be thought of as a generalization of the method in [3] to find optimal canonical strategies in the serial computation setting. As we pointed out before in the Introduction (also see Fig. 2), our strategies under PCP already have the potential to outperform some naive and intuitive parallelization methods. On the other hand, the PCP model is somewhat restrictive, and we introduce our second model CCP. The CCP model is quite complex and we have been unable to find a method for finding optimal canonical strategies under the CCP model. Instead, we develop an algorithm that can parallelize any given canonical strategy and compute its cost under the CCP model. Our experiments have shown that taking strategies which are optimal under PCP and parallelizing them under the CCP model provides very good results, and so we use PCP optimal strategies as a starting point in our search for well-performing CCP strategies.

One might also question why we restrict to looking only at canonical strategies rather than the more general well-formed strategies. The analysis in [3] shows that an optimal strategy is always canonical in the serial setting, and so intuitively one would expect that this fact carries over to the parallel setting. Furthermore, the simple structure that canonical strategies provide offer a much simpler analysis of their cost in the parallel setting when compared to the set of well-formed strategies. In our analysis, we computed all strategies having $n \leq 5$ leaves and, while there were sometimes non-canonical strategies having minimal parallelized cost (with an ad-hoc assignment of processors), the minimal cost strategies could each be achieved through a canonical strategy which was parallelized under the CCP model.

As an example, we single out the three strategies S_3, S_4, and S_5 shown in Fig. 3. S_3 is clearly non-canonical, while S_4 and S_5 are canonical strategies. When taking $p = q = 1$, the costs of S_3, S_4 and S_5 using one processor are $13, 12$, and 13, respectively, and S_4 is an optimal strategy in this scenario. When using two processors we cannot apply PCP or CCP parallelization to S_3 since it is non-canonical, but exhausting all possible parallelizations of S_3 one finds that the minimal cost of S_3 using two processors is 9. Using Theorem 2 and Algorithm 1 in the sections to follow, we find that when using two processors S_4 has a PCP cost and a CCP cost of 10, while S_5 has a PCP cost of 10 and a CCP cost of 9. We illustrate these minimal costs in Fig. 3. We label each edge in each strategy with an integer i indicating that the corresponding computation is performed at the i'th iteration. In particular, if two edges are labeled with the same integer, then the corresponding computations are performed in parallel. This shows that finding minimal parallelized cost non-canonical strategies is sometimes possible, but we seem to be able to do just as well by searching through canonical strategies using the CCP model. In other words, we propose some systematic and efficient methods for constructing strategies for parallel implementation of isogeny based systems. Searching for parallel strategies in an ad-hoc way (e.g. exhausting all (non-)canonical strategies, or trying all possible parallelizations of a strategy) may yield better results but it may not be feasible for cryptographically interesting parameters.

3   Parallelization Models

In this section we detail the parallelization models we consider in this paper. We begin with a couple definitions. Recall from Sect. 1 that T_n is the weighted directed graph whose vertices are lattice points of the unit triangular equilateral lattice between the x-axis, the line $y = \sqrt{3}x$, and the line $y = -\sqrt{3}(x - n + 1)$.

Definition 1. *For $1 \leq i \leq n$ we define $\mathcal{L}_i$ and $\mathcal{R}_i$ as the positive and negative slope diagonals, respectively, of T_n containing the point $(i - 1, 0)$.*

Definition 2. *Let S be a set of edges in T_n. We define*

- *the i-th q-bin of S, denoted $Q_i(S)$, for $1 \leq i \leq n - 1$ as the set of negative slope edges in S lying between the lines $\mathcal{L}_i$ and $\mathcal{L}_{i+1}$;*
- *the i-th p-bin of S, denoted $P_i(S)$, for $0 \leq i \leq n - 1$ as the set of positive slope edges in S lying on the line $\mathcal{L}_{i+1}$.*

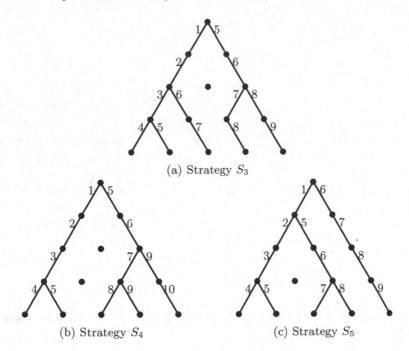

(a) Strategy S_3

(b) Strategy S_4 (c) Strategy S_5

Fig. 3. Three strategies for comparing our methods against ad-hoc methods. We assume $p = q = 1$. S_3 is non-canonical with a serial computation cost of 12, and its minimum cost with 2 processors is 9. S_4 and S_5 are canonical strategies with serial costs 12 and 13, respectively. Using our algorithms, both S_4 and S_5 can be extracted as two optimal strategies with cost 10 under the PCP model using two processors. S_5 can further be refined under the CCP model, and its cost reduces to 9 matching the parallelized cost of S_3. Each edge in each strategy is labeled with an integer i indicating that the corresponding computation is performed at the i'th iteration. If two edges are labeled with the same integer, then the corresponding computations are performed in parallel. Labels were assigned in S_4 using the PCP or CPC models (they are equivalent in this case), and in S_5 using the CCP model. (We thank the anonymous reviewer of PQC 2018 for pointing out the example of strategy S_3. The example was very enlightening in our analysis and led us to consider the CCP model.)

We note that $|P_0(S)| = n - 1$ and that $P_{n-1}(S)$ is always empty for canonical strategies S, but these are included for algorithmic purposes. See Fig. 4 for an example of these definitions on a strategy with $n = 4$. We then have that the serial cost of a strategy S is $C_{p,q}(S) = \sum_{i=0}^{n-1} |P_i(S)| p + \sum_{i=1}^{n-1} |Q_i(S)| q$.

We adopt the definition of a *strategy* as in [3]; more precisely, we work entirely with canonical strategies unless otherwise stated. Throughout this paper, S will denote a strategy on n leaves, and K will be the total number of processing

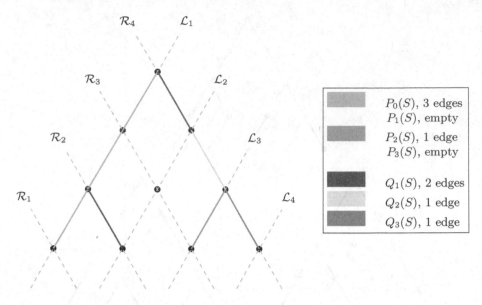

Fig. 4. An example of the lines $\mathcal{L}_i$ and $\mathcal{R}_i$ and the bins $P_i(S)$ and $Q_i(S)$ on a strategy S with $n = 4$.

units deployed for parallelization. *Evaluating* a strategy S involves the following computations:

(1) computation of elliptic curves E_i for $i = 0, 1, \ldots, n$.
(2) the evaluation of $[\ell]$ at varying points on varying curves: for $P \in E_i$ for some i, we must compute $\ell \cdot P$. These evaluations correspond to edges in $P_i(S)$.
(3) the evaluation of isogenies at varying points on varying curves: for $P \in E_{i-1}$ and $\phi_i : E_{i-1} \to E_i$ for some i, we must compute $\phi_i(P)$. These evaluations correspond to edges in $Q_i(S)$.

Item (1) is done through Vélu's formulas and its variants [9]. What kind of parallelization might one consider among the above computations? The following theorem limits the possibilities.

Theorem 1. *Let S be a canonical strategy with $n \geq 3$ leaves and let a and b be distinct edges in the bins $P_i(S)$ and $P_j(S)$, respectively. Then a and b cannot be parallelized together.*

Proof. We use induction on n. There are only two strategies with $n = 3$ leaves, pictured in Fig. 2. If $i = j = 0$, we may assume the path in $P_0(S)$ is ab; in this case the input to the computation corresponding to b is the output of the computation corresponding to a, and so these computations must be done sequentially. If $i \neq j$ then we may assume $i = 0$ and $j = 1$; the elliptic curve E_1 cannot be defined until the point corresponding to the leaf at the vertex $(0, 0)$ is reached, and so a must be evaluated before b.

Suppose the theorem holds for all canonical strategies having less than n leaves. Let S' and S'' be the left and right branches of S, respectively (these exist since S is well-formed and canonical). Let r, r' and r'' be the roots of S, S' and S'', respectively. Then $S = S' \cup S'' \cup rr' \cup rr''$. See Fig. 5 for a visual depiction. Then a and b each lie in either rr', S', or S''. To begin evaluating S'' we must have S' completely evaluated first, and to begin evaluating S' we must have rr' completely evaluated first. We can therefore assume that a and b both lie in the same set: either rr', S' or S''. The edges in rr' must be done successively, and so we get no parallelization in this case. Our induction hypothesis takes care of the S' and S'' cases.

In light of this theorem, we focus on grouping together edges $a_1, \ldots, a_K$ in *batches* to be parallelized, with all a_i lying in some $Q_{j_i}(S)$ with the possible exception that one edge is in some $P_m(S)$. In this work we consider two different but similar parallelization models: the per-curve parallel model and the consecutive-curve parallel model. The per-curve model is simpler and gives very nice theoretic results. Under this model we prove the existence of optimal strategies and show how to extract them efficiently. However, this model suffers from a fairly large amount of idle processors, which motivates the consecutive-curve model. We found that the consecutive-curve model gives overall lower costs as we will see in Sect. 5.

Parallelization Model 1. *The Per-Curve Parallel (PCP) model imposes a parallelization on a canonical strategy S in which the only computations that we allow to be parallelized are isogeny evaluations which involve the same isogeny. That is, we evaluate $P_0(S)$ in serial, then evaluate $Q_1(S)$ in parallel, then evaluate $P_1(S)$ in serial, then evaluate $Q_2(S)$ in parallel, etc.*

It should now be clear why the name "Per-Curve Parallel" was chosen: parallelization is applied on a per-curve basis.

Remark 1. Fix an index i with $1 \leq i \leq n - 1$. For a canonical strategy S we remark that all edges in $P_i(S)$ form a connected path whose target is a leaf of S. This means there is some point $P_i \in E_i$ and some non-negative integer t_i such that the vertices on the path of $P_i(S)$ correspond exactly to the points $P_i, [\ell] P_i, [\ell]^2 P_i, \ldots, [\ell]^{t_i} P_i$ and such that P_i is reached through an isogeny evaluation (and not a multiplication by ℓ). The isogeny evaluation which produces P_i will be the bottom-most edge in $Q_i(S)$, for otherwise S would not be well-formed. As a consequence, the bottom-most edge in $Q_i(S)$ must be evaluated in order to begin the evaluations in $P_i(S)$. When $i = 0$, P_i is the root of S and $t_0 = n - 1$.

Parallelization Model 2. *The Consecutive-Curve Parallel (CCP) model imposes a parallelization on a canonical strategy S in which:*

1. *we apply parallelization among $P_i(S) \cup Q_i(S)$ for $i = 1, 2, \ldots, n-1$ and among $Q_i(S) \cup Q_{i-1}(S)$ for $i = 2, \ldots, n - 1$. In the former case due to Theorem 1, we parallelize 1 edge from $P_i(S)$ with $K - 1$ edges from $Q_i(S)$ when possible.*

In the latter case we parallelize K edges from $Q_i(S)$ together when possible, and if the last of these batches has idle processors then we apply these processors to the bottom-most edges in the next bin $Q_{i+1}(S)$ whenever possible (so that the last few edges from $Q_{i-1}(S)$ will be parallelized with the first few edges in $Q_i(S)$).

2. *before any edges in $P_i(S)$ can be evaluated, the bottom-most edge in $Q_i(S)$ must be evaluated (see Remark 1)*

The name "Consecutive-Curve Parallel" comes from parallelizing $Q_i(S)$ with $Q_{i+1}(S)$. The CCP model is more involved than the PCP model, but it significantly reduces the number of idle processors overall since all evaluations in $P_i(S)$ are done in serial in the PCP model. In addition, any "leftover" processors from evaluating $Q_i(S)$ are tasked with beginning evaluations in $Q_{i+1}(S)$, which the PCP model doesn't account for.

In the single processor $K = 1$ setting, the notion of cost $C_{p,q}(S)$ of a strategy S is well defined as in Eq. 3. When $K > 1$, the cost of the strategy depends upon the model used. We will write $C_{p,q}^K(S)$ for the parallelized cost of the strategy S, with the choice of parallelization model understood from context.

Naturally, for given n and K we seek parallelized strategies S having n leaves which are *optimal* under the PCP and CCP parallelization models, by which we mean that $C_{p,q}^K(S)$ is minimal among all possible strategies S with the same parameters p, q, K under the corresponding model.

4 Parallelized Strategies and Their Optimality

Let S be a strategy and (p, q) a measure. We expand our definition of measure to include a parallelization parameter K denoting the total number of processing units deployed for the evaluation of the strategy. We also write $|S|$ for the number of leaves in the strategy S.

In this section we are concerned with finding optimal strategies in the PCP and CCP models. In the PCP case, we will see in Subsect. 4.1 that we are able to determine explicitly what these optimal strategies are and easily find their cost. In the CCP model, we give an algorithm that will determine the cost of a given strategy under this model.

4.1 Constructing Optimal Strategies Under PCP

Here we restrict exclusively to the PCP model unless otherwise stated. In the terminology of Definition 2, recall that the PCP model performs calculations on a per-bin basis. That is, we perform the computations in $P_i(S)$ in serial, then we parallelize the edges in $Q_i(S)$, and repeat for the next index $i + 1$. One would expect the parallelized cost of S to be

$$\sum_{i=0}^{n} |P_i(S)| p + \sum_{i=1}^{n-1} \left\lceil \frac{|Q_i(S)|}{K} \right\rceil q \tag{4}$$

when parallelizing computations on each curve individually.

Definition 3. *Let S be a set of edges in T_n and let K be the total number of processors deployed. For $k \in [1, \ldots, K]$, we let $C^{k/K}(S)$ denote the parallelized cost of S when exactly k out of the total K processors are available for the first batch of computations within each bin $Q_i(S)$, with the understanding that the other processors are occupied with other computations during this first batch.*

This relates to our previous cost definition by the equality $C^K(S) = C^{K/K}(S)$. As we will see shortly, we can compute this cost by examining the cost of the left and right branches of S, but will have "one less processor" at the start of computations in the left branch of S, which motivates the above definition. We make this precise now by stating and proving our main result for this section.

Theorem 2. *Let S be a canonical strategy with $n > 2$ leaves. Let S' and S'' be, respectively, the left and right branches of S, having i and $n - i$ leaves, respectively. Fix k with $1 \leq k \leq K$. Then under the PCP model we have*

$$C_{p,q}^{k/K}(S) = \begin{cases} C_{p,q}^{k-1/K}(S') + C_{p,q}^{k/K}(S'') + (n-i)p + q & \text{if } k > 1 \\ C_{p,q}^{K/K}(S') + C_{p,q}^{k/K}(S'') + (n-i)p + iq & \text{if } k = 1 \end{cases}$$

Proof. Let r, r' and r'' be the roots of S, S' and S'', respectively. Let $\hat{r}$ be the vertex on the edge rr'' lying at the intersection of $\mathcal{L}_i$ and $\mathcal{R}_n$ (so that $\hat{r}r''$ is just a single edge). Then because S is canonical we can write $S = S' \cup S'' \cup rr' \cup r\hat{r} \cup \hat{r}r''$. See Fig. 5.

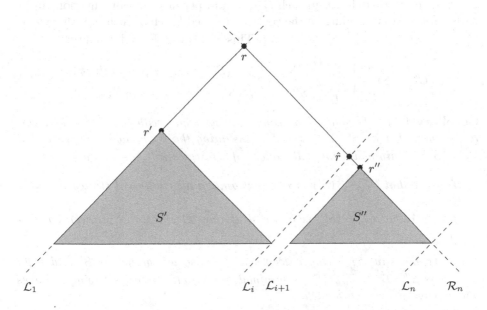

Fig. 5. Separating a strategy into its branches and connecting paths.

Then since

- the negative slope edges in $S' \cup r\hat{r}$ are exactly those in $\bigcup_{j=1}^{i-1} Q_j(S)$,
- the negative slope edge $\hat{r}r''$ is exactly $Q_i(S)$, and
- the negative slope edges in S'' are exactly $\bigcup_{j=i+1}^{n-1} Q_j(S)$,

these portions of S can be parallelized separately and we get

$$C^{k/K}(S) = C^{k/K}(S' \cup r\hat{r}) + C^{k/K}(S'') + C^{k/K}(rr') + C^{k/K}(\hat{r}r''). \quad (5)$$

As $\hat{r}r''$ is a single edge, we get $C_{p,q}^{k/K}(\hat{r}r'') = q$. Since rr' contains exactly $|S''|$ many positive slope edges, no parallelization can be applied and so $C_{p,q}^{k/K}(rr') = |S''|p = (n-i)p$. We now have

$$C_{p,q}^{k/K}(S) = C_{p,q}^{k/K}(S' \cup r\hat{r}) + C_{p,q}^{k/K}(S'') + (n-i)p + q.$$

We examine $C^{k/K}(S' \cup r\hat{r})$. Let $q_1, \ldots, q_{i-1}$ be the edges making up $r\hat{r}$, with $q_j \in Q_j(S)$ for $1 \le j \le i-1$. If $k > 1$, then the computation of the edge q_j must put in the first batch of bin $Q_j(S')$, occupying one of the processors on the first batch of computations. This leaves $k-1$ available processors on the first batch in S', and so $C^{k/K}(S' \cup r\hat{r}) = C^{k-1/K}(S')$. If $k = 1$, then only one processor is available for our first computation; we use this processor on the edges q_j to fill a batch within each $Q_j(S)$, and perform the computations in S' with a fresh batch. Adding on the cost of the filled batch in each bin, this yields $C_{p,q}^{k/K}(S' \cup r\hat{r}) = C_{p,q}^{K/K}(S') + (i-1)q$. This yields the desired equality:

$$C_{p,q}^{k/K}(S) = \begin{cases} C_{p,q}^{k-1/K}(S') + C_{p,q}^{k/K}(S'') + (n-i)p + q & \text{if } k > 1 \\ C_{p,q}^{K/K}(S') + C_{p,q}^{k/K}(S'') + (n-i)p + iq & \text{if } k = 1 \end{cases} \quad (6)$$

Corollary 1. *Let S' and S'' be canonical strategies with i and $n-i$ leaves, respectively. Fix k with $1 \le k \le K$. Assuming the PCP model, suppose that $C_{p,q}^{k/K}(S'')$ is minimal among all choices of strategies with $n-i$ leaves, and:*

- *if $k > 1$ that $C_{p,q}^{k-1/K}(S')$ is minimal among all choices of strategies with i leaves,*
- *if $k = 1$ that $C_{p,q}^{K/K}(S')$ is minimal among all choices of strategies with i leaves.*

Let S be the strategy having n leaves and whose left branch is S' and right branch is S''. Then $C_{p,q}^{k/K}(S)$ is minimal among all strategies having n leaves with i leaves in the left branch.

Proof. Let $\hat{S}$ be any canonical strategy with n leaves having i leaves in the left branch. Let $\hat{S}'$ and $\hat{S}''$ be the left and right branches of $\hat{S}$, respectively. We must show that $C^{k/K}(\hat{S}) \ge C^{k/K}(S)$.

Suppose $k > 1$. Both S and $\hat{S}$ satisfy the hypotheses of Theorem 2, and by the minimality assumptions on S' and S'' we have

$$\begin{aligned}
C^{k/K}(\hat{S}) &= C_{p,q}^{k-1/K}(\hat{S}') + C_{p,q}^{k/K}(\hat{S}'') + (n-i)p + q \\
&\geq C_{p,q}^{k-1/K}(S') + C_{p,q}^{k/K}(S'') + (n-i)p + q \\
&= C^{k/K}(S).
\end{aligned}$$

When $k = 1$, a similar sequence of (in)equalities can be derived by using the $k = 1$ case of Eq. 6.

The above corollary tells us that we can construct optimal strategies inductively from smaller optimal strategies, just as in the serial setting. We now define a function which computes the cost of an optimal parallelized strategy with n leaves, and can be used to construct such a strategy. Due to Theorem 2, we need only take a minimum over all possible partitions of the leaves of a strategy to find an optimal strategy.

Definition 4. *For a measure* (p, q, K), *we define a* cost *function* $C_{p,q}^{k/K}(n)$, *or* $C^{k/K}(n)$ *when* p, q *are clear from context, for* $n \in \mathbb{N}$ *and* $0 \leq k \leq K$ *recursively as*

1. $C_{p,q}^{k/K}(1) = 0$ *for* $k > 0$
2. $C_{p,q}^{k/K}(2) = p + q$ *for* $k > 0$
3. $C_{p,q}^{0/K}(n) = C_{p,q}^{K/K}(n) + (n-1)q$ *for all* n
4. $C_{p,q}^{k/K}(n) = \min\limits_{i \in [1, n-1]} \{ C_{p,q}^{k-1/K}(i) + C_{p,q}^{k/K}(n-i) + (n-i)p + q \}$ *for* $k > 0, n > 2$.

We define $C_{p,q}^{K}(n)$ *to be* $C_{p,q}^{K/K}(n)$.

Property (3.) is justified as beginning a new batch of computations since we've "run out" of processors. The cost of the new batch is calculated in $C_{p,q}^{K/K}(n)$, and we add the cost of $(n-1)q$ for the completed batch.

Note that when $K = 1$, we have

$$\begin{aligned}
C_{p,q}^{1}(n) &= \min\limits_{i \in [1, n-1]} \{ C_{p,q}^{0/1}(i) + C_{p,q}^{1/1}(n-i) + (n-i)p + q \} \\
&= \min\limits_{i \in [1, n-1]} \{ C_{p,q}^{1}(i) + C_{p,q}^{1}(n-i) + (n-i)p + iq \}
\end{aligned}$$

and so $C_{p,q}^{1}(n) = C_{p,q}(n)$ for all p, q, n, as one would expect when using one processor.

If one computes $C^{K}(n)$ for $n \leq K + 1$, then property (3.) above will never be used. In this case, we can derive an explicit formula for $C^{K}(n)$.

Theorem 3. *For all* $n \leq K + 1$, *we have* $C_{p,q}^{K}(n) = (n-1)(p+q)$ *for all* p, q.

Proof. We use induction on n. The base case $n = 1$ is given by property (1.) in the definition above.

Assume the theorem holds for all $i < n$. Then

$$
\begin{aligned}
C_{p,q}^K(n) &= \min_{i \in [1,n-1]} \{C_{p,q}^{K-1/K}(i) + C_{p,q}^K(n-i) + (n-i)p + q\} \\
&= \min_{i \in [1,n-1]} \{(i-1)(p+q) + (n-i-1)(p+q) + (n-i)p + q\} \\
&= \min_{i \in [1,n-1]} \{(2n-i-2)p + (n-1)q\} \\
&= (n-1)(p+q)
\end{aligned}
$$

where the second equality follows from the inductive hypothesis since $i \leq n-1 \leq k$ and $n-i < n \leq k+1$.

Note that in the proof above the minimum always occurs when i is maximized, meaning that the partition on the leaves used in the optimal strategy is $(n-1, 1)$. This yields an entirely isogeny-based strategy. We should also point out that the number $(n-1)(p+q)$ is a universal lower bound on the parallelized cost of any strategy with n leaves, independent of which method of parallelization is used; this is because the edges on the lines $\mathcal{L}_1$ and $\mathcal{R}_n$ can never be parallelized together in any way, and adding the cost of these edges gives $(n-1)(p+q)$.

4.2   Searching for Optimal Strategies Under CCP

The CCP model is more difficult to work with than the PCP model. In our analysis, we were unable to find any direct or recursive formula which gives the cost of a strategy S having n leaves using K processors under the CCP model. As a consequence, we settle for an algorithm which computes the cost of a given strategy S under fixed parameters p, q, K.

We give a high level overview of the algorithm before jumping into the specific details. Recall that the CCP model iterates through the Q bins of a strategy S, attempting to apply parallelization to $P_i(S) \cup Q_i(S)$ and $Q_i(S) \cup Q_{i-1}(S)$. The edges in $Q_i(S)$ are evaluated in a bottom-to-top fashion. We define a variable $leftover_i$ as the number of processors not initially assigned an edge on the last batch of isogeny computations in $Q_i(S)$. The i-th iteration of the algorithm will consider a picture resembling that of Fig. 6.

In order to gain access to the bin $P_i(S)$, the bottom-most edge in $Q_i(S)$ must first be evaluated. If $leftover_i$ is zero (meaning that the last batch in the previous iteration was actually full), we must perform one batch of entirely isogeny computations to access the $P_i(S)$ bin. Following this the $P_i(S)$ edges cannot be parallelized among themselves due to Theorem 1, and so in order to avoid idle processors we group one edge from $P_i(S)$ with $K-1$ edges from $Q_i(S)$. Two cases arise when grouping edges in this way: the bin $P_i(S)$ is exhausted first, or $Q_i(S)$ is.

The latter case is handled simply: we compute any edges remaining in $P_i(S)$ in serial and begin anew on the next index. In the former case we will have

$$
u_i = |Q_i(S)| - v_i - |P_i(S)|(K-1)
$$

edges from $Q_i(S)$ left to compute, where v_i is either *leftover$_i$* if *leftover$_i$* $\neq 0$ or K otherwise. We group these edges into batches of size K; if $(u_i \bmod K)$ is nonzero then the remaining $K - (u_i \bmod K)$ many processors can be tasked with performing computations in the next Q bin $Q_{i+1}(S)$, which determines *leftover$_{i+1}$*.

All of the above assumes that $|Q_i(S)| - K \geq 0$, and special care must be taken when this is not the case (when the bin size is very small). Recall that we group the few top-most edges in $Q_{i-1}(S)$ with the few bottom-most edges in $Q_i(S)$, but we cannot put any edges on the same $\mathcal{R}_j$ line into the same batch. In this situation we put as many of the bottom-most edges of $Q_i(S)$ into the batch as possible (and so possibly having idle processors in this batch).

This methodology gives rise to Algorithm 1. We now explain each step of this algorithm.

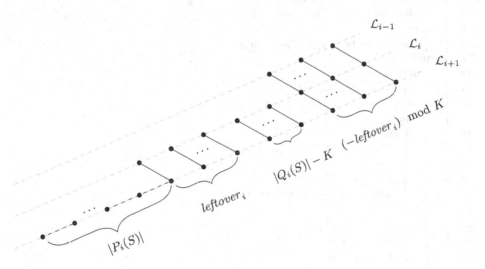

Fig. 6. The i-th q and p bins of a strategy S, together with the edges placed into the last batch of isogeny computations in bin $Q_{i-1}(S)$. Other edges in $Q_{i-1}(S)$ which are not in the last batch have been omitted from this figure, as they will have been parallelized on the previous iteration of the algorithm.

Details of Algorithm 1: We keep track of variables r, s, and t representing the p, q, and $\max\{p, q\}$ costs, respectively. The algorithm begins by initializing s and t to 0, and r to $n - 1$ to account for the $n - 1$ multiplications by ℓ occurring on curve 0. We then iterate through each $Q_i(S)$ bin with the index i, determining the cost while accounting for processors which were idle in the final batch of isogeny evaluations being carried over to the next bin through the variable *leftover*. Recall that we perform isogeny evaluations in a bottom-to-top fashion, and so the lowest positioned edges in each bin have priority on being grouped into batches first.

Algorithm 1. CCP-Cost

Parameters: p, q, K

Input   : Integer vectors $Q = (Q_1, \ldots, Q_{n-1})$ and $P = (P_1, \ldots, P_{n-2})$
     referencing some canonical strategy S having n leaves, where
  -- Q_i is the number of negative slope edges in S between curves $i - 1$ and i.
  -- P_i is the number of positive slope edges in S on curve i.

with the first curve being curve 0.

Output: A tuple (r, s, t), where r, s, and t are respectively the number of p, q,
     and $\max\{p, q\}$ computations when parallelizing S with K processors

```
 1  s, t, leftover ← 0, r ← n − 1
 2  for i = 1 to n − 1 do
 3  │    binSize ← Qi − K
 4  │    if binSize < 0 then
 5  │    │    binSize ← K − leftover
 6  │    │    leftover ← Qi − (K − leftover)
 7  │    else
 8  │    │    binSize ← binSize + K − leftover
 9  │    end
10  │    if leftover == 0 then
11  │    │    binSize ← binSize − K
12  │    │    s ← s + 1
13  │    if binSize > 0 then
14  │    │    if binSize ≥ (K − 1)Pi then
15  │    │    │    binSize ← binSize − (K − 1)Pi
16  │    │    │    t ← t + Pi
17  │    │    │    s ← s + ⌈binSize/K⌉
18  │    │    │    leftover ← (−binSize)  mod K
19  │    │    else
20  │    │    │    t ← t + ⌈binSize/(K−1)⌉
21  │    │    │    r ← r + Pi − ⌈binSize/(K−1)⌉
22  │    │    │    leftover ← 0
23  │    │    end
24  │    else
25  │    │    r ← r + Pi
26  │    │    leftover ← 0
27  │    end
28  end
29  return (r, s, t)
```

Lines 3–9: Together lines 3–9 account for subtracting off a number of edges from the $Q_i(S)$ bin according to the number of leftover processors from the previous iteration. However we cannot group together an edge in $Q_{i-1}(S)$ with an edge in $Q_i(S)$ which lie on the same line $\mathcal{R}_j$, since these would have to be done sequentially. We note that

> if an edge on line $\mathcal{R}_j$ is present in bin $Q_{i-1}(S)$ in a canonical strategy, then either the target of this edge is a leaf or the edge on line $\mathcal{R}_j$ is present in bin $Q_i(S)$.

Suppose that there are *leftover* many idle processors on the last batch of isogeny computations in bin $Q_{i-1}(S)$. Then the first batch of isogeny computations in bin $Q_i(S)$ will include the $K - leftover$ isogeny computations from bin $Q_{i-1}(S)$. From the note above, we cannot include the last $K - leftover$ edges in the first batch of bin $Q_i(S)$ since they lie on the same $\mathcal{R}$ lines as the leftover edges in bin $Q_{i-1}(S)$ already being included in the first batch. We subtract these $K - leftover$ edges from the bin to prevent them from being included in the batch. Then we subtract off the first *leftover* many edges from the bin to include them in the first batch, which gives

$$\text{binSize} \leftarrow Q_i - (K - \text{leftover}) - \text{leftover} = Q_i - K$$

in line 3.

If the first *leftover* edges and the last $k - leftover$ edges in $Q_i(S)$ overlap, then this will cause the size of the bin to become negative, which is checked for in line 4. In this case, we reset the bin size to the $K - leftover$ edges that we took out in line 5. We also change *leftover* in line 6 to the actual amount of edges of $Q_i(S)$ which were included in this first batch.

If there was no overlap between these two groups, we simple add back in the edges we previously took out in line 8.

Lines 10–13: if there were no leftover processors from isogeny evaluations in bin $Q_{i-1}(S)$, then we must perform a single batch of isogeny evaluations in bin $Q_i(S)$ in order to access the bin $P_i(S)$.

Lines 13–27: Line 13 checks if there are still remaining edges in the bin after the execution of line 10. If there are no edges left, we perform the multiplications by ℓ in bin $P_i(S)$ in serial in the Else clause of line 24. In this case, there is no remainder on the isogeny evaluations since we must compute all edges in $P_i(S)$ before beginning on $Q_{i+1}(S)$.

When there are edges left in $Q_i(S)$, we group them in "mixed" batches consisting of one single edge from $P_i(S)$ with $K - 1$ edges from $Q_i(S)$. Line 14 checks if there are enough remaining edges in $Q_i(S)$ to do $|P_i(S)|$ many of these mixed batches. If there are (lines 15–18), we add their cost and then group any remaining edges from $Q_i(S)$ into batches of size K. If there aren't enough edges in $P_i(S)$ (lines 20–22), we perform as many mixed batches as we can and then compute the remaining $P_i(S)$ edges in serial.

We make a note that when $K = 1$, line 16 should be changed to $r \leftarrow r + P_i$ since each batch consists of only a single positive slope edge.

5   An Analysis of Parallelized Strategies and Comparisons

In this section, we detail how we search for optimal strategies and report on their costs and the speedup percentages that they provide.

The PCP model is quite easy to work with due to the convenient recursive formula in Definition 4. This formula tells us exactly the cost of the best possible strategy having n leaves using K processors in the PCP model. A simple recursive script will compute $C_{p,q}^K(n)$ for any n and K in the PCP setting and allow us to determine which strategies are universally optimal under this model.

The CCP model is much more problematic. In Sect. 4.2 we gave an algorithm for determining the cost of a given strategy under the CCP model. Since the number of canonical strategies having n leaves is equal to the n-th Catalan number, it is infeasible to use this algorithm to compute the cost of every canonical strategy for large n. In our analysis we focused on the cryptographically interesting parameters (n, p, q) taken from [8] of $(239, 27.8, 17)$ and $(186, 25.8, 22.8)$ for K in the set $\{2, 3, 4, 7, 8, n - 2, n - 1\}$ (where $n = 186$ is derived from $372/2$, in which the division by 2 comes from using 4-isogenies as a composition of two 2-isogenies). These parameters together provide 124-bits of quantum security for SIDH. For each set of parameters we looked at three sets of strategies: *all canonical strategies*, *serially optimal strategies*, and *PCP optimal strategies*.

For our parameters the sets of *serially optimal strategies* have size $183,579,396$ for $n = 239$ and $1,623,160$ for $n = 186$. For the $n = 186$ case we are able to parallelize *every* strategy in the set using Algorithm 1, and the minimal costs for a few values of K are shown in Table 1.

Table 1. Data for parameters $n = 186, p = 25.8, q = 22.8$. Row PCP: optimal PCP costs over all canonical strategies. Row CCP S.O.: best CCP costs over all 1,623,160 serially optimal strategies. Row CCP A.C.: best CCP costs among 5,000,000 randomly sampled canonical strategies. Row CCP P.O.: best CCP costs among 5,000,000 randomly sampled PCP optimal strategies. Percent speedup is over the optimal serial cost of 34256.4.

	K	2	3	4	5	6	7	8
PCP	Cost	25942.2	22521.6	20373.0	19197.0	17941.2	16978.8	16617.0
	% speedup	24.27	34.26	40.53	43.96	47.63	50.44	51.49
CCP S.O.	Cost	24247.2	21784.8	20941.2	20781.6	20781.6	20781.6	20781.6
	% speedup	29.22	36.41	38.87	39.34	39.34	39.34	39.34
CCP A.C.	Cost	25440.6	22200.6	20880.6	19825.2	19606.2	19218.6	18739.2
	% speedup	25.73	35.19	39.05	42.13	42.77	43.90	45.30
CCP P.O.	Cost	23890.2	20515.2	18252.6	17555.4	16482.0	16021.2	15294.6
	% speedup	30.26	40.11	46.72	48.75	51.89	53.23	55.35

For the other two sets we chose a uniformly random subset of $5,000,000$ strategies for each parameter tuple. In each case we computed the cost of all strategies in the subset using Algorithm 1 and reported on the minimal cost found within that set. The results from the *all canonical strategies* set were not as good as one might hope as shown in the "CCP A.C." row of Table 1 for $n = 186$, and this is likely because there are simply too many strategies which don't parallelize very well. Better results might come from this set upon further statistical or theoretical analysis. For example, choosing a different distribution than uniform on the partitioning of the leaves of a strategy based on (n, p, q, K) might give better results. We found no obvious correlation between the parameters and the optimal partitions, but due to Theorem 3 we know that the initial partitioning tends toward $(n - 1, 1)$ as $K \to n$.

In contrast the results from the *PCP optimal strategies* sets were typically the best of all. The data we found is shown in Table 2 with comparisons to the optimal PCP strategy cost and the optimal serial cost. As expected the CCP model performs no worse than the PCP model in this case, and often performs much better with speedups of up to 10% over the best PCP strategy. Of particular interest are the cases of $K = 2, 4, 8$, where we find speedups over the serially optimal strategy of $28.40\%, 43.85\%, 52.43\%$ respectively in the $n = 239$ case and $30.26\%, 46.72\%, 55.35\%$ respectively in the $n = 186$ case, and each of these cases we have noticeably lower costs than what we get with the PCP method. We have no reason to suspect that the strategies we found (which were chosen uniformly at random from a subset of canonical strategies) are optimal strategies under the CCP model, and so further speedups than this may be possible.

It's worth pointing out that as far as we are aware of in the literature the method of choosing a serially optimal strategy and applying parallelization to it has been the only approach used to apply parallelization to the SIDH protocol, such as in [8]. Table 1 shows at least for $n = 186$ that this is not an optimal approach, as the best serially optimal strategy (row CCP S.O.) is outperformed by a PCP optimal strategy parallelized under CCP (row CCP P.O.) for every parameter value K.

Tables 1 and 2 suggest that if one were to implement a parallelized version of SIDH, they would fix a parameter tuple (n, p, q, K) and find a strategy which performs well under the CCP model. We found such efficient strategies by sampling randomly from the optimal PCP strategies and saving the ones with minimal costs under Algorithm 1, but other methods of finding efficient strategies under CCP may exist. Once a strategy has been chosen to be used with CCP, Algorithm 1 also tells the user how to group the operations for parallelization to achieve the desired cost.

6 Concluding Remarks

We have introduced two models of parallelization, Per-Curve Parallel (PCP) and Consecutive-Curve Parallel (CCP), for computation of an isogeny of large degree, which can be used in supersingular isogeny-based Diffie-Hellman key exchange. For the PCP model we gave a recursive formula for the cost an optimal strategy, and for the CCP model we gave Algorithm 1 which computes the cost of a given strategy under CCP. Intuitively we expect that strategies parallelized under the CCP model should perform better than under the PCP model, and our experiments reflect this as well. Furthermore, our constructions (cost formula for PCP and algorithm for CCP) tell the user explicitly how to parallelize the operations within the strategy. In the case of the PCP model, our cost formula also tells the user how to construct a strategy which is optimal under the PCP model.

Table 2. Comparison of PCP and CCP strategy costs. Above is the costs for parameters $n = 239, p = 27.8, q = 17$, while below is the costs for $n = 186, p = 25.8, q = 22.8$. For each parameter set (n, p, q, K) in the CCP rows, we chose $5,000,000$ strategies uniformly at random from the set of strategies which were optimal under the PCP model; these strategies were parallelized under the CCP model and the minimum cost from this set is reported in the table.

$n = 239$	K	1	2	3	4	7	8	237	238
Serial	Cost	41653.8	–	–	–	–	–	–	–
PCP	Cost	41653.8	31886.0	27858.0	25328.8	21572.6	20851.2	10679.4	10662.4
	% speedup over Serial	0	23.45	33.12	39.19	48.21	49.94	74.36	74.40
CCP	Cost	41653.8	29931.0	25835.0	23390.8	20399.6	19814.2	10679.4	10662.4
	% speedup over Serial	0	28.14	37.98	43.85	51.03	52.43	74.36	74.40
	% speedup over PCP	0	6.13	7.26	7.65	5.44	4.97	0	0
$n = 186$	K	1	2	3	4	7	8	184	185
Serial	Cost	34256.4	–	–	–	–	–	–	–
PCP	Cost	34256.4	25942.2	22521.6	20373.0	16978.8	16617.0	9013.8	8991.0
	% speedup over Serial	0	24.27	34.26	40.53	50.44	51.50	73.69	73.75
CCP	Cost	34256.4	23890.2	20515.2	18252.6	16021.2	15294.6	9013.8	8991.0
	% speedup over Serial	0	30.26	40.11	46.72	53.23	55.35	73.69	73.75
	% speedup over PCP	0	7.91	8.91	10.40	5.64	7.96	0	0

For cryptographically interesting parameters at the 124-bit quantum security level, our experiments with the PCP model found strategies which yield up to 24% (for $K = 2$), 40% (for $K = 4$), and 51% (for $K = 8$) theoretical speed ups over the optimal serial strategies for $n = 186$, and up to 23% (for $K = 2$), 39% (for $K = 4$), and 50% (for $K = 8$) theoretical speed ups over the optimal serial strategies for $n = 239$. Furthermore for the CCP model we were able to find strategies which yield up to 30% (for $K = 2$), 47% (for $K = 4$), and 55% (for $K = 8$) theoretical speed ups over the optimal serial strategies for $n = 186$, and up to 28% (for $K = 2$), 44% (for $K = 4$), and 52% (for $K = 8$) theoretical speed ups over the optimal serial strategies for $n = 239$. In the CCP case the costs that we found were only the best strategies that we happened to come across, and so further speedups may be possible in this case. Our results and comparison are purely theoretical as we do not take into account implementation related (scheduling, sycronization, etc.) costs. It would be interesting to see the practical impact of our methods in a side-channel protected implementation.

Acknowledgements. The authors would like to thank our reviewers for their comments and corrections. Research reported in this paper was supported by the Army Research Office under the award number W911NF-17-1-0311. The content is solely the responsibility of the authors and does not necessarily represent the official views of the Army Research Office.

References

1. Costello, C., Hisil, H.: A simple and compact algorithm for SIDH with arbitrary degree isogenies. In: Takagi, T., Peyrin, T. (eds.) ASIACRYPT 2017. LNCS, vol. 10625, pp. 303–329. Springer, Cham (2017). https://doi.org/10.1007/978-3-319-70697-9_11

2. Costello, C., Longa, P., Naehrig, M.: Efficient algorithms for supersingular isogeny diffie-hellman. In: Robshaw, M., Katz, J. (eds.) CRYPTO 2016. LNCS, vol. 9814, pp. 572–601. Springer, Heidelberg (2016). https://doi.org/10.1007/978-3-662-53018-4_21

3. De Feo, L., Jao, D., Plût, J.: Towards quantum-resistant cryptosystems from supersingular elliptic curve isogenies. J. Math. Cryptol. 8(3), 209–247 (2014)

4. Faz-Hernández, A., López, J., Ochoa-Jiménez, E., Rodríguez-Henríquez, F.: A faster software implementation of the supersingular isogeny Diffie-Hellman key exchange protocol. IEEE Trans. Comput. 2017

5. Galbraith, S.D., Petit, C., Silva, J.: Identification protocols and signature schemes based on supersingular isogeny problems. In: Takagi, T., Peyrin, T. (eds.) ASIACRYPT 2017. LNCS, vol. 10624, pp. 3–33. Springer, Cham (2017). https://doi.org/10.1007/978-3-319-70694-8_1

6. Galbraith, S.D., Vercauteren, F.: Computational problems in supersingular elliptic curve isogenies. Quantum Inf. Process. 17(10), 265 (2018)

7. Koziel, B., Azarderakhsh, R., Kermani, M., Jao, D.: Post-quantum cryptography on FPGA based on Isogenies on elliptic curves. IEEE Trans. Circuits Syst. 64, 86–99 (2017)

8. Koziel, B., Azarderakhsh, R., Mozaffari-Kermani, M.: fast hardware architectures for supersingular isogeny diffie-hellman key exchange on FPGA. In: Dunkelman, O., Sanadhya, S.K. (eds.) INDOCRYPT 2016. LNCS, vol. 10095, pp. 191–206. Springer, Cham (2016). https://doi.org/10.1007/978-3-319-49890-4_11

9. Moody, D., Shumow, D.: Analogues of Velu's formulas for isogenies on alternate models of elliptic curves. Math. Comput. 85(300), 1929–1951 (2016)

More Efficient Lattice PRFs from Keyed Pseudorandom Synthesizers

Hart Montgomery[✉]

Fujitsu Laboratories of America, Sunnyvale, USA
hmontgomery@us.fujitsu.com

Abstract. We develop new constructions of lattice-based PRFs using keyed pseudorandom synthesizers. We generalize all of the known 'basic' parallel lattice-based PRFs–those of [BPR12], [BLMR13], and [BP14]–to build highly parallel lattice-based PRFs with smaller modulus (and thus better reductions from worst-case lattice problems) while still maintaining computational efficiency asymptotically equal to the fastest known lattice-based PRFs at only the cost of larger key sizes.

In particular, we build several parallel (in NC^2) lattice-based PRFs with modulus independent of the number of PRF input bits based on both standard LWE and ring LWE. Our modulus for these PRFs is just $O\left(m^{f(m)}\right)$ for lattice dimension m and any function $f(m) \in \omega(1)$. The only known parallel construction of a lattice-based PRF with such a small modulus is a construction from Banerjee's thesis [Ban15], and some of our parallel PRFs with equivalently small modulus have smaller key sizes and are very slightly faster (when using FFT multiplication). These PRFs also asymptotically match the computational efficiency of the most efficient PRFs built from *any* LWE- or ring LWE-based assumptions known today, respectively, and concretely require less computation per output than any known parallel lattice-based PRFs (again when using FFT multiplication).

We additionally use our techniques to build other efficient PRFs with very low circuit complexity (but higher modulus) which improve known results on highly parallel lattice PRFs. For instance, for input length λ, we show that there exists a ring LWE-based PRF in NC^1 with modulus proportional to m^{λ^c} for any $c \in (0,1)$. Constructions from lattices with this circuit depth were only previously known from larger moduli.

Keywords: Lattices · Pseudorandom functions
Learning with errors · Pseudorandom synthesizers

1 Introduction

Pseudorandom functions, first defined by Goldreich, Goldwasser, and Micali [GGM84], are one of the most fundamental building blocks in cryptography. They are used for a wide variety of cryptographic applications, including

---

The full version of this paper is available on the IACR cryptology eprint archive.

© Springer Nature Switzerland AG 2018
D. Chakraborty and T. Iwata (Eds.): INDOCRYPT 2018, LNCS 11356, pp. 190–211, 2018.
https://doi.org/10.1007/978-3-030-05378-9_11

encryption, message integrity, signatures, key derivation, user authentication, and much more. PRFs are important in computational complexity as well since they can be used to build lower bounds in learning theory.

In a nutshell, a PRF is a function that is indistinguishable from a truly random function[1]. The most efficient PRFs are built from block ciphers like AES and security is based on ad-hoc *interactive* assumptions. It is a longstanding open problem to construct PRFs that are efficient as these block ciphers from *offline* assumptions like factoring or the decisional Diffie-Hellman problem. The history of PRFs based on standard, offline assumptions is long and filled with many interesting constructions [NR97]. For a full treatment of PRFs and their applications, we highly recommend reading [BR17]. In this work, however, we specifically focus on lattice-based PRFs.

While there are many desirable properties of good PRFs, three that immediately come to mind are speed, parallelization, and cryptographic hardness. Speed speaks for itself: all other things equal, faster PRFs are better. Parallelization is also another desired quality: it means that PRFs can practically be computed more quickly and has interesting implications for complexity theory [BFKL94]. Of course, PRFs that are harder to break are also more desirable. Throughout this paper we will examine all of these PRF qualities.

Learning with Errors. In this work, we base our PRFs on the hardness of the *learning with errors* (LWE) problem [Reg05], which is the most commonly used lattice problem in cryptography[2]. Informally, the LWE problem is, for a uniformly random fixed key $s \in \mathbb{Z}_q^n$, random samples $a_i \leftarrow \mathbb{Z}_q^n$, and discrete Gaussian noise terms δ_i, to distinguish from random the distribution consisting of samples of the form $(a_i, a_i^\top \cdot s + \delta_i \mod q)$.

Regev [Reg05] showed that solving the LWE problem is as hard as finding approximate solutions to certain worst-case lattice problems. The quality of the approximate solution (and thus the hardness of the problem solved) was proportional to the ratio of the modulus q to the width of the Gaussian noise terms. Most LWE-based cryptosystems today rely on the hardness of an LWE instance with a small, polynomial Gaussian noise distribution, so the hardness of the scheme is typically directly tied to the modulus q. Thus, decreasing the modulus of LWE-based cryptosystems is an important goal across many areas of lattice cryptography[3].

1.1 Lattice-Based PRFs

It has been known how to build completely sequential (and thus high depth) PRFs from LWE by using generic constructions like [GGM84] since the original LWE result [Reg05] was published. For instance, it is possible to build a very

---

[1] We give a precise definition in Sect. 2.
[2] Please see Sect. 2 for a comprehensive definition of and discussion on the LWE problem.
[3] For a full treatment of lattice and LWE complexity, we strongly recommend [MG12].

simple lattice-based PRF using the [GGM84] construction by treating LWE as a PRG. This simple construction also has the added benefit of a polynomially sized modulus q. However, these PRFs from generic constructions are maximally sequential and very inefficient, since LWE noise (i.e. Gaussians) has to be sampled at every step in the generic construction.

The study of PRFs based on lattice problems truly began in 2011, when Banerjee, Peikert, and Rosen [BPR12] invented the learning with rounding (LWR) problem, reduced to it from LWE, and showed that it could be used to build efficient and highly parallel PRFs. The authors built three new PRFs using the new rounding technique: one using the GGM construction, one using pseudorandom synthesizers, and one direct construction. The ring-based direct construction had the nice property that it could be implemented in NC^1, even if it was slightly less efficient than the generic constructions.

In a follow-up work, Boneh, Lewi, Montgomery, and Raghunathan [BLMR13] invented the first key homomorphic PRF in the standard model (from any assumption) using lattices. While their PRF was not extremely efficient, key homomorphic PRFs have a wide variety of applications, and their techniques (in particular, the use of LWE samples with low noise) turned out to be useful in other applications.

Most recently, Banerjee and Peikert [BP14] developed a general family of key homomorphic PRFs that dramatically improved upon the PRFs in [BLMR13] and even were (for certain choices of parameters) competitive with the non-key homomorphic PRFs of [BPR12] in terms of performance. The authors used a clever tree structure and rigorous analysis to carefully schedule 'bit decomposition' that allowed for good performance while still managing to retain key homomorphism.

In his Ph.D. thesis, Banerjee [Ban15] further improved the pseudorandom synthesizer construction technique from [BPR12], which allowed for tighter asymptotic constructions than previously known[4]. Around the same time, Döttling and Schröder [DS15] showed how to use their general technique of on-the-fly adaptation to also build LWE-based PRFs with relatively small moduli from low-depth circuits.

Concurrent Work. Very recently, and in a work concurrent with (and independent from) ours, Jager, Kurek, and Pan [JKP18] introduce all-prefix universal hash functions and show how to use these in conjunction with the augmented cascade construction [BMR10] to build efficient lattice-based PRFs with slightly superpolynomial modulus. Their LWE-based PRF can be thought of as a much more efficient version of [DS15]. We do not fully analyze this construction here, but it is likely more efficient than ours (although it does not have quite as small of a modulus as some of our constructions).

---

[4] To our knowledge this result has not been formally published in conference proceedings.

Application-Focused Lattice PRFs. Lattice PRFs have also been used for a number application specific PRFs, including puncturable PRFs [GGM84], constrained PRFs [BW13], [DKW16] (including key homomorphic constrained PRFs [BFP+15], [BV15]), PRFs secure against related key attacks [LMR14], and PRFs that hide constraints or functions [CC17], [BKM17], [BTVW17], [KW17], [PS17]. It is not known how to achieve many of these results from standard, non-lattice assumptions. Moreover, many of these works utilize very strong versions of the LWE assumption. It is our hope that the techniques introduced in this paper can be used to improve the efficiency and assumptions of some of these works.

1.2   Pseudorandom Synthesizers and Lattices

Pseudorandom synthesizers were first invented by Naor and Reingold in their famous work [NR95] as a way to construct PRFs with low circuit depth. The first synthesizer PRF constructions from lattices were introduced in [BPR12].

It can be cumbersome to define synthesizer PRFs in a way that is immediately understandable, so we present an 8-bit version of the synthesizer PRF from [BPR12]. Let the matrices $\mathbf{S}_{i,b} \in \mathbb{Z}_q^{m \times m}$ for $i \in [1, ..., 8]$ and $b \in \{0, 1\}$ be sampled uniformly at random. The original lattice-based synthesizer construction of [BPR12] had the following form on an 8-bit input $\mathbf{x} = x_1...x_8$:

$$\left\lfloor \left\lfloor \lfloor \mathbf{S}_{1,x_1} \cdot \mathbf{S}_{2,x_2} \rceil_{p_2} \lfloor \mathbf{S}_{3,x_3} \cdot \mathbf{S}_{4,x_4} \rceil_{p_2} \right\rceil_{p_1} \cdot \left\lfloor \lfloor \mathbf{S}_{5,x_5} \cdot \mathbf{S}_{6,x_6} \rceil_{p_2} \lfloor \mathbf{S}_{7,x_7} \cdot \mathbf{S}_{8,x_8} \rceil_{p_2} \right\rceil_{p_1} \right\rceil_{p_0}$$

Note that this construction has the unfortunate requirement that $q >> p_2 >> p_1 >> p_0$. In his thesis [Ban15], Banerjee showed how to eliminate this 'tower of moduli' requirement from this synthesizer construction by using rectangular matrices. To illustrate this, suppose we set our modulus q and our rounding parameter p such that $q = p^2$. Let the matrices $\mathbf{S}_{i,b}$ for $i \in [1, ..., 4]$ and $b \in \{0, 1\}$ now be defined such that $\mathbf{S}_{i,b} \in \mathbb{Z}_q^{m \times 2m}$. We can take the product of the transpose of one of these matrices with another and round in the following way:

$$\left\lfloor \left[\mathbf{S}_{i,b_i}^{\mathsf{T}} \right] \cdot \left[\mathbf{S}_{i+1,b_{i+1}} \right] \right\rceil_p = \mathbf{T} \in \mathbb{Z}_p^{2m \times 2m}$$

for some matrix $\mathbf{T}$ that will be indistinguishable from random by the hardness of LWR. In addition, note that $\mathbf{T}$ has enough entropy to produce a new, uniformly random matrix $\mathbf{S}' \in \mathbb{Z}_q^{m \times 2m}$. In fact, we can just set

$$\mathbf{S}' = \mathbf{T} \cdot \begin{bmatrix} \mathbf{I}_m \\ q\mathbf{I}_m \end{bmatrix}$$

to trivially extract this randomness. Suppose we now consider a 4-bit input $\mathbf{x} = x_1...x_4$: if we put this all together, we can present a four-bit version of the

PRF based on the improved synthesizer from [Ban15] in the following way:

$$\left\lfloor \left(\left[\begin{bmatrix} \mathbf{S}_{1,x_1}{}^{\mathsf{T}} \end{bmatrix} \cdot \begin{bmatrix} \mathbf{S}_{2,x_2} \end{bmatrix} \right]_p \cdot \begin{bmatrix} \mathbf{I}_m \\ q\mathbf{I}_m \end{bmatrix} \right) \cdot \right.$$

$$\left. \left(\left[\begin{bmatrix} \mathbf{S}_{3,x_3}{}^{\mathsf{T}} \end{bmatrix} \cdot \begin{bmatrix} \mathbf{S}_{4,x_4} \end{bmatrix} \right]_p \cdot \begin{bmatrix} \mathbf{I}_m \\ q\mathbf{I}_m \end{bmatrix} \right)^{\mathsf{T}} \right\rfloor_p \cdot \begin{bmatrix} \mathbf{I}_m \\ q\mathbf{I}_m \end{bmatrix}$$

This new synthesizer construction from [Ban15] was the first lattice-based PRF construction where the modulus q was independent of the input length λ of the PRF.

1.3 Our Contributions

In this paper, we introduce a new, general technique that we use to build new lattice-based PRFs by applying a pseudorandom synthesizer structure [NR95] to the three main generic PRF constructions of [BPR12], [BLMR13], and [BP14]. While our constructions are not key homomorphic, they are either as efficient or more efficient and have as small or smaller modulus (and thus better reductions to worst-case lattice problems) than existing lattice-based PRFs. For lattice-based PRFs with any degree of parallelism, only the synthesizer-based PRF from Banerjee's thesis [Ban15] matches the most efficient of our constructions asymptotically, and our constructions are (slightly) more efficient in practice assuming we use fast Fourier transform multiplication[5].

In order to illustrate our construction technique, we will start by considering the [BLMR13] PRF F_{BLMR}. Recall that F_{BLMR} uses two public matrices $\mathbf{A}_0, \mathbf{A}_1 \in \mathbb{Z}_2^{m \times m}$ where the entries of these matrices are sampled uniformly at random from $\{0,1\}$ such that $\mathbf{A}_0$ and $\mathbf{A}_1$ are full-rank. The dimension m is derived from the security parameter, and the key for the PRF is a single vector $\mathbf{k} \in \mathbb{Z}_q^m$ and its input domain is $\{0,1\}^\lambda$. We also need an integer modulus q and a rounding parameter p. The PRF at the point $\mathbf{x} = x_1 \cdots x_\lambda \in \{0,1\}^\lambda$ is defined as

$$F_{BLMR}(\mathbf{k}, x) = \left\lceil \prod_{i=1}^{\lambda} \mathbf{A}_{x_i} \cdot \mathbf{k} \right\rfloor_p \tag{1.1}$$

where $\lceil \cdot \rfloor_p$ denotes the standard rounding operation[6]. F_{BLMR} is both key homomorphic and massively parallelizable (in NC^2) which is quite desirable. However the construction is quite inefficient, and the modulus q required for the PRF to be secure is enormous ($\log q$ scales linearly with the input length λ), meaning

---

[5] See [Fat06] and especially [KSN+04] for a discussion of integer multiplication algorithms.

[6] $\lceil \cdot \rfloor_p : \mathbb{Z}_q \to \mathbb{Z}_p$ as $\lceil x \rfloor_p = i$, where $i \cdot \lfloor q/p \rfloor$ is the largest multiple of $\lfloor q/p \rfloor$ that does not exceed x.

that the worst-case lattice problems that we can reduce to the PRF require quite strong assumptions. The work of [BP14] aims to alleviate some of these problems by cleverly inserting some 'bit decomposition' operations into the evaluation of the PRF, getting an overall tree structure that results in more efficient PRFs with better modulus.

Many years ago, the cascade construction [BCK96] and the augmented cascade construction [BMR10] were used to build more efficient PRFs by adding 'key material' at every layer of the PRF. This generally increased the key size of the resulting PRFs, but increased efficiency and (sometimes) allowed for weaker assumptions. To this point, no such ideas have been applied to PRFs based on lattice assumptions. While we cannot directly utilize these constructions for lattices, we can apply their core idea–add 'key material' at every layer of the construction–to build more efficient PRFs.

To this end, suppose we view F_{BLMR} in tree form and add an additional secret key at every layer of the tree (up to $\log \lambda$ total). In order to do this efficiently, we use a pseudorandom synthesizer construction [NR95]. Pseudorandom synthesizers, which we define in Sect. 3, are efficient ways to construct parallelizable PRFs.

In our work, we construct a keyed synthesizer S_ℓ which has a square matrix $\mathbf{S} \in \mathbb{Z}_q^{m \times m}$ with entries sampled uniformly at random over $\mathbb{Z}_q$ as a key. Our synthesizer is parameterized by a parallelization factor ℓ and uses ℓ 'lists' of binary random matrices $\mathbf{A}_{i,x_i} \in \mathbb{Z}_2^{m \times m}$, where $i \in [1, \ell]$ is the list indicator and $x_i \in [1, k_i]$ is the index of a particular matrix $\mathbf{A}_{i,x_i}$ in the list i. Each output block of our synthesizer S_ℓ looks like the following:

$$S_\ell(x_1, ... x_\ell) \stackrel{\text{def}}{=} \left[\left[\prod_{i=1}^{\ell} \mathbf{A}_{i,x_i} \right] \mathbf{S} \right]_p$$

S_ℓ looks very much like F_{BLMR} with input length ℓ, although there are some key differences that we need in order for the synthesizer construction to efficiently work. We rigorously define and prove the security of this synthesiser S_ℓ later in the paper, and show how the pseudorandom synthesizer construction of [NR95] can be used to turn various versions of this synthesizer into PRFs. The proof of security borrows elements from the proofs of [BLMR13] and especially [BP14].

Now suppose that we set our rounding parameter $p = 2$. This will turn out to be a practical parameter choice. We next select binary matrices $\mathbf{A}_{i,b} \in \mathbb{Z}_2^{m \times m}$ uniformly at random for $i \in [1, ..., 8]$ and $b \in \{0, 1\}$ and keys $\mathbf{S}_1, \mathbf{S}_2, \mathbf{S}_3 \in \mathbb{Z}_q^{m \times m}$ uniformly at random. Our synthesizer S_2 can be used to build what we call the PRF F^2 which, on 8-bit input $\mathbf{x} = x_1...x_8$, gives us the following construction:

$$\left\lfloor \lfloor \lfloor \mathbf{A}_{1,x_1} \mathbf{A}_{2,x_2} \mathbf{S}_1 \rfloor_2 \cdot \lfloor \mathbf{A}_{3,x_3} \mathbf{A}_{4,x_4} \mathbf{S}_1 \rfloor_2 \mathbf{S}_2 \rfloor_2 \cdot \right. \tag{1.2}$$

$$\left. \lfloor \lfloor \mathbf{A}_{5,x_5} \mathbf{A}_{6,x_6} \mathbf{S}_1 \rfloor_2 \cdot \lfloor \mathbf{A}_{7,x_7} \mathbf{A}_{8,x_8} \mathbf{S}_1 \rfloor_2 \mathbf{S}_2 \rfloor_2 \mathbf{S}_3 \right\rfloor_2$$

Note that each rounded subset product (i.e. $\lfloor \mathbf{A}_{1,x_1} \mathbf{A}_{2,x_2} \mathbf{S}_1 \rfloor_2$) evaluates to a new random-looking matrix over $\mathbb{Z}_2^{m \times m}$ (assuming the nonuniform LWE assumption from [BLMR13]), so after one stage of (parallel) evaluation, the above 8-bit PRF looks like the following:

$$\left\lfloor \left\lfloor \tilde{\mathbf{A}}_{x_1,x_2} \tilde{\mathbf{A}}_{x_3,x_4} \mathbf{S}_2 \right\rceil_2 \cdot \left\lfloor \tilde{\mathbf{A}}_{x_5,x_6} \tilde{\mathbf{A}}_{x_7,x_8} \mathbf{S}_2 \right\rceil_2 \mathbf{S}_3 \right\rceil_2$$

for random-looking matrices $\tilde{\mathbf{A}}_{x_i,x_{i+1}} \in \mathbb{Z}_2^{m \times m}$ that depend on the bits x_i and x_{i+1}. We can also generalize and pick higher values of ℓ, some of which will have interesting ramifications. Below we show a (abbreviated) 16-bit construction of a PRF F^4 using S_4:

$$\left\lfloor \left\lfloor \mathbf{A}_{1,x_1} \mathbf{A}_{2,x_2} \mathbf{A}_{3,x_3} \mathbf{A}_{4,x_4} \mathbf{S}_1 \right\rceil_2 \cdot \left\lfloor \mathbf{A}_{5,x_5} \mathbf{A}_{6,x_6} \mathbf{A}_{7,x_7} \mathbf{A}_{8,x_8} \mathbf{S}_1 \right\rceil_2 \cdot \left\lfloor \cdot \right\rceil_2 \cdot \left\lfloor \cdot \right\rceil_2 \cdot \mathbf{S}_2 \right\rceil_2$$

The inputs to each layer of our PRFs are new, random-looking binary matrices, which allows our synthesizer to compose nicely. Right away, it should be obvious that our synthesizer offers some advantages over the basic synthesizer construction of [BPR12]. Most obviously, our rounding parameter and modulus can be independent of the PRF length. However, some of the comparisons are a bit more nuanced. While we can build PRFs from (almost) any choice of ℓ– including PRFs that include synthesizers with different choices of ℓ–we examine one particular choice in Sect. 5 which we briefly discuss here.

PRF F^2 from Synthesizer S_2. Our PRF F^2, which we showed for 8 bits in Eq. 1.2, is one of the simplest PRFs we can build, but also one of the most efficient. F^2 has modulus $O\left(m^{\omega(1)}\right)^7$ which is currently the (asymptotically) smallest known modulus for any lattice PRF that uses rounding. We note that this modulus is independent of the input length λ of the PRF. F^2 is as asymptotically efficient in terms of output per work as the naive pseudorandom synthesizer of [BPR12], which is currently the most efficient known PRF in this regard from standard lattices. Additionally, F^2 can be computed (practically, even) in circuit class NC^2, meaning that it is highly parallelizable. The only drawback of F^2 is the relatively large key size.

F^2 is the second (after that in [Ban15]) known lattice PRF with modulus $m^{\omega(1)}$ independent of the number of input bits of the PRF with *any* sublinear circuit depth and also happens to also be one of the most efficient known lattice PRF (from standard lattices) in terms of output per work.

Synthesizing [BPR12]. We can also apply pseudorandom synthesizers to other lattice PRF constructions. The logical place to continue is, of course, the original lattice PRF construction: [BPR12]. There are substantial differences between [BLMR13] and [BPR12], the largest of which is that F_{BLMR} uses a LWE sample subset-product structure while the direct PRF F_{BPR} from [BPR12] uses a key subset-product structure. However, it turns out we can still build interesting PRFs by synthesizing F_{BPR} with a few minor tricks.

---

[7] We use this as shorthand for $O\left(m^{f(m)}\right)$ for any function $f(m) \in \omega(1)$. This is technically incorrect, but a nice convenience and is common in LWR literature.

In this vein, we next construct a keyed synthesizer K_ℓ inspired by (and almost identical to) F_{BPR} which has a square matrix $\mathbf{A} \in \mathbb{Z}_q^{m \times m}$ with entries sampled uniformly at random over $\mathbb{Z}_q$ as a key. K_ℓ is parameterized by a parallelization factor ℓ and uses ℓ 'lists' of random matrices $\mathbf{S}_{i,x_i} \in \mathbb{Z}_p^{m \times m}$ for some superpolynomially large p, where $i \in [1, \ell]$ is the list indicator and $x_i \in [1, k_i]$ is the index of a particular matrix $\mathbf{S}_{i,x_i}$ in the list i. Each output block of K_ℓ looks like the following:

$$K_\ell (x_1, ... x_\ell) \stackrel{\text{def}}{=} \left\lfloor \mathbf{A} \left[\prod_{i=1}^{\ell} \mathbf{S}_{i,x_i} \right] \right\rceil_p \tag{1.3}$$

We chose to make the keys uniformly random in order to make the synthesizer compose properly when used to construct PRFs[8]. Of course, proving synthesizer security requires that LWE problem is hard with superpolynomially large uniform key and noise, but this is a very straightforward (and probably known in folklore) result. We prove this as a part of our analysis. In the meantime, we again demonstrate how such a synthesizer would look by spelling out an 8-bit version of K_ℓ.

Suppose we select 'public samples' $\mathbf{A}_1, \mathbf{A}_2, \mathbf{A}_3 \in \mathbb{Z}_q^{m \times m}$ uniformly at random and 'secret' matrices $\mathbf{S}_{i,b} \in \mathbb{Z}_p^{m \times m}$ uniformly at random for $i \in [1, ..., 8]$ and $b \in \{0, 1\}$. K_2 can be used to build a PRF that we refer to as P^2, which, on 8-bit input $\mathbf{x} = x_1 ... x_8$, gives us the following construction (where all computations are performed $\mod q$):

$$\left\lfloor \mathbf{A}_1 \left\lfloor \mathbf{A}_2 \lfloor \mathbf{A}_3 \mathbf{S}_{1,x_1} \mathbf{S}_{2,x_2} \rceil_p \cdot \lfloor \mathbf{A}_3 \mathbf{S}_{3,x_3} \mathbf{S}_{4,x_4} \rceil_p \right\rceil_p \right. \cdot \tag{1.4}$$

$$\left. \left\lfloor \mathbf{A}_2 \lfloor \mathbf{A}_3 \mathbf{S}_{5,x_5} \mathbf{S}_{6,x_6} \rceil_p \cdot \lfloor \mathbf{A}_3 \mathbf{S}_{7,x_7} \mathbf{S}_{8,x_8} \rceil_p \right\rceil_p \right\rceil_p$$

As is evident from the equations, one can view K_ℓ as a sort of key-sample flip-flop with S_ℓ. However, this relationship is not exact, since the hardness results of LWE with different distributions of keys and samples are not equivalent.

Moving to Rings. While PRFs derived from K_ℓ (which we will call P^ℓ) are a little bit more complicated and slightly less efficient than those built from S_ℓ (although this doesn't show up asymptotically under $\tilde{O}(\cdot)$ notation), they have one huge advantage over constructions from S_ℓ: they admit ring instantiations. As in [BPR12], we can almost immediately derive a ring form of K_ℓ, which we call $K_{R,\ell}$, and a corresponding PRF $P^{R,\ell}$. These ring PRFs allow us to match the efficiency of all previously known ring LWE-based PRFs while maintaining a slightly superpolynomial modulus at the cost of only more key size.

PRFs P^2 and $P^{R,2}$ from Synthesizers K_2 and $K_{R,2}$. Our PRF P^2, which we showed for 8 bits in Eq. 1.4, turns out to have parameters almost exactly

---

[8] There are other choices available for the key distribution here–perhaps even more efficient ones.

asymptotically equivalent to F^2, including modulus $O\left(m^{\omega(1)}\right)$. The only difference is that P^2 has large secret keys, while F^2 has large public parameters.

We see substantial improvements when we move to rings. The PRF $P^{R,2}$ is the second (after that in [Ban15]) known PRF based on the hardness of ring LWE with modulus $m^{\omega(1)}$ (where m is now the degree of the polynomial of the ring R) with *any* sublinear circuit depth. In addition, $P^{R,2}$ matches the most efficient known ring LWE-based PRFs (like those of [BP14]) in terms of asymptotic computational efficiency. Again, the only drawback is larger key sizes.

Synthesizing [BP14]. We can also build what we call tree-based synthesizers, which are based on and look almost identical to the PRFs from [BP14]. We call these synthesizers T_ℓ and $T_{R,\ell}$ for the standard and ring-based versions, respectively. While the tree constructions are a bit too complicated to explain here, we note that we can get parameters asymptotically equivalent to what we have achieved earlier for simple PRFs based on T_2 and $T_{R,2}$ (which we call B^2 and $B^{R,2}$, respectively).

The main advantage of these tree-based synthesizers is that we can potentially build many more interesting PRFs than we otherwise could with the simpler synthesizers. We have yet to fully explore the potential of these synthesizers, but we think that there might be many interesting applications.

Moving to Higher ℓ. In our PRF constructions, ℓ essentially acts as a parallelization parameter–it can be thought of as a 'locality' parameter for the synthesizer. While we do not seem to gain anything in the integer lattice setting from setting ℓ to be anything higher than a constant (other than the case where $\ell = \lambda$ and we gain key homomorphic properties for certain PRFs), we can achieve some theoretically interesting results from a higher ℓ in the ring setting.

In [BPR12], the authors showed how to construct a PRF in NC^1 using ring LWE. We generalize this PRF with our synthesizer $K_{R,\ell}$ and show two interesting choices of PRF to examine with higher ℓ. We first consider the PRF $P^{R,\lambda^{\frac{1}{\log \lambda}}}$ which is built using the synthesizer $K_{R,\lambda^{\frac{1}{\sqrt{\log \lambda}}}}$. This PRF has modulus $m^{\omega(1)\left(\lambda^{\frac{1}{\sqrt{\log \lambda}}}\right)}$, which is clearly large but is still smaller than m^{λ^c} for any constant c. The synthesizer construction tree of this PRF also has depth $\sqrt{\log n}$, so we can build this PRF in overall circuit depth of $O\left((\log n)^{\frac{3}{2}}\right)$, giving us a PRF in the unorthodox class $NC^{1.5}$.

We finally consider the PRF $P^{R,\lambda^{\lambda^c}}$ for some constant $c \in (0,1)$ which is based upon the synthesizer $K_{R,\lambda^{\lambda^c}}$. This PRF has modulus $m^{\omega(1)\lambda^c}$ and synthesizer construction with constant depth, meaning that it can be built in NC^1. This PRF is interesting because it is the PRF with the smallest modulus that we can build in NC^1 using our techniques. This lets us build lattice PRFs in NC^1 with any subexponential modulus (assuming λ is polynomial in m), but we still do not know how to break this subexponential barrier (or if it is even possible). The existence of PRFs in NC^1 has many interesting implications in complexity theory [BFKL94], so building them from maximally hard assumptions is an important problem.

Comparison with Previous Work. In Table 1 on the adjacent page, we compare our new PRFs with those of the relevant previous works. We borrow the table format from [BP14].

The General View. While our synthesizers look very similar to existing PRF constructions, the PRF constructions themselves can be viewed as generalizations of those in [BPR12], [BLMR13], and [BP14]. In fact, setting $\ell = \lambda$ for the synthesizers S_ℓ, K_ℓ, and T_ℓ result in PRFs that are almost identical to those in [BLMR13], [BPR12], and [BP14], respectively. However, we do lose the key homomorphic properties of [BLMR13] and [BP14] when we set $\ell < \lambda$.

Concrete Instantiations and Parameters. If we want to instantiate an actual PRF, we need to look beyond the asymptotics. It is relatively straightforward to see that, except for the highly sequential GGM-based and [BP14] sequential constructions, the constructions of PRFs from synthesizers with small values of ℓ and the construction from Banerjee's thesis [Ban15] are the most efficient overall PRF constructions (although the key sizes are larger) for large input lengths: these have at most a small constant (either 2 or 4) times the number of multiplications as the more direct constructions with a substantially smaller modulus. So, we choose to analyze concretely the synthesizer constructions here.

If we fix a particular 'lattice security dimension' m and a particular subexponential parameter (derived from how we set the modulus in any learning with rounding reduction) r_p, we can examine how some of the schemes work practically. We show these concrete metrics in Table 2. While we have not implemented these PRFs or closely examined the speed for various multiplication algorithms, it seems like the PRF $P_{R,2}$ is a strong candidate to be the fastest parallel PRF from lattices known today.

Theoretical Implications. A lofty goal in lattice-based cryptography is to build a PRF based on the hardness of LWE with polynomial modulus q. While we obviously do not achieve that in this work, we seemingly make progress towards this goal. In particular, previous lattice-based PRFs typically relied on long subset-products of matrices multiplied by a secret key. In this work, we show that only a 2-subset product of matrices multiplied by a key is generally sufficient to build a PRF. This substantially generalizes the requirements seemingly needed to build a lattice-based PRF with polynomial modulus. We hope that this result can be used as a stepping stone towards such PRFs.

1.4   Paper Outline

Unfortunately, due to the space constraints for the conference version of the paper, we do not have space to include a substantial amount of content. However, we hope that the main body of the paper provides a clear and concise explanation

Table 1. Comparison with Previous Work: The parameters are with respect to PRFs with input length λ and reductions to worst-case lattice problems in dimension m. We let τ denote the exponent of matrix multiplication. Brackets $[\cdot]$ denote parameters of a ring-LWE construction that are better than those of integer lattices (when ring LWE schemes are possible). We ignore constants, lower-order terms, and logarithmic factors. *Parallel Circuit Complexity* refers to the parallel circuit complexity class of the PRF, when applicable. *Parallel Matrix Comp.* refers to the parallel matrix complexity of a PRF in terms of matrix multiplication operations, which is a much better measure of practical parallelizability than circuit depth. PP refers to public parameters, again when applicable. ★ refers to polynomially-sized parameters that are too big to fit nicely in the table (and are relatively unimportant anyway). ✠ refers to parameters that are dependent on a (unspecified) universal hash function.

Reference	Modulus	Parallel circuit complexity	Parallel matrix or Ring Comp.
[BPR12] GGM	$m^{\omega(1)}$	–	λ
[BPR12] synth	$m^{\log \lambda}$	NC^2	$\log \lambda$
[BPR12] direct	m^{λ}	NC^2 $[NC^1]$	$\log \lambda$
[BLMR13]	m^{λ}	NC^2	$\log \lambda$
[BP14] sequen	$m^{\omega(1)}$	–	λ
[BP14] balanced	$m^{\log \lambda}$	NC^2	$\log \lambda$
[Ban15] synth	$m^{\omega(1)}$	NC^2	$\log \lambda$
[DS15]	$m^{\log \lambda}$	$NC^{1+o(1)}$	$\log \log \lambda$
This Work: F^2	$m^{\omega(1)}$	NC^2	$\log \lambda$
This Work: P^2	$m^{\omega(1)}$	NC^2	$\log \lambda$
This Work: B^2	$m^{\omega(1)}$	NC^2	$\log \lambda$
This Work: $P^{\lambda^{\frac{1}{\sqrt{\log \lambda}}}}$	$m^{\omega(1)\lambda^{\left(\lambda^{\frac{1}{\sqrt{\log \lambda}}}\right)}}$	NC^2 $[NC^{1.5}]$	$\log \lambda$
This Work: P^{λ^c}	$m^{\omega(1)\lambda^c}$	NC^2 $[NC^1]$	$\log \lambda$

Reference	Key size	PP size	Time/out	Out
[BPR12] GGM	m	m^2 $[m]$	λm $[\lambda]$	m
[BPR12] synth	λm^2 $[\lambda m]$	0 [0]	$\lambda m^{\tau-2}$ $[\lambda]$	m^2 $[m]$
[BPR12] direct	$\lambda^3 m^2$ $[\lambda^2 m]$	0 [0]	$\lambda^3 m$ $[\lambda^2]$	λm $[\lambda m]$
[BLMR13]	$\lambda^2 m$	$\lambda^3 m^2$	$\lambda^3 m$	λm
[BP14] sequen	m	m^2 $[m]$	$\lambda m^{\tau-1}$ $[\lambda]$	m
[BP14] balanced	m	m^2 $[m]$	$\lambda m^{\tau-1}$ $[\lambda]$	m
[Ban15] synth	λm^2 $[\lambda m]$	0 [0]	$\lambda m^{\tau-2}$ $[\lambda]$	m^2 $[m]$
[DS15]	m^2 $[m]$	✠	✠	m
This Work: F^2	m^2	λm^2	$\lambda m^{\tau-2}$	m^2
This Work: P^2	λm^2 $[\lambda m]$	m^2 $[m]$	$\lambda m^{\tau-2}$ $[\lambda]$	m^2 $[m]$
This Work: B^2	m^2 $[m]$	λm^2 $[\lambda m]$	$\lambda m^{\tau-2}$ $[\lambda]$	m^2 $[m]$
This Work: $P^{\lambda^{\frac{1}{\sqrt{\log \lambda}}}}$	★	★	★	★
This Work: P^{λ^c}	★	★	★	★

Some Comments: We note that the large keys of some of the PRFs (including ours) can be computed using a lattice-based PRG in practical cases, making this less of an issue in practice. In addition, we note that it is trivial to modify the (standard LWE-based) PRFs from [BP14] to have exactly the same time per output and output sizes as our PRFs F^2, P^2, and B^2 by expanding the secret to be a full-rank matrix rather than a vector. The authors of [BP14] mention this as an optimization in their work.

Table 2. Practical Comparison with Previous Work: In this table we consider the practical implementation of the most efficient parallel lattice-based PRFs. We compare our PRFs to that from [Ban15] in terms of concrete efficiency. The parameters are with respect to PRFs with input length λ and reductions from worst-case lattice problems in dimension m. Brackets [·] denote parameters of a ring-LWE construction that are better than those of integer lattices (when ring LWE schemes are possible). For simplicity, we only state higher-order terms (i.e., we ignore polynomial terms when superpolynomial terms exist, and we ignore constants when polynomial terms exist). The term r_p–short for 'rounding parameter'–refers to the (superpolynomially large) value induced by the LWE $\to$ LWR reduction that we have denoted in Table 1 as $m^{\omega(1)}$.

Reference	Modulus	Matrix/Ring dimension	Key size (Matrices/Ring elements)
[Ban15] synth	r_p^2	m	$8\lambda \times \mathbb{Z}_q^{m\times m} [R_q]$
This Work: F^2	r_p	$m\log q$	$4\lambda \times \mathbb{Z}_2^{m\log q\times m\log q}, \log\lambda \times \mathbb{Z}_q^{m\log q\times m\log q}$
This Work: P^2	r_p^3	m	$4\lambda \times \mathbb{Z}_p^{m\times m} [R_p], \log\lambda \times \mathbb{Z}_q^{m\times m} [R_q]$

Reference	Matrix/Ring multiplies	Matrix/Ring product computation
[Ban15] synth	$4(\lambda-1)$	$\mathcal{U}\left(\mathbb{Z}_q^{m\times m}\right) \times \mathcal{U}\left(\mathbb{Z}_q^{m\times m}\right) [\mathcal{U}(R_q) \times \mathcal{U}(R_q)]$
This Work: F^2	$2(\lambda-1)$	$\mathcal{U}\left(\mathbb{Z}_2^{m\log q\times m\log q}\right) \times \mathcal{U}\left(\mathbb{Z}_q^{m\log q\times m\log q}\right)$
This Work: P^2	$2(\lambda-1)$	$\mathcal{U}\left(\mathbb{Z}_{r_p}^{m\times m}\right) \times \mathcal{U}\left(\mathbb{Z}_q^{m\times m}\right) [\mathcal{U}(R_{r_p}) \times \mathcal{U}(R_q)]$

Some Comments: The 'matrix/ring product computation' shows the distributions of what matrices or rings we are multiplying in the PRFs above. All of the operations are computed modulo q, but, for our PRFs listed here, some of the matrices or rings are nonuniform (and thus can be multiplied more quickly). The fastest PRF is most likely either P^2 for regular LWE or the ring version of P^2 for ring LWE (this is definitely the case if we use fast Fourier transform (FFT) multiplication over field elements, but less clear for other multiplication algorithms–the construction from [Ban15] or F^2 may be faster for asymptotically slow modular multiplication algorithms).

of the results. As we have mentioned before, we encourage interested readers to find the full version of our paper on the IACR cryptology eprint archive.

The rest of the paper proceeds as follows: we begin by defining some basic cryptographic notation and facts about PRFs and lattice problems in Sect. 2. A reader knowledgeable in lattices and PRFs can safely skip this section. In Sect. 3, we define pseudorandom synthesizers and state results from [NR95] on constructions of PRFs from pseudorandom synthesizers. Our definitions are phrased a little differently than those in [NR95], but the meaning is identical.

In Sect. 4 we formally define our first pseudorandom synthesizer S_ℓ which is based on F_{BLMR} and give an overview of the proof of security. We then give formal analysis of the PRF F^2 which we build from S_2 in Sect. 5. Finally, in Sect. 6 we conclude and state what we consider are interesting and important open problems in the area.

2   Preliminaries

We start by discussing some basic background material for the paper. A reader who is familiar with the basic cryptographic concepts in each subsection can safely skip the respective subsections.

2.1   Notation

For a random variable X we denote by $x \leftarrow X$ the process of sampling a value x according to the distribution of X. Similarly, for a finite set S we denote by $x \leftarrow S$ the process of sampling a value x according to the uniform distribution over S. We sometimes also use $\mathcal{U}(S)$ to denote the uniform distribution over a set S. We typically use bold lowercase letters (i.e. $\mathbf{a}$) to denote vectors and bold uppercase letters (i.e. $\mathbf{A}$) to denote matrices.

For two bit-strings x and y (or vectors $\mathbf{x}$ and $\mathbf{y}$) we denote by $x\|y$ their concatenation. A non-negative function $f : \mathbb{N} \to \mathbb{R}$ is negligible if it vanishes faster than any inverse polynomial. We denote by $\mathrm{Rk}_i(\mathbb{Z}_p^{a \times b})$ the set of all $a \times b$ matrices over $\mathbb{Z}_p$ of rank i.

We unfortunately do not have space for a comprehensive preliminaries section in the conference proceedings format. For a full preliminaries, please refer to the full version of our paper[9].

3   Pseudorandom Synthesizers

In this section we introduce pseudorandom synthesizers. Pseudorandom synthesizers were invented by Naor and Reingold in their seminal work [NR95]. Since previous general-purpose PRF constructions were entirely sequential [GGM84] (i.e. had circuit depth at least linear in the number of input bits), which was both theoretically and practically inefficient, Naor and Reingold developed a new technique for building highly parallel PRF constructions which they called pseudorandom synthesizers.

We spend a little bit more time on this than usual because we present the material in a different way than [NR95] or Omer Reingold's thesis [Rei]. Rather than using synthesizer ensembles, we opt for the more modern game-based definitions where keys are chosen randomly (rather than functions are selected randomly from an ensemble). This means that, in addition to a traditional synthesizer, we need to define a keyed synthesizer as well. The definitional changes require us to change the way the definitions of synthesizers are presented, but we note that the content remains exactly the same.

We additionally generalize some of the definitions to cover alternative constructions that are mentioned in [NR95] (and shown to be secure), but not covered by the main definition. We start by defining a basic pseudorandom synthesizer.

---

[9] Available on the IACR cryptology eprint archive.

Pseudorandom Synthesizer Basics. A pseudorandom synthesizer is, in rough terms, a two-input function $S(\cdot, \cdot)$ parameterized by two integers m and n such that on random inputs $(x_1, ..., x_m) \in \mathcal{X}$ and $(w_1, ..., w_n) \in \mathcal{W}$, the matrix $\mathbf{M}$ of all mn values of $S(x_i, w_j) = \mathbf{M}_{ij}$ is indistinguishable from random.

Let m and n be integers. In precise terms, a pseudorandom synthesizer is an efficiently computable function $S : \mathcal{X} \times \mathcal{W} \to \mathcal{Y}$ parameterized by m and n, where $\mathcal{X}$ and $\mathcal{W}$ are the two input domains and $\mathcal{Y}$ is the range. In this paper, we sometimes allow the synthesizer to take additional public parameters pp and use $S_{pp} : \mathcal{X} \times \mathcal{W} \to \mathcal{Y}$ to denote such a synthesizer.

Security for a synthesizer is defined using two experiments between a challenger and an adversary $\mathcal{A}$. For $b \in \{0, 1\}$ the challenger in Exp_b works as follows.

1. When $b = 0$ the challenger sets $f(\cdot, \cdot) \overset{\mathrm{def}}{=} S(\cdot, \cdot)$.
2. When $b = 1$ the challenger chooses a random function $f : \mathcal{X} \times \mathcal{W} \to \mathcal{Y}$.
3. The challenger samples input values $(x_1, ..., x_m) \leftarrow \mathcal{X}$ and $(w_1, ..., w_n) \leftarrow \mathcal{W}$ and sends the values $f(x_i, w_j)$ for all $i \in [1, m]$ and $j \in [1, n]$ to the adversary. Eventually the adversary outputs a bit $b' \in \{0, 1\}$.

For $b \in \{0, 1\}$ let W_b be the probability that $\mathcal{A}$ outputs 1 in Exp_b.

Definition 1. *A synthesizer $S : \mathcal{X} \times \mathcal{W} \to \mathcal{Y}$ is secure if for all efficient adversaries $\mathcal{A}$ the quantity*

$$\mathsf{SYNTH}_{adv}[\mathcal{A}, F] \overset{\mathrm{def}}{=} |W_0 - W_1|$$

is negligible.

The above definition is what most papers that present pseudorandom synthesizers use. However, as we earlier alluded, the work of [NR95] allows for substantially more generality. First, we note that we can also use *keyed* synthesizers. A keyed pseudorandom synthesizer (KPS) is, in rough terms, a *keyed* two-input function $S(k, \cdot, \cdot)$ parameterized by two integers m and n such that on sets of random inputs $(x_1, ..., x_m) \in \mathcal{X}$ and $(w_1, ..., w_n) \in \mathcal{W}$, the matrix $\mathbf{M}$ of all mn values of $S(k, x_i, w_j) = \mathbf{M}_{ij}$ is indistinguishable from random. We note that on successive queries to the KPS, the same key is used but new input values x_i and y_i are chosen.

Additionally, we note that it is not necessary that our synthesizer S be a function with two inputs and one output. S could have three (or more) inputs, as long as we can still prove security. Once again, Naor and Reingold show a proof of security for this case as well in [NR95]. Thus, we overload S so that it can take more than two inputs. We next present a modified definition of a pseudorandom synthesizer that takes all of these extra considerations into account.

Keyed Pseudorandom Synthesizer Definition. Let ℓ be an integer, and let $n_1, ..., n_\ell$ be integers as well. In precise terms, a pseudorandom synthesizer is an efficiently computable function $S : \mathcal{K} \times \mathcal{X}_1 \times ... \times \mathcal{X}_\ell \to \mathcal{Y}$ parameterized by ℓ, $n_1, ... , n_\ell$ where $\mathcal{K}$ is the keyspace, $\mathcal{X}_1, ... , \mathcal{X}_\ell$ are the ℓ input domains and $\mathcal{Y}$ is the range. In this paper, we sometimes allow the synthesizer to take additional public parameters pp and use $S_{pp} : \mathcal{K} \times \mathcal{X}_1 \times ... \times \mathcal{X}_\ell \to \mathcal{Y}$ to denote such a keyed synthesizer.

Security for a synthesizer is defined using two experiments between a challenger and an adversary $\mathcal{A}$. For $b \in \{0,1\}$ the challenger in Exp_b works as follows.

1. When $b = 0$ the challenger selects a random key $k \leftarrow \mathcal{K}$ and sets $f(\cdot, \cdot) \stackrel{\text{def}}{=} S(k, \cdot, ..., \cdot)$.
2. When $b = 1$ the challenger chooses a random function $f : \mathcal{X}_1 \times ... \times \mathcal{X}_\ell \to \mathcal{Y}$.
3. The challenger samples input values $(x_{1,1}, ..., x_{1,n}) \leftarrow \mathcal{X}_1$, $(x_{2,1}, ..., x_{2,n}) \leftarrow \mathcal{X}_2, ... , (x_{\ell,1}, ..., x_{\ell,n}) \leftarrow \mathcal{X}_\ell$ and sends the values $f(x_{1,i_1}, x_{2,i_2}..., x_{\ell,i_\ell})$ for all $i_1 \in [1, n_1] , ... , i_\ell \in [1, n_\ell]$ to the adversary. If the number of possible values of f is superpolynomial, the adversary is allowed to adaptively query f on inputs of the form $(i_1, ..., i_\ell)$ of its choice. The challenger repeats this process an arbitrary polynomial number of times. Eventually the adversary outputs a bit $b' \in \{0,1\}$.

For $b \in \{0,1\}$ let W_b be the probability that $\mathcal{A}$ outputs 1 in Exp_b.

Definition 2. *A synthesizer $S : \mathcal{K} \times \mathcal{X}_1 \times ... \times \mathcal{X}_\ell \to \mathcal{Y}$ is secure if for all efficient adversaries $\mathcal{A}$ the quantity*

$$\mathsf{SYNTH}_{adv}[\mathcal{A}, F] \stackrel{\text{def}}{=} |W_0 - W_1|$$

is negligible.

3.1   Building PRFs from Synthesizers

In this section we explain how to build pseudorandom functions from synthesizers using the main theorem from [NR95]. We use different terminology but the content of the theorem statement remains the same.

ℓ-Admissible Synthesizers. In order to build synthesizers that combine ℓ inputs into one, we need to make sure that the overall bit length of our input is appropriate for our synthesizer length ℓ. To see how this might go wrong, suppose we are trying to construct a 4-bit PRF from a 3-way synthesizer. If we combine inputs 1, 2, and 3, we will get another input $1'$. But we will only have input 4 to combine with it. If our synthesizer only works on three inputs (and not two—some synthesizers might work on both two or three inputs), we will be stuck and unable to finish our PRF! The authors of [NR95] do not explicitly mention such an idea, but it is implicit in their work.

Definition 3. *We say that a number λ is ℓ-admissible if the following procedure outputs one:*

1. *While $\lambda \geq \ell$:*
   (a) Write $\lambda = \ell k + r$ where $r \in [0, \ell]$
   (b) Set $\lambda = \frac{\lambda - r}{\ell} + r$.

2. Output λ

We note that any λ is 2-admissible. For larger values of ℓ, the situation is slightly more complicated. In practice, this admissibility fact won't be too much of an issue–we can just pick λ to be a multiple of ℓ, for instance–but we need it for our synthesizer definition to be complete.

Definition 4.   *Squeeze Function SQ^ℓ_{sk}: Let X be some set and sk some secret key. Let k and ℓ be an integers, and let $k \bmod \ell = r$. For every function $S^\ell_{sk} : X^\ell \to X$, and every sequence of inputs $L = \{x_1, ..., x_k\}$ where $x_i \in X$ we define the squeeze $SQ_{sk}(L)$ to be the sequence $L' = \left\{ x'_1, ..., x'_{\lfloor \frac{k}{\ell} \rfloor}, x'_{\lfloor \frac{k}{\ell} \rfloor + 1}, ..., x'_{\lfloor \frac{k}{\ell} \rfloor + r} \right\}$ where $x'_i = S_{sk}\left(x_{\ell i - (\ell - 1)}, ...x_{\ell i - 1}, x_{\ell i} \right)$ for $i \le \lfloor \frac{k}{\ell} \rfloor$, and if $k \ne 0 \bmod \ell$, then for each $i \ge \lfloor \frac{k}{\ell} \rfloor$ we set $x'_i = x_{(\ell - 1) \lfloor \frac{k}{\ell} \rfloor + i}$.*

Definition 5.   *Let ℓ be an integer, and let λ be an ℓ-admissible integer. Let $S^\ell_{sk} : X^\ell \to X$ be a family of keyed pseudorandom synthesizers with key generation algorithm KeyGen. We define a pseudorandom function F in the following way:*
Key Generation:

1. *For $j \in [1, ..., \lceil \log_\ell \lambda \rceil]$, sample $sk_i \leftarrow KeyGen$.*
2. *For $i \in [1, ..., \lambda]$ and $b \in [0, 1]$, sample $x_{i,b} \leftarrow X$.*

Evaluation: For some bit string $\mathbf{i} \in \mathbb{Z}_2^\lambda = \{i_1 i_2 ... i_\lambda\}$ we have

$$F_{pp}(\mathbf{i}) = SQ^\ell_{sk_1}\left(SQ^\ell_{sk_2}\left(... SQ^\ell_{sk_{\lceil \log_\ell \lambda \rceil}} \{x_{1,i_1}, x_{2,i_2}, ..., x_{\lambda, i_\lambda}\} ... \right) \right)$$

We next state the main theorem from [NR95], which proves that any adversary that can distinguish the PRF construction in Definition 5 from random can be used to distinguish the output of the synthesizer S^ℓ_{sk} from random. We paraphrase the theorem slightly to accommodate our definitions, which, as we have mentioned numerous times, are slightly different from those in [NR95].

Theorem 1.   *Let ℓ and λ be integers. Let S^ℓ_{sk} be a pseudorandom synthesizer as defined in Definition 2, and let F_{pp} be the function defined in Definition 5. Any adversary that can distinguish F_{pp} from a truly random function with advantage ε can be used to distinguish S^ℓ_{sk} from random with advantage $\frac{\varepsilon}{\log_\ell \lambda}$.*

Proof. This theorem is almost exactly Theorem 5.1 of [NR95] and the proof can be found there.

4   Sample Subset-Product Pseudorandom Synthesizer

In this section we define a new pseudorandom synthesizer based on the LWE assumption and prove that it is secure. Our synthesizer S_ℓ very closely resembles the PRF from [BLMR13]. We choose to present this synthesizer first because it is the simplest and most intuitive construction and has the easiest composition into PRFs.

4.1　Synthesizer Definition

Definition 6.　*Let m, q, p, and ℓ be integers such that $p \leq q$ and $\frac{2^{\lceil \log q \rceil} - q}{q}$ is negligible. Let $k_1, ..., k_\ell \in \mathbb{Z}$ be positive integers. For $i \in [1, \ell]$ and $j \in [1, k_i]$ let $\mathbf{A}_{i,j} \in \mathbb{Z}_2^{m \times m}$ be a uniformly distributed binary matrix. Let $\mathbf{S} \in \mathbb{Z}_q^{m \times m}$ be a matrix sampled uniformly at random.*

We define the synthesizer $S_\ell : \mathbb{Z}_q^{m \times m} \times \left[\left(\mathbb{Z}_2^{m \times m} \right)^{k_1} \times ... \times \left(\mathbb{Z}_2^{m \times m} \right)^{k_\ell} \right] \rightarrow \mathbb{Z}_p^{k_1 m \times ... \times k_\ell m}$ in the following way: for each $m \times m$ block of output of S_ℓ, define

$$S_\ell (x_1, ...x_\ell) \stackrel{\text{def}}{=} \left[\left[\prod_{i=1}^{\ell} \mathbf{A}_{i,x_i} \right] \mathbf{S} \right]_p \tag{4.1}$$

where $x_i \in [1, k_i]$.

We now offer some comments on our synthesizer S_{LWE}. First, note that as long as p is even, we can 'chain' this synthesizer. In other words, if this is the case, we can modify the output of each (i, j)-block of the synthesizer $S_{\text{LWE}(i,j)}$ to be a random matrix over $\mathbb{Z}_2^{m \times m}$ by just computing the output modulo two. Later in the paper (when we select parameters and analyze the overall PRF's performance) we will comment more on this.

We defer the security proof of this construction to the full version of the paper. We note that, while the proof of [BP14] could be applied obliquely to get the same result, the implied parameters are not as good as we can achieve with a direct proof.

5　Constructions of PRFs from S_ℓ

In this section we show how our synthesizer S_ℓ can be used to build PRFs. We also show some optimizations that we can achieve by slightly modifying the overall synthesizer construction (in a way that doesn't affect security).

We start by stating an overall theorem about the security of PRFs constructed from our pseudorandom synthesizers using the synthesizer construction of [NR95]. This theorem follows almost immediately from applying the synthesizer construction security theorem of [NR95] as we stated in Theorem 1 to our theorem in the full version of the paper proving our synthesizers S_ℓ secure.

Theorem 2.　*Let m, n, q, p, λ, and ℓ be integers. In words, m will be our lattice dimension, n will be the dimension of the LWE problem we reduce to, q is our modulus, p is our rounding parameter, λ is our PRF length, and ℓ is our synthesizer parameter. Let $\psi \in \mathbb{Z}$ be a B-bounded noise distribution. We additionally require that $q \geq 2m^{\ell + \omega(1)} Bp$ and $\frac{2^k - q}{q}$ is negligible for some integer k. Let $p = 2$.*

Let the PRF F^ℓ defined by applying the synthesizer construction defined in Definition 5 to the synthesizer S_ℓ defined in Definition 6. Let Q be the number of queries an adversary makes to F^ℓ.

Any adversary that can distinguish F^ℓ from random with advantage ε can be used to solve the $(q, n, \psi, \mathcal{U}_q^n, U_q^n)$-LWE problem (the standard version of LWE) with advantage $\frac{\varepsilon}{2mQ\log\lambda}$.

As long as we follow the parameter choices implied by Theorem 2, we can build PRFs from S_ℓ for any choice of ℓ. In fact, we could mix and match different values of ℓ in a single PRF, but we do not see any logical reason to do so (perhaps odd hardware constraints could make such a thing useful).

In the rest of the section, we examine the PRF F^2, which we build from S_2.

5.1   The Synthesizer S_2

We start by analyzing the simplest, and yet one of the most efficient, synthesizers: S_2. From Theorem 2, we know that F^2–the PRF built using S_2–is secure as long as $q \geq 2m^{2+\omega(1)}Bp = 2m^{\omega(1)}Bp$ and $\frac{2^k - q}{q}$ is negligible for some integer k. We let $p = 2$.

Let's evaluate our PRF construction F^2. We start by noting that, due to our choice of parameters, $q \geq 4m^{2+\omega(1)}Bp = 2m^{\omega(1)}Bp$, and thus

$$\log q = O\left(\omega\left(1\right)\left(\log m\right)\right) = O\left(\omega\left(1\right)\left(\log\left(n\log q\right)\right)\right) =$$

$$O\left(\omega\left(1\right)\left(\log\left(n\right)\log\log\left(q\right)\right)\right) = O\left(\omega\left(1\right)\log n\right)$$

Efficiency Calculations. At level i of the synthesizer tree, we do $2 \cdot 2^{\log(\lambda)-i}$ matrix multiplications and $2^{\log(\lambda)-i}$ matrix rounding operations. If we sum over all levels of the tree, we perform $4\lambda - 1$ matrix multiplications and $2\lambda - 1$ matrix rounding operations. Since $p = 2$, the output of our synthesizer is a random-looking binary matrix that can be immediately used at the next level, and we do not have to spend any computational time reformatting this output.

Our matrices are of dimension size $m = n\log q$. If we set τ to be the exponent corresponding to optimal matrix multiplication, this means our PRF can be computed in time $\tilde{O}(\lambda n^\tau)$. In addition, note that we output m^2 bits, so our operations per output bit is $\tilde{O}(\lambda n^{\tau-2})$.

Parallel Complexity. We subdivide parallel complexity into two categories: complexity in terms of matrix operations if this is what we are only allowed (i.e. multiply, add, and round) and absolute complexity. Our PRF F^2 clearly has $O(\log\lambda)$ matrix operation complexity, since the longest potential path from root to leaf on our synthesizer tree has $2\log(\lambda)$ multiplies and $\log(\lambda)$ rounding operations. Since matrix multiplication is in NC^1 [RW04] and our synthesizer tree has depth $\log\lambda$, this means F^2 is in NC^2.

Key and Public Parameter Size. The one area that our construction F^2 does not do well on is key size. For our key, we need $\log\lambda$ uniform matrices in $\mathbb{Z}_q^{m\times m}$ as well as 2λ matrices in $\mathbb{Z}_2^{m\times m}$. We note that these additional binary matrices can be made public without any loss of security (since they are what would traditionally be the public matrices of an LWE instance). This gives us secret key sizes of $\log(\lambda)(m^2\log q)$ and public parameter sizes of λm^2.

6   Conclusion and Open Problems

In this paper, we showed how to build more efficient lattice-based PRFs using keyed pseudorandom synthesizers. We constructed PRFs with only slightly superpolynomial modulus (independent of the number of input bits) that match the efficiency of the otherwise most efficient known constructions. We also show how to build other PRFs that imply interesting results on the parallel circuit complexity of lattice PRFs.

6.1   Open Problems

We conclude the paper by stating two open problems. We think these problems are very important for lattice cryptography in general, as well as (obviously) PRFs.

LWR with Polynomial Modulus. Recall that the learning with rounding problem states that, informally, it is hard to distinguish samples of the form $\left(\mathbf{a}_i, \lceil \mathbf{a}_i{}^\mathsf{T}\mathbf{s} \rfloor_p\right)$ from random for uniformly random samples $\mathbf{a}_i$ and secret key $\mathbf{s}$. For an unbounded number of samples (what is needed for a PRF reduction), the hardness of LWR is only known when the modulus q of the problem instance is superpolynomially large. This doesn't seem natural to us, and we think attempting to prove that LWR is hard for a polynomial modulus and unbounded samples (or showing evidence that this is not true, although we consider this unlikely) is a worthwhile endeavor. While there have been several papers establishing the hardness of LWR with polynomial hardness and a *fixed* numbers of samples [AKPW13], [BGM+16], [BLL+15], [AA16], the security of LWR with a polynomial modulus and unbounded polynomial samples is still unknown.

Earlier this year, Montgomery [Mon18] showed how to build a (highly non-standard) variant of learning with rounding with polynomial modulus and unbounded samples. However, at a first glance, this new LWR variant does not seem to be able to be used with our techniques (or any others, for that matter) to build PRFs with polynomial modulus. Constructing a variant of LWR that can be used to build PRFs with polynomial modulus (or showing a security proof with polynomial modulus and unbounded samples for the regular version of LWR) remains an important open problem in our opinion.

Subset Product LWE with Polynomial Modulus. Let $\mathbf{A}_{i,b_i} \in \mathbb{Z}_q^{m \times m}$ for $i \in [1, ...\lambda]$ and $b_i \in \{0, 1\}$ be matrices selected uniformly at random, and let $\mathbf{s} \in \mathbb{Z}_q^m$ also be sampled uniformly from random. We call the following function subset product LWE:

$$F\left(x_1...x_\lambda\right) = \prod_{i=1}^{\lambda} \mathbf{A}_{i,x_i}\mathbf{s} + \boldsymbol{\delta}_{\mathbf{x}}$$

where $\boldsymbol{\delta}_{\mathbf{x}}$ is a noise vector selected independently at random for each input $\mathbf{x}$.

Currently, for polynomial λ, this function F can only be shown to be hard for modulus on the order of m^λ. Like LWR, we see no real reason why this problem should not be hard for a polynomial modulus: there are no known attacks, and the large modulus seems to be a relic of the hybrid argument in the proofs. We think attempting to prove this function secure for smaller modulus (and thus achieve better lattice hardness results) is a great candidate for future research.

References

[AA16]   Alperin-Sheriff, J., Apon, D.: Dimension-preserving reductions from LWE to LWR. IACR Cryptology ePrint Archive, 2016:589 (2016)

[AKPW13] Alwen, J., Krenn, S., Pietrzak, K., Wichs, D.: Learning with rounding, revisited. In: Canetti, R., Garay, J.A. (eds.) CRYPTO 2013. LNCS, vol. 8042, pp. 57–74. Springer, Heidelberg (2013). https://doi.org/10.1007/978-3-642-40041-4_4

[Ban15]   Banerjee, A.: New constructions of cryptographic pseudorandom functions. Ph.D. thesis (2015). https://smartech.gatech.edu/bitstream/handle/1853/53916/BANERJEE-DISSERTATION-2015.pdf?sequence=1&isAllowed=y

[BCK96]   Bellare, M., Canetti, R., Krawczyk, H.: Pseudorandom functions revisited: the cascade construction and its concrete security. In 37th Annual Symposium on Foundations of Computer Science, Burlington, Vermont, pp. 514–523. IEEE Computer Society Press (1996)

[BFKL94] Blum, A., Furst, M., Kearns, M., Lipton, R.J.: Cryptographic primitives based on hard learning problems. In: Stinson, D.R. (ed.) CRYPTO 1993. LNCS, vol. 773, pp. 278–291. Springer, Heidelberg (1994). https://doi.org/10.1007/3-540-48329-2_24

[BFP+15] Banerjee, A., Fuchsbauer, G., Peikert, C., Pietrzak, K., Stevens, S.: Key-homomorphic constrained pseudorandom functions. In: Dodis, Y., Nielsen, J.B. (eds.) TCC 2015. LNCS, vol. 9015, pp. 31–60. Springer, Heidelberg (2015). https://doi.org/10.1007/978-3-662-46497-7_2

[BGM+16] Bogdanov, A., Guo, S., Masny, D., Richelson, S., Rosen, A.: On the hardness of learning with rounding over small modulus. In: Kushilevitz, E., Malkin, T. (eds.) TCC 2016. LNCS, vol. 9562, pp. 209–224. Springer, Heidelberg (2016). https://doi.org/10.1007/978-3-662-49096-9_9

[BKM17]   Boneh, D., Kim, S., Montgomery, H.: Private puncturable PRFs from standard lattice assumptions. In: Coron, J.-S., Nielsen, J.B. (eds.) EUROCRYPT 2017. LNCS, vol. 10210, pp. 415–445. Springer, Cham (2017). https://doi.org/10.1007/978-3-319-56620-7_15

[BLL+15] Bai, S., Langlois, A., Lepoint, T., Stehlé, D., Steinfeld, R.: Improved security proofs in lattice-based cryptography: using the Rényi divergence rather than the statistical distance. In: Iwata, T., Cheon, J.H. (eds.) ASIACRYPT 2015. LNCS, vol. 9452, pp. 3–24. Springer, Heidelberg (2015). https://doi.org/10.1007/978-3-662-48797-6_1

[BLMR13] Boneh, D., Lewi, K., Montgomery, H., Raghunathan, A.: Key homomorphic PRFs and their applications. In: Canetti, R., Garay, J.A. (eds.) CRYPTO 2013. LNCS, vol. 8042, pp. 410–428. Springer, Heidelberg (2013). https://doi.org/10.1007/978-3-642-40041-4_23

[BMR10] Boneh, D., Montgomery, H.W., Raghunathan, A.: Algebraic pseudoran-
dom functions with improved efficiency from the augmented cascade. In:
Al-Shaer, E., Keromytis, A.D., Shmatikov, V. (eds.) ACM CCS 10: 17th
Conference on Computer and Communications Security, pp. 131–140.
ACM Press, New York (2010)

[BP14] Banerjee, A., Peikert, C.: New and improved key-homomorphic pseudorandom functions. In: Garay, J.A., Gennaro, R. (eds.) CRYPTO 2014.
LNCS, vol. 8616, pp. 353–370. Springer, Heidelberg (2014). https://doi.
org/10.1007/978-3-662-44371-2_20

[BPR12] Banerjee, A., Peikert, C., Rosen, A.: Pseudorandom functions and lattices. In: Pointcheval, D., Johansson, T. (eds.) EUROCRYPT 2012.
LNCS, vol. 7237, pp. 719–737. Springer, Heidelberg (2012). https://doi.
org/10.1007/978-3-642-29011-4_42

[BR17] Bogdanov, A., Rosen, A.: Pseudorandom functions: three decades later.
Tutorials on the Foundations of Cryptography. ISC, pp. 79–158. Springer,
Cham (2017). https://doi.org/10.1007/978-3-319-57048-8_3

[BTVW17] Brakerski, Z., Tsabary, R., Vaikuntanathan, V., Wee, H.: Private constrained PRFs (and more) from LWE. In: Kalai, Y., Reyzin, L. (eds.) TCC
2017. LNCS, vol. 10677, pp. 264–302. Springer, Cham (2017). https://doi.
org/10.1007/978-3-319-70500-2_10

[BV15] Brakerski, Z., Vaikuntanathan, V.: Constrained key-homomorphic PRFs
from standard lattice assumptions - Or: how to secretly embed a circuit
in your PRF. In: Dodis, Y., Nielsen, J.B. (eds.) TCC 2015. LNCS, vol.
9015, pp. 1–30. Springer, Heidelberg (2015). https://doi.org/10.1007/978-
3-662-46497-7_1

[BW13] Boneh, D., Waters, B.: Constrained pseudorandom functions and their
applications. In: Sako, K., Sarkar, P. (eds.) ASIACRYPT 2013. LNCS,
vol. 8270, pp. 280–300. Springer, Heidelberg (2013). https://doi.org/10.
1007/978-3-642-42045-0_15

[CC17] Canetti, R., Chen, Y.: Constraint-hiding constrained PRFs for NC^1 from
LWE. In: Coron, J.-S., Nielsen, J.B. (eds.) EUROCRYPT 2017. LNCS,
vol. 10210, pp. 446–476. Springer, Cham (2017). https://doi.org/10.1007/
978-3-319-56620-7_16

[DKW16] Deshpande, A., Koppula, V., Waters, B.: Constrained pseudorandom
functions for unconstrained inputs. In: Fischlin, M., Coron, J.-S. (eds.)
EUROCRYPT 2016. LNCS, vol. 9666, pp. 124–153. Springer, Heidelberg
(2016). https://doi.org/10.1007/978-3-662-49896-5_5

[DS15] Döttling, N., Schröder, D.: Efficient pseudorandom functions via on-the-
fly adaptation. In: Gennaro, R., Robshaw, M. (eds.) CRYPTO 2015.
LNCS, vol. 9215, pp. 329–350. Springer, Heidelberg (2015). https://doi.
org/10.1007/978-3-662-47989-6_16

[Fat06] Fateman, R.J.: When is FFT multiplication of arbitrary-precision polynomials practical? University of California, Berkeley (2006)

[GGM84] Goldreich, O., Goldwasser, S., Micali, S.: How to construct random functions (extended abstract). In: 25th Annual Symposium on Foundations
of Computer Science, Singer Island, Florida, 24–26 October 1984, pp.
464–479. IEEE Computer Society Press (1984)

[JKP18] Jager, T., Kurek, R., Pan, J.: Simple and more efficient PRFs with tight
security from LWE and matrix-DDH. Cryptology ePrint Archive, Report
2018/826 (2018). https://eprint.iacr.org/2018/826

[KSN+04] Knuth, D.E., Saitou, H., Nagao, T., Matui, S., Matui, T., Yamauchi, H.:
          of Book: The Art of Computer Programming.-Volume 2, Seminumerical
          Algorithms (Japanese Edition), vol. 2. ASCII (2004)

[KW17]   Kim, S., Wu, D.J.: Watermarking cryptographic functionalities from stan-
          dard lattice assumptions. In: Katz, J., Shacham, H. (eds.) CRYPTO 2017.
          LNCS, vol. 10401, pp. 503–536. Springer, Cham (2017). https://doi.org/
          10.1007/978-3-319-63688-7_17

[LMR14]   Lewi, K., Montgomery, H., Raghunathan, A.: Improved constructions of
          PRFs secure against related-key attacks. In: Boureanu, I., Owesarski, P.,
          Vaudenay, S. (eds.) ACNS 2014. LNCS, vol. 8479, pp. 44–61. Springer,
          Cham (2014). https://doi.org/10.1007/978-3-319-07536-5_4

[MG12]   Micciancio, D., Goldwasser, S.: Complexity of Lattice Problems: A Cryp-
          tographic Perspective, vol. 671. Springer, New York (2012). https://doi.
          org/10.1007/978-1-4615-0897-7

[Mon18]   Montgomery, H.: A nonstandard variant of learning with rounding with
          polynomial modulus and unbounded samples. Cryptology ePrint Archive,
          Report 2018/100 (2018). https://eprint.iacr.org/2018/100

[NR95]   Naor, M., Reingold, O.: Synthesizers and their application to the parallel
          construction of pseudo-random functions. In: 36th Annual Symposium on
          Foundations of Computer Science, Milwaukee, Wisconsin, 23–25 October
          1995, pp. 170–181. IEEE Computer Society Press (1995)

[NR97]   Naor, M., Reingold, O.: Number-theoretic constructions of efficient
          pseudo-random functions. In: 38th Annual Symposium on Foundations
          of Computer Science, Miami Beach, Florida, 19–22 October 1997, pp.
          458–467. IEEE Computer Society Press (1997)

[PS17]   Peikert, C., Shiehian, S.: Privately constraining and programming PRFs,
          the LWE way. Cryptology ePrint Archive, Report 2017/1094 (2017).
          https://eprint.iacr.org/2017/1094

[Reg05]   Regev, O.: On lattices, learning with errors, random linear codes, and
          cryptography. In: Gabow, H.N., Fagin, R. (eds.) 37th Annual ACM Sym-
          posium on Theory of Computing, pp. 84–93. ACM Press, New York (2005)

[Rei]   Reingold, O.: Pseudorandom synthesizers, functions, and permutations

[RW04]   Rudich, S., Wigderson, A.: Computational Complexity Theory, vol. 10.
          American Mathematical Soc., Providence (2004)

Asymmetric Key Cryptography and Cryptanalysis

A Las Vegas Algorithm to Solve the Elliptic Curve Discrete Logarithm Problem

Ayan Mahalanobis[1]([⊠]), Vivek Mohan Mallick[1], and Ansari Abdullah[2]

[1] IISER Pune, Pune, India
ayan.mahalanobis@gmail.com, vmallick@iiserpune.ac.in
[2] Savitribai Phule Pune University, Pune, India
abdullah0096@gmail.com

Abstract. In this paper, we describe a new Las Vegas algorithm to solve the elliptic curve discrete logarithm problem. The algorithm depends on a property of the group of rational points on an elliptic curve and is thus not a generic algorithm. The algorithm that we describe has some similarities with the most powerful index-calculus algorithm for the discrete logarithm problem over a finite field. The algorithm has no restriction on the finite field over which the elliptic curve is defined.

Keyword: Elliptic curve discrete logarithm problem

1 Introduction

Public-key cryptography is a backbone of this modern society. Many of the public-key cryptosystems depend on the *discrete logarithm problem* as their cryptographic primitive. Of all the groups used in a discrete logarithm based protocol, the group of *rational points of an elliptic curve* is the most popular. In this paper, we describe a **Las Vegas algorithm** to solve the elliptic curve discrete logarithm problem.

There are two kinds of attack on the discrete logarithm problem. One is generic. This kind of attack works in any group. Examples of such attacks are the baby-step giant-step attack [8, Proposition 2.22] and Pollard's rho [8, Sect. 4.5]. The other kind of attack depends on the group used. Example of such an attack is the index-calculus attack [8, Sect. 3.8] on the multiplicative group of a finite field. An attack similar to index-calculus for elliptic curves, known as xedni calculus, was developed by Silverman [9,13]. However, it was found to be no better than exhaustive search. Another similar work in the direction of ours is Semaev [11] which has given rise to index-calculus algorithms for elliptic curves. A curious reader can consult Amadori et al. [1] for a list of references. Our approach to solve the elliptic curve discrete logarithm problem is completely different and has **no restriction on the finite field** on which the elliptic curve is defined.

© Springer Nature Switzerland AG 2018
D. Chakraborty and T. Iwata (Eds.): INDOCRYPT 2018, LNCS 11356, pp. 215–227, 2018.
https://doi.org/10.1007/978-3-030-05378-9_12

The main algorithm is divided into two algorithms. The first one reduces the elliptic curve discrete logarithm problem to a problem in linear algebra. We call the linear algebra problem, Problem L. This reduction is a Las Vegas algorithm with **probability of success** 0.6 and is **polynomial** in both time and space complexity. The second half of the algorithm is solving Problem L. This is the current bottle-neck of the whole algorithm and better algorithms to solve Problem L will produce better algorithms to solve elliptic curve discrete logarithm problem. The success of the main algorithm is $0.6 \times (\log p)^2 / p$ where every pass is polynomial in time and space complexity. This also shows that our algorithm is worse than Pollard's rho or other square root attack for sufficiently large p.

1.1   Notations

All elliptic curves in this paper are non-singular curves defined over a finite field
    of arbitrary characteristic.
All curves are projective plane curves. We do not deal with the affine case though
    that can be achieved with minor modification.
The group of rational points of the elliptic curve is assumed to be of prime
    order [8, Remark 2.33] and we reserve p for that prime.
We denote by $\mathbb{P}^2 (\mathbb{F})$ the projective plane over the field $\mathbb{F}$.

1.2   The Central Idea Behind Our Attack

Let G be a cyclic group of prime order p. Let P be a non-identity element and $Q(= mP)$ belong to G. The *discrete logarithm problem* is to compute the m mod p. One way to find m is the following: fix a positive integer k; for $i = 1, 2, \ldots, k$ find positive integers n_i, $1 \leq n_i < p$ such that $\sum_{i=1}^{k} n_i = m$ mod p. The last equality is hard to compute because we do not know m. However we can decide whether

$$\sum_{i=1}^{k} n_i P = Q \tag{1}$$

and based on that we can decide if $\sum_{i=1}^{k} n_i = m$ mod p. Once the equality holds, we have found m and the discrete logarithm problem is solved.

The number of possible choices of n_i for a given k that can solve the discrete logarithm problem is the number of partitions of m into k parts modulo a prime p. The applicability of the above method depends on, how fast can one decide on the equality in the above equation and on the probability, how likely is it that a given set of positive integers n_i sums to m mod p?

2   The Elliptic Curve Discrete Logarithm Problem

The elliptic curve discrete logarithm problem (ECDLP) is an important problem in modern public-key cryptography. This paper is about a new probabilistic algorithm to solve this problem. We denote by $\mathcal{E}(\mathbb{F}_q)$ the group of rational points of the elliptic curve $\mathcal{E}$ over $\mathbb{F}_q$. It is well known that there is an isomorphism $\mathcal{E}(\mathbb{F}_q) \to \mathrm{Pic}^0(\mathcal{E})$ given by $P \mapsto [P] - [\mathcal{O}]$ [10, Proposition I.4.10].

Theorem 1. *Let $\mathcal{E}$ be an elliptic curve over $\mathbb{F}_q$ and $P_1, P_2, \ldots, P_k$ be points on that curve, where $k = 3n'$ for some positive integer n'. Then $\sum_{i=1}^{k} P_i = \mathcal{O}$ if and only if there is a curve $\mathcal{C}$ over $\mathbb{F}_q$ of degree n' that passes through these points. Multiplicities are intersection multiplicities.*

Proof. Assume that $\sum_{i=1}^{k} P_i = \mathcal{O}$ in $\mathbb{F}_q$ and then it is such in the algebraic closure $\bar{\mathbb{F}}_q$. From the above isomorphism, $\sum_{i=1}^{k} P_i \mapsto \sum_{i=1}^{k} [P_i] - k[\mathcal{O}]$. Then $\sum_{i=1}^{k} [P_i] - k[\mathcal{O}]$ is zero in the Picard group $\mathrm{Pic}_{\bar{\mathbb{F}}_q}^0(\mathcal{E})$. Then there is a rational function $\dfrac{\phi}{z^{n'}}$ over $\mathbb{P}^2(\bar{\mathbb{F}}_q)$ such that

$$\sum_{i=1}^{k} [P_i] - k[\mathcal{O}] = \mathrm{div}\left(\frac{\phi}{z^{n'}}\right) \tag{2}$$

Bezout's theorem justifies that $\deg(\phi) = n'$, since ϕ is zero on $P_1, P_2, \ldots, P_k$. We now claim, there is ψ over $\mathbb{F}_q$ which is also of degree n' and passes through $P_1, P_2, \ldots, P_k$. First thing to note is that there is a finite extension of $\mathbb{F}_q$, $\mathbb{F}_{q^N}$(say) in which all the coefficients of ϕ lies and $\gcd(q, N) = 1$. Let G be the Galois group of $\mathbb{F}_{q^N}$ over $\mathbb{F}_q$ and define

$$\psi = \sum_{\sigma \in G} \phi^\sigma. \tag{3}$$

Clearly $\deg(\psi) = n'$. Note that, since P_i for $i = 1, 2, \ldots, k$ is in $\mathbb{F}_q$ is invariant under σ. Furthermore, σ being a field automorphism, P_i is a zero of ϕ^σ for all $\sigma \in G$. This proves that P_i are zeros of ψ and then Bezout's theorem shows that these are the all possible zeros of ψ on $\mathcal{E}$. The only thing left to show is that ψ is over $\mathbb{F}_q$. To see that, lets write $\phi = \sum_{i+j+k=n'} a_{ijk} x^i y^j z^k$. Then $\psi = \sum_{i+j+k=n'} \sum_{\sigma \in G} a_{ijk}^\sigma x^i y^j z^k$. However, it is well known that $\sum_{\sigma \in G} a^\sigma \in \mathbb{F}_q$ for all $a \in \mathbb{F}_{q^N}$.

Conversely, suppose we are given a curve $\mathcal{C}$ of degree n' that passes through $P_1, P_2, \ldots, P_k$. Then consider the rational function $\mathcal{C}/z^{n'}$. Then this function has zeros on P_i, $i = 1, 2, \ldots, k$ and a pole of order k at $\mathcal{O}$. The above isomorphism says $\sum_{i=1}^{k} P_i = \mathcal{O}$. $\qquad\square$

2.1   How to Use the Above Theorem in Our Algorithm

We choose k such that $k = 3n'$ for some positive integer n'. Then we choose random points $P_1, P_2, \ldots, P_s$ and $Q_1, Q_2, \ldots, Q_t$ such that $s + t = k$ from $\mathcal{E}$ and check if there is a homogeneous curve of degree n' that passes through these points where $P_i = n_i P$ and $Q_j = -n'_j Q$ for some integers n_i and n'_j. If there is a curve, the discrete logarithm problem is solved. Otherwise repeat the process by choosing a new set of points $P_1, P_2, \ldots, P_s$ and $Q_1, Q_2, \ldots, Q_t$. To choose these points P_i and Q_j, we choose random integers n_i, n'_j and compute $n_i P$ and $-n'_j Q$. We choose n_i and n'_j to be distinct from the ones chosen before. This gives rise to distinct points P_i and Q_j on $\mathcal{E}$.

The only question remains, how do we say if there is a homogeneous curve of degree n' passing through these selected points? One can answer this question using linear algebra.

Let $C = \sum_{i+j+k=n'} a_{ijk} x^i y^j z^k$ be a *complete* homogeneous curve of degree n'. We assume that an ordering of i, j, k is fixed throughout this paper and C is presented according to that ordering. By complete we mean that the curve has all the possible monomials of degree n'. We need to check if P_i for $i = 1, 2, \ldots, s$ and Q_j for $j = 1, 2, \ldots, t$ satisfy the curve C. Note that, there is no need to compute the values of a_{ijk}, just mere existence will solve the discrete logarithm problem.

Let P be a point on $\mathcal{E}$. We denote by $\overline{P}$ the value of C when the values of x, y, z in P is substituted in C. In other words, $\overline{P}$ is a linear combination of a_{ijk} with the fixed ordering. Similarly for Qs. We now form a matrix $\mathcal{M}$ where the rows of $\mathcal{M}$ are $\overline{P_i}$ for $i = 1, 2, \ldots, s$ and $\overline{Q_j}$ for $j = 1, 2, \ldots, t$. If this matrix has a non-zero left-kernel, we have solved the discrete logarithm problem. By *left-kernel* we mean the kernel of $\mathcal{M}^{\mathrm{T}}$, the transpose of $\mathcal{M}$.

2.2   Why Look at the Left-Kernel Instead of the Kernel

In this paper, we will use the left-kernel more often than the (right) kernel of $\mathcal{M}$. We denote the left-kernel by $\mathcal{K}$ and kernel by $\mathcal{K}'$. We first prove the following theorem:

Theorem 2. *The following are equivalent:*

(a) $\mathcal{K} = 0$.
(b) $\mathcal{K}'$ *only contain curves that are a multiple of $\mathcal{E}$.*

Proof. The proof uses a simple counting argument. First recall the well-known fact that the number of monomials of degree d is $\binom{d+2}{2}$. Furthermore, notice two things – all multiples of $\mathcal{E}$ belongs to $\mathcal{K}'$ and the dimension of that vector-space (multiples of $\mathcal{E}$) is $\binom{n'-1}{2} = \dfrac{(n'-2)(n'-1)}{2}$, where n' is as defined earlier.

Now, $\mathcal{M}$ is as defined earlier, has $3n'$ rows and $\dfrac{(n'+1)(n'+2)}{2}$ columns. Then $\mathcal{K} = 0$ means that the row-rank of $\mathcal{M}$ is $3n'$. So the dimension of the $\mathcal{K}'$ is

$$\frac{(n'+1)(n'+2)}{2} - 3n' = \frac{(n'-2)(n'-1)}{2}.$$

This proves (a) implies (b).

Conversely, if $\mathcal{K}'$ contains all the curves that are a multiple of $\mathcal{E}$ then its dimension is at least $\dfrac{(n'-2)(n'-1)}{2}$, then the rank is $3n'$, making $\mathcal{K} = 0$.   □

It is easy to see, while working with the above theorem $\mathcal{M}$ cannot repeat any row. So from now onward we would assume that $\mathcal{M}$ has no repeating rows. For all practical purposes this means that we are working with distinct partitions (also known as unique partitions). By distinct partition we mean those partitions which has no repeating parts.

A question that becomes significantly important later is, instead of choosing k points from the elliptic curve what happens if we choose $k + l$ points for some positive integer l. The answer to the question lies in the following theorem.

Theorem 3. *If $l \geq 1$, the dimension of the left kernel of $\mathcal{M}$ is l.*

Proof. First assume $l \geq 1$. In this case, any non-trivial element of $\mathcal{K}'$ will define a curve which passes through more than $3n'$ points of the elliptic curve. Since the elliptic curve is irreducible, it must be a component of the curve. Thus the equation defining the curve must be divisible by the equation defining the elliptic curve. Thus, the dimension of $\mathcal{K}'$ is the dimension of all degree n' homogeneous polynomials which are divisible by the elliptic curve. This is the dimension of all degree $n' - 3$ homogeneous polynomials. Thus, we get

$$\dim(\mathcal{K}') = \frac{(n' - 2)(n' - 1)}{2}.$$

On the other hand, by rank-nullity theorem, it follows:

$$\dim(\mathcal{K}') + \dim(\text{image}(\mathcal{M})) = \frac{(n'+2)(n'+1)}{2}$$
$$\dim(\mathcal{K}) + \dim(\text{image}(\mathcal{M}^{\mathrm{T}})) = 3n' + l.$$

Thus, since row rank and the column rank of a matrix are equal,

$$\dim(\mathcal{K}) = 3n' + l - \frac{(n' + 2)(n' + 1)}{2} + \dim(\mathcal{K}') = l.$$

Corollary 1. *Assume that $\mathcal{M}$ has $3n' + l$ rows, computed from the same number of points of the elliptic curve $\mathcal{E}$. If there is a curve $\mathcal{C}$ intersecting $\mathcal{E}$ non-trivially in $3n'$ points among $3n' + l$ points, then there is a vector v in $\mathcal{K}$ with at least l zeros. Conversely, if there is a vector v in $\mathcal{K}$ with at least l zeros, then there is a curve $\mathcal{C}$ passing through those $3n'$ points that correspond to the non-zero entries of v in $\mathcal{M}$.*

Proof. Assume that there is a non-trivial curve $\mathcal{C}$ intersecting $\mathcal{E}$ in $3n'$ points. Then construct the matrix $\mathcal{M}'$ whose rows are the points of intersection. Then from the earlier theorem we see that $\mathcal{K}$ for this matrix $\mathcal{M}'$ is non-zero. In all the vectors of $\mathcal{K}$ if we put zeros in the place where where we deleted rows then those are element of the left kernel of $\mathcal{M}$. It is clear that these vectors will have at least l zeros.

Conversely, if there is a vector with at least l zeros in $\mathcal{K}$, then by deleting l zeros from the vector and corresponding rows from $\mathcal{M}$ we have the required result from the theorem above. □

Algorithm 1. Reducing ECDLP to a linear algebra problem (Problem L)

Data: Two points P and Q, such that $mP = Q$
Result: m
Select a positive integers, n' and $l = 3n'$. Initialize a matrix with $3n' + l$ rows
and $\binom{n'+2}{2}$ columns. Initialize a vector $\mathcal{I}$ of length $3n' - 1$ and another vector $\mathcal{J}$
of length $l + 1$. Initialize integers $A, B = 0$.
repeat
    for $i = 1$ *to* $3n' - 1$ **do**
        repeat
            choose a random integer r in the range $[1, p)$
        until r *is not in* $\mathcal{I}$
        $\mathcal{I}[i] \leftarrow r$
        compute rP
        compute $\overline{rP}$
        insert $\overline{rP}$ as the ith row of the matrix $\mathcal{M}$
    end
    for $i = 1$ *to* $l + 1$ **do**
        repeat
            choose a random integer r in the range $[1, p)$
        until r *is not in* $\mathcal{J}$
        $\mathcal{J}[i] \leftarrow r$
        compute $-rQ$
        compute $\overline{-rQ}$
        insert $\overline{-rQ}$ as the $(3n' + i - 1)^{\text{th}}$ row of the matrix $\mathcal{M}$
    end
    compute $\mathcal{K}$ as the left-kernel of $\mathcal{M}$
until $\mathcal{K}$ *has a vector* v *with* l *zeros (Problem L)*
for $i = 1$ *to* $3n' - 1$ **do**
    if $v[i] \neq 0$ **then**
        $A = A + \mathcal{I}[i]$
    end

end
for $i = 3n'$ *to* $3n' + l$ **do**
    if $v[i] \neq 0$ **then**
        $B = B + \mathcal{J}[i - 3n' + 1]$
    end

end
return $A \times B^{-1} \bmod p$

The algorithm that we present in this paper has two parts. One reduces it to a
problem in linear algebra and the other solves that linear algebra problem which
we call Problem L. The first algorithm, Algorithm 1, is Las Vegas in nature with
high success probability and is polynomial in both time and space complexity.

3   The Main Algorithm – Reducing ECDLP to a Linear Algebra Problem (Problem L)

Why Is This Algorithm Better Than Exhaustive Search. In the exhaustive search we would have picked a random set of $3n'$ points and then checked to see if the sum of those points is Q. In the above algorithm we are taking a set of $3n' + l$ points and then checking all possible $3n'$ subsets of this set simultaneously. There are $\binom{3n'+l}{l}$ such subsets. This is one of the main advantage of our algorithm.

Probability of Success of the Above Algorithm. To compute the probability, we need to understand the number of unique partitions of an integer m modulo a prime p. For our definition of partition, order of the parts does not matter. The number of partitions is proved in the following theorem:

Theorem 4. *Let* $2 < k \leq p/2$ *be an integer. The number of k unique partitions of m modulo a odd prime p is at least*

$$\frac{(p-1)(p-2)\ldots(p-k+2)(p-2k+1)}{k!}.$$

Proof. The argument is a straight forward counting argument. We think of k parts as k boxes. Then the first box can be filled with $p-1$ choices, second with $p-2$ choices and so on. The last but one, $k-1$ box can be filled with $p-k+1$ choices. When all $k-1$ boxes are filled then there is only one choice for the last box, it is m minus the sum of the other boxes. So it seems that the count is $(p-1)(p-2)\ldots(p-k+1)$ choices.

However there is a problem, the choice in the last box might not be different from the first $k-1$ choices. To remove that possibility we remove $k-1$ choices from the last but one box. Furthermore, the choice in the last box can not be zero that removes one more choice from the last but one box.

In some pathological cases, the above argument might remove more choices than necessary. Thus the above formula gives a lower bound for the number of k distinct partitions.

Since order does not matter, we divide by $k!$.   □

Consider the event, m is fixed, we pick k integers less than $p/2$. What is the probability that those numbers form a partition of m. From the above theorem, number of favorable events is

$$\frac{(p-1)(p-2)\ldots(p-k+2)(p-2k+1)}{k!}$$

and the total number of events is $\binom{p}{k}$. Since for all practical purposes k is much smaller than p, we approximate the probability to be $\frac{1}{p}$.

Now we look at the probability of success of our algorithm. In our algorithm we choose $3n'$ points from $3n' + l$ points. This can be done in $\binom{3n'+l}{l}$ ways. Then the probability of success of the algorithm is $1 - \left(1 - \frac{1}{p}\right)^{\binom{3n'+l}{l}}$.

Let us first look at the $\left(1 - \frac{1}{p}\right)^p$. It is well known that $\left(1 - \frac{1}{p}\right)^p$ tends to $\frac{1}{e}$ when p tends to infinity. So if we can make $\binom{3n'+l}{l}$ close to p, then we can claim the asymptotic probability of our algorithm is $1 - \frac{1}{e}$ which is greater than $\frac{1}{2}$.

Since we are dealing with matrices, it is probably the best that we try to keep the size of it as small as possible. Note that the binomial coefficient is the biggest when it is of the form $\binom{2n}{n}$ for some positive integer n. Furthermore, from Stirling's approximation it follows that for large enough n, $\binom{2n}{n} \approx \frac{4^n}{\sqrt{\pi n}}$.

So, when we take $3n' = l$ and such that $\binom{3n'+l}{l} = p$ then l is the solution to the equation $l = O(1) + \log l + \log p$.

To understand the time complexity of this algorithm (without the linear algebra problem), the major work done is finding the kernel of a matrix. Using Gaussian elimination, there is an algorithm to compute the kernel which is cubic in time complexity. Thus we have proved the following theorem:

Theorem 5. *When p tends to infinity, the probability of success of the above algorithm is approximately $1 - \frac{1}{e} \approx 0.6321$. The size of the matrix required to reach this probability is $O(\log p)$. This makes our algorithm polynomial in both time and space complexity.*

3.1   Few Comments

Accidentally Solving the Discrete Logarithm Problem. It might happen, that while computing rP and rQ in our algorithm, it turns out that for some r_1 and r_2, $r_1 P = r_2 Q$. In that case, we have solved the discrete logarithm problem. We should check for such accidents. However, in a real life situation, the possibility of an accident is virtually zero, so we ignored that in our algorithm completely.

On the Number of Ps and Qs in Our Algorithm. The algorithm will take as input P and Q and produce different Ps and Qs and the produce a vector v with l many zeros. If all of these l zeros fall either in the place of Ps or Qs exclusively, then we have not solved the discrete logarithm problem. To avoid this, we have chosen Ps and Qs of roughly same size, with one more P than Q. This way the vector v will have atleast one non-zero in the place of both P and Q.

Allowing, Detecting and Using Multiple Intersection Points in Our Algorithm. One obvious idea to make our algorithm slightly faster: allow multiplicities of intersection between the curve C and the elliptic curve $\mathcal{E}$. This will increase the computational complexity. Since the elliptic curve is smooth at the points one is interested in, one observes that with high probability the multiplicity of intersection will coincide with the multiplicity of the point in C. This reduces to checking if various partial derivatives are zero. This can easily be done by introducing extra rows in the matrix $\mathcal{M}$. Then the algorithm reduces to finding vectors with zeroes in a particular pattern. This is same as asking for special type of solutions in Problem L. However, this has to be implemented efficiently as probability of such an event occurring is around $1/p$ for large primes p.

3.2   Choosing Ps and Qs Uniformly Random

We raise an obvious question, can one choose the set of n_i and n'_j (see Sect. 2.1) which give rise to P_i and Q_j respectively in such a way that the probability of solving the discrete logarithm problem is higher than the uniformly random selection? In this paper we choose P_i and Q_j uniformly random.

4   Dealing with the Linear Algebra Problem

This paper provides an efficient algorithm to reduce the elliptic curve discrete logarithm problem to a problem in linear algebra. We call it the Problem L.

At this stage we draw the attention of the reader to some similarities that emerge between the most powerful attack on the discrete logarithm problem over finite fields, the index-calculus algorithm, and our algorithm. In an index-calculus algorithm, the discrete logarithm problem is reduced to a linear algebra problem. Similar is the case with our algorithm. However, in our case, the linear algebra problem is of a different genre and not much is known about this problem. In this paper, we have not been able to solve the linear algebra problem completely. However, we made some progress and we report on that in this section.

Problem L. Let W be a l-dimensional subspace of a n-dimensional vector space V. The vectors in the vectors space are presented as linear sum of some fixed basis of V. The problem is to determine, if W contains a vector with l zeros. If there is one such vector, find that vector.

This problem is connected with the earlier algorithm in a very straightforward way. We need to determine if the left-kernel of the matrix $\mathcal{M}$ contains a vector with l zeros and that is where Problem L must be solved efficiently for the overall algorithm to run efficiently. As is customary, we would assume that the kernel $\mathcal{K}$ is presented as a matrix of size $l \times (3n' + l)$, where each row is an element of the basis of $\mathcal{K}$. Recall that we chose $3n' = l$.

A algorithm that we developed, uses Gaussian elimination algorithm multiple times to solve Problem L. In particular we use the row operations from the Gaussian elimination algorithm. Abusing our notations slightly, we denote the basis matrix of $\mathcal{K}$ by $\mathcal{K}$ as well. Now we can think of $\mathcal{K}$ to be made up of two blocks of $l \times l$ matrix. Our idea is to do Gaussian elimination to reduce each of these blocks to a diagonal matrix one after the other. The reason that we do that is, when the first block has been reduced to diagonal, every row of the matrix has at least $l - 1$ zeros. So we are looking for another zero in some row. The row reduction that produced the diagonal matrix in the first block might also have produced that extra zero and we are done. However, if this is not the case, we go on to diagonalize the second block and check for that extra zero like we did for the first block.

Algorithm 2. Multiple Gaussian elimination algorithm

Data: The basis matrix $\mathcal{K}$

Result: Determine if Problem L is solved. If yes, output the vector that solves Problem L

for $i=1$ to 2 **do**

    row reduce block i to a lower triangular block

    check all rows of the new matrix to check if any one has l zeros

    if *there is a row with l zeros* **then**

        | STOP and return the row

    end

    row reduce the lower-triangular block to a diagonal block

    check all rows of the new matrix to check if any one has l zeros

    if *there is a row with l zeros* **then**

        | STOP and return the row

    end

end

STOP (Problem L not solved)

---

5 Complexity, Implementation and Conclusion

5.1 Complexity

We describe the complexity of the whole algorithm in this section. First note that the whole algorithm is the composition of two algorithms, one is Algorithm 1, which has success probability 0.6 and the other is the linear algebra problem. It is easy to see from conditional probability that the probability of success of the whole algorithm is the product of the probability of success of Algorithms 1 and 2.

Let us now calculate the probability of Algorithm 2 under the condition that Algorithm 1 is successful. In other words, we know that Algorithm 1 has found a $\mathcal{K}$ whose span contains a vector with l zeros. What is the probability that Algorithm 2 will find it?

Notice that Algorithm 2 can only find zero if they are in certain positions and the number of such positions is l^2. Total number of ways that there can be l zeros in a vector of size $3n' + l$ is $\binom{3n'+l}{l}$. In our setting we have already assumed that $\binom{3n'+l}{l} \approx p$. Then the probability of success of the whole algorithm is

$$0.6 \times \frac{(\log p)^2}{p}.$$

Which is a significant improvement over exhaustive search!

One thing to notice, the probability of success is $1 - \left(1 - \frac{1}{p}\right)^{\binom{3n'+l}{l}}$ and in the probability estimate we have $\binom{3n'+l}{l}$ in the denominator. Furthermore, one observes that in this paper we have taken $\binom{3n'+l}{l}$ to approximately equal the prime p. One can now question our choice and argue, if we took $\binom{3n'+l}{l}$ to be much smaller than p, we might get a better algorithm. Alas, this is not the case, $1 - \left(1 - \frac{1}{p}\right)^{p^{\frac{1}{n}}}$ tends to 0 as p tends to infinity for $n \geq 2$.

5.2   Implementation

The aim of our implementation is to determine an average number of tries required by the Las Vegas algorithm to solve a elliptic curve discrete logarithm problem. However to generate real life data that solves discrete logarithm problems is very time consuming. So we set a cut off, if the number of steps taken by the program is more than $\sqrt{p}$ where p is the order of P, we stop the program. This way we deal with the black swan situation that normally happens in any Las Vegas algorithm. In each step we generate points on the elliptic curve, form the matrix $\mathcal{M}$ and then the left kernel $\mathcal{K}$, perform two row-operations on $\mathcal{K}$ and see if there are l zeros in a row after each row-reduction of $\mathcal{K}$. The algorithm was implemented using NTL [12].

The left-kernel $\mathcal{K}$ of M is a matrix of size $l \times 2l$. Thus $\mathcal{K}$ consists of two matrices of size $l \times l$ stacked sideways. Both these matrices in $\mathcal{K}$ are row-reduced to a diagonal matrix using only row operations in our experiment. The first attempt reduces columns 1 to l in $\mathcal{K}$ and checks for a row with l zeros. If a row with l zeros is present, the discrete logarithm is solved. If the first attempt fails to reduce $\mathcal{K}$ which contain a row with l zeros, second row reduction is applied. The second attempt reduces columns $l+1$ to $2l$ and checks for a row with l zeros. If this reduced form contains a row with l zeros DLP is solved. If the first as well as the second row reduction does not yield a row with l zeros the algorithm is restarted with a fresh choice of random points on the elliptic curve.

The Las Vegas algorithm was executed 80 times and try-count, the number of tries, for each execution was recorded. The data is presented in Table 1. Each

Table 1. No. of steps required to solve ECDLP using our algorithm

Field Size	$\sqrt{p}$	Solved in less than $\sqrt{p}$ tries	Can't solve in less than $\sqrt{p}$ tries
2^{17}	256	80	00
2^{19}	512	80	00
2^{23}	2048	77	03
2^{29}	16384	39	41
2^{31}	32768	26	54
2^{37}	262144	07	73

of the 80 executions resulted in either a value for try-count if the DLP was solved before the try-count reached $\sqrt{p}$ or no value for try-count if DLP was not solved in less than $\sqrt{p}$ steps.

5.3 Conclusion

We conclude this paper by saying that we have found a new genre of attack against the elliptic curve discrete logarithm problem. This attack has some similarities with the well-known index-calculus algorithm. In an index-calculus algorithm, the discrete logarithm problem is reduced to a problem in linear algebra and then the linear algebra problem is solved. However, the similarities are only skin deep as our linear algebra problem is completely new.

Acknowledgment. We are indebted to the anonymous referees for their careful reading of the manuscript and detail comments. Due to lack of time, we were not able to incorporate the new research directions suggested. However, those comments have certainly piqued our interest and we thank the referees for those.

References

1. Amadori, A., Pintore, F., Sala, M.: On the discrete logarithm problem for prime-field elliptic curve. Finite Fields Appl. **51**, 168–182 (2018)
2. Bernstein, D.J., Lange, T.: Non-uniform cracks in the concrete: the power of free precomputation. In: Sako, K., Sarkar, P. (eds.) ASIACRYPT 2013. LNCS, vol. 8270, pp. 321–340. Springer, Heidelberg (2013). https://doi.org/10.1007/978-3-642-42045-0_17
3. Fulton, W.: Algebraic Curves (2008, self-published)
4. Galbraith, S., Gaudry, P.: Recent progress on the elliptic curve discrete logarithm problem. Des. Codes Crypt. **78**, 51–78 (2016)
5. Galbraith, S.D., Gebregiyorgis, S.W.: Summation polynomial algorithms for elliptic curves in characteristic two. In: Meier, W., Mukhopadhyay, D. (eds.) INDOCRYPT 2014. LNCS, vol. 8885, pp. 409–427. Springer, Cham (2014). https://doi.org/10.1007/978-3-319-13039-2_24
6. Gaudry, P.: Index calculus for abeian varieties of small dimension and the elliptic curve discrete logarithm problem. J. Symbolic Comput. **44**, 1690–1702 (2009)
7. Harris, J.: Algebraic Geometry. Springer, New York (1992). https://doi.org/10.1007/978-1-4757-2189-8
8. Silverman, J.H., Pipher, J., Hoffstein, J.: An Introduction to Mathematical Cryptography. UTM. Springer, New York (2008). https://doi.org/10.1007/978-0-387-77993-5
9. Jacobson, M.J., Koblitz, N., Silverman, J.H., Stein, A., Teske, E.: Analysis of the xedni calculus attack. Des. Codes Crypt. **20**(1), 41–64 (2000)
10. Milne, J.S.: Elliptic Curves. BookSurge Publishers (2006)
11. Semaev, I.: Summation polynomials and the discrete logarithm problem on elliptic curves (2004). https://eprint.iacr.org/2004/031
12. Shoup, V.: NTL: a library for doing number theory (2016). http://www.shoup.net/ntl

13. Silverman, J.H.: The xedni calculus and the elliptic curve discrete logarithm problem. Des. Codes Crypt. **20**(1), 5–20 (2000)
14. Wiener, M.J., Zuccherato, R.J.: Faster Attacks on Elliptic Curve Cryptosystems. In: Tavares, S., Meijer, H. (eds.) SAC 1998. LNCS, vol. 1556, pp. 190–200. Springer, Heidelberg (1999). https://doi.org/10.1007/3-540-48892-8_15

Pairing-Friendly Twisted Hessian Curves

Chitchanok Chuengsatiansup[1($\boxtimes$)] and Chloe Martindale[2]

[1] INRIA and ENS de Lyon, 46 Allée d'Italie, 69364 Lyon Cedex 07, France
chitchanok.chuengsatiansup@ens-lyon.fr
[2] Department of Mathematics and Computer Science,
Technische Universiteit Eindhoven,
P.O. Box 513, 5600 MB Eindhoven, The Netherlands
c.r.martindale@tue.nl

Abstract. This paper presents efficient formulas to compute Miller doubling and Miller addition utilizing degree-3 twists on curves with j-invariant 0 written in Hessian form. We give the formulas for both odd and even embedding degrees and for pairings on both $\mathbb{G}_1 \times \mathbb{G}_2$ and $\mathbb{G}_2 \times \mathbb{G}_1$. We propose the use of embedding degrees 15 and 21 for 128-bit and 192-bit security respectively in light of the NFS attacks and their variants. We give a comprehensive comparison with other curve models; our formulas give the fastest known pairing computation for embedding degrees 15, 21, and 24.

Keywords: Twisted Hessian curves · Pairing-friendly curves
Ate pairing · Degree-3 twists · Explicit formulas

1 Introduction

Pairings on elliptic curves have various applications in cryptography, ranging from very basic key exchange protocols, such as one round tripartite Diffie–Hellman [29,30], to complicated protocols, such as identity-based encryption [8, 22,26,47]. Pairings also help to improve currently existing protocols, such as signature schemes, to have shortest possible signatures [9].

Curves that are suitable for pairings are called *pairing-friendly curves*, and these curves must satisfy specific properties. It is extremely rare that a randomly generated elliptic curve is pairing-friendly, so pairing-friendly curves have to be generated in a specific way. Examples of famous and commonly used pairing-friendly curves include Barreto-Naehrig curves [5] (BN curves), Barreto-Lynn-Scott curves [4] (BLS curves), and Kachisa-Schaefer-Scott curves [33] (KSS curves).

Chitchanok Chuengsatiansup acknowledges the support of Bpifrance in the context of the national projet RISQ (P141580). Chloe Martindale was supported by the Commission of the European Communities through the Horizon 2020 program under CHIST-ERA USEIT (NWO project 651.002.004).

© Springer Nature Switzerland AG 2018
D. Chakraborty and T. Iwata (Eds.): INDOCRYPT 2018, LNCS 11356, pp. 228–247, 2018.
https://doi.org/10.1007/978-3-030-05378-9_13

The performance of pairing-based cryptography relies on elliptic-curve-point arithmetic, computation of line functions and pairing algorithms. A pairing is a bilinear map from two elliptic curve groups $\mathbb{G}_1$ and $\mathbb{G}_2$ to a target group $\mathbb{G}_T$. To achieve a good performance, as well as having an efficient pairing algorithm, it is desirable to have a fast elliptic-curve-point arithmetic in both $\mathbb{G}_1$ and $\mathbb{G}_2$.

The security of pairings depends mainly on the cost of solving the discrete logarithm problem (DLP) in the three groups previously mentioned, namely, $\mathbb{G}_1$, $\mathbb{G}_2$, and $\mathbb{G}_T$. Since one can attack pairing-based protocols by attacking any of these three groups, the cost of solving DLP must be sufficiently high in all of these three groups.

1.1   Choice of Curves and Embedding Degrees

One way to improve the performance of pairings is to improve the performance of the underlying point arithmetic. Many authors have studied efficient point arithmetic via the representation of elliptic curves in a specific model, for example, Hessian form [32,50] and Edwards form [7,17].

Pairings based on Edwards curves, along with examples of pairing-friendly Edwards curves, were proposed by Arene, Lange, Naehrig and Ritzenthaler [1]. They found that the computation of line functions necessary to compute the pairing is much more complicated than if the curves were written in Weierstrass form. In other words, even though Edwards curves allow faster point arithmetic, this gain is somewhat outweighed by the slower computation of line functions. Li, Wu, and Zhang [40] proposed the use of quartic and sextic twists for Edwards curves, improving the efficiency of both the point arithmetic and the computation of the line functions.

Pairings based on Hessian curves with even embedding degrees were proposed by Gu, Gu and Xie [23]. They provided a geometric interpretation of the group law on Hessian curves along with an algorithm for computing Tate pairing on elliptic curves in Hessian form. However, no pairing-friendly curves in Hessian form were given.

Bos, Costello and Naehrig [10] investigated the possibility of using a model of a curve (such as Edwards or Hessian) allowing for fast point arithmetic and transforming to Weierstrass form for the actual computation of the pairing. They found that for every elliptic curve E in the BN-12, BLS-12, and KSS-18 families of pairing-friendly curves, if E is isomorphic over $\mathbb{F}_q$ to a curve in Hessian or Edwards form, then it is not isomorphic over $\mathbb{F}_{q^k}$ to a curve in Hessian or Edwards form, where k is the embedding degree. This implies that the point arithmetic has to be performed on curves in Weierstrass form — not all curves can be written in special forms such as Hessian or Edwards form. This idea of using different curve models comes at a cost of at least one conversion between other curve models into Weierstrass form.

In this article we study the efficiency of curves in Hessian form for pairing computations. Hessian curves with j-invariant 0 have degree-3 twists that can also be written in Hessian form. This means that we can take full advantage of speed-up techniques for point arithmetic and pairing computations that move arithmetic to subfields via the twist, e.g., as studied for Edwards curves in [40], without the expensive curve conversion to Weierstrass form. We use the families proposed by [20], in which we could find three families that can be written in Hessian form.

Regardless of which model of elliptic curve was being studied, most of the previous articles on this topic were considering even embedding degrees. One of the main advantages of even embedding degrees is the applicability of a denominator elimination technique in the pairing computation (avoiding a field inversion) which does not directly apply to odd embedding degrees. Examples of pairing algorithms for curves in Weierstrass form with odd embedding degree include the work by Lin, Zhao, Zhang and Wang in [41], by Mrabet, Guillermin and Ionica in [43], and by Fouotsa, Mrabet and Pecha in [19].

1.2 Attacks on Solving DLP over Finite Fields

Due to recent advances in number field sieve (NFS) techniques for attacking the discrete logarithm problem for pairing-friendly elliptic curves over finite fields [2,3,31,35] (NFS attacks and their variants), it is necessary to re-evaluate the security of pairing-friendly curves. In [18], Fotiadis and Konstantinou propose countering these attacks by using families with a higher ρ-value. In this paper, we investigate the feasibility of an alternative method: increasing the embedding degree. This has the advantage of keeping the low ρ-value of previously proposed families, but it is disadvantaged by the less efficient pairing computations. This article attempts to analyze the use of Hessian curves in combating this. Previous research on computing pairings with Hessian curves addressed only even embedding degrees, and in order to make use of degree-3 twists the embedding degree should be divisible by 6. Prior to the NFS attacks and their variants, the favoured embedding degree for 128-bit security was 12, so that to increase the embedding degree while making use of cubic twists the next candidate is 15. However, as 15 is odd the formulas of [23] do not apply; for this reason one focus of this article is to provide formulas for embedding degree 15. Similarly, the pre-NFS favourite embedding degree for 192-bit security was 18, which we propose to increase to 21. Observe further that for 192-bit security, the families of [18] all require the embedding degree to be greater than 21.

1.3 Our Contributions

We present formulas for computing pairings on both $\mathbb{G}_1 \times \mathbb{G}_2$ and $\mathbb{G}_2 \times \mathbb{G}_1$ for a curve given in Hessian form that admits degree-3 twists. These formulas exploit the degree-3 twists where possible: in moving the point arithmetic in $\mathbb{F}_{q^k}$ to $\mathbb{F}_{q^{k/3}}$ and performing the computations for the line functions in $\mathbb{F}_{q^{k/3}}$ in place of $\mathbb{F}_{q^k}$. For efficient curve arithmetic (before applying the use of twists) we refer to Bernstein, Chuengsatiansup, Kohel, and Lange [6].

We analyze the efficiency of the pairing computation in each case, focussing on the embedding degrees that should correspond to 128- and 192-bit security.

Our analysis shows that for embedding degree 12, Hessian curves are outperformed by twisted Edwards curves, but for embedding degrees 15, 21, and 24 our formulas give the most efficient known pairing computation. We do not consider 18 as we do not know of any curve constructions for this case. As explained above, our main focus is on odd embedding degrees, as we propose the use of $k = 15$ and $k = 21$ as a countermeasure against the NFS attacks and their variants.

We also give concrete constructions of pairing-friendly Hessian curves for both embedding degrees and a proof-of-concept implementation of the optimal ate pairing for these cases.

2   Background on Pairings

Let E be an elliptic curve defined over a finite field $\mathbb{F}_q$ where q is a prime. Let r be the largest prime factor of $n = \#E(\mathbb{F}_q) = q + 1 - t$ where t is the trace of Frobenius. The *embedding degree* with respect to r is defined to be the smallest positive integer k such that $r|(q^k - 1)$. Let $\mu_r \subseteq \mathbb{F}_{q^k}^*$ be the group of r-th roots of unity. For $m \in \mathbb{Z}$ and $P \in E[r]$, let $f_{m,P}$ be a function with divisor $\operatorname{div}(f_{m,P}) = m(P) - ([m]P) - (m-1)(\mathcal{O})$, where $\mathcal{O}$ denotes the neutral element of E. The reduced Tate pairing is defined as

$$\tau_r : E(\mathbb{F}_{q^k})[r] \times E(\mathbb{F}_{q^k})/[r]E(\mathbb{F}_{q^k}) \longrightarrow \mu_r$$
$$(P,Q) \qquad\qquad \mapsto f_{r,P}(Q)^{\frac{q^k-1}{r}}.$$

We address the computation of the reduced Tate pairing restricted to $\mathbb{G}_1 \times \mathbb{G}_2$, where

$$\mathbb{G}_1 = E[r] \cap \ker(\phi_q - [1]) \text{ and } \mathbb{G}_2 = E[r] \cap \ker(\phi_q - [q]) \subseteq E(\mathbb{F}_{q^k}).$$

Here ϕ_q denotes the q-power Frobenius morphism on E. We denote the restriction of τ_r to $\mathbb{G}_1 \times \mathbb{G}_2$ by

$$e_r : \mathbb{G}_1 \times \mathbb{G}_2 \longrightarrow \mu_r.$$

Let $T = t - 1$. We define the *ate pairing* a_T by restricting the Tate pairing to $\mathbb{G}_2 \times \mathbb{G}_1$ so that

$$a_T : \mathbb{G}_2 \times \mathbb{G}_1 \longrightarrow \mu_r$$
$$(P,Q) \quad \mapsto f_{T,P}(Q)^{\frac{q^k-1}{r}}.$$

Note that in addition to $\mathbb{G}_1$ and $\mathbb{G}_2$ being switched, the subscript r (i.e., the number of loops) is also changed to T.

Algorithm 1 shows Miller's algorithm to compute the reduced Tate pairing or the ate pairing. Let $m \in \{r, T\}$ and represent the binary format of m by $(m_{n-1}, \ldots, m_1, m_0)_2$. For any two points R, S on E denote by $l_{R,S}$ the line passing through R and S, and by v_R the line passing through R and $-R$. We further define $\ell_{2R} = l_{R,R}/v_{2R}$ and $\ell_{R,P} = l_{R,P}/v_{R+P}$. Miller's algorithm outputs the Tate pairing if $m = r$, $P \in \mathbb{G}_1$, and $Q \in \mathbb{G}_2$, and outputs the ate pairing if $m = T$, $P \in \mathbb{G}_2$, and $Q \in \mathbb{G}_1$.

Algorithm 1 Miller's algorithm

Require: $m = (m_{n-1}, \ldots, m_1, m_0)_2$ and $P, Q \in E[r]$ with $P \neq Q$
1: Initialize $R = P$ and $f = 1$
2: **for** $i := n - 2$ **down to** 0 **do**
3:     $f \leftarrow f^2 \cdot \ell_{2R}(Q)$
4:     $R \leftarrow 2R$
5:     **if** $m_i = 1$ **then**
6:         $f \leftarrow f \cdot \ell_{R,P}(Q)$
7:         $R \leftarrow R + P$
8: $f \leftarrow f^{(q^k - 1)/r}$

---

3   Curve Constructions

Even though every elliptic curve can be written in Weierstrass form, only those that contain points of order 3 can be written in (twisted) Hessian form. Almost all methods to generate pairing-friendly curves are for generating pairing-friendly Weierstrass curves, so we find pairing-friendly Hessian curves by searching through constructions of pairing-friendly Weierstrass curves for curves that have points of order 3, and converting those into Hessian form. The families that we present below are guaranteed to have points of order 3.

In order to give fast formulas for curve arithmetic, it is desirable for the pairing-friendly curves that we consider to have *twists*. Recall that a *degree-d twist* of an elliptic curve $E/\mathbb{F}_q$ is an elliptic curve $E'/\mathbb{F}_{q^e}$ that is isomorphic to E over a degree-d extension of $\mathbb{F}_{q^e}$ but not over any smaller field. Recall also (e.g., [49]) that the only degrees of twists that occur for elliptic curves are $d \in \{2, 3, 4, 6\}$ such that $d|k$, and that degree 3 and 6 twists occur only for elliptic curves with j-invariant 0. We concentrate in this article on twists of degree 3, partly motivated by our aforementioned interest in embedding degrees $k = 15$ and 21. Twisted Hessian curves with j-invariant 0 are of the form

$$\mathcal{H}_a : aX^3 + Y^3 + Z^3 = 0.$$

Suppose that $a \in \mathbb{F}_q$ is a non-cube such that for $\omega \in \mathbb{F}_{q^3}$ with $a = \omega^3$, the element ω generates $\mathbb{F}_{q^k}$ as a $\mathbb{F}_{q^{k/3}}$-vector space. Then $\mathcal{H}_a$ is a degree-3 twist of $\mathcal{H}_1$; the two curves are isomorphic via

$$\varphi: \quad \mathcal{H}_a \quad \rightarrow \quad \mathcal{H}_1 \tag{1}$$
$$(X : Y : Z) \mapsto (\omega X : Y : Z).$$

In particular, if $R' \in \mathcal{H}_a(\mathbb{F}_{q^{k/3}})$, then $\varphi(R') \in \mathbb{G}_2$. Analogously to [4], we choose the $\mathbb{G}_2$ input point for the pairing from $\varphi(\mathcal{H}_a(\mathbb{F}_{q^{k/3}}))$. The simplicity of the twist isomorphism allows us to do many calculations in $\mathbb{F}_{q^{k/3}}$ instead of $\mathbb{F}_{q^k}$, as explained in detail on a case-by-case basis in Sect. 4.

3.1   Degree Six Twists of Hessian Curves

In this article we include, for completeness, formulas for computing pairings of Hessian curves with even embedding degree. As we want to make use of the natural twist of degree 3, the embedding degrees that we consider are also divisible by 3, so that we are in fact considering embedding degrees divisible by 6.

As mentioned above, degree-6 twists only occur for elliptic curves with j-invariant 0. Let a and ω be as in the previous section and let $\alpha \in \mathbb{F}_{q^2}$ generate $\mathbb{F}_{q^{k/3}}$ as a $\mathbb{F}_{q^{k/6}}$-vector space. Then

$$\mathbb{F}_{q^k} = \mathbb{F}_{q^{k/6}} + \alpha \mathbb{F}_{q^{k/6}} + \omega \mathbb{F}_{q^{k/6}} + \alpha\omega \mathbb{F}_{q^{k/6}} + \omega^2 \mathbb{F}_{q^{k/6}} + \alpha\omega^2 \mathbb{F}_{q^{k/6}}.$$

Define the triangular elliptic curve $T/\mathbb{F}_q : \alpha^2 VW(V + aW) = U^3$. Then we can adapt the isomorphism of [6, Theorem 5.3] to see that T is a degree-2 twist of $\mathcal{H}_a$ via the isomorphism

$$\begin{array}{rcl} \psi: \quad T & \to & \mathcal{H}_a \\ (U : V : W) & \mapsto & (U : \beta(\alpha V - 54W) : \beta(-\alpha V + 54\zeta_3^2 W)), \end{array} \quad (2)$$

where $\beta = \zeta_3 - \zeta_3^2$ and $\zeta_3 \in \mathbb{F}_q$ is a primitive cube root of unity. In particular, the triangular elliptic curve T is a degree-6 twist of $\mathcal{H}_1$ via the composition $\varphi \circ \psi$, where φ is as given in Eq. 1.

3.2 Checking for Points of Order 3

Let $E/\mathbb{F}_q$ be an elliptic curve. There is a Hessian model of E if and only if $E(\mathbb{F}_q)$ contains a point of order 3. To apply the formulas in the following sections we require both E and the degree-3 twist of E that we consider to have order 3. Recall that $\#E(\mathbb{F}_q) = q + 1 - t$, where t is the trace of Frobenius; by [24] the two non-trivial degree-3 twists E' satisfy:

$$\begin{array}{lll} \#E'(\mathbb{F}_q) = q + 1 - (3f - t)/2 & \quad \text{with} & t^2 - 4q = -3f^2, \\ \#E'(\mathbb{F}_q) = q + 1 - (-3f - t)/2 & \quad \text{with} & t^2 - 4q = -3f^2. \end{array}$$

It is also necessary that, for the twist E' that we use, $\#E'(\mathbb{F}_q)$ is divisible by r (recall that r was the largest prime factor of $\#E(\mathbb{F}_q)$); exactly one of the two possible twists satisfies this condition. So to choose a family for which the elliptic curve E can be rewritten in Hessian form together with a degree-3 twist, it suffices to check that 3 divides $q + 1 - t$ and that $3r$ divides $q + 1 - (\pm 3f - t)/2$ (for one choice of sign).

3.3 Generating Curves

Recall that E is an elliptic curve defined over a finite field $\mathbb{F}_q$ where q is prime, and r is the largest prime factor of $\#E(\mathbb{F}_q)$. The embedding degree k is the smallest integer k such that $r|q^k - 1$. Constructions of parametric families of pairing-friendly curves give an elliptic curve E with integral coefficients and polynomials $q(x)$ and $r(x)$, where for each x_0 such that $q(x_0)$ is prime and $r(x_0)$ has a large prime factor, the reduction of E mod $q(x_0)$ is a pairing-friendly curve with parameters $q = q(x_0)$ and $r = r(x_0)$.

Cyclotomic families are families of curves where the underlying field K is a cyclotomic field, the size r of the largest prime-order subgroup of the group of $\mathbb{F}_q$-point is a cyclotomic polynomial, and the field K contains $\sqrt{-D}$ for some small discriminant D. We searched through [20] and found three cyclotomic-family constructions that satisfy the conditions outlined in the previous section; for each family $D = 3$. The following constructions generate pairing-friendly Weierstrass curves which have a (twisted) Hessian model [6, Sect. 5]. Note that twists of these curves (see Sect. 3.2) are also expressible in twisted Hessian form. We denote the cyclotomic polynomial of degree n by $\Phi_n(x)$.

Recall the L-notation: $L_N[\ell, c] = \exp\left((c+o(1))(\ln N)^\ell(\ln \ln N)^{1-\ell}\right)$. The best complexity for NFS attacks up until recently was $L_{q^k}[1/3, 1.923]$, but now due to work of [35] the best complexity for composite k is reduced to $L_{q^k}[1/3, 1.526]$. In particular, with the earlier figure, a 256-bit prime q together with embedding degree $k = 12$ gave a security complexity of 139 bits, but that has now been brought down to 110 bits. To compensate, a pairing implementation using embedding degree 12 aiming for 128-bit security would have to increase the size of the base field to about 364 bits. We propose increasing the embedding degree instead to $k = 15$, for which the base field does not have to increased so dramatically; see details below. Similarly, with the earlier figure, a 384-bit prime q together with embedding degree $k = 18$ gave a security complexity of 194 bits, but that has now been brought down to 154 bits. To compensate, a pairing implementation using embedding degree 18 aiming for 192-bit security now requires $\log(q) \approx 653$, giving $k \log(q) \approx 11754$. We propose increasing the embedding degree instead to $k = 21$ or $k = 24$, for which the base field does not have to be increased so dramatically; see details below.

Construction 1: $k \equiv 3 \pmod{18}$. This construction follows Construction 6.6 in [20]. Pairing-friendly curves with embedding degree $k \equiv 3 \pmod{18}$ can be constructed using the following polynomials:

$$r(x) = \Phi_{2k}(x),$$
$$t(x) = x^{k/3+1} + 1,$$
$$q(x) = \tfrac{1}{3}(x^2 - x + 1)(x^{2k/3} - x^{k/3} + 1) + x^{k/3+1}.$$

For this construction, the resulting curves and their twists all have points of order 3. However, there is no such x_0 for which both $q(x_0)$ and $r(x_0)$ are prime. This means that $r(x_0)$ factors, and the largest prime-order subgroup of $E(\mathbb{F}_q)$ actually has less than $r(x_0)$ elements. Recall that the discriminant $D = 3$: the curves are defined by an equation of the form $y^2 = x^3 + b$ and have cubic twists. The ρ-value of this family is $\rho = (2k/3+2)/\varphi(k)$ where φ is the Euler φ-function. For $k = 21$ this gives $\rho = 4/3$. To get 192-bit security we have to take r about 420 bits, for which we get $\log(q) \approx 560$ and $k \log(q) \approx 11760$.

Construction 2: $k \equiv 9, 15 \pmod{18}$. This construction follows Construction 6.6 in [20]. Pairing-friendly curves with embedding degree $k \equiv 9, 15 \pmod{18}$ can be constructed using the following polynomials:

$$r(x) = \Phi_{2k}(x),$$
$$t(x) = -x^{k/3+1} + x + 1,$$
$$q(x) = \tfrac{1}{3}(x+1)^2(x^{2k/3} - x^{k/3} + 1) - x^{2k/3+1}.$$

This satisfies all the same properties as Construction 1. For $k = 15$ the ρ-value is $\rho = 3/2$. To get 128-bit security we have to take $\log(r) \approx 256$. Then $\log(q) \approx 384$, and $k \log(q) \approx 5760$. This actually gives 143-bit security; a family with a lower ρ-value would be more efficient.

Construction 3: $k \equiv 0 \pmod 6$ and $18 \nmid k$. This construction follows Construction 6.6 in [20]. Pairing-friendly curves with embedding degree $k \equiv 0 \pmod 6$ where $18 \nmid k$ can be constructed using the following polynomials:

$$r(x) = \Phi_k(x),$$
$$t(x) = x + 1,$$
$$q(x) = \tfrac{1}{3}(x-1)^2(x^{k/3} - x^{k/6} + 1) + x.$$

For this construction, the resulting curves and their twists all have points of order 3. There also exists x_0 such that both $q(x_0)$ and $r(x_0)$ are prime. The curves generated by this construction admit sextic twists. The ρ-value for this construction is given by $\rho = (k/3 + 2)/\varphi(k)$ where φ is the Euler φ-function. For $k = 12$ this gives $\rho = 3/2$ and for $k = 24$ this gives $\rho = 5/4$. To get 192-bit security with $k = 24$ we need $\log(r) \approx 392$, for which $\log(q) \approx 490$ and $k \log(q) \approx 11760$.

For all the constructions outlined above, the curves are given in Weierstrass form as $v^2 = u^3 + b$. To convert a pairing-friendly Weierstrass curve of the above form that has a point (u_3, v_3) of order 3 into twisted Hessian form, we refer to [6]. The authors give explicit transformations showing that there is a Hessian model of the above curve given by $aX^3 + Y^3 + Z^3 = 0$, where $a = 27(u_3^6/v_3^3 - 2v_3)$. Let **m**, **s** and $\mathbf{m}_c$ denote field multiplication, field squaring and field multiplication by a small constant respectively. They compute the total cost for the whole conversion to be $9\mathbf{m} + 2\mathbf{s} + 5\mathbf{m}_c$ plus one inversion and one cube root computation.

4   Computation of Line Functions

Each iteration of Miller's loop (Algorithm 1) includes a *Miller doubling* step and some of the iterations also include a *Miller addition* step. The Miller doubling step has four costly parts: computing the double of a point R on the curve, computing the Miller function $\ell_{R,R} = l_{R,R}/v_{2R}$, squaring an element $f \in \mathbb{F}_{q^k}$, and multiplying f^2 by $\ell_{R,R}$. The Miller addition step has three costly parts: computing the sum of two points P and R on the curve, computing the Miller function $\ell_{P,R} = l_{P,R}/v_{P+R}$, and multiplying an element $f \in \mathbb{F}_{q^k}$ by $\ell_{P,R}$. We attempt in the following sections to optimize each of these parts for Hessian curves $\mathcal{H}/\mathbb{F}_q : X^3 + Y^3 + Z^3 = 0$ of j-invariant 0 for pairings on both $\mathbb{G}_1 \times \mathbb{G}_2$ (such as the Tate pairing) and $\mathbb{G}_2 \times \mathbb{G}_1$ (such as the ate pairing).

4.1 Denominator Elimination

It is, of course, desirable to avoid the field inversion that results from dividing by $v_{P_1+P_2}(Q)$, with $P_2 = R$ and $P_1 \in \{R, P\}$, which we can do (to some extent). For curves in (twisted) Hessian form, the neutral group element is given by $(0 : -1 : 1)$, and negation by $-(x, y) = (x/y, 1/y)$ (in affine coordinates). This means that the line $v_{P_1+P_2}$ passing through $P_3 = P_1 + P_2$ and $(0 : -1 : 1)$ has a more complicated form than for many other popular curve shapes (such as short Weierstrass or Edwards). Namely, writing $(X_3 : Y_3 : Z_3) = P_3$ and $(X_Q : Y_Q : Z_Q)$, we have

$$v_{P_3}(Q) : (Z_3 + Y_3)X_Q - (Z_Q + Y_Q)X_3.$$

When considering pairings on $\mathbb{G}_1 \times \mathbb{G}_2$, we have that $P_3 \in \mathbb{G}_1$ and $Q \in \mathbb{G}_2$, and when considering pairings on $\mathbb{G}_2 \times \mathbb{G}_1$, we have that $P_3 \in \mathbb{G}_2$ and $Q \in \mathbb{G}_1$. As $v_{P_3}(Q) = v_Q(P_3)$, exactly the same arguments apply to $\mathbb{G}_1 \times \mathbb{G}_2$ as to $\mathbb{G}_2 \times \mathbb{G}_1$ in this case; say for simplicity that $P_3 \in \mathbb{G}_1$ and $Q \in \mathbb{G}_2$. Suppose that we have chosen Q such that there exists $Q' \in \mathcal{H}_a(\mathbb{F}_{q^{k/3}})$ for which $Q = \varphi(Q')$, where φ is the cubic twist isomorphism from Eq. 1.

Even Embedding Degrees. The following is essentially a rephrasing of the denominator elimination technique presented in [23] (although they do not mention pairings on $\mathbb{G}_2 \times \mathbb{G}_1$).

Assume now that $6|k$. In particular, by the discussion in Sect. 3.1, the triangular curve $T : \alpha^2 VW(V + \omega^3 W) = U^3$, with α and ω as in Sect. 3.1, defines a quadratic twist of $\mathcal{H}_{\omega^3}$ via the isomorphism ψ of Eq. 2. We choose our point $Q' \in \mathcal{H}_{\omega^3}(\mathbb{F}_{q^{k/3}})$ from the image under ψ of $T(\mathbb{F}_{q^{k/6}})$, so that there exist $U, V, W \in \mathbb{F}_{q^{k/6}}$ for which $Q' = (U : \beta(\alpha V - 54W) : \beta(-\alpha V + 54\zeta_3^2 W))$, where $\beta = \zeta_3 - \zeta_3^2$ and $\zeta_3 \in \mathbb{F}_q$ is a primitive cube root of unity. Evaluation of v_{P_3} at $Q = \varphi(Q')$ then gives

$$v_{2R}(Q) : (Z_3 + Y_3)U\omega - 54\beta(\zeta_3^2 - 1)WX_3 \in \mathbb{F}_{q^{k/2}}.$$

This value will go to 1 in the final exponentiation step of Miller's algorithm (Algorithm 1), so without loss of generality we can set it to 1 throughout the computation.

Odd Embedding Degrees. Unfortunately the denominator elimination technique of [23] does not apply to this case; instead we extend ideas of [41,43]. Observe that $\frac{1}{x-y} = \frac{x^2+xy+y^2}{x^3-y^3}$. Let $Q' = (X_{Q'}, Y_{Q'}, Z_{Q'})$. Plugging $x = (Z_3 + Y_3)X_{Q'}\omega$ and $y = (Z_{Q'} + Y_{Q'})X_3$ in $\frac{1}{v_{P_3}(Q)}$ with $Q = \varphi(Q')$, we get that the denominator

$x^3 - y^3$ is in $\mathbb{F}_{q^{k/3}}$ so will go to 1 in the final exponentiation, hence can be set to 1 for the whole computation. That is, we replace $\frac{1}{v_{P_3}(Q)}$ by the numerator

$$n_{P_3}(Q) = ((Z_3 + Y_3)X_{Q'})^2\omega^2 + (Z_3 + Y_3)X_{Q'}(1 + Y_{Q'})X_3\omega + ((1 + Y_{Q'})X_3)^2,$$

and we replace the Miller function $\ell_{P_1,P_2}(Q)$ by $n_{P_3}(Q)\cdot l_{P_1,P_2}(Q)$. The numerator $n_{P_3}(Q)$ can be computed with cost $\frac{2k}{3}\mathbf{m} + \frac{1}{9}\mathbf{M} + \frac{2}{9}\mathbf{S}$ via

$$u = (Z_3 + Y_3)X_{Q'}; \quad v = (1 + Y_{Q'})X_3; \quad n = u^2\omega^2 + (u \cdot v)\omega + v^2.$$

4.2 Miller Doubling

Let $R = (X_1 : Y_1 : Z_1) \in \mathcal{H}_b(K)$ for $b \in \{1, a\}$. The fastest known formulas to compute $2R = (X_3 : Y_3 : Z_3)$ (due to [6]) are as follows:

$$T = Y_1^2; \qquad A = Y_1 \cdot T; \qquad S = Z_1^2; \qquad B = Z_1 \cdot S;$$
$$X_3 = X_1 \cdot (A - B); \quad Y_3 = -Z_1 \cdot (2A + B); \quad Z_3 = Y_1 \cdot (A + 2B).$$

The cost for point doubling with the above formulas is $5\mathbf{m} + 2\mathbf{s}$ in K.

In all that follows we denote multiplication and squaring in $\mathbb{F}_q$ by $\mathbf{m}$ and $\mathbf{s}$ respectively, and multiplication and squaring in $\mathbb{F}_{q^k}$ by $\mathbf{M}$ and $\mathbf{S}$ respectively. We also assume always that $3|k$.

Pairings on $\mathbb{G}_1 \times \mathbb{G}_2$. The Miller doubling function is given by

$$\ell_{R,R}(Q) = l_{R,R}(Q)/v_{2R}(Q).$$

For pairings on $\mathbb{G}_1 \times \mathbb{G}_2$ the input points are $P \in \mathbb{G}_1$ and $Q \in \mathbb{G}_2$, and R will be a multiple of P.

We first address the computation of $l_{R,R}(Q)$. This line is the tangent line to $\mathcal{H}_1$ at R evaluated at Q, which is given by $l_{R,R}(Q) : X_1^2 X_Q + TY_Q + S$, where $R = (X_1 : Y_1 : Z_1)$ and $T = Y_1^2$, and $S = Z_1^2$ are the values that were computed in the point doubling computation. Set $Q' = (X_{Q'} : Y_{Q'} : 1)$ and $Q = \varphi(Q')$, where $\varphi : \mathcal{H}_a \to \mathcal{H}_1$ is the twist isomorphism Eq. 1 (this is possible as $3|k$). Then we can write $l_{R,R}(Q)$ as

$$l_{R,R}(Q) : (SY_{Q'} + T) + aX_{Q'}X_1^2\omega,$$

which can be computed with cost $\frac{2k}{3}\mathbf{m} + \mathbf{s}$ via

$$U = X_1^2; \quad V = SY_{Q'}; \quad W = \eta U; \quad l_{R,R}(Q) = V + T + W\omega,$$

where $\eta = aX_{Q'}$ and can be precomputed. We now split into cases.

Even Embedding Degrees. By Sect. 4.1, we can set the denominator of the Miller doubling function to 1, so that the computation of the line function $l_{R,R}(Q)$ is in fact the computation of the whole Miller (doubling) function $\ell_{R,R}(Q)$.

Furthermore, a general element of $\mathbb{F}_{q^k}$ considered as element of the $\mathbb{F}_{q^{k/3}}$-vector space generated by ω will be of the form $c_1\omega + c_2\omega^2 + c_3\omega^3$, but for $\ell_{2R}(Q)$ we have that $c_2 = 0$. In particular, the multiplication of $\ell_{2R}(Q)$ with f^2 in Step 3 of Algorithm 1 will not be the full cost of a general multiplication in $\mathbb{F}_{q^k}$ (that is, approximately $k^2\mathbf{m}$), but by schoolbook multiplication will cost 6 multiplications in $\mathbb{F}_{q^{k/3}}$, which amounts to $6\left(\frac{k}{3}\right)^2\mathbf{m} = \frac{2}{3}\mathbf{M}$. Putting together all of the above, the Miller doubling step for even embedding degrees costs

$$\left(5 + \frac{2k}{3}\right)\mathbf{m} + 3\mathbf{s} + \frac{2}{3}\mathbf{M} + \mathbf{S}.$$

Odd Embedding Degrees. By Sect. 4.1, we have $\ell_{2R}(Q) = n_{2R}(Q)\cdot l_{R,R}(Q)$, where $n_{2R}(Q)$ is as given in Sect. 4.1. Putting the above together, the Miller doubling step for odd embedding degrees costs

$$\left(5 + \frac{4k}{3}\right)\mathbf{m} + 3\mathbf{s} + \frac{16}{9}\mathbf{M} + \frac{11}{9}\mathbf{S}.$$

Pairings on $\mathbb{G}_2 \times \mathbb{G}_1$. In this case, the input points are $P \in \mathbb{G}_2$ and $Q \in \mathbb{G}_1$, and R will be a multiple of P. We choose $P = (X_P : Y_P : 1) \in \varphi(\mathcal{H}_a(\mathbb{F}_{p^{k/3}}))$, where φ is the twist isomorphism given in Eq. 1. As $R = (X_1 : Y_1 : Z_1)$ is a multiple of P, it is also in the image of $\mathcal{H}_a(\mathbb{F}_{p^{k/3}})$ under φ; let $R' \in \mathcal{H}_a(\mathbb{F}_{p^{k/3}})$ be the pre-image of R under φ. As $2R = 2\varphi(R') = \varphi(2R')$, we can perform the doubling operation on the cubic twist $\mathcal{H}_a$, so that the operation count occurs in $\mathbb{F}_{q^{k/3}}$. That is, point doubling can be performed using 5 multiplications and 2 squarings in $\mathbb{F}_{q^{k/3}}$, which amounts to $\frac{5}{9}\mathbf{M} + \frac{2}{9}\mathbf{S}$. For even embedding degrees this can be done slightly faster, which we address below.

As for pairings on $\mathbb{G}_1 \times \mathbb{G}_2$, we address the computations of the line function

$$l_{R,R}(Q) : X_1^2 X_Q + TY_Q + S, \tag{3}$$

where $T = Y_1^2$ and $S = Z_1^2$, in order to compute the Miller doubling function.

Even Embedding Degrees. Assume now that $6|k$. As described in Sect. 4.1 we choose the input point from $\mathbb{G}_2$, in this case $P = \varphi(P')$, such that P' is in the image of the quadratic twist isomorphism ψ given in Sect. 3.1. This implies that $R' = \varphi^{-1}(R)$, as a multiple of P', also lies in this image, so that there exist $U_1, V_1, W_1 \in \mathbb{F}_{q^{k/6}}$ for which

$$R' = (X_1' : Y_1' : Z_1') = (U_1 : \beta(\alpha V_1 - 54W_1) : \beta(-\alpha V_1 + 54\zeta_3^2 W_1)), \tag{4}$$

where $\beta = \zeta_3 - \zeta_3^2$ and $\zeta_3 \in \mathbb{F}_q$ is a primitive cube root of unity. Here ω and α are as in Sect. 3.1. We also have $X_1' \in \mathbb{F}_{q^{k/6}}$ and $Y_1', Z_1' \in \mathbb{F}_{q^{k/3}}$.

This gives us a small saving in the point doubling calculation. In the preamble we stated that all the point doubling arithmetic is performed in $\mathbb{F}_{q^{k/3}}$. However, the final step in the computation of X_3' (the X-coordinate of $2R'$) is not a full multiplication in $\mathbb{F}_{q^{k/3}}$ but a multiplication of a $\mathbb{F}_{q^{k/6}}$-element X'_1 with a $\mathbb{F}_{q^{k/3}}$-element $(A - B)$, costing $2 \left(\frac{k}{6}\right)^2 \mathbf{m} = \frac{1}{18}\mathbf{M}$ using schoolbook multiplication instead of $\frac{1}{9}\mathbf{M}$. So we save $\frac{1}{18}\mathbf{M}$ on the point doubling for even embedding degrees, resulting in $\frac{1}{2}\mathbf{M} + \frac{2}{9}\mathbf{S}$.

As shown in Sect. 4.1, the Miller doubling function $\ell_{R,R}(Q)$ is just given by the line function $l_{R,R}(Q)$ in this case, the computation of which we now address. As above we have that $R = (X_1 : Y_1 : Z_1) = (X'_1\omega : Y'_1 : Z'_1)$ so that Eq. 3 becomes

$$l_{R,R}(Q) : (X'_1)^2 X_Q\omega^2 + TY_Q + S.$$

The values S and T are computed during the point doubling computation and lie in $\mathbb{F}_{q^{k/3}}$, so the computation of $\ell_{R,R}(Q) = l_{R,R}(Q)$ costs an additional squaring in $\mathbb{F}_{q^{k/6}}$, multiplication of a $\mathbb{F}_{q^{k/6}}$-element with a $\mathbb{F}_q$-element, and multiplication of a $\mathbb{F}_{q^{k/3}}$-element with a $\mathbb{F}_q$-element, giving $\frac{k}{2}\mathbf{m} + \frac{1}{36}\mathbf{S}$ via

$$c_1 = (X'_1)^2; \quad c_2 = c_1 X_Q; \quad c_3 = TY_Q.$$

Additionally, the formula for $\ell_{R,R}(Q)$ considered as an element of the $\mathbb{F}_{q^{k/6}}$-vector space generated by ω and α has no coefficient of ω, $\alpha\omega$, or $\alpha\omega^2$. Therefore the multiplication of $\ell_{R,R}(Q)$ with a general element (i.e., f^2) of $\mathbb{F}_{q^k}$ costs only $3 \cdot 6 \left(\frac{k}{6}\right)^2 \mathbf{m} = \frac{1}{2}\mathbf{M}$ with schoolbook arithmetic.

Putting the above together, the full Miller doubling step for even embedding degrees costs

$$\frac{k}{2}\mathbf{m} + \mathbf{M} + \frac{5}{4}\mathbf{S}.$$

Odd Embedding Degrees. By Sect. 4.1, the Miller doubling function $\ell_{R,R}(Q)$ is given by $\ell_{R,R}(Q) = n_{2R}(Q) \cdot l_{R,R}(Q)$, where $n_{2R}(Q)$ is as given in Sect. 4.1. As described above for even embedding degrees, we have that

$$l_{R,R}(Q) : (X'_1)^2 X_Q\omega^2 + TY_Q + S,$$

where $S, T \in \mathbb{F}_{q^{k/3}}$ and are computed during the point doubling computation. In the case of odd embedding degrees, we have that $X'_1 \in \mathbb{F}_{q^{k/3}}$, so that the cost of commutating $l_{R,R}(Q)$ via c_1, c_2, and c_3 as above is $\frac{2k}{3}\mathbf{m} + \frac{1}{9}\mathbf{S}$. Putting the above together, the whole Miller doubling step for odd embedding degrees costs

$$\frac{4k}{3}\mathbf{m} + \frac{7}{3}\mathbf{M} + \frac{14}{9}\mathbf{S}.$$

4.3 Miller Addition

Let $P_1 = P = (X_1 : Y_1 : 1)$ and $P_2 = R = (X_2 : Y_2 : Z_2) \in \mathcal{H}_b(K)$ for $b \in \{1, a\}$. The fastest known formulas to compute $P_1 + P_2 = P_3 = (X_3 : Y_3 : Z_3)$ for $P_1 \neq P_2$ (due to [25]) are as follows:

$$A = X_1 \cdot Z_2; \qquad C = Y_1 \cdot X_2; \qquad D = Y_1 \cdot Y_2; \qquad F = \eta \cdot X_2;$$
$$G = (D + Z_2) \cdot (A - C); \quad H = (D - Z_2) \cdot (A + C); \quad X_3 = G - H;$$

$$J = (D + F) \cdot (A - Y_2); \quad K = (D - F) \cdot (A + Y_2); \quad Y_3 = K - J;$$
$$Z_3 = J + K - G - H - 2(Z_2 - F) \cdot (C + Y_2),$$

where $\eta = aX_1$ can be precomputed. The cost for point addition with the above formulas is $9\mathbf{m}$ in K.

Pairings on $\mathbb{G}_1 \times \mathbb{G}_2$. The Miller addition function is given by

$$\ell_{P_1,P_2}(Q) = l_{P_1,P_2}(Q)/v_{P_1+P_2}(Q).$$

For pairings on $\mathbb{G}_1 \times \mathbb{G}_2$ the input points are $P \in \mathbb{G}_1$ and $Q \in \mathbb{G}_2$, and for addition we have that $P_1 = P = (X_1 : Y_1 : 1)$ and $P_2 = R = (X_2 : Y_2 : Z_2)$ is a multiple of P.

The line $l_{P_1,P_2}(Q)$ is the line passing through P and R evaluated at Q. As above we write $Q = (\omega X'_Q : Y'_Q : 1)$ with $Q' = (X_{Q'}, Y_{Q'} : 1) \in \mathcal{H}_a(\mathbb{F}_{q^{k/3}})$. Then

$$l_{P,R}(Q) : (E - Y_2)X_1 + (Y_{Q'} - Y_1)(A - X_2) - (E - Y_2)X_{Q'}\omega,$$

where $E = Y_1 Z_2$, and where A is the value that was computed during the computation of $P+R$. In particular, the cost of computing $l_{P,R}(Q)$ is $\left(2 + \frac{2k}{3}\right)\mathbf{m}$ via

$$E = Y_1 \cdot Z_2; \quad L = (E - Y_2) \cdot X_1; \quad M = (Y_{Q'} - Y_1) \cdot (A - X_2);$$
$$N = (E - Y_2) \cdot X_{Q'}; \quad l_{P,R}(Q) = L + M - N\omega.$$

Even Embedding Degrees. By Sect. 4.1, the Miller addition function $\ell_{P_1,P_2}(Q)$ is just given by $l_{P_1,P_2}(Q)$ in this case. Also, exactly as for the Miller doubling function, multiplying a general element of $\mathbb{F}_{q^k}$ with $l_{P,R}(Q)$ costs only $\frac{2}{3}\mathbf{M}$. Putting together all of the above, the entire Miller addition step costs

$$\left(11 + \frac{2k}{3}\right)\mathbf{m} + \frac{2}{3}\mathbf{M}.$$

Odd Embedding Degrees. By Sect. 4.1, the Miller addition function $\ell_{P_1,P_2}(Q)$ is given by $\ell_{P_1,P_2}(Q) = n_{P_1+P_2}(Q) \cdot l_{P_1,P_2}(Q)$, where $n_{P_1+P_2}(Q)$ is as given in Sect. 4.1. Putting together all of the above, the entire Miller addition step costs

$$\left(11 + \frac{4k}{3}\right)\mathbf{m} + \frac{16}{9}\mathbf{M} + \frac{2}{9}\mathbf{S}.$$

Pairings on $\mathbb{G}_2 \times \mathbb{G}_1$. For pairings on $\mathbb{G}_2 \times \mathbb{G}_1$ the input points $P \in \mathbb{G}_2$ and $Q \in \mathbb{G}_1$, and in the Miller addition function $\ell_{P_1,P_2}(Q)$ we have that $P_1 = P = (X_1 : Y_1 : 1)$ and $P_2 = R = (X_2 : Y_2 : Z_2)$, which is some multiple of P. In exactly the same way as discussed for the Miller doubling function, the point addition can be performed in the group $\mathcal{H}_a(\mathbb{F}_{q^{k/3}})$ in place of $\mathcal{H}(\mathbb{F}_{q^k})$, so that the operation count occurs in $\mathbb{F}_{q^{k/3}}$. That is, point addition can be performed using 9 multiplications in $\mathbb{F}_{q^{k/3}}$, which amounts to $1\mathbf{M}$. For even embedding degrees this can be done faster, which we address below. As for pairings on $\mathbb{G}_1 \times \mathbb{G}_2$, we will need to compute the line function

$$l_{P,R}(Q) : -(E - Y_2)X_Q + (E - Y_2)X_1 + (Y_Q - Y_1)(A - X_2),$$

where $E = Y_1Z_2$ and $A = X_1Z_2$. Let $P = \varphi(P')$ and $R = \varphi(R')$ be the images of $P' = (X_1', Y_1', 1)$ and $R' = (X_2', Y_2', Z_2') \in \mathcal{H}_a(\mathbb{F}_{q^{k/3}})$ respectively under the twist isomorphism φ of Eq. 1. Then

$$l_{P,R}(Q) : -(E' - Y_2')X_Q + (C' - Y_2'X_1' + Y_Q(A' - X_2'))\omega,$$

where $E' = Y_1'Z_2'$, $A = A'\omega$, $C = C'\omega$ and $A' = X_1'Z_2'$ and $C' = Y_1'X_2'$ are the values that were computed during the point addition. This can be computed in $\frac{2k}{3}\mathbf{m} + \frac{2}{9}\mathbf{M}$ via

$$E' = Y_1' \cdot Z_2'; \quad d_1 = Y_2' \cdot X_1'; \quad d_2 = (E'-Y_2') \cdot X_Q; \quad d_3 = (A'-X_2') \cdot Y_Q.$$

Even Embedding Degrees. Suppose now that $6|k$. As described already for Miller doubling, we may choose $U_2, V_2, W_2 \in \mathbb{F}_{q^{k/6}}$ such that

$$R' = (U_2 : \beta(\alpha V_2 - 54W_2) : \beta(-\alpha V_2 + 54\zeta_3^2 W_2)),$$

where $\beta = \zeta_3 - \zeta_3^2$ and $\zeta_3 \in \mathbb{F}_q$ is a primitive cube root of unity (c.f. Eq. 4). Note that we do not apply this to P because we want to make use of the mixed addition with $Z_1 = 1$.

This gives us a small saving in the point addition calculation: the computations of C and of F now cost $\frac{1}{18}\mathbf{M}$ each instead of $\frac{1}{9}\mathbf{M}$ each, saving $\frac{1}{18}\mathbf{M}$; the cost for point addition is therefore $\frac{8}{9}\mathbf{M}$.

As shown in Sect. 4.1, the Miller addition function $\ell_{P,R}(Q)$ is just given by the line function $l_{P,R}(Q)$ in this case. Multiplication of a general element in $\mathbb{F}_{q^k}$ with $\ell_{P,R}(Q)$ costs only $\frac{2}{3}\mathbf{M}$ as $\ell_{P,R}(Q)$ has no coefficient of ω^2. Putting together all of the above, we get the cost for the whole Miller addition step

$$\frac{2k}{3}\mathbf{m} + \frac{16}{9}\mathbf{M}.$$

Odd Embedding Degrees. By Sect. 4.1, the Miller addition function $\ell_{P,R}(Q)$ is given by $n_{P+R}(Q) \cdot l_{P,R}(Q)$ in this case, where $n_{P+R}(Q)$ is as given in Sect. 4.1. Putting together all of the above, we get the cost for the full Miller addition step

$$\frac{4k}{3}\mathbf{m} + 3\mathbf{M} + \frac{2}{9}\mathbf{S}.$$

5   Comparison

As this paper primarily concerns cubic twists, we only discuss results for embedding degrees that are divisible by 3. To our knowledge, most of the previous work on the optimization of operation counts for one iteration of Miller's loop concentrated on pairings for $\mathbb{G}_1 \times \mathbb{G}_2$. To properly compare different results, we need to take into account the number of iterations of Miller's loop, which differs greatly between $\mathbb{G}_1 \times \mathbb{G}_2$ and $\mathbb{G}_2 \times \mathbb{G}_1$.

For pairings on $\mathbb{G}_1 \times \mathbb{G}_2$, the lowest number of iterations occurs for the twisted ate pairing when twists are available, or the reduced Tate pairing when twists are not available. In this paper, we explicitly address the first case, so the twisted ate pairing gives the minimal number of iterations. Let t be the trace of Frobenius, let $T = t - 1$, and let d be the degree of the twist. The number of iterations of Miller's loop for the twisted ate pairing is given by $\log(T_e)$, where $T_e \equiv T^e$ (mod r) and $1 < e|d$. Also T is a d-th root of unity in $\mathbb{F}_r$, so when $d = 6$ the smallest value of $\log(T_e)$ is $\log(T_2) \approx \log(r)/3$, and when $d = 3$ the smallest value of $\log(T_e)$ is $\log(T_3) \approx \log(r)$. For more details on the twisted ate pairing see [24].

For pairings on $\mathbb{G}_2 \times \mathbb{G}_1$, the lowest number of iterations occurs for the optimal ate pairing. The best-case-scenario (which can in principle occur for any embedding degree) is $\log(r)/\varphi(k)$ iterations of Miller's loop, where φ is the Euler φ-function. This scenario takes x as the input for the Miller's algorithm (e.g., in place of $r = r(x)$ as in Tate). For more details on the optimal ate pairing see [51].

We compared previous results in this area for Weierstrass curves with Jacobian coordinates [1,27], Weierstrass curves with projective coordinates [15], Edwards curves [1], Edwards curves with sextic twists [40], and Hessian curves with quadratic twists [23]. Most of these papers considered only pairings on $\mathbb{G}_1 \times \mathbb{G}_2$ (many of them were written before Vercauteren's paper [51] on optimal pairings) and only even embedding degree (to avoid dealing with denominators).

5.1   Comparing Results for $\mathbb{G}_2 \times \mathbb{G}_1$

The only other paper containing operation counts for pairings on $\mathbb{G}_2 \times \mathbb{G}_1$ and embedding degree divisible by 3, to our knowledge, is [15], which considers projective Weierstrass coordinates. In that paper they look at even embedding degrees, so we only compare our results for the optimal ate pairing when $k = 12$ and 24 (c.f. Construction 3). Assume for simplicity that $\mathbf{s} \approx 0.8\mathbf{m}$. The formulas presented in [15] give an operation count of

$$\frac{41}{36}\mathbf{M} + \frac{41}{36}\mathbf{S} \approx \begin{cases} 295.2\mathbf{m} \ \ k = 12 \\ 1180.8\mathbf{m} \ \ k = 24 \end{cases} \text{ for Miller doubling and}$$

$$\frac{4}{3}\mathbf{M} + \frac{1}{18}\mathbf{S} \approx \begin{cases} 198.4\mathbf{m} \ \ k = 12 \\ 793.6\mathbf{m} \ \ k = 24 \end{cases} \text{ for Miller addition.}$$

The formulas presented in this paper give an operation count of

$$\frac{k}{2}\mathbf{m} + \mathbf{M} + \frac{5}{4}\mathbf{S} \approx \begin{cases} 294.0\mathbf{m} \ \ k = 12 \\ 1164.0\mathbf{m} \ \ k = 24 \end{cases} \text{ for Miller doubling and}$$

$$\frac{2k}{3}\mathbf{m} + \frac{16}{9}\mathbf{M} \approx \begin{cases} 264.0\mathbf{m} \ \ k = 12 \\ 1040.0\mathbf{m} \ \ k = 24 \end{cases} \text{ for Miller addition.}$$

As the formulas for Hessian form are faster for doubling but slower for adding (with respect to projective Weierstrass form), there is a trade-off to assess. Suppose that we wish to compute the optimal ate pairing and that we have an example for which the input for Miller's algorithm is x. The pairing can then be computed in $\log(x) = \log(r)/\varphi(k)$ iterations of Miller's loop — this amounts to $O(\log(x))$ Miller doubling steps, $O(\mathsf{Ham}(x))$ Miller addition steps, where $\mathsf{Ham}(x)$ denotes the Hamming weight of x, and the final exponentiation. When $k = 12$, the formulas presented in [15] compute the pairing in $\approx 295.2 \cdot O(\log(x)) + 198.4 \cdot O(\mathsf{Ham}(x))$ multiplications in $\mathbb{F}_q$ and an exponentiation, and the formulas presented in this paper compute the pairing in $\approx 294.0 \cdot O(\log(x)) + 264.0 \cdot O(\mathsf{Ham}(x))$ multiplications in $\mathbb{F}_q$ and an exponentiation. That is, the formulas using Hessian curves outperform the projective Weierstrass curves only for an x-value such that $\log(x) > 54.67 \cdot \mathsf{Ham}(x)$. When $k = 24$, the formulas using Hessian curves outperform the projective Weierstrass curves for an x-value such that $\log(x) > 14.67 \cdot \mathsf{Ham}(x)$.

5.2   Comparing Results for $\mathbb{G}_1 \times \mathbb{G}_2$

Comparing the aforementioned papers $[1, 15, 23, 27, 40]$, and our results, we see that the fastest curve model for embedding degree divisible by 6 together with a $\mathbb{G}_1 \times \mathbb{G}_2$ pairing is the Edwards form with sextic twists [40] using $\left(\frac{4k}{3} + 4\right)\mathbf{m} + 7\mathbf{s} + \frac{1}{3}\mathbf{M} + \mathbf{S}$ for one Miller doubling step and $\left(\frac{4k}{3} + 12\right)\mathbf{m} + \frac{1}{3}\mathbf{M}$ for one Miller addition step. The fastest curve model for odd embedding degree divisible by 3 together with a $\mathbb{G}_1 \times \mathbb{G}_2$ pairing is the projective Weierstrass form [15] using $(k + 6)\mathbf{m} + 7\mathbf{s} + \mathbf{M} + \mathbf{S}$ for one Miller doubling step and $(k + 13)\mathbf{m} + 3\mathbf{s} + \mathbf{M}$ for one Miller addition step.

5.3   Comparing $\mathbb{G}_1 \times \mathbb{G}_2$ and $\mathbb{G}_2 \times \mathbb{G}_1$

In the following table we compare the operation counts from the *most efficient* curve shape for each subcase (optimal ate vs. twisted ate and even vs. odd) in what we hope is a meaningful way: we give the number of $\mathbb{F}_q$-multiplications per Miller doubling/addition multiplied by $\frac{1}{\log(r)} \times$ the number of iterations. We call these numbers DBLc (for doubling compare) and ADDc (for addition compare). We assume here that $\mathbf{s} = 0.8\mathbf{m}$ for simplicity. The most efficient option for each subcase is as follows.

For embedding degree 12, [40] is clearly the most efficient. For embedding degrees 15 and 21, our results are clearly the most efficient. For embedding degree 24, doubling is more efficient in Hessian form with optimal ate while addition is more efficient in Edwards form with twisted ate. We could assess this trade-off in a similar way to the trade-off that was required to compare results for even embedding degrees for optimal ate pairings; our results will outperform those of [40] when the Hamming weight of x is sufficiently low compared to $\log(x)$.

Not included in Table 1 are the precomputation costs (which are relatively low for our constructions) and the final exponentiation costs (which are roughly

Table 1. Best operation counts for DBLc and ADDc for each embedding degree and type of pairing

k	Pairing	Model	# iterations	DBLc	ADDc
12	Twisted ate	Edwards [40]	$\log(r)/3$	62.9	25.3
	Optimal ate	Projective [15]	$\log(r)/4$	73.8	49.6
	Optimal ate	Hessian (this paper)	$\log(r)/4$	73.5	66.0
15	Twisted ate	Projective [15]	$\log(r)$	431.6	255.4
	Optimal ate	Hessian (this paper)	$\log(r)/8$	103.1	120.0
21	Twisted ate	Projective [15]	$\log(r)$	826.4	477.4
	Optimal ate	Hessian (this paper)	$\log(r)/12$	133.8	155.9
24	Twisted ate	Edwards [40]	$\log(r)/3$	231.5	78.7
	Optimal ate	Projective [15]	$\log(r)/8$	147.5	99.2
	Optimal ate	Hessian (this paper)	$\log(r)/8$	140.7	134.0

uniform across all curve shapes). A significant part of the precomputation cost for many models is the conversion between curve models, which is not necessary for our constructions. (Recall that for BN, BLS, and KSS, this conversion is always necessary if one wants to take advantage of the fast point arithmetic on Hessian or Edwards curves, as proven in [10].)

6   Concluding Remarks

This paper presents efficient formulas to compute Miller doubling and Miller addition on curves of j-invariant 0 with embedding degree divisible by 3 when written in Hessian form. This paper presents formulas for both pairings of the form $\mathbb{G}_1 \times \mathbb{G}_2$ and $\mathbb{G}_2 \times \mathbb{G}_1$ and compares the efficiency of these formulas to the best known formulas of previous research. We present the first formulas for pairings on $\mathbb{G}_2 \times \mathbb{G}_1$ that utilize twists of degree 3 in the case of odd embedding degrees, and the first formulas that utilize twists of degree 3 for Hessian curves in all cases. Our formulas for embedding degrees 15, 21, and (subject to trade-offs) 24 are the most efficient among known choices.

Curves generated by the methods used in this paper (originally due to [20]) are guaranteed to have twists of degree 3 and have embedding degree $k \equiv 3, 9, 15$ (mod 18) or $k \equiv 0$ (mod 6) where $18 \nmid k$. We suggest updating the use of embedding degree 12 to 15 for 128-bit security and 18 to 21 for 192-bit security in light of the NFS attacks and their variants. This allows us to keep the relatively small primes for the base field and a low ρ-value. We additionally suggest including $k = 24$ in any future (more precise) comparisons, as our results show that this may be competitive with $k = 21$ (since the ρ-value for $k = 24$ is lower than that of $k = 21$).

In future work, we plan to study precisely how the NFS attacks and their variants apply to our constructions in order to be able to properly evaluate the security and propose concrete parameters. A comparison between the larger embedding degrees (but low ρ-value) that we suggest in this paper and the higher ρ-value (but small embedding degrees) suggested in [18] would be very interesting, but we leave this for future work. It would also be interesting to evaluate the performance of other curve models with degree 3 twists on $\mathbb{G}_2 \times \mathbb{G}_1$ pairings. We also consider the optimized implementation as future work.

References

1. Arene, C., Lange, T., Naehrig, M., Ritzenthaler, C.: Faster computation of the Tate pairing. IACR Cryptology ePrint Archive, 2009:155 (2009). http://eprint.iacr.org/2009/155
2. Barbulescu, R., Gaudry, P., Guillevic, A., Morain, F.: Improving NFS for the discrete logarithm problem in non-prime finite fields. In: Eurocrypt 2015 [44], pp. 129–155 (2015)
3. Barbulescu, R., Gaudry, P., Kleinjung, T.: The tower number field sieve. In: Asiacrypt 2015 [28], pp. 31–55 (2015)
4. Barreto, P.S.L.M., Lynn, B., Scott, M.: On the selection of pairing-friendly groups. In: SAC 2003 [42], pp. 17–25 (2003)
5. Barreto, P.S.L.M., Naehrig, M.: Pairing-friendly elliptic curves of prime order. In: SAC 2005 [45], pp. 319–331 (2006). http://cryptosith.org/papers/pfcpo.pdf
6. Bernstein, D.J., Chuengsatiansup, C., Kohel, D., Lange, T.: Twisted Hessian curves. In: LATINCRYPT 2015 [39], pp. 269–294 (2015). http://cr.yp.to/papers.html#hessian
7. Bernstein, D.J., Lange, T.: Faster addition and doubling on elliptic curves. In: Asiacrypt 2007 [37], pp. 29–50 (2007). http://cr.yp.to/newelliptic/newelliptic-20070906.pdf
8. Boneh, D., Franklin, M.K.: Identity-based encryption from the Weil pairing. In: CRYPTO 2001 [34], pp. 213–229 (2001). http://www.iacr.org/archive/crypto2001/21390212.pdf
9. Boneh, D., Lynn, B., Shacham, H.: Short signatures from the Weil pairing. J. Cryptol. **17**(4), 297–319 (2004). http://crypto.stanford.edu/~dabo/pubs/papers/weilsigs.ps
10. Bos, J.W., Costello, C., Naehrig, M.: Exponentiating in pairing groups. In: SAC 2013 [38] (2013). https://eprint.iacr.org/2013/458.pdf
11. Bosma, W. (ed.): ANTS 2000. LNCS, vol. 1838. Springer, Heidelberg (2000). https://doi.org/10.1007/10722028
12. Cao, Z., Zhang, F. (eds.): Pairing 2013. LNCS, vol. 8365. Springer, Cham (2014). https://doi.org/10.1007/978-3-319-04873-4
13. Koç, Ç.K., Naccache, D., Paar, C. (eds.): CHES 2001. LNCS, vol. 2162. Springer, Heidelberg (2001). https://doi.org/10.1007/3-540-44709-1
14. Chowdhury, D.R., Rijmen, V., Das, A. (eds.): INDOCRYPT 2008. LNCS, vol. 5365. Springer, Heidelberg (2008). https://doi.org/10.1007/978-3-540-89754-5
15. Costello, C., Hisil, H., Boyd, C., González Nieto, J.M., Wong, K.K.-H.: Faster pairings on special Weierstrass curves. In: Pairing 2009 [48], pp. 89–101 (2009)
16. Cramer, R. (ed.): EUROCRYPT 2005. LNCS, vol. 3494. Springer, Heidelberg (2005). https://doi.org/10.1007/b136415
17. Edwards, H.M.: A normal form for elliptic curves. Bulletin Am. Mathe. Soc. **44**, 393–422 (2007). http://www.ams.org/bull/2007-44-03/S0273-0979-07-01153-6/home.html

18. Fotiadis, G., Konstantinou, E.: TNFS resistant families of pairing-friendly elliptic curves. J. Theor. Comput. Sci. (2018, to appear)
19. Fouotsa, E., El Mrabet, N., Pecha, A.: Optimal ate pairing on elliptic curves with embedding degree 9, 15 and 27. IACR Cryptology ePrint Archive, 2016:1187 (2016). http://eprint.iacr.org/2016/1187
20. Freeman, D., Scott, M., Teske, E.: A taxonomy of pairing-friendly elliptic curves. J. Cryptol. **23**(2), 224–280 (2010). http://eprint.iacr.org/2006/372/
21. Galbraith, S.D., Paterson, K.G. (eds.): Pairing 2008. LNCS, vol. 5209. Springer, Heidelberg (2008). https://doi.org/10.1007/978-3-540-85538-5
22. Gentry, C., Silverberg, A.: Hierarchical ID-based cryptography. In: Asiacrypt 2002 [52], pp. 548–566 (2002). http://www.cs.ucdavis.edu/~franklin/ecs228/pubs/extra_pubs/hibe.pdf
23. Gu, H., Gu, D., Xie, W.L.: Efficient pairing computation on elliptic curves in Hessian form. In: Rhee, K.-H., Nyang, D.H. (eds.) ICISC 2010. LNCS, vol. 6829, pp. 169–176. Springer, Heidelberg (2011). https://doi.org/10.1007/978-3-642-24209-0_11
24. Hess, F., Smart, N.P., Vercauteren, F.: The Eta pairing revisited. IEEE Trans. Inf. Theor. **52**(10), 4595–4602 (2006). http://eprint.iacr.org/2006/110
25. Hışıl, H.: Elliptic curves, group law, and efficient computation. Ph.D. thesis, Queensland University of Technology (2010)
26. Horwitz, J., Lynn, B.: Toward hierarchical identity-based encryption. In: Eurocrypt 2002 [36], pp. 466–481 (2002). http://theory.stanford.edu/~horwitz/pubs/hibe.pdf
27. Ionica, S., Joux, A.: Another approach to pairing computation in Edwards coordinates. In: INDOCRYPT 2008 [14], pp. 400–413 (2008)
28. Iwata, T., Cheon, J.H. (eds.): ASIACRYPT 2015. LNCS, vol. 9452. Springer, Heidelberg (2015). https://doi.org/10.1007/978-3-662-48797-6
29. Joux, A.: A one round protocol for tripartite Diffie-Hellman. In: ANTS-IV [11], pp. 385–393 (2000). http://cgi.di.uoa.gr/~aggelos/crypto/page4/assets/joux-tripartite.pdf
30. Joux, A.: A one round protocol for tripartite Diffie-Hellman. J. Cryptol. **17**(4), 263–276 (2004)
31. Joux, A., Pierrot, C.: The special number field sieve in $\mathbb{F}_{p^n}$, application to pairing-friendly constructions. In: Pairing 2013 [12], pp. 45–61 (2013)
32. Joye, M., Quisquater, J.-J.: Hessian elliptic curves and side-channel attacks. In: CHES 2001 [13], pp. 402–410 (2001). http://joye.site88.net/
33. Kachisa, E.J., Schaefer, E.F., Scott, M.: Constructing Brezing-Weng pairing-friendly elliptic curves using elements in the cyclotomic field. In: Pairing 2008 [21], pp. 126–135 (2008)
34. Kilian, J. (ed.): CRYPTO 2001. LNCS, vol. 2139. Springer, Heidelberg (2001). https://doi.org/10.1007/3-540-44647-8
35. Kim, T., Barbulescu, R.: Extended tower number field sieve: a new complexity for the medium prime case. In: CRYPTO 2016 [46], pp. 543–571 (2016)
36. Knudsen, L.R. (ed.): EUROCRYPT 2002. LNCS, vol. 2332. Springer, Heidelberg (2002). https://doi.org/10.1007/3-540-46035-7
37. Kurosawa, K. (ed.): ASIACRYPT 2007. LNCS, vol. 4833. Springer, Heidelberg (2007). https://doi.org/10.1007/978-3-540-76900-2
38. Lange, T., Lauter, K., Lisoněk, P. (eds.): SAC 2013. LNCS, vol. 8282. Springer, Heidelberg (2014). https://doi.org/10.1007/978-3-662-43414-7
39. Lauter, K., Rodríguez-Henríquez, F. (eds.): LATINCRYPT 2015. LNCS, vol. 9230. Springer, Cham (2015). https://doi.org/10.1007/978-3-319-22174-8

40. Li, L., Wu, H., Zhang, F.: Pairing computation on Edwards curves with high-degree twists. In: Lin, D., Xu, S., Yung, M. (eds.) Inscrypt 2013. LNCS, vol. 8567, pp. 185–200. Springer, Cham (2014). https://doi.org/10.1007/978-3-319-12087-4_12

41. Lin, X., Zhao, C., Zhang, F., Wang, Y.: Computing the ate pairing on elliptic curves with embedding degree k = 9. IEICE Trans. **91-A(9)**, 2387–2393 (2008)

42. Matsui, M., Zuccherato, R.J. (eds.): SAC 2003. LNCS, vol. 3006. Springer, Heidelberg (2004). https://doi.org/10.1007/b96837

43. El Mrabet, N., Guillermin, N., Ionica, S.: A study of pairing computation for elliptic curves with embedding degree 15. IACR Cryptology ePrint Archive, 2009:370 (2009). http://eprint.iacr.org/2009/370

44. Oswald, E., Fischlin, M. (eds.): EUROCRYPT 2015. LNCS, vol. 9056. Springer, Heidelberg (2015). https://doi.org/10.1007/978-3-662-46800-5

45. Preneel, B., Tavares, S. (eds.): SAC 2005. LNCS, vol. 3897. Springer, Heidelberg (2006). https://doi.org/10.1007/11693383

46. Robshaw, M., Katz, J. (eds.): CRYPTO 2016. LNCS, vol. 9814. Springer, Heidelberg (2016). https://doi.org/10.1007/978-3-662-53018-4

47. Sahai, A., Waters, B.: Fuzzy identity-based encryption. In: Eurocrypt 2005 [16], pp. 457–473 (2005). http://eprint.iacr.org/2004/086/

48. Shacham, H., Waters, B. (eds.): Pairing 2009. LNCS, vol. 5671. Springer, Heidelberg (2009). https://doi.org/10.1007/978-3-642-03298-1

49. Silverman, J.H.: The Arithmetic of Elliptic Curves. GTM, vol. 106. Springer, New York (2009). https://doi.org/10.1007/978-0-387-09494-6

50. Smart, N.P.: The Hessian form of an Hessian curve. In: CHES 2001 [13], pp. 118–125 (2001)

51. Vercauteren, F.: Optimal pairings. IEEE Trans. Inf. Theor. **56**(1), 455–461 (2010)

52. Zheng, Y. (ed.): ASIACRYPT 2002. LNCS, vol. 2501. Springer, Heidelberg (2002). https://doi.org/10.1007/3-540-36178-2

A Family of FDH Signature Schemes Based on the Quadratic Residuosity Assumption

Giuseppe Ateniese[1], Katharina Fech[2], and Bernardo Magri[2(✉)]

[1] Stevens Institute of Technology, Hoboken, USA
[2] Friedrich-Alexander-Universität Erlangen-Nürnberg, Erlangen, Germany
{katharina.fech,bernardo.magri}@fau.de

Abstract. Signature schemes are arguably the most crucial cryptographic primitive, and devising *tight* security proofs for signature schemes is an important endeavour, as it immediately impacts the feasibility of deployment in real world applications. Hash-then-sign signature schemes in the Random Oracle Model, such as RSA-FDH, and Rabin-Williams variants are among the fastest schemes to date, but that unfortunately do not enjoy tight security proofs based on the one-wayness of their trapdoor function; instead, all known tight proofs rely on variants of the (non-standard) Φ-Hiding assumption. As our main contribution, we introduce a family of hash-then-sign signature schemes, inspired by a lossy trapdoor function from Freeman et al. (JoC' 13), that is tightly secure under the Quadratic Residuosity assumption. Our first scheme has the property of having *unique* signatures, while the second scheme is deterministic with an extremely fast signature verification, requiring at most 3 modular multiplications.

Keywords: Digital signatures · Full domain hash
Tight security proof · Quadratic residuosity · Lossy trapdoor function

1 Introduction

After the beginning of public-key cryptography [13] many new computational problems were devised, and along with them came cryptographic schemes based on the difficulty of solving those problems. At first, asymptotic security analysis was enough to claim the robustness of a given scheme, but it was realized later that a more precise analysis was required to measure the security of a scheme under a realistic scenario. A security proof is built upon computational complexity theory, using polynomial-time reductions from a well established hard problem to the problem of solving (or breaking) the cryptographic scheme. If this reduction is possible, we can say that breaking the cryptographic scheme is as difficult as solving the well established hard problem (up to a polynomial). If this polynomial is of a high degree, it can degrade the security of the scheme

© Springer Nature Switzerland AG 2018
D. Chakraborty and T. Iwata (Eds.): INDOCRYPT 2018, LNCS 11356, pp. 248–262, 2018.
https://doi.org/10.1007/978-3-030-05378-9_14

considerably, even rendering it useless for practical applications. Bellare and Rogaway [4] started dealing with security reductions that explicitly stated the polynomial factors involved in those reductions, making it possible to build *tight* reductions, in which the polynomial is a small constant.

1.1 Hash-then-Sign Signature Schemes

In 1993, Bellare and Rogaway [3] introduced the Full Domain Hash (FDH) signature scheme based on RSA (RSA-FDH), where the message is hashed to the full domain of the underlying trapdoor function before being signed (also known as "hash-then-sign" schemes). The security proof presented in [3] for RSA-FDH was not tight, making the actual scheme potentially impractical for an acceptable level of security. Fortunately, probabilistic FDH (PFDH) schemes, which prepend a short random string to the message, already allow for tight proofs. In particular, Katz and Wang [22] showed that even a single bit of randomness is enough for achieving tight proofs.

Signature schemes that behave deterministically are usually more efficient and easier to implement, what makes them invaluable for practical applications. Moreover, it is a fact that signature schemes secure in the Random Oracle Model (ROM) are much more practical than schemes secure in the standard model [6,8,10,16,34], therefore, in this paper we only focus on FDH schemes with deterministic signatures in the ROM.

We mainly categorize signature schemes into four distinct classes, namely probabilistic, derandomized, deterministic and unique, that we describe next.

- Probabilistic schemes utilize randomness during the signing process; signatures are always different (with high probability) even if the same message is signed twice with the same signing key. Some examples of probabilistic schemes are PSS [4], Schnorr [28], El-Gamal [14], and Bitcoin's ECDSA.
- Derandomized schemes are probabilistic schemes that demonstrate a deterministic behavior but still requires an internal use of randomness. It is folklore that any randomized signature scheme can be turned into a deterministic one; merely generate the random coins used during the signing algorithm through a pseudo-random function (PRF) that takes the message as input. Then, the random coins used to sign a particular message will be fixed, therefore producing a deterministic signature for each message. Unfortunately, in some cases, the derandomization process can lead to several vulnerabilities [23]. Signature schemes in the derandomized category include the Derandomized Rabin-Williams (DRW) scheme, where the signature is a square root selected uniformly at random out of four possibilities, and returned systematically (by using the PRF "trick").
- Deterministic schemes always produce the same signature for each message without relying on randomness (or derandomization) for signing, but the verification algorithm accepts more than 1 valid signature per message (for each key pair).

– Lastly, unique schemes are deterministic schemes where the verification algorithm only accepts as valid the only signature ever produced by its signing algorithm (for each message and key). Schemes in this category are the Absolute Principal Rabin-Williams (APRW) scheme, and the RSA-FDH (since RSA [27] defines a permutation over $\mathbb{Z}_n^*$).

In Table 1 we show a quick comparison between FDH signature schemes.

Table 1. Comparison of different hash-then-sign signature schemes.

	Assumption	Derandomized?	Unique?	Tight?
DRW [5]	Factoring	✓	✗	✓
APRW [30]	2-Φ/4-Hiding	✗	✓	✓
RSA-FDH [20]	Φ-Hiding	✗	✓	✓
BLS [7]	EC-CDH	✗	✓	✗
Katz-Wang [22]	RSA	✓	✗	✓
Our scheme Π_u (Sect. 3.1)	Quadratic residuosity	✗	✓	✓
Our scheme Π_d (Sect. 3.2)	Quadratic residuosity	✗	✗	✓

1.2   Previous Work

A seminal impossibility result by Coron [12] states that any FDH signature scheme with *unique* signatures could not hope to have a tight security proof. Kakvi and Kiltz [20] clarified that Coron's impossibility result only holds when the trapdoor permutation is certified. They also presented a tight security proof for RSA-FDH based on the Φ-Hiding assumption [9].

Bernstein [5] studied all variants of Rabin-Williams signatures and devised an ingenious tight proof for the DRW scheme (which he calls "fixed unstructured"), where it releases systematically one of the four square roots that is initially selected at random. Bernstein also provides a non-tight security proof for APRW (the unique signature version of the scheme) and left as an open problem finding a tight proof for it. Seurin [30] first showed that the Rabin function is lossy and then presented a tight security proof for APRW, but under a new assumption dubbed 2-Φ/4-Hiding assumption.

Unique signatures received renewed attention lately, as Bader et al. [2] extended the seminal meta-reduction of Coron [12] by showing that any security proof for unique signatures based on static assumptions or in the security of the underlying trapdoor permutation must lose a factor of q_s in its security reduction, where q_s is the number of signature queries asked by the adversary. Later, Guo et al. [18] clarified that the authors of [2] implicitly assumed in their meta-reduction that the simulator is only allowed to extract information from the adversary's forgeries when trying to invert the underlying trapdoor permutation; [18] circumvents the impossibility of [2] by allowing the simulator (in addition) to extract information from the adversary's hash queries. In [18] the authors present a unique signature scheme based on Computational Diffie-Hellman (CDH) with a tight security proof, with the drawback that the size of a signature is logarithmic in the number of hash queries asked by the adversary. Shacham [31] improves on the results of [18] and presents a version of the unique scheme of [18] with succinct signatures, where each signature consists of 2 group elements. Unfortunately, the scheme of [31] is still not as fast as RSA-FDH or any Rabin-Williams variant.

Thus, to summarize: All the unique schemes with tight security proofs from the assumption that the underlying trapdoor function (or permutation) is one-way are not efficient. On the other hand, efficient unique schemes such as RSA-FDH and APRW have a tight security proof that relies on the lossiness of the trapdoor function and are based on variants of the Φ-Hiding assumption. Seurin (cf. Theorem 5 in [30]) noted that it is very unlikely that FDH-RSA and APRW will have a tight security reduction from, respectively, inverting RSA or factoring. It is evident that the state of affairs is a bit confusing. FDH-RSA and Rabin-Williams signatures with non-tight proofs were criticized as being potentially impractical due to the large size of the parameters involved. Their tight proofs, however, rely on new assumptions that appear to be markedly stronger than factoring [19,29]. How should these results be interpreted in practice? Should we trust these new assumptions and keep parameters short or should we use large parameters to account for possible cryptanalytic attacks on these new assumptions?

1.3   State of Affairs

What is wrong with randomness? Generating cryptographically-strong random or pseudo-random numbers (RNG or PRNG) has always been a challenging endeavor. Several devices are even unable to generate random numbers that are good enough for cryptographic purposes. For instance, smart cards and sensors are not usually capable of collecting enough entropy. Some are susceptible to *reset* attacks where the PRNG is brought back to previous states. A reset attack can be devastating for signature schemes since it could be possible even to recover the signing keys of the user [26]. The same attack can be applied to virtualized systems where the adversary can take snapshots of a virtual machine and later replay them with distinct messages to recover the signing key. When possible, probabilistic schemes should be avoided in these circumstances.

What is wrong with derandomization? Despite showing a deterministic behavior, derandomized schemes still require randomness to sign messages. Therefore, it is crucial to have a sound derandomization process; otherwise, it can be a source of vulnerabilities [17,23]. For instance, a simple fault attack during the derandomization leads to a full key recovery attack in the derandomized Rabin-Williams scheme (by outputting 2 different square roots of the same message), while deterministic schemes are immune to such attacks.

What is wrong with the Φ-Hiding assumption? The Φ-Hiding assumption appears to be much stronger than factoring, and it does not hold in some cases, as shown in [19,29]. The RSA-FDH scheme is tightly secure under the Φ-Hiding assumption, while the APRW scheme is tightly secure under a new assumption dubbed 2-Φ/4-Hiding assumption [30]. As reported by Seurin [30], the 2-Φ/4-Hiding assumption is clearly stronger than quadratic residuosity (on which our schemes rely instead): When $n \equiv 1 \mod 4$, the 2-Φ/4-Hiding problem is equivalent to the problem of establishing whether -1 is a square in $\mathbb{Z}_n^*$; thus, it's enough to provide $y = -x^2 \mod n$, for a random $x \in \mathbb{Z}_n^*$, to a quadratic residuosity solver to violate the 2-Φ/4-Hiding assumption.

A Case for Unique Signatures. Ateniese et al. [1] shows a generic subversion attack against virtually all probabilistic and deterministic signature schemes that leads to the complete recovery of the signing key. The intuition behind the attack is that the adversary builds a subverted signing algorithm that leaks bits of the signing key through the produced signatures; this is only possible because the signature contains randomness that is used to "disguise" the parts of the signing key that is being leaked. Deterministic schemes are also susceptible to such attacks since the bits of the signing key can still be leaked through the choice of the signature that is returned among the possible options. On the other hand, [1] shows that unique signature schemes are *secure* against the class of subversion attacks that satisfies the verifiability condition[1]. When used in tandem with a cryptographic reverse firewall [25] unique signature schemes are secure against *all* classes of subversion attacks [1]. Therefore, unique signatures are recommended for settings where the generation of randomness is problematic, and subversion attacks are a concern.

1.4 Our Contribution

Our contribution is a family of FDH signature schemes in the ROM with tight security proofs to the Quadratic Residuosity (QR) assumption[2]. The family consists of a unique scheme and a deterministic scheme, both based on a variation of a lossy function from [15]. To argue tight security for the unique signature scheme, we leverage the results of Kakvi and Kiltz [20] that show a generic proof for any unique scheme based on a lossy trapdoor function. As far as we could ascertain, this is the first unique signature scheme tightly secure under the quadratic residuosity assumption (and non-tightly secure under factoring). Besides, the reduction is tighter than the one in [30], i.e., our unique scheme is closer to quadratic residuosity than principal Rabin-Williams is to the 2-Φ/4-Hiding assumption.

---

[1] The verifiability condition informally says that *all* signatures produced by the signing algorithm must be valid for the corresponding verification key.

[2] Arguably, the next best assumption after factoring is quadratic residuosity, which has been extensively studied, at least as much as the RSA assumption.

The efficiency of the schemes in our family is comparable to that of the Rabin-Williams family, which are considered the fastest (for signature verification) signature schemes ever devised [5]. The unique scheme does require the computation of a Jacobi symbol (as the unique variant of Rabin-Williams also does) but we believe such a computation carries an unfair stigma. In reality, computing Jacobi symbols can be performed very efficiently [24, 32] (in particular in $O(n^2 / \log n)$ as reported in [24]), and can be parallelized [24] to harness recent multicore and/or distributed platforms. Nevertheless, for applications where the verification process has to be even faster, we provide a deterministic signature scheme that does not require the computation of Jacobi symbols.

2   Preliminaries

2.1   Basic Notations

When A is a deterministic algorithm, we write $y := A(x)$ to denote a run of A on input x and output y; if A is a randomized algorithm then $y \leftarrow A(x; r)$ denotes a run of A on input x and randomness r; when it is clear from context we simply write $y \leftarrow A(x)$. An algorithm A is probabilistic polynomial-time (PPT) if A is randomized and for any input $x, r \in \{0, 1\}^*$ the computation of $A(x; r)$ terminates in at most $poly(|x|)$ steps. We denote with $\kappa \in \mathbb{N}$ the security parameter. A function $\nu : \mathbb{N} \to [0, 1]$ is negligible in the security parameter (or simply negligible) if it vanishes faster than the inverse of any polynomial in κ, i.e., $\nu(\kappa) = \kappa^{-\omega(1)}$. For a random variable $\mathbf{X}$, we write $\mathbb{P}[\mathbf{X} = x]$ for the probability that $\mathbf{X}$ takes on a particular value $x \in \mathcal{X}$ (where $\mathcal{X}$ is the set where $\mathbf{X}$ is defined).

2.2   Number Theory

We denote by $\mathbb{J}_n$ the set of all $x \in \mathbb{Z}_n^*$ with Jacobi symbol 1, by $\bar{\mathbb{J}}_n$ the set of all $x \in \mathbb{Z}_n^*$ with Jacobi symbol -1, and by $\mathbb{QR}_n$ the set of all quadratic residues of $\mathbb{Z}_n^*$. For $n \in \mathbb{Z}$, we call n a Williams integer if $n = pq$ for primes p and q of the form $p \equiv 3 \mod 8$ and $q \equiv 7 \mod 8$. Our results rely on the following lemmas from [15, 33].

Lemma 1. *Let $n = pq$ be a Williams integer and let $x \in \mathbb{QR}_n$. The equation $x \equiv y^2 \mod n$ takes four distinct values, namely $\{\pm y_0, \pm y_1\}$, where*

(i) for $b \in \{0, 1\}$, we have that y_b and $-y_b$ are both either in $\mathbb{J}_n$ or $\bar{\mathbb{J}}_n$,
(ii) $y_0 \in \mathbb{J}_n$ if and only if $y_1 \in \bar{\mathbb{J}}_n$.

Lemma 2. *Let $n = pq$ be a Williams integer, then $2 \in \bar{\mathbb{J}}_n$.*

Lemma 3. *For $n, x, y \in \mathbb{Z}$, where $x \not\equiv \pm y \mod n$, if $x^2 \equiv y^2 \mod n$ then $\gcd(n, x - y)$ gives a non-trivial factor of n.*

2.3   Signature Schemes

A signature scheme is a triple of algorithms $\Pi = (\mathsf{KGen}, \mathsf{Sign}, \mathsf{Vrfy})$ specified as follows:

- KGen takes as input the security parameter κ and outputs a verification/ signing key pair $(vk, sk) \in \mathcal{VK} \times \mathcal{SK}$, where $\mathcal{VK} := \mathcal{VK}_\kappa$ and $\mathcal{SK} := \mathcal{SK}_\kappa$ denote the sets of all verification and secret keys produced by $\mathsf{KGen}(1^\kappa)$.
- Sign takes as input the signing key $sk \in \mathcal{SK}$, a message $m \in \mathcal{M}$ and random coins $r \in \mathcal{R}$, and outputs a signature $\sigma \in \Sigma$.
- Vrfy takes as input the verification key $vk \in \mathcal{VK}$ and a pair (m, σ), and outputs a decision bit that equals 1 iff σ is a valid signature for message m under the key vk.

The correctness of a signature scheme informally says that verifying honestly generated signatures always works.

Definition 1 (Correctness). *Let* $\Pi = (\mathsf{KGen}, \mathsf{Sign}, \mathsf{Vrfy})$ *be a signature scheme. We say that* Π *satisfies (perfect) correctness if for all* (vk, sk) *output by* KGen, *and all* $m \in \mathcal{M}$,

$$\mathbb{P}\left[\mathsf{Vrfy}(vk, (m, \mathsf{Sign}(sk, m))) = 1\right] = 1,$$

where the probability is taken over the randomness of the signing algorithm.

The standard notion of security for a signature scheme demands that no PPT adversary given access to a signing oracle returning signatures for arbitrary messages, can forge a signature on a "fresh" message (not asked to the signing oracle).

Definition 2 (Existential unforgeability). *Let* $\Pi = (\mathsf{KGen}, \mathsf{Sign}, \mathsf{Vrfy})$ *be a signature scheme. We say that* Π *is* (t, q, ε)-*existentially unforgeable under chosen-message attacks if for all adversaries* A *running in time* t *it holds:*

$$\mathbb{P}\left[\mathsf{Vrfy}(vk, (m^*, \sigma^*)) = 1 \wedge m^* \notin \mathcal{Q} : \begin{array}{l} (vk, sk) \leftarrow \mathsf{KGen}(1^\kappa); \\ (m^*, \sigma^*) \leftarrow \mathsf{A}^{\mathsf{Sign}(sk, \cdot)}(vk) \end{array}\right] \leq \varepsilon,$$

where $\mathcal{Q} = \{m_1, \ldots, m_q\}$ *denotes the set of queries to the signing oracle. If for all* $t, q = poly(\kappa)$ *there exists* $\varepsilon(\kappa) = negl(\kappa)$ *such that* Π *is* (t, q, ε)-*existentially unforgeable under chosen-message attacks (EUF-CMA for short), then we simply say* Π *is EUF-CMA.*

We define the so-called *unique* signatures next. Informally, a signature scheme is unique if, for any message, there is only a single signature that verifies w.r.t. an honestly generated verification key.

Definition 3 (Uniqueness). *Let* Π *be a signature scheme. We say that* Π *satisfies uniqueness if for all* vk *output by* KGen, *and all* $m \in \mathcal{M}$, *there exists a single value* $\sigma \in \Sigma$ *such that* $\mathsf{Vrfy}(vk, (m, \sigma)) = 1$.

3   The Signature Scheme Family

In this section, we describe the two components of our hash-and-sign family of signature schemes. Our family is a variant of the Rabin-Williams family [5], and is inspired by a *lossy* trapdoor function from [15]. We first describe the unique signature scheme based on QR in Sect. 3.1, followed by the deterministic scheme based on QR in Sect. 3.2.

3.1   Unique Scheme Π_u

Let the functions $h, j : \mathbb{Z}_n \to \{0, 1\}$ be defined as

$$h(x) = \begin{cases} 1, & \text{if } x > n/2, \\ 0, & \text{otherwise,} \end{cases}$$

$$j(x) = \begin{cases} 1, & \text{if } x \in \bar{\mathbb{J}}_n, \\ 0, & \text{otherwise.} \end{cases}$$

We build the unique signature scheme $\Pi_u = (\mathsf{KGen}, \mathsf{Sign}, \mathsf{Vrfy})$ as follows:

- $(vk, sk) \leftarrow \mathsf{KGen}(1^\kappa)$: The key generation algorithm takes as input the security parameter 1^κ and produces a pair of corresponding verification and signing keys. The signing key sk is composed of two randomly sampled $\kappa/2$-bit primes p and q of the form $p \equiv 3 \mod 8$ and $q \equiv 7 \mod 8$. The verification key vk is defined by $n := pq$ and a randomly sampled parameter $s \in \mathbb{J}_n \setminus \mathbb{QR}_n$.
- $\sigma := \mathsf{Sign}(sk, m)$: Set $b := 0$ and hash the message m to obtain $x := H(m)$, where $H : \{0, 1\}^* \to \mathbb{Z}_n^*$ is a collision-resistant hash function. Compute $x' := x \cdot 2^{j(x)} \mod n$ and iff $x' \notin \mathbb{QR}_n$ set $b := 1$ and compute $x' := x' \cdot s \mod n$ with the public parameter s. Now that $x' \in \mathbb{QR}_n$ we use the signing key to compute the four modular square roots of x' and select the single root y such that $j(y) = j(x)$ and $h(y) = b$ (according to Lemma 1); set $\sigma := y$ and output σ.
- $b := \mathsf{Vrfy}(vk, m, \sigma)$: If $\sigma \notin \{1, \ldots, n-1\}$ then output 0, otherwise output $H(m) = \sigma^2 \cdot 2^{-j(\sigma)} \cdot s^{-h(\sigma)} \mod n$.

On uniqueness. We note that for a signature scheme to be considered unique, it is necessary, but not sufficient, that the signing algorithm always returns the same signature when the same message is signed more than once. To fully characterize a unique signature scheme, the verification algorithm needs (for each verification key vk) to reject as invalid all the signatures for a particular message m, except the only signature for m that is ever returned by the signing algorithm. It is easy to see that the scheme Π_u above satisfies these requirements, as for each key a single signature σ is ever produced for some message m, and only σ is ever accepted as a signature for m.

3.2   Deterministic Scheme Π_d

In order to achieve even better efficiency, we construct additionally the deterministic variant Π_d of the previous signature scheme. We define the deterministic scheme $\Pi_d = (\mathsf{KGen}', \mathsf{Sign}', \mathsf{Vrfy}')$, where the algorithms KGen' and Sign' are exactly the same as KGen and Sign in Π_u, and the verification algorithm Vrfy' is described below:

- $b := \mathsf{Vrfy}'(vk, m, \sigma)$: If $\sigma \notin \{1, \ldots, n-1\}$ then output 0, otherwise output $(H(m) = \sigma^2 \cdot s^{-h(\sigma)} \mod n) \vee (H(m) = \sigma^2 \cdot s^{-h(\sigma)} \cdot 2^{-1} \mod n)$.

Note that although the signing algorithm will always return a unique signature for each message, the verification algorithm does accept 2 different signatures for a message. The main advantage of the deterministic scheme over the unique scheme is efficiency; while the unique scheme requires computation of a Jacobi symbol in the signature verification, the deterministic scheme only needs to perform 3 modular multiplications (in the worst case).

4   Security Analysis

In this section, we analyze the security of the signature schemes presented in Sect. 3. We first present a security proof for Π_u based on the hardness of factoring, and then a *tight* security proof based on QR. To achieve the latter, we leverage the results of Kakvi and Kiltz [21] on unique signatures based on *lossy* functions. Later we also present a tight security proof for the Π_d signature scheme based on QR.

4.1   Security of Π_u Based on Factoring

Theorem 1. *If the Integer Factorization Problem (IFP) is (t, ε)-hard, then the unique signature scheme Π_u is $(t', q_h, q_s, \varepsilon')$-secure, with*

$$t = t' + (q_h + q_s + 1) \cdot O(\kappa^2) \quad and \quad \varepsilon = \frac{\varepsilon'}{4 \cdot (q_h + q_s + 1)}.$$

Proof. Let A be an adversary that $(t', q_h, q_s, \varepsilon')$-breaks Π_u. We build a reduction R that uses A as a subroutine and (t, ε)-breaks the IFP.

The reduction R receives a modulus $n = pq$ from the challenger, and its objective is to factor n. Instead of sampling $s \in \mathbb{J}_n \setminus \mathbb{QR}_n$, which R is not able to, it simply samples an $s \in \mathbb{J}_n$. When $s \in \mathbb{QR}_n$ the reduction aborts, what happens with probability $1/2$. The reduction R sends $vk := (n, s)$ to A. We allow the adversary A to make two types of oracle queries, namely hash and sign queries, that R must answer with the same distribution as a real signing oracle would. The reduction R maintains a list $\mathcal{L} := \emptyset$ of hash queries and a counter i that is initialized by 0. R chooses a random $\ell \in \{1, ..., q\}$, where $q := q_h + q_s$, and answers the queries as follows:

- **Hash queries:** Upon a hash query for message m check if $m \in \mathcal{L}$; if yes, then return x from the triple $(m, x, y) \in \mathcal{L}$, otherwise proceed as follows. Increment the counter i, and if $i \neq \ell$ the reduction R chooses a random $y_i \in \mathbb{Z}_n^*$ and sets $x_i = y_i^2 \cdot 2^{-j(y_i)} \cdot s^{-h(y_i)} \mod n$. However, when $i = \ell$, reduction R chooses random values $y_i \in \mathbb{Z}_n^*$, $\alpha, \beta \in \{0, 1\}$ and sets $x_i = y_i^2 \cdot 2^{-\alpha} \cdot s^{-\beta} \mod n$. Store the triple (m_i, x_i, y_i) in the list $\mathcal{L}$ and return x_i.
- **Sign queries:** When A makes a sign query for a message m, reduction R checks if there exists a triple $(m, x, y) \in \mathcal{L}$; if not, R simply makes the corresponding hash query itself. Return y as the signature of message m.

The adversary A eventually outputs a forgery (m_i, σ_i), and we assume wlog that $(m_i, x_i, y_i) \in \mathcal{L}$. If $i = l$ we have that both $y' = \sigma_i \cdot 2^{-j(\sigma)} \cdot s^{-h(\sigma)} \mod n$ and y_i are square roots of y_i^2. With probability $1/2$, the roots y' and y_i are not the complement of each other, and in that case we can factor n by computing $\gcd(n, y' - y_i)$, due to Lemma 3. The running time for R is the running time of the adversary A plus all the oracle queries. $\square$

The reduction R is required to answer all the oracle queries that A makes; in particular, R needs to produce valid signatures to all the messages queried by A without knowing the signing key. Before every signature query for message m is made, a corresponding hash query for m needs to be made to the reduction R; the reduction first samples a random $y \in \mathbb{Z}_n^*$ and returns $H(m) := y^2 \cdot 2^{-j(y)} \cdot s^{-h(y)} \mod n$ as the answer to the hash query. To answer a signature query for message m, the reduction R returns y as a valid signature for m.

In order to factor, R selects an index $\ell \in \{1, \ldots, q\}$ during initialization, and for the ℓ-th hash query made by A the reduction R replies with $x_\ell = y_\ell^2 \cdot 2^{-\alpha} \cdot s^{-\beta} \mod n$ for $y_\ell \in \mathbb{Z}_n^*$, $\alpha, \beta \in \{0, 1\}$. The reduction is then able to factor with probability $1/2$ if A produces a pair (m_ℓ, σ_ℓ) as a forgery for the message m_ℓ.

We note that the above security reduction can be further improved to roughly $\varepsilon = \varepsilon'/4q_s$ by applying a technique by Coron [11].

4.2 Tight Security of Π_u Based on QR

The unique signature scheme Π_u of Sect. 3.1 is a variant of a lossy trapdoor function based on QR from [15]. In fact, the changes made to our scheme were carefully crafted so the scheme would still maintain its lossiness; the main difference is that n is a Williams integer so that $2 \in \bar{\mathbb{J}}_n$.

To instantiate the lossy version of our scheme, KGen needs to be modified to sample the public parameter $s \in \mathbb{QR}_n$, in contrast to the injective version, where $s \in \mathbb{J}_n \setminus \mathbb{QR}_n$. Note that the only difference between the lossy and the injective version of the scheme is the domain of s; in both cases $s \in \mathbb{J}_n$, but in the lossy version $s \in \mathbb{QR}_n$, while in the injective version $s \notin \mathbb{QR}_n$. Distinguishing among these two cases is precisely the QR assumption, so an adversary that is able to distinguish must solve the QR problem. Since the lossy version of the scheme is 2-to-1 [15] and the injective version is a permutation in $\{1,\ldots,n\}$, the scheme has lossiness of 1-bit.

For the tight security proof of the scheme Π_u we leverage the generic result of Kakvi and Kiltz [21] for unique signatures based on lossy functions, that intuitively states that any unique signature scheme based on a lossy function has a tight security reduction based on the lossiness of the function. From that, we achieve the following result.

Theorem 2. *If the Quadratic Residuosity assumption is $(t_{QR}, \varepsilon_{QR})$-hard, then for any q_h, q_s the unique signature scheme Π_u is $(t, q_h, q_s, \varepsilon)$-EUF-CMA secure in the random oracle model with*

$$t = t_{QR} - q_h \cdot \mathcal{O}(\kappa^2) \quad and \quad \varepsilon = 3 \cdot \varepsilon_{QR}.$$

4.3   Tight Security of Π_d Based on QR

In this section, we build a reduction from breaking the security of the Π_d scheme to breaking the security of the Π_u scheme. Since the Π_u scheme has tight security to the QR problem, then Π_d has also tight security to the QR problem.

Theorem 3. *If the Π_u scheme is $(t', q_h, q_s, \varepsilon')$-EUF-CMA secure, then the deterministic signature scheme Π_d is $(t, q_h, q_s, \varepsilon)$-EUF-CMA secure, with $t = t'$, and $\varepsilon = 2 \cdot \varepsilon'$.*

Proof. Assume there exists an adversary A that $(t, q_h, q_s, \varepsilon)$-breaks the security of Π_d. Then, we build another adversary A' that $(t', q_h, q_s, \varepsilon')$-breaks the security of Π_u.

Adversary A':
- Receive the verification key $vk := (n, s)$ from the challenger and send it to A.
- Upon any hash or signature query from A, forward the query to its corresponding oracle and send the reply to A.
- Eventually, receive a forgery (m, σ) from A. Sample a random bit b and return the pair $(m, \sigma \cdot 2^b)$ to the challenger as a forgery for message m.

For the analysis, we note that the simulation performed by A' is perfect since the hash and signature oracles from both schemes are exactly the same. By assumption, A produces a valid forgery (m, σ) with non-negligible probability, and in that case, $(m, \sigma \cdot 2^b)$ is a valid forgery for A' when $\sigma \cdot 2^b$ has the same Jacobi symbol as $H(m)$, what happens with probability $1/2$ when b is sampled at random. Therefore, if A breaks the security of Π_d with probability ε, then A' breaks the security of Π_u with probability $\varepsilon/2$.   □

5   Performance

When the factors p and q are known, calculating the Jacobi symbol of an element is very efficient since it is enough to compute two Legendre symbols. In particular, for $x \in \mathbb{Z}_n^*$, $x \in \mathbb{J}_n$ if $x^{(p-1)/2} \mod p = x^{(q-1)/2} \mod q$, otherwise $x \in \overline{\mathbb{J}}_n$.[3] The signature σ is the unique square root y of the square x such that $j(y) = j(x)$ and $y > n/2$ iff $x > n/2$. Computing such a square root is very efficient thanks to the Chinese remainder theorem.

In general, when p and q are known, the computation of the Jacobi symbol and a square root share several calculations and can be optimized when performed simultaneously. Since computing Jacobi symbols when p and q are unknown is computationally more expensive than other modular operations, we recommend the deterministic version Π_{d} of our scheme for applications where unique signatures are not necessary.

While in the unique signature scheme the computation of a Jacobi symbol (for signature verification) is necessary, in the deterministic scheme it is enough to compute $t := \sigma^2 \cdot s^{-h(\sigma)} \mod n$ and then check whether any of $H(m) = t$ or $H(m) = t \cdot 2^{-1} \mod n$ holds to consider the signature σ as valid.

A note on efficiency. Our Π_{u} scheme has comparable speed to the unique signature scheme from the Rabin-Williams family, denoted by APRW* in [30]. The running time of the verification algorithm is dominated by the computation of a Jacobi symbol in both schemes. Our deterministic scheme Π_{d} is very efficient, requiring at most 3 modular multiplications for signature verification.

6   Conclusions

We presented a family of FDH signature schemes with tight security based on a standard assumption (QR). The schemes are as efficient as other variants of Rabin-Williams which hold the record for fastest signature verification schemes [5]. A tight security proof for the APRW scheme was presented only recently by Seurin [30], and his proof is based on the lossiness of the APRW function, which is based on a new assumption called 2-Φ/4-Hiding, that is a variation of the Φ-Hiding problem [9]. Unlike QR, the Φ-Hiding problem is a new and poorly understood assumption as remarked in [19,29].

In practice, since the security of our signature scheme is based on the QR assumption, in comparison to RSA-FDH and APRW, it is possible to safely employ smaller parameters for comparable levels of security, which leads to even better efficiency.

---

[3] We do not consider cases where the Jacobi or Legendre symbols are 0 since they happen with negligible probability.

References

1. Ateniese, G., Magri, B., Venturi, D.: Subversion-resilient signature schemes. In: Ray, I., Li, N., Kruegel, C. (eds.) 22nd Conference on Computer and Communications Security – ACM CCS 2015, pp. 364–375. ACM Press (2015)
2. Bader, C., Jager, T., Li, Y., Schäge, S.: On the impossibility of tight cryptographic reductions. In: Fischlin, M., Coron, J.-S. (eds.) EUROCRYPT 2016. LNCS, vol. 9666, pp. 273–304. Springer, Heidelberg (2016). https://doi.org/10.1007/978-3-662-49896-5_10
3. Bellare, M., Rogaway, P.: Random oracles are practical: a paradigm for designing efficient protocols. In: Ashby, V. (ed.) ACM CCS 93: 1st Conference on Computer and Communications Security, pp. 62–73. ACM Press, November 1993
4. Bellare, M., Rogaway, P.: The exact security of digital signatures: how to sign with RSA and Rabin. In: Maurer, U. (ed.) EUROCRYPT 1996. LNCS, vol. 1070, pp. 399–416. Springer, Heidelberg (1996). https://doi.org/10.1007/3-540-68339-9_34
5. Bernstein, D.J.: Proving tight security for Rabin-Williams signatures. In: Smart, N. (ed.) EUROCRYPT 2008. LNCS, vol. 4965, pp. 70–87. Springer, Heidelberg (2008). https://doi.org/10.1007/978-3-540-78967-3_5
6. Boneh, D., Boyen, X.: Short signatures without random oracles. In: Cachin, C., Camenisch, J.L. (eds.) EUROCRYPT 2004. LNCS, vol. 3027, pp. 56–73. Springer, Heidelberg (2004). https://doi.org/10.1007/978-3-540-24676-3_4
7. Boneh, D., Lynn, B., Shacham, H.: Short signatures from the Weil pairing. J. Cryptol. 17(4), 297–319 (2004)
8. Boneh, D., Shen, E., Waters, B.: Strongly unforgeable signatures based on computational Diffie-Hellman. In: Yung, M., Dodis, Y., Kiayias, A., Malkin, T. (eds.) PKC 2006. LNCS, vol. 3958, pp. 229–240. Springer, Heidelberg (2006). https://doi.org/10.1007/11745853_15
9. Cachin, C., Micali, S., Stadler, M.: Computationally Private Information Retrieval with Polylogarithmic Communication. In: Stern, J. (ed.) EUROCRYPT 1999. LNCS, vol. 1592, pp. 402–414. Springer, Heidelberg (1999). https://doi.org/10.1007/3-540-48910-X_28
10. Chevallier-Mames, B., Joye, M.: A practical and tightly secure signature scheme without hash function. In: Abe, M. (ed.) CT-RSA 2007. LNCS, vol. 4377, pp. 339–356. Springer, Heidelberg (2006). https://doi.org/10.1007/11967668_22
11. Coron, J.-S.: On the exact security of full domain hash. In: Bellare, M. (ed.) CRYPTO 2000. LNCS, vol. 1880, pp. 229–235. Springer, Heidelberg (2000). https://doi.org/10.1007/3-540-44598-6_14
12. Coron, J.-S.: Optimal security proofs for PSS and other signature schemes. In: Knudsen, L.R. (ed.) EUROCRYPT 2002. LNCS, vol. 2332, pp. 272–287. Springer, Heidelberg (2002). https://doi.org/10.1007/3-540-46035-7_18
13. Diffie, W., Hellman, M.: New directions in cryptography. IEEE Trans. Inf. Theory 22(6), 644–654 (1976)
14. ElGamal, T.: A public key cryptosystem and a signature scheme based on discrete logarithms. IEEE Trans. Inf. Theory 31, 469–472 (1985)
15. Freeman, D.M., Goldreich, O., Kiltz, E., Rosen, A., Segev, G.: More constructions of lossy and correlation-secure trapdoor functions. J. Crypt. 26(1), 39–74 (2013)
16. Gennaro, R., Halevi, S., Rabin, T.: Secure hash-and-sign signatures without the random oracle. In: Stern, J. (ed.) EUROCRYPT 1999. LNCS, vol. 1592, pp. 123–139. Springer, Heidelberg (1999). https://doi.org/10.1007/3-540-48910-X_9

17. Granboulan, L.: How to repair ESIGN. In: Cimato, S., Persiano, G., Galdi, C. (eds.) SCN 2002. LNCS, vol. 2576, pp. 234–240. Springer, Heidelberg (2003). https://doi.org/10.1007/3-540-36413-7_17

18. Guo, F., Chen, R., Susilo, W., Lai, J., Yang, G., Mu, Y.: Optimal security reductions for unique signatures: bypassing impossibilities with a counterexample. In: Katz, J., Shacham, H. (eds.) CRYPTO 2017. LNCS, vol. 10402, pp. 517–547. Springer, Cham (2017). https://doi.org/10.1007/978-3-319-63715-0_18

19. Herrmann, M.: Improved cryptanalysis of the multi-prime ϖ - hiding assumption. In: Nitaj, A., Pointcheval, D. (eds.) AFRICACRYPT 2011. LNCS, vol. 6737, pp. 92–99. Springer, Heidelberg (2011). https://doi.org/10.1007/978-3-642-21969-6_6

20. Kakvi, S.A., Kiltz, E.: Optimal security proofs for full domain hash, revisited. In: Pointcheval, D., Johansson, T. (eds.) EUROCRYPT 2012. LNCS, vol. 7237, pp. 537–553. Springer, Heidelberg (2012). https://doi.org/10.1007/978-3-642-29011-4_32

21. Kakvi, S.A., Kiltz, E.: Optimal security proofs for full domain hash, revisited. J. Crypt. **31**(1), 276–306 (2018)

22. Katz, J., Wang, N.: Efficiency improvements for signature schemes with tight security reductions. In: Sushil, J., Vijayalakshmi, A., Trent, J. (eds.) ACM CCS 03: 10th Conference on Computer and Communications Security – ACM CCS 2003, pp. 155–164. ACM Press, October 2003

23. Leurent, G., Nguyen, P.Q.: How risky is the random-oracle model? In: Halevi, S. (ed.) CRYPTO 2009. LNCS, vol. 5677, pp. 445–464. Springer, Heidelberg (2009). https://doi.org/10.1007/978-3-642-03356-8_26

24. Eikenberry, S.M., Sorenson, J.P.: Efficient algorithms for computing the Jacobi symbol. J. Symb. Comput. **26**, 509–523 (1998)

25. Mironov, I., Stephens-Davidowitz, N.: Cryptographic reverse firewalls. In: Oswald, E., Fischlin, M. (eds.) EUROCRYPT 2015. LNCS, vol. 9057, pp. 657–686. Springer, Heidelberg (2015). https://doi.org/10.1007/978-3-662-46803-6_22

26. Ristenpart, T., Yilek, S.: When good randomness goes bad: virtual machine reset vulnerabilities and hedging deployed cryptography. In: ISOC Network and Distributed System Security Symposium - NDSS 2010. The Internet Society, February/March 2010

27. Rivest, R.L., Shamir, A., Adleman, L.M.: A method for obtaining digital signature and public-key cryptosystems. Commun. ACM **21**(2), 120–126 (1978)

28. Schnorr, C.P.: Efficient identification and signatures for smart cards. In: Brassard, Gilles (ed.) CRYPTO 1989. LNCS, vol. 435, pp. 239–252. Springer, New York (1990). https://doi.org/10.1007/0-387-34805-0_22

29. Schridde, C., Freisleben, B.: On the validity of the phi-hiding assumption in cryptographic protocols. In: Pieprzyk, J. (ed.) ASIACRYPT 2008. LNCS, vol. 5350, pp. 344–354. Springer, Heidelberg (2008). https://doi.org/10.1007/978-3-540-89255-7_21

30. Seurin, Y.: On the lossiness of the rabin trapdoor function. In: Krawczyk, H. (ed.) PKC 2014. LNCS, vol. 8383, pp. 380–398. Springer, Heidelberg (2014). https://doi.org/10.1007/978-3-642-54631-0_22

31. Shacham, H.: Short unique signatures from RSA with a tight security reduction (in the random oracle model). In: 22nd Financial Cryptography and Data Security (2018)

32. Shallit, J., Sorenson, J.: A binary algorithm for the Jacobi symbol. ACM SIGSAM Bull. **27**, 4–11 (1993)
33. Shoup, V.: A Computational Introduction to Number Theory and Algebra. Cambridge University Press, New York (2009)
34. Waters, B.: Efficient identity-based encryption without random oracles. In: Cramer, R. (ed.) EUROCRYPT 2005. LNCS, vol. 3494, pp. 114–127. Springer, Heidelberg (2005). https://doi.org/10.1007/11426639_7

Symmetric Key Cryptanalysis

Using MILP in Analysis of Feistel Structures and Improving Type II GFS by Switching Mechanism

Mahdi Sajadieh[1]([⊠]) and Mohammad Vaziri[2]([⊠])

[1] Department of Electrical Engineering, Khorasgan Branch,
Islamic Azad University, Isfahan, Iran
m.sajadieh@khuisf.ac.ir

[2] Department of Mathematics, Iran University of Science
and Technology (IUST), Tehran, Iran
mohammad.vaziri67@gmail.com

Abstract. Some features of Feistel structures have caused them to be considered as an efficient structure for design of block ciphers. Although several structures are proposed relied on Feistel structure, the type-II generalized Feistel structures (GFS) based on SP-functions are more prominent. Because of difference cancellation, which occurs in Feistel structures, their resistance against differential and linear attack is not as expected. In order to improve the immunity of Feistel structures against differential and linear attack, two methods are proposed. One of them is using multiple MDS matrices, and the other is using changing permutations of sub-blocks.

In this paper by using mixed-integer linear programming (MILP) and summation representation method, a technique to count the active S-boxes is proposed. Moreover in some cases, the results proposed by Shibutani at SAC 2010 are improved. Also multiple MDS matrices are applied to GFS, and by relying on a proposed approach, the new inequalities related to using multiple MDS matrices are extracted, and results of using the multiple MDS matrices in type II GFS are evaluated. Finally results related to linear cryptanalysis are presented. Our results show that using multiple MDS matrices leads to 22% and 19% improvement in differential cryptanalysis of standard and improved 8 sub-blocks structures, respectively, after 18 rounds.

Keywords: MILP · Generalized Feistel structure
Switching mechanism · Differential cryptanalysis · Linear cryptanalysis

1 Introduction

Nowadays, security is one of the most important components of information transition, and cryptography is an inseparable part of security. Block ciphers are one of the most important tools, which are used in cryptography. These ciphers must be resistant against the existing security cryptanalysis such as differential and linear cryptanalysis.

© Springer Nature Switzerland AG 2018
D. Chakraborty and T. Iwata (Eds.): INDOCRYPT 2018, LNCS 11356, pp. 265–281, 2018.
https://doi.org/10.1007/978-3-030-05378-9_15

Feistel structures form a significant category of block ciphers, which have been under several evaluation so far. Perhaps CAMELLIA [1] and CLEFIA [13] are the most important block ciphers that are designed based on these structures. The CLEFIA block cipher uses four sub-blocks Feistel structure with switching mechanism [11]. In switching mechanism multiple MDS matrices with specified properties are used. Using switching mechanism in CLEFIA provides *1.3* times more active S-boxes rather than the structure with one matrix. Also as mentioned in [11,12], for two sub-blocks Feistel structure with multiple MDS matrices, the total number of active S-boxes is *1.2* times higher than two sub-blocks Feistel structure with one MDS matrix.

A lot of methods have so far been proposed to count the number of active S-boxes of Feistel structures. The first method for Feistel structures with SPN round functions is proposed in [5]. This method is able to offer a lower bound for the number of differentially and linearly active S-boxes with branch number β. In [10], a method is proposed to calculate the minimum number of active S-boxes of block cipher Camellia, and the existing bound is improved for this block cipher. Also in [8,14], the number of active S-boxes is obtained by changing the standard method, and proposing a particular algorithm. Although employing multiple MDS matrices in GFS is discussed in [3,9], accurate results are not reported.

Probably, using mixed-integer linear programming method in calculating the number of active S-boxes of block ciphers is one of the most important existing methods. This method is discussed in several papers such as [2,6]. In order to evaluate word-oriented block ciphers, however, a comprehensive method is proposed in [6]. Also in [7] due to the better performance of the features of MDS matrices, a method is proposed.

In this paper by using linear programming and the proposed idea in [7], first a method to count the number of differentially and linearly active S-boxes in Feistel structures is presented, and the obtained results are compared with results in [8]. Other major contribution of the paper is referring to inequalities that are extracted from imposing switching mechanism, and results that are obtained by imposing switching mechanism on generalized Feistel structures are presented. Moreover, the results for the best 8 sub-blocks Feistel structures that employ 2 and 4 multiple MDS matrices are reported. Based on our researchs, the results of using $\frac{l}{4}$ MDS matrices and $\frac{l}{2}$ MDS matrices in differential cryptanalysis of l sub-blocks structures, are fairly close. Finally we analyze the switching mechanism in linear attack.

The rest of the paper is organized as follows. In Sect. 2 we review the details of GFS structures, give a brief description about MILP, and explain about summation representation, which is used in our MILP method. In Sect. 3 first we present our method to calculate the minimum number of differentially active S-boxes of *2* sub-blocks Feistel structures, and we generalize this method for structures with higher number of sub-blocks. In Sect. 4 inequalities which describe switching mechanism are proposed, and the results of imposing the switching properties on generalized Feistel structures are presented. In Sect. 5 by expanding the proposed method, linear cryptanalysis is evaluated. Finally, we conclude in Sect. 6.

2   Preliminaries

In this section we clarify what type of Feistel structures exactly we aim to evaluate, and then give a definition of MILP. Moreover, we point out what is the difference between our method and the well-known MILP method, which is proposed in [6].

2.1   GFS Structures

In GFS, a plaintext is divided to l sub-blocks, where l is an even integer. If $(X_0, X_1, ..., X_{l-1})$ represents the l divided sub-blocks of a state with size of lmn-bit, a single round of l sub-blocks GFS follows a permutation over $(\{0,1\}^{mn})^l$, as:

$$(X_0, X_1, ..., X_{l-1}) \rightarrow \pi(X_0, F_0(X_0) \oplus X_1, X_2, F_1(X_2) \oplus X_3, ..., F_{l/2-1}(X_{l-2}) \oplus X_{l-1}) \quad (1)$$

In relation (1) $F_i : \{0,1\}^{mn} \rightarrow \{0,1\}^{mn}$ is the i-th round function, and $\pi : (\{0,1\}^{mn})^l \rightarrow (\{0,1\}^{mn})^l$ is a deterministic permutation over l sub-blocks. Figure 1, illustrates the relation (1), where possible connections of output sub-blocks of the state and input sub-blocks of next state are denoted by dotted lines. Throughout the paper, we consider each round function be an SP-function, and each sub-block is consisted of n S-boxes with size of m bits. Therefore, it is easy to verify that a GFS with l sub-blocks is an lmn-bit block cipher.

In this paper we assume that π is a word-based permutation. In the rest of the paper, GFS_l^{std} is interpretted as a standard type-II GFS [8] with l sub-blocks, where $\pi(X_0, X_1, ..., X_{l-1}) = (X_1, X_2, ..., X_{l-1}, X_0)$, and GFS_l^{imp} is interpretted as an improved type-II GFS with l sub-blocks as pointed out by the authors of [14]. For instance, the permutation in $GFS_6^{imp}(No.1)$ is as $\pi(X_0, X_1, ..., X_5) = (X_3, X_0, X_1, X_4, X_5, X_2)$, and the permutation in $GFS_8^{imp}(No.1)$, which is one of the most important evaluated structures in this paper, is as $\pi(X_0, X_1, ..., X_7) = (X_3, X_0, X_1, X_4, X_7, X_2, X_5, X_6)$.

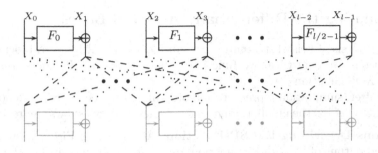

Fig. 1. GFS with l sub-blocks

2.2 MILP

In linear programming, a linear objective function $f(x_1, x_2, ..., x_n)$ is optimized, subject to a given set of linear constraints, which are produced by decision variables $x_i, 1 \le i \le n$. If certain decision variables are restricted to be integer values, such programs are called mixed-integer linear programming.

2.3 Summation Representation

As mentioned above, each round function in GFS contains an mn-bit block as an input, and each bijective S-box is m bits (n parallel m-bit S-boxes), and also $\mathbf{P}$ is an $n \times n$ matrix with m-bit elements, where we assume that β is the branch number of this matrix. In order to count the number of active S-boxes, the truncated method [7] is used. Therefore, in this case, the S-box does not have any effect on truncated difference or mask. Because of using branch number, the place of elements does not care to be zero or not, and just the number of them is important. Hence for every n truncated vector bits, the summation of elements of that vector are allocated (i.e. we replace an integer number between 0 and n instead of a vector with size n). From now on, we call this method "summation representation" [7]. We emphasize that in summation representation, 2^n possible representation reduces to $n + 1$ possible representation. For instance, in relation (2) the truncated representation and summation representation are shown for a vector as an input of F-function with 4 8-bit elements:

$$\begin{pmatrix} 6 \\ 15 \\ 0 \\ 158 \end{pmatrix} \xrightarrow{truncated} \begin{pmatrix} 1 \\ 1 \\ 0 \\ 1 \end{pmatrix} \xrightarrow{summation} 3 \qquad (2)$$

In order to count the differentially and linearly active S-boxes of word-oriented block ciphers, the truncated method is used in [6]. In contrast, we use summation method to count the differentially and linearly active S-boxes of word-oriented block ciphers. Throughout this paper all of the inputs and outputs are shown in a summation representation. It is worth mentioning that, our method can be exploited for both structures with MDS and non MDS matrices. However, in this paper we assume that all of applied matrices are MDS.

3 Counting the Differentially Active S-Boxes

In cryptanalysis of Feistel structure (two sub-blocks or multiple sub-blocks) with SP-functions, we deal with two functions. One of them is SP-function and the other is XOR function.

Hereafter, summation representation of a difference vector "$\mathbf{x}$" is denoted by "x^c", where "x^c" is the number of non zero elements of "$\mathbf{x}$" shows the results for.

Equations Describing the SP-Function. According to Fig. 2, assume that input and output of the i-th SP-function are x_i^c and z_i^c, respectively, where both of them are an integer number between 0 and n.

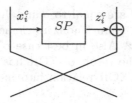

Fig. 2. Summation variables related to the SP-function of Feistel structure

The branch number of matrix **P** is β. Therefore, we have:

$$\begin{cases} z_i^c = 0 & \text{if } x_i^c = 0 \\ x_i^c + z_i^c \geq \beta \text{ otherwise} \end{cases} \qquad (3)$$

The function is conditional. Considering [6], we need to introduce a new binary dummy variable b_i to convert the condition into inequality, where $b_i \in \{0, 1\}$. Then we have:

$$\begin{cases} x_i^c + z_i^c \geq \beta b_i \\ b_i \leq x_i^c \leq nb_i \\ b_i \leq z_i^c \leq nb_i \end{cases} . \qquad (4)$$

Note that we assumed that an employed matrix be MDS, and this leads to $\beta = n + 1$. In this case if x_i^c be nonzero, certainly z_i^c is nonzero, since the maximum amount of x_i^c is n, and $x_i^c + z_i^c \geq n + 1$ causes that $z_i^c \geq 1$. Therefore inequality $b_i \leq z_i^c$ is redundant and it could be eliminated. As a rule, for an SP-function, inequalities are turned as follows:

$$\begin{cases} 0 \leq x_i^c \leq n \\ 0 \leq z_i^c \leq n \\ x_i^c + z_i^c \geq (n + 1)b_i \\ b_i \leq x_i^c \leq nb_i \\ z_i^c \leq nb_i \end{cases} \qquad (5)$$

Equations Describing the XOR Operation. For describing the XOR operation consider Fig. 3. To evaluate XOR operation in summation structure, regard to $\mathbf{y_i} = \mathbf{x_i} \oplus \mathbf{z_i}$, it is clear that the maximum amount of y_i^c is equal to the summation of two inputs. Also the minimum amount of y_i^c won't be less than the difference of the absolute value of two inputs.

Fig. 3. Summation variables related to XOR operation of Feistel structure

For instance, if x_i^c and z_i^c are equal to 3 and 1, respectively, the maximum amount that z_i^c can eliminate from 3 nonzero elements of x_i^c is 1, and the minimum amount of y_i^c will be 2. Also in best case z_i^c is nonzero in a place that x_i^c is zero and in this case the result of XOR has 4 nonzero elements. Under this notation, for converting XOR relation into inequality, the following three inequalities are obtained:

$$\begin{cases} x_i^c + z_i^c \geq y_i^c \\ \|x_i^c - z_i^c\| \leq y_i^c \end{cases} \implies \begin{cases} x_i^c + z_i^c \geq y_i^c \\ x_i^c - z_i^c \leq y_i^c \\ z_i^c - x_i^c \leq y_i^c \end{cases} \tag{6}$$

Therefore, for each round of Feistel structure with an SP-type F-function, where matrix $\mathbf{P}$ is an MDS matrix, we need 4 variables and 11 inequalities. More precisely 8 inequalities are derived from SP-function, and 3 inequalities are derived from XOR operation. Needless to say, if we wanted to describe such a structure, which contains n S-boxes in its F-functions, with prior well-known MILP model, we needed to define $4n$ variables. Also $2n + 1$ inequalities were needed to describe the SP-function, and $4n$ inequalities were needed to describe the XOR operation. Besides that, we need just 1 binary dummy variable for SP-function in our model, whereas we need 1 and n binary dummy variables for SP-function and XOR operation in prior model, respectively.

We know that, counting the number of nonzero inputs of SP-functions is equivalent to count the number of active S-boxes. According to the way of defining variables in our method, x_i variable denotes the number of active S-boxes in the i-th SP-function. Therefore, to calculate the minimum number of active S-boxes, the summation of x_i variables must be minimized.

3.1 Evaluating Two Sub-blocks Feistel Structure

Figure 4 shows the details of a two sub-blocks Feistel structure starting from the first round.

According to indices of variables in Fig. 4, and inequalities (5) and (6), for each round with an MDS matrix we have:

$$\begin{cases} 0 \leq x_i^c \leq n \\ 0 \leq z_i^c \leq n \\ x_i^c + z_i^c \geq (n+1)b_i \qquad and \qquad (for\ i \geq 1) \\ b_i \leq x_i^c \leq nb_i \\ z_i^c \leq nb_i \end{cases} \begin{cases} x_{i-2}^c + z_{i-1}^c \geq x_i^c \\ x_{i-2}^c - z_{i-1}^c \leq x_i^c \\ z_{i-1}^c - x_{i-2}^c \leq x_i^c \end{cases} \tag{7}$$

It is worth noting that, the variable corresponding plain text $(x_{-1}^c + x_0^c)$ must be nonzero. Thus the inequality $x_{-1}^c + x_0^c \geq 1$ must be added. Finally by organizing inequalities system and calculating the minimum amount of $\sum_{j=0}^{n-1} x_j^c$ and solving it by IBM ILOG CPLEX [4], the minimum number of active S-boxes for $n = 4$ and $n = 8$ for r rounds with branch number β are obtained as $\lfloor \frac{r}{4} \rfloor (\beta + 1) + (r) mod 4 - 1$. Table 1 shows our results for two sub-blocks Feistel with $n = 4$ and $n = 8$ and $\beta = n + 1$, which are corresponded with [8].

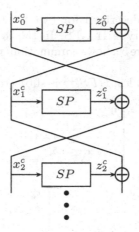

Fig. 4. The way of defining summation variables in two sub-blocks Feistel structure

Table 1. Minimum number of active S-boxes of two sub-blocks Feistel

Round	Feistel with $n = 4$	Feistel with $n = 8$
1	0	0
2	1	1
3	2	2
4	5	9
5	6	10
6	7	11
7	8	12
8	11	19
9	12	20
10	13	21
11	14	22
12	17	29
13	18	30
14	19	31
15	20	32
16	23	39
17	24	40
18	25	41

3.2 Evaluating Generalized Feistel Structures

The process that has been described for two sub-blocks Feistel structure can be expanded to type I and type II GFS. In the following the inequalities are described for GFS_8^{std}. Figure 5 shows summation variables for the first three rounds of GFS_8^{std}.

According to the above rules, GFS_8^{std} is subjected to:

$$\begin{cases} 0 \leq x_i^c \leq n \\ 0 \leq z_i^c \leq n \\ x_i^c + z_i^c \geq (n+1)b_i \\ b_i \leq x_i^c \leq nb_i \\ z_i^c \leq nb_i \end{cases} \tag{8}$$

$$(for\ 4 \leq i \leq 7) \begin{cases} x_{i-8}^c + z_{i-4}^c \geq x_i^c \\ x_{i-8}^c - z_{i-4}^c \leq x_i^c \\ z_{i-8}^c - x_{i-4}^c \leq x_i^c \end{cases} (for\ i \geq 8) \begin{cases} x_{i-8+(i+1)mod4-(i)mod4}^c + z_{i-4}^c \geq x_i \\ x_{i-8+(i+1)mod4-(i)mod4}^c - z_{i-4}^c \leq x_i^c \\ z_{i-4}^c - x_{i-8+(i+1)mod4-(i)mod4}^c \leq x_i^c \end{cases} \tag{9}$$

Our results for $n = 4$ are summarized for standard and improved generalized Feistel structures from $l = 4$ sub-blocks till $l = 16$ sub-blocks in Tables 2 and 3, respectively. In these tables results are compared with [8]. In these tables our different results are bold.

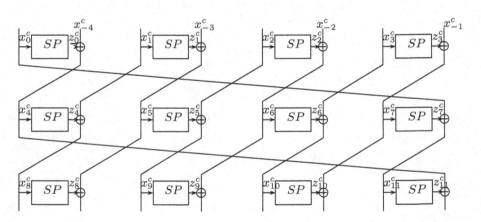

Fig. 5. The way of defining summation variables in standard eight sub-blocks Feistel structure

Table 2. The minimum number of active S-boxes in GFS_l^{std} with n = 4, the columns marked by "*" are our results

Round	GFS_4^{std}		GFS_6^{std}		GFS_8^{std}		GFS_{10}^{std}		GFS_{12}^{std}		GFS_{14}^{std}		GFS_{16}^{std}	
	*	[8]	*	[8]	*	[8]	*	[8]	*	[8]	*	[8]	*	[8]
1	0	0	0	0	0	0	0	0	0	0	0	0	0	0
2	1	1	1	1	1	1	1	1	1	1	1	1	1	1
3	2	2	2	2	2	2	2	2	2	2	2	2	2	2
4	6	6	6	6	6	6	6	6	6	6	6	6	6	6
5	8	8	8	8	8	8	8	8	8	8	8	8	8	8
6	12	12	12	12	12	12	12	12	12	12	12	12	12	12
7	12	12	14	14	14	14	14	14	14	14	14	14	14	14
8	13	13	18	18	18	18	18	18	18	18	18	18	18	18
9	14	14	21	21	21	21	21	21	21	21	21	21	21	21
10	18	18	25	25	25	25	25	25	25	25	25	25	25	25
11	20	20	27	27	28	28	28	28	28	28	28	28	28	28
12	24	24	30	30	36	36	36	36	36	36	36	36	36	36
13	24	24	31	31	36	36	39	39	39	39	39	39	39	39
14	25	25	35	35	37	37	43	43	43	43	43	43	43	43
15	26	26	37	37	38	38	47	47	47	47	47	47	47	47
16	30	30	41	41	42	42	54	54	54	54	54	54	54	54
17	32	32	43	43	44	44	58	58	58	58	58	58	52	52
18	36	36	47	47	48	48	**62**	58	62	62	62	62	62	62

4   Evaluating Switching Mechanism

In switching mechanism instead of using one matrix, multiple matrices are used in a way that the number of differentially and linearly active S-boxes will be significantly more than the case of using one matrix. In this section, at first inequalities related to switching properties for two sub-blocks structure are described, and then for four sub-blocks. Finally inequalities for six and eight sub-blocks structure are listed.

In Fig. 6 switching mechanism is imposed on two sub-blocks Feistel structure. In this structure, two MDS matrices $\mathbf{M_1}$ and $\mathbf{M_2}$ are used, where the branch number of matrix $[\mathbf{M_1}\ \mathbf{M_2}]_{n \times 2n}$ is $n + 1$. The matrices $\mathbf{M_1}$ and $\mathbf{M_2}$ should be allocated to round functions, in a way that avoid difference cancellation. For more details, we refer to [11]. In order to count the number of active S-boxes of this block cipher, some inequalities must be added to prior corresponded model, which has only one matrix in its structure.

Table 3. The minimum number of active S-boxes in GFS_l^{imp} with n = 4, the columns marked by "*" are our results

Round	GFS_6^{imp}		GFS_8^{imp}		GFS_{10}^{imp}		GFS_{12}^{imp}		GFS_{14}^{imp}		GFS_{16}^{imp}	
	*	[8]	*	[8]	*	[8]	*	[8]	*	[8]	*	[8]
1	0	0	0	0	0	0	0	0	0	0	0	0
2	1	1	1	1	1	1	1	1	1	1	1	1
3	2	2	2	2	2	2	2	2	2	2	2	2
4	6	6	6	6	6	6	6	6	6	6	6	6
5	8	8	8	8	8	8	8	8	8	8	8	8
6	12	12	12	12	12	12	12	12	12	12	12	12
7	14	14	14	14	14	14	14	14	14	14	14	14
8	**23**	22	23	23	**26**	23	18	18	**26**	23	**26**	23
9	24	24	26	26	29	29	21	21	29	29	31	31
10	26	26	29	29	**35**	34	29	29	37	37	**43**	40
11	28	28	32	32	36	36	32	32	40	40	48	48
12	32	32	39	39	**43**	45	**42**	39	**52**	49	**57**	54
13	**34**	33	**42**	40	44	44	45	45	54	54	60	60
14	38	38	**45**	44	48	48	**54**	53	**64**	60	**66**	63
15	40	40	46	46	50	50	57	57	**66**	63	**69**	70
16	**48**	46	50	50	54	54	**61**	60	**77**	71	76	76
17	48	48	52	52	56	56	64	64	**82**	76	78	78
18	50	50	56	56	**68**	65	**70**	68	**84**	83	87	87

According to Fig. 6, relations between inputs and outputs of five consecutive rounds are as follows:

$$\begin{cases} x_i = z_{i-1} \oplus x_{i-2} \\ x_{i-2} = z_{i-3} \oplus x_{i-4} \end{cases} \implies \quad x_i = z_{i-1} \oplus z_{i-3} \oplus x_{i-4} \tag{10}$$

More precisely, according to the effect of S-box on truncated method, the above relation can be described as follows:

$$[M_1\ M_2] \begin{bmatrix} x_{i-1} \\ x_{i-3} \end{bmatrix} = x_i \oplus x_{i-4} \quad or \quad [M_2\ M_1] \begin{bmatrix} x_{i-1} \\ x_{i-3} \end{bmatrix} = x_i \oplus x_{i-4} \tag{11}$$

Now converting switching mechanism into inequalities contains two steps: the first step refers to the way of interpreting the relation (10), and the second step refers to guaranteeing at least one of amounts x_{i-1} and x_{i-3} must be nonzero. In the following, the above two steps are elaborated, respectively.

Firstly, according to feature of switching, the matrix $[M_1\ M_2]_{n \times 2n}$ has branch number $n + 1$. Thus, if in a relation $[M_1\ M_2] \begin{bmatrix} a \\ b \end{bmatrix} = c \oplus d$, at least

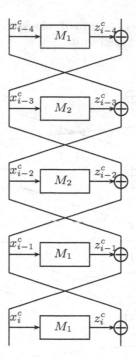

Fig. 6. The way of defining summation variables in two sub-blocks Feistel structure, imposed by switching mechanism

one of the amounts **a** and **b** are nonzero, the relation $a^c + b^c + c^c + d^c \geq n+1$ is established. To be more specific, $a^c + b^c + \|\mathbf{c} \oplus \mathbf{d}\| \geq n+1$ is correct, and since it is easy to verify that $c^c + d^c \geq \|\mathbf{c} \oplus \mathbf{d}\|$, consequently the relation is described as mentioned. It is remarkable that, this relation corresponds with proposed lemma in [12]. If all the variables a^c, b^c, c^c, d^c are zero, a paradox occurs in the inequity. In order to avoid this paradox, a new binary dummy variable needs to be defined.

Secondly, at least one of amounts **a** and **b** are supposed to be nonzero. Towards this end, the addition of a^c and b^c must be greater or equal to 1. Also, it is obvious that the addition of a^c and b^c is not more than $2n$. As a result, the relation $1 \leq a^c + b^c \leq 2n$ is attained. It is easy to verify that the paradox in prior step appears again. In order to overcome the aforementioned problem, the same dummy variable, which is defined in previous step, is used.

With all these taken to account, by defining the new binary dummy variable called bb_i, the description of switching properties for five consecutive rounds of two sub-blocks Feistel structure is as follows:

$$x_i^c + x_{i-1}^c + x_{i-3}^c + x_{i-4}^c \geq (n+1)bb_i$$
$$bb_i \leq x_{i-1}^c + x_{i-3}^c \leq 2nbb_i \tag{12}$$

Therefore, for each five consecutive rounds, inequalities related to switching mechanism must be added.

Figure 7 shows the structure of four sub-blocks type II GFS (CLEFIA), which two matrices M_1 and M_2 are used in it:

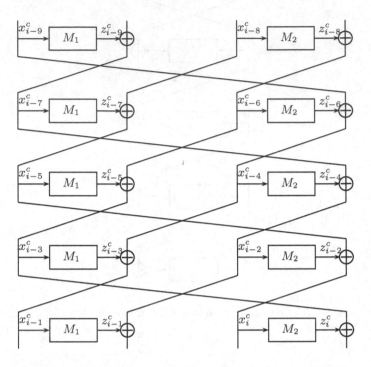

Fig. 7. The way of defining summation variables in CLEFIA

As mentioned above, in CLEFIA two MDS matrices are used. Patterning the process that was done for two sub-blocks structure, for adding inequalities related to switching mechanism, relation between inputs and outputs of five consecutive rounds is as follows:

$$\begin{cases} x_{i-1} = z_{i-3} \oplus z_{i-6} \oplus x_{i-9} \\ x_i = z_{i-2} \oplus z_{i-7} \oplus x_{i-8} \end{cases} \tag{13}$$

By following the same process which was done for two sub-blocks structure, we have:

$$x_{i-1}^c + x_{i-3}^c + x_{i-6}^c + x_{i-9}^c \geq (n+1)bb_{i-1}$$
$$bb_{i-1} \leq x_{i-3}^c + x_{i-6}^c \leq 2nbb_{i-1}$$

$$\tag{14}$$

$$x_i^c + x_{i-2}^c + x_{i-7}^c + x_{i-8}^c \geq (n+1)bb_i$$
$$bb_i \leq x_{i-2}^c + x_{i-7}^c \leq 2nbb_i$$

Therefore, for each five consecutive rounds, the inequalities of switching feature must be added. We stress that, the obtained results exactly match with [13].

In six sub-blocks type II GFS which consists of standard and improved structure [14], in order to have the best performance, three MDS matrices $\mathbf{M_1}$, $\mathbf{M_2}$ and $\mathbf{M_3}$ must be used. More matrices have not considerable influence. Inequalities that are obtained for switching feature in GFS_6^{std} are generalized of CLEFIA. For $GFS_6^{imp}(No.1)$, one of the relations between inputs and outputs for each seven consecutive rounds is as follows:

$$x_{i-2} = z_{i-5} \oplus z_{i-9} \oplus z_{i-16} \oplus x_{i-20} \qquad (15)$$

And finally we have:

$$x_{i-2}^c + x_{i-5}^c + x_{i-9}^c + x_{i-16}^c + x_{i-20}^c \geq (n+1)bb_{i-2}$$
$$bb_{i-2} \leq x_{i-5}^c + x_{i-9}^c + x_{i-16}^c \leq 3nbb_{i-2} \qquad (16)$$

Therefore, for each seven consecutive rounds the inequalities of switching feature must be added.

In eight sub-blocks type II GFS that consists of standard and improved structure [14], four MDS matrices $\mathbf{M_1}$, $\mathbf{M_2}$, $\mathbf{M_3}$ and $\mathbf{M_4}$ are recommended to apply. Inequalities that are obtained for switching feature in GFS_8^{std} are generalized of prior structure, and for $GFS_8^{imp}(No.1)$, one of the relations between inputs and outputs for each nine consecutive rounds is as follows:

$$x_{i-3}^c + x_{i-7}^c + x_{i-13}^c + x_{i-20}^c + x_{i-30}^c + x_{i-35}^c \geq (n+1)bb_{i-3}$$
$$bb_{i-3} \leq x_{i-7}^c + x_{i-13}^c + x_{i-20}^c + x_{i-30}^c \leq 4nbb_{i-3} \qquad (17)$$

Table 4, shows the results for standard and improved generalized Feistel structures with $n = 4$, by considering switching properties. We point out that in l sub-blocks structures, in order to have more powerful structure $\frac{l}{2}$ different MDS matrices must be applied. However, this process negatively impacts the costs on generalized Feistel structures with larger sub-blocks. Fortunately, although using less matrices in structures with larger sub-blocks makes them a bit weaker, such these structures are still efficient, and instead lead to reduce the costs significantly. The obtained results for GFS_8^{imp} with 2 MDS matrices and GFS_{12}^{imp} with 3 MDS matrices, which are listed in Table 4, are enough to emphasis our claim.

5   Counting the Linearly Active S-Boxes

It is well known that, because of duality between differential and linear attack, the method of counting the linearly active S-boxes is identical to differentially active S-boxes in many regards. As shown in [5,12], counting the linearly active S-boxes could be calculated by using the simple transformation in Fig. 8. We emphasize that, in linear cryptanalysis, Feistel structures with SP-functions convert to Feistel structures with PS-functions, and x^c denotes the summation representation of vector $\mathbf{\Gamma.x}$.

Table 4. Minimum number of differentially active S-boxes of generalized Feistel structures imposed by switching properties with n = 4

Round	2 MDS matrices			3 MDS matrices			4 MDS matrices		5 MDS matrices		6 MDS matrices	
	Feistel	CLEFIA	GFS_8^{imp}	GFS_6^{std}	GFS_6^{imp}	GFS_{12}^{imp}	GFS_8^{std}	GFS_8^{imp}	GFS_{10}^{std}	GFS_{10}^{imp}	GFS_{12}^{std}	GFS_{12}^{imp}
1	0	0	0	0	0	0	0	0	0	0	0	0
2	1	1	1	1	1	1	1	1	1	1	1	1
3	2	2	2	2	2	2	2	2	2	2	2	2
4	5	6	6	6	6	6	6	6	6	6	6	6
5	6	8	8	8	8	8	8	8	8	8	8	8
6	10	12	12	12	12	12	12	12	12	12	12	12
7	10	14	14	14	14	14	14	14	14	14	14	14
8	11	18	26	18	25	26	18	23	18	26	18	26
9	12	20	29	21	27	29	21	26	21	29	21	29
10	15	22	34	25	31	41	26	32	25	37	25	37
11	16	24	37	28	33	44	31	36	28	40	28	42
12	20	28	42	34	36	56	36	44	36	49	36	50
13	20	30	43	37	38	61	41	46	39	51	39	54
14	21	34	47	38	43	67	48	49	47	54	45	65
15	22	36	49	42	46	69	50	52	51	56	50	72
16	25	38	53	44	52	72	53	56	58	62	56	77
17	26	40	55	48	55	75	56	60	64	66	62	79
18	30	44	64	50	59	79	59	67	68	70	67	85
19	30	46	67	54	61	81	62	69	72	78	75	88
20	31	50	72	57	64	92	66	77	74	83	85	..
21	32	52	74	61	67	..	69	80	78	88	88	..
22	35	55	83	64	72	..	73	86	80	95	92	..
23	36	56	84	69	76	..	76	88	84	97	94	..
24	40	59	88	73	81	..	80	91	87	..	..	..

Fig. 8. transforming differential vectors to linear masks

Regardless of switching mechanism, if the matrix **M** is an MDS matrix, the number of linearly active S-boxes for both standard and improved structures will be equal to differentially active S-boxes. In case of using switching mechanism, in order to clarify, consider Fig. 9 as a special example.

The inequalities related to switching mechanism in linear cryptanalysis readily can be extracted from Theorem 3 in [12]. The relation (18) shows one of 4 relations between inputs and outputs of three consecutive rounds, in Fig. 9:

$$\Gamma.\mathbf{x_5} = (\mathbf{M_2^T})^{-1}\Gamma.\mathbf{x_1} \oplus (\mathbf{M_1^T})^{-1}\Gamma.\mathbf{x_8} = \left[(\mathbf{M_1^T})^{-1} \ (\mathbf{M_2^T})^{-1}\right] \begin{bmatrix} \Gamma.\mathbf{x_1} \\ \Gamma.\mathbf{x_8} \end{bmatrix} \quad (18)$$

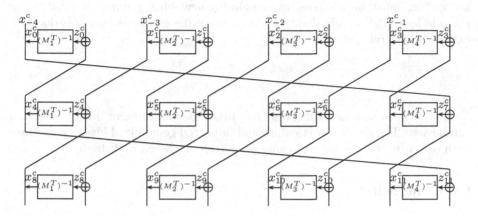

Fig. 9. Defining summation variables in GFS_8^{std} imposed by switching mechanism

Table 5. Minimum number of linearly active S-boxes of standard and improved generalized Feistel structures imposed by switching properties with $n = 4$

Round	Feistel	CLEFIA	GFS_6^{std}	GFS_6^{imp}	GFS_8^{std}	GFS_8^{imp}	GFS_{10}^{std}	GFS_{10}^{imp}	GFS_{12}^{std}	GFS_{12}^{imp}
1	0	0	0	0	0	0		0	0	0
2	1	1	1	1	1	1	1	1	1	1
3	5	5	5	5	5	5	5	5	5	5
4	5	8	8	8	8	8	8	8	8	8
5	7	10	10	10	10	10	10	11	10	11
6	10	15	16	16	16	16	16	16	16	16
7	11	16	18	22	18	22	18	22	18	22
8	12	19	24	27	24	30	24	30	24	30
9	15	21	26	30	26	32	26	38	26	38
10	16	24	32	33	34	38	34	43	34	43
11	17	26	35	35	39	43	39	50	39	51
12	20	31	37	38	45	49	45	53	45	59
13	21	32	40	40	48	51	50	55	50	65
14	22	35	42	46	51	54	58	58	58	72
15	25	37	47	52	53	56	63	61	63	74
16	26	40	50	56	56	62	69	66	69	77
17	27	42	55	60	58	66	73	72	74	79
18	30	47	58	63	64	72	75	78	85	85
19	31	48	63	65	66	77	78	86	92	91
20	32	51	67	68	72	82	80	91	99	99
21	35	53	69	71	74	88	86	97	101	107
22	36	56	72	76	82	94	88	103	104	112
23	37	58	74	82	87	97	94	105	106	120
24	40	63	79	86	93	100	96	108	112	..

Without loss of generality, under the assumption that at least one of amounts x_1 and x_8 must be nonzero, by defining a new binary dummy variable bb_i, the relations (19) are obtained, based on the same process which was done for extracting the relations (12).

$$x_1^c + x_5^c + x_8^c \geq (n+1)bb_i$$
$$bb_i \leq x_1^c + x_8^c \leq 2nbb_i \tag{19}$$

Due to obtained inequalities in (19), other inequalities can be obtained in a similar way. The results for standard and improved generalized Feistel structures with $n = 4$ by considering switching properties are listed in Table 5.

6 Conclusion

In this paper, by relying on MILP and summation representation, we introduced an approach to calculate the number of differentially and linearly active S-boxes until 24 rounds. We first explained, how XOR relation and SP-function can be converted to inequalities. Then we listed the tables related to standard, and improved generalized Feistel structures. Moreover, we clarified the way of constructing inequalities related to employing multiple MDS matrices in generalized Feistel structures type II, and presented the results. Finally, we confirmed the effect of switching mechanism on linear cryptanalysis. Due to obtained results for linear cryptanalysis, it is clear that switching is more effective on linear cryptanalysis. Since, the effect of switching on each GFS starts from the third round, in linear cryptanalysis. Aside from the fact that our method does not apply for structures such as AES (because of $shiftrow$ operation), our approach significantly reduces the number of inequalities for other structures compared with the previous approach based on MILP.

We would like to point out that, employing the multiple MDS matrices in improved 8 sub-blocks structures leads to enhance the number of active S-boxes almost 20% for 18 rounds, and creates a structure so close to RIJNDAEL-256 (RIJNDAEL-256 has 105 differentially active S-boxes for 12 rounds). For larger blocks, switching can not diffuse until 18 rounds, in differential cryptanalysis.

Besides that, in differential cryptanalysis, we have confirmed that in improved 8 sub-blocks structure, if we apply 2 different MDS matrices, only 3 differentially active S-boxes is lower than applying 4 matrices after 24 rounds (91 for 4 matrices and 88 for 2 matrices). By doing so, we not only benefit from switching features, but also apply fewer resources. It is worth mentioning that, our approach can be generalized for other Feistel structures, and is usable in designing future block ciphers.

References

1. Aoki, K., et al.: *Camellia*: a 128-bit block cipher suitable for multiple platforms — design and analysis. In: Stinson, D.R., Tavares, S. (eds.) SAC 2000. LNCS, vol. 2012, pp. 39–56. Springer, Heidelberg (2001). https://doi.org/10.1007/3-540-44983-3_4

2. Beierle, C., et al.: The SKINNY family of block ciphers and its low-latency variant MANTIS. In: Robshaw, M., Katz, J. (eds.) CRYPTO 2016, Part II. LNCS, vol. 9815, pp. 123–153. Springer, Heidelberg (2016). https://doi.org/10.1007/978-3-662-53008-5_5

3. Bogdanov, A.: On unbalanced feistel networks with contracting MDS diffusion. Des. Codes Crypt. **59**(1), 3558 (2011)

4. IBM: IBM ILOG CPLEX Optimizer. http://www.ibm.com/software/integration/optimization/cplex-optimizer/

5. Kanda, M.: Practical security evaluation against differential and linear cryptanalyses for feistel ciphers with SPN round function. In: Stinson, D.R., Tavares, S. (eds.) SAC 2000. LNCS, vol. 2012, pp. 324–338. Springer, Heidelberg (2001). https://doi.org/10.1007/3-540-44983-3_24

6. Mouha, N., Wang, Q., Gu, D., Preneel, B.: Differential and linear cryptanalysis using mixed-integer linear programming. In: Wu, C.-K., Yung, M., Lin, D. (eds.) Inscrypt 2011. LNCS, vol. 7537, pp. 57–76. Springer, Heidelberg (2012). https://doi.org/10.1007/978-3-642-34704-7_5

7. Sajadieh, M., Mirzaei, A., Mala, H., Rijmen, V.: A new counting method to bound the number of active s-boxes in Rijndael and 3D. Des. Codes Crypt. **83**(2), 327–343 (2017)

8. Shibutani, K.: On the diffusion of generalized feistel structures regarding differential and linear cryptanalysis. In: Biryukov, A., Gong, G., Stinson, D.R. (eds.) SAC 2010. LNCS, vol. 6544, pp. 211–228. Springer, Heidelberg (2011). https://doi.org/10.1007/978-3-642-19574-7_15

9. Shirai, T., Araki, K.: On generalized feistel structures using the diffusion switching mechanism. IEICE Trans. Fundam. Electron. Commun. Comput. Sci. **E91-A(8)**, 2120–2129 (2008)

10. Shirai, T., Kanamaru, S., Abe, G.: Improved upper bounds of differential and linear characteristic probability for camellia. In: Daemen, J., Rijmen, V. (eds.) FSE 2002. LNCS, vol. 2365, pp. 128–142. Springer, Heidelberg (2002). https://doi.org/10.1007/3-540-45661-9_10

11. Shirai, T., Shibutani, K.: Improving immunity of feistel ciphers against differential cryptanalysis by using multiple MDS matrices. In: Roy, B., Meier, W. (eds.) FSE 2004. LNCS, vol. 3017, pp. 260–278. Springer, Heidelberg (2004). https://doi.org/10.1007/978-3-540-25937-4_17

12. Shirai, T., Shibutani, K.: On feistel structures using a diffusion switching mechanism. In: Robshaw, M. (ed.) FSE 2006. LNCS, vol. 4047, pp. 41–56. Springer, Heidelberg (2006). https://doi.org/10.1007/11799313_4

13. Shirai, T., Shibutani, K., Akishita, T., Moriai, S., Iwata, T.: The 128-Bit block-cipher CLEFIA (Extended Abstract). In: Biryukov, A. (ed.) FSE 2007. LNCS, vol. 4593, pp. 181–195. Springer, Heidelberg (2007). https://doi.org/10.1007/978-3-540-74619-5_12

14. Suzaki, T., Minematsu, K.: Improving the generalized feistel. In: Hong, S., Iwata, T. (eds.) FSE 2010. LNCS, vol. 6147, pp. 19–39. Springer, Heidelberg (2010). https://doi.org/10.1007/978-3-642-13858-4_2

Tools in Analyzing Linear Approximation for Boolean Functions Related to FLIP

Subhamoy Maitra[1], Bimal Mandal[1], Thor Martinsen[2], Dibyendu Roy[3(✉)], and Pantelimon Stănică[2]

[1] Indian Statistical Institute, Kolkata, India
subho@isical.ac.in, bimalmandal90@gmail.com
[2] Naval Postgraduate School, Monterey, USA
{tmartins,pstanica}@nps.edu
[3] National Institute of Science Education and Research (HBNI), Bhubaneswar, India
roydibyendu.rd@gmail.com

Abstract. For cryptographic purposes, we generally study the characteristics of a Boolean function in n-variables with the inherent assumption that each of the n-bit inputs take the value 0 or 1, independently and randomly with probability $1/2$. However, in the context of the FLIP stream cipher proposed by Méaux et al. (Eurocrypt 2016), this type of analysis warrants a different approach. To this end, Carlet et al. (IACR Trans. Symm. Crypto. 2018) recently presented a detailed analysis of Boolean functions with restricted inputs (mostly considering inputs with weight $\frac{n}{2}$) and provided certain bounds on linear approximation, which are related to restricted nonlinearity. The Boolean function used in the FLIP cipher reveals that it is actually a direct sum of several Boolean functions on a small number of inputs. Thus, with a different approach, we start a study in order to understand how the inputs to the composite function are distributed on the smaller functions. In this direction, we obtain several results that summarize the exact biases related to such Boolean functions. Finally, for the nonlinear filter function of FLIP, we obtain the lower bound on the restricted Walsh–Hadamard transform (i.e., upper bound on restricted nonlinearity). Our techniques provide a general theoretical framework to study such functions and better than previously published estimations of the biases, which is directly linked to the security parameters of the stream cipher.

Keywords: Bias · Boolean function · FLIP
Homomorphic encryption · Restricted domain · Stream cipher

1 Introduction

The search for practical solutions to efficient homomorphic encryption schemes, ushered in a new paradigm in stream cipher design, and received serious attention, recently. One important step in this direction has appeared in [1]. Shortly thereafter, the papers [5,6] started analyzing the constituent Boolean function(s)

© Springer Nature Switzerland AG 2018
D. Chakraborty and T. Iwata (Eds.): INDOCRYPT 2018, LNCS 11356, pp. 282–303, 2018.
https://doi.org/10.1007/978-3-030-05378-9_16

in the FLIP stream cipher. An initial version of this cipher was cryptanalyzed in [3]. In the FLIP stream cipher, the keystream bit is computed by using one nonlinear filter function, which takes input from a restricted domain. Recently, in [2,7], the properties of the Boolean functions in such a restricted domain [6] were studied in detail.

In this paper we consider a different approach. It is evident that for the implementation of efficient homomorphic encryption schemes, the underlying stream cipher must be simple. This requires Boolean functions with simple Algebraic Normal Form (ANF) having many linear and low degree terms connected by simple $\mathbb{F}_2$ addition. Further, the existing Boolean functions in the FLIP cipher require each variable to be part of only one subfunction in the ANF. Given such restrictions, it is evident that such functions will not have good cryptographic properties. Thus, one requires a large number of variables, and consequently, we want to get the required security with the least possible number of inputs. The study of such functions is much easier using the standard Walsh–Hadamard transform if we consider that the inputs appear independently and uniformly at random. However, this is not the case here, since only the inputs of a specific weight play a role. Quite involved mathematical techniques have been exploited in [2,7] to study such functions. The analysis of FLIP, as a consequence of these works, requires more attention, as specific numerical bounds on both sides are not available.

Let us now discuss the issue from a more technical viewpoint (we refer to the notations later in this section). Carlet et al. [2] observed that different properties of a Boolean function F defined over $\mathbb{F}_2^n$ degrade significantly when the inputs come from a restricted subset $E \subset \mathbb{F}_2^n$. In the case of the FLIP stream cipher, the inputs of the nonlinear filter function remain a 0/1 string of length n with weight $\frac{n}{2}$ for all rounds. So, the nonlinear filter function always takes input from a restricted subset $E \subset \mathbb{F}_2^n$. Based on this observation, Carlet et al. [2] studied several properties of a Boolean function in a restricted domain. Mesnager et al. [7] further analyzed Boolean functions on restricted domains and proposed a lower bound of the bias, although, the numerical computation of the upper bound of the bias is practically not possible by that technique. The papers [2,7] consider the properties of the complete function F (see Sect. 1.2), given that the input is of fixed weight.

In this paper, we concentrate on this issue and first notice that if $\mathbf{x} = \mathbf{x}_1\|\mathbf{x}_2\| \cdots \|\mathbf{x}_n$ (concatenation) and $\mathbf{x} \in E_{n,k}$ (the definition of $E_{n,k}$ is provided in Sect. 1.1) then $\mathbf{x}_1, \mathbf{x}_2, \ldots, \mathbf{x}_n$ does not follow a uniform distribution. This observation motivates us to study the restricted Walsh–Hadamard transform by considering the exact probability distribution of $\mathbf{x}_1, \mathbf{x}_2, \ldots, \mathbf{x}_n$. In fact, it is worth mentioning now that if the input $\mathbf{x}_i$ of a Boolean function f_i does not follow a uniform distribution, then the original properties of f_i (assuming uniform distribution) changes significantly. Further, by considering the actual distribution of the input (rather than a uniform one), we expect to achieve a tighter bound for the bias given the nonlinear filter function used in the FLIP stream cipher. Naturally a tighter bound will provide much better approximation for the security parameters of the FLIP cipher.

CONTRIBUTION AND ORGANIZATION. Our approach considers how the inputs to the composite function are distributed on the smaller functions. In this direction, we present some tools to start our analysis in Sect. 2. Then our main motivation is to obtain more accurate linear approximations of nonlinear Boolean functions when the inputs are restricted, which we discuss in Sect. 3. However, the formulae that we arrive at are quite complicated to be directly compared with equally complicated expressions of [2,7]. Thus, in this direction, numerical data will provide better understanding of these results, as we discuss in Sect. 4. For that we refer to the $n = 530$ variable Boolean function that has been considered in [2]. Straightforward analysis of the Walsh–Hadamard spectrum shows that when we consider that the inputs are uniform, such a function has maximum absolute Walsh–Hadamard transform value in $[2^{-79}, 2^{-78}]$. Thus, the bias to a linear function looks quite low. However, our analysis shows that when the inputs are taken of weight $\frac{n}{2} = 265$, then the restricted Walsh–Hadamard transform is much higher. The maximum absolute value is in $[2^{-18.49}, 2^{-13.59}]$. We obtain the upper bound by considering the idea of [2] and the lower bound is obtained from our detailed analysis in this paper. That is, our work complements the work of [2] to bound the maximum absolute restricted Walsh–Hadamard transform value of a function on large number of variables used in the FLIP stream cipher.

Before proceeding further let us present some background material.

1.1 Boolean Functions

Let $\mathbb{F}_2$ and $\mathbb{F}_2^n$ be the prime binary field, respectively, the extension field over $\mathbb{F}_2$ of degree n. Let $\mathbb{F}_2^n = \{\mathbf{x} = (x_1, x_2, \ldots, x_n) : x_i \in \mathbb{F}_2, \text{for all } 1 \leq i \leq n\}$ be the vector space over $\mathbb{F}_2$ of dimension n. We denote the concatenation of two (or more) binary strings $\mathbf{x}', \mathbf{x}''$ by $\mathbf{x}' || \mathbf{x}''$. The cardinality of a set S is denoted by $|S|$. Any function $f : \mathbb{F}_2^n \longrightarrow \mathbb{F}_2$ is said to be a Boolean function in n-variables, whose set is denoted by $\mathcal{B}_n$. These functions can be represented in a unique way (called the *Algebraic Normal Form* (ANF) of f) as

$$f(\mathbf{x}) = \sum_{\mathbf{a} \in \mathbb{F}_2^n} \mu_{\mathbf{a}} \left(\prod_{i=1}^{n} x_i^{a_i} \right), \text{ for all } \mathbf{x} \in \mathbb{F}_2^n, \text{where } \mu_{\mathbf{a}} \in \mathbb{F}_2.$$

The *Hamming weight* of $\mathbf{x} \in \mathbb{F}_2^n$ is defined as $wt(\mathbf{x}) = \sum_{i=1}^{n} x_i$, where the sum is over $\mathbb{Z}$, the ring of integers. The *algebraic degree* of a Boolean function $f \in \mathcal{B}_n$ is defined as $\deg(f) = \max_{\mathbf{a} \in \mathbb{F}_2^n} \{wt(\mathbf{a}) : \mu_{\mathbf{a}} \neq 0\}$. Let $E_{n,i} = \{\mathbf{x} \in \mathbb{F}_2^n : wt(\mathbf{x}) = i\}$, for all $0 \leq i \leq n$. The *support* of $f \in \mathcal{B}_n$ is defined as $\text{supp}(f) = \{\mathbf{x} \in \mathbb{F}_2^n : f(\mathbf{x}) = 1\}$. A Boolean function is said to be *balanced* if the cardinality of its support set is $|\text{supp}(f)| = 2^{n-1}$. If the algebraic degree of a Boolean function $f \in \mathcal{B}_n$ is at most 1 then f is an affine function, and its set is $\mathcal{A}_n = \{l_{\mathbf{a},\varepsilon} : \mathbf{a} \in \mathbb{F}_2^n, \varepsilon \in \mathbb{F}_2\}$, where $l_{\mathbf{a},\varepsilon}(\mathbf{x}) = \mathbf{a} \cdot \mathbf{x} + \varepsilon$, for all $\mathbf{x} \in \mathbb{F}_2^n$. If $\varepsilon = 0$, then $l_{\mathbf{a},0}$ is a linear function. The *Hamming distance* between any $f, g \in \mathcal{B}_n$ is defined by $d_H(f, g) = |\{\mathbf{x} \in \mathbb{F}_2^n : f(\mathbf{x}) \neq g(\mathbf{x})\}|$. The *correlation* between two Boolean functions $f, g \in \mathcal{B}_n$ is defined by

$$\mathrm{corr}(f, g) = \left| \frac{|\{\mathbf{x} : f(\mathbf{x}) = g(\mathbf{x})\}| - |\{\mathbf{x} : f(\mathbf{x}) \neq g(\mathbf{x})\}|}{2^n} \right|.$$

To measure the correlation between an n-variable Boolean function f and a linear function $l_{\mathbf{a},0}$, we use the *Walsh–Hadamard transform*, defined by

$$\mathcal{W}_f(\mathbf{a}) = \frac{1}{2^n} \sum_{\mathbf{x} \in \mathbb{F}_2^n} (-1)^{f(\mathbf{x}) + \mathbf{a} \cdot \mathbf{x}}.$$

We observe that the absolute value of the Walsh–Hadamard transform of $f \in \mathcal{B}_n$ at a fixed point $\mathbf{a} \in \mathbb{F}_2^n$ provides us the correlation between the Boolean function f and the linear function $l_{\mathbf{a},0}$, i.e., $\mathrm{corr}(f, l_{\mathbf{a},0}) = |\mathcal{W}_f(\mathbf{a})|$, for all $\mathbf{a} \in \mathbb{F}_2^n$. The multiset $[\mathcal{W}_f(\mathbf{a}) : \mathbf{a} \in \mathbb{F}_2^n]$, which is the *Walsh–Hadamard spectrum* of f, provides us the correlation between the Boolean function f and all possible linear functions. From the Parseval's identity for arbitrary $f \in \mathcal{B}_n$,

$$\sum_{\mathbf{a} \in \mathbb{F}_2^n} \mathcal{W}_f(\mathbf{a})^2 = 1,$$

we obtain $\max_{\mathbf{a} \in \mathbb{F}_2^n} |\mathcal{W}_f(\mathbf{a})| \geq \frac{1}{2^{n/2}}$.

A Boolean function $f \in \mathcal{B}_n$ (n even) is said to be bent if and only if the correlation between f and $\{l_{\mathbf{a},0} | \mathbf{a} \in \mathbb{F}_2^n\}$ is $\frac{1}{2^{n/2}}$, i.e., $\mathrm{corr}(f, l_{\mathbf{a},0}) = |\mathcal{W}_f(\mathbf{a})| = 2^{-\frac{n}{2}}$, for all $\mathbf{a} \in \mathbb{F}_2^n$.

Now if we assume that an n-variable Boolean function $f \in \mathcal{B}_n$ takes input from a restricted domain, then to calculate the correlation between f and a linear function $l_{\mathbf{a},0}$, we need to consider the inputs $\mathbf{x}$ only from a restricted domain. Here we assume that f takes inputs of weight k, i.e., $\mathbf{x} \in E_{n,k} := \{\mathbf{x} \in \mathbb{F}_2^n : wt(\mathbf{x}) = k\}$. Certainly, $|E_{n,k}| = \binom{n}{k}$. Under this assumption, the (restricted domain) correlation between the Boolean function f and a linear function $l_{\mathbf{a},0}$ is

$$\mathrm{corr}^{(k)}(f, l_{\mathbf{a},0}) = \left| \frac{|\{\mathbf{x} : f(\mathbf{x}) = l_{\mathbf{a},0}(\mathbf{x})\}| - |\{\mathbf{x} : f(\mathbf{x}) \neq l_{\mathbf{a},0}(\mathbf{x})\}|}{|E_{n,k}|} \right|.$$

Further, to calculate this correlation, we shall define the Walsh–Hadamard transform $\mathcal{W}_f^{(k)}(\mathbf{a})$ of a Boolean function f in a restricted domain $E_{n,k}$, $0 \leq k \leq n$, by

$$\mathcal{W}_f^{(k)}(\mathbf{a}) = \frac{1}{|E_{n,k}|} \sum_{\mathbf{x} \in E_{n,k}} (-1)^{f(\mathbf{x}) + \mathbf{a} \cdot \mathbf{x}}.$$

Here we define two more notations, which are used throughout the article.

Definition 1. *Let* $\mathbf{x} = (x_1, \ldots, x_n) \in E_{n,k}$ *and* $n = n_1 + n_2$, *and* $\mathbf{x}' = (x_1, \ldots, x_{n_1})$, $\mathbf{x}'' = (x_{n_1+1}, \ldots, x_n)$. *Then* $E_{n_1,i}^{n=n_1+n_2,k} = \{\mathbf{x}' \in \mathbb{F}_2^{n_1} \mid \mathbf{x} \in E_{n,k}, n = n_1 + n_2$ *and* $wt(\mathbf{x}') = i\}$ *and* $E_{n_2,j}^{n=n_1+n_2,k} = \{\mathbf{x}'' \in \mathbb{F}_2^{n_2} \mid \mathbf{x} \in E_{n,k}, n = n_1 + n_2$ *and* $wt(\mathbf{x}'') = j\}$.

Certainly, we can continue the splitting process, and if $n = n_1 + n_2 + n_3$ then $E_{n_1,i}^{n=n_1+n_2+n_3,k}$, $E_{n_2,j}^{n=n_1+n_2+n_3,k}$ and $E_{n_3,r}^{n=n_1+n_2+n_3,k}$ can be inferred from the above definition. More generally it can be extended to $n = n_1 + n_2 + \cdots + n_q$.

1.2   Design Specification of the FLIP Stream Cipher

In this section we describe the design specification of the FLIP family of stream ciphers (initially, presented in [5]). The main motivation behind this proposal was to construct a fully homomorphic encryption (FHE) scheme with the limited error growth using a symmetric key primitive. Since block ciphers are based on complicated round functions, it seems to be difficult to construct such a FHE. After this proposal, Duval et al. [3] came up with an attack on the FLIP ciphers. Shortly thereafter, Méaux et al. [6] modified the design specification of FLIP and proposed the final modified version of the FLIP stream cipher.

The FLIP cipher is based on three components: one register of length n, one pseudorandom number generator (PRNG), one nonlinear filter function F involving n-variables.

The cipher stores the secret key K of length n into the register and a PRNG is initialized with the initialization vector IV. In each clock, the PRNG generates a number which corresponds to a permutation. This pseudorandom permutation permutes the state bits of the register. Finally, the nonlinear filter function takes the current state as input to generate keystream bits. The pictorial description of the FLIP stream cipher is described in Fig. 1.

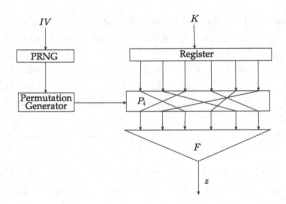

Fig. 1. Design specification of FLIP

The nonlinear filter function $F = f_1 + f_2 + f_3$ has three component functions: f_1 is a linear function, f_2 is a quadratic bent function and f_3 is a special type of triangular function. The ANFs of these functions are described below:

– **L-type function.** A Boolean function L_n in n-variables is said to be of L-type if it is of the following form $L_n(x_0, x_1, \ldots, x_{n-1}) = \sum_{i=0}^{n-1} x_i$.

- **Q-type function.** The algebraic normal form of the Q-type bent function Q_{2n} in $2n$-variables is $Q_{2n}(x_0, x_1, \ldots, x_{2n-1}) = \sum_{i=0}^{n-1} x_{2i} x_{2i+1}$.
- **T-type function.** For a positive integer n, the algebraic normal form of the n-th T-type triangular function T_n in $\frac{n(n+1)}{2}$ variables is

$$T_n(x_0, x_1, \ldots, x_{\frac{n(n+1)}{2}-1}) = \sum_{i=1}^{n} \prod_{j=0}^{i-1} x_{j+\sum_{\ell=0}^{i-1} \ell}.$$

Thus, the nonlinear filter function F in n-variables is a direct sum of three Boolean functions f_1, f_2 and f_3 involving n_1, n_2 and n_3 variables (such that $n = n_1 + n_2 + n_3$), respectively, where the algebraic normal form of these functions are as follows:

- $f_1(x_0, x_1, \ldots, x_{n_1-1}) = L_{n_1}(x_0, x_1, \ldots, x_{n_1-1})$.
- $f_2(x_{n_1}, x_{n_1+1}, \ldots, x_{n_1+n_2-1}) = Q_{n_2}(x_{n_1}, x_{n_1+1}, \ldots, x_{n_1+n_2-1})$.
- $f_3(x_{n_1+n_2}, x_{n_1+n_2+1}, \ldots, x_{n_1+n_2+n_3-1})$ is the direct sum of r triangular function T_k, where each T_k involves independent variables.

The final algebraic normal form of the nonlinear filter function F is

$$F = L_{n_1} + Q_{n_2} + \sum_{i=1}^{r} T_k.$$

The function that we concentrate on in this paper is the one with the notation FLIP$(42, 128, {}^8\Delta^9)$ as described in [6]. This means that $n_1 = 42, n_2 = 2 \cdot 64 = 128, n_3 = 8 \cdot (1 + 2 + \cdots + 9) = 360$. That is there are 42 terms in the linear functions (L-type), 64 many quadratic terms in the quadratic functions (Q-type) and further there are eight T-type functions each having terms of degree 1 to 9, i.e., each one having 45 many variables.

The designers of the FLIP stream ciphers suggested that the weight of the secret key of length n must be $\frac{n}{2}$. In each round, one pseudorandom permutation is applied on the register, which permutes the index of the secret key bits. The nonlinear filter function F takes the updated state of the register as input to produce an output bit. It is then clear that the weight of the state of the cipher in each round remains fixed (i.e., $\frac{n}{2}$). From the expression of the keystream we can formally write $F(S_n^t) = z_t$, where $wt(S_n^t) = \frac{n}{2}$.

At Crypto 2016, Duval et al. [3] proposed an attack on the old version of the FLIP stream cipher as introduced in [5]. The attack complexities for two instances of FLIP, namely, $n = 192$ ($n_1 = 47, n_2 = 40, n_3 = 105$) and $n = 400$ ($n_1 = 87, n_2 = 82, n_3 = 231$), are 2^{54}, respectively, 2^{68}. The previously described modified design has then been proposed by Méaux et al. [6] to counter this attack.

2   Tools for Our Analysis

Here we first review the existing techniques and then move forward to our new ideas. One may note that given the simple structure of the Boolean function in the FLIP cipher, it is not hard to study the nonlinearity under the framework of Walsh–Hadamard transform. One can easily verify that, in uniform domain $2^{-79} < \max_{\mathbf{a} \in \mathbb{F}_2^{530}} |\mathcal{W}_f(\mathbf{a})| < 2^{-78}$.

However, the scenario is completely different in restricted domain. We will actually see at the end of this paper, due to the restriction on inputs, this maximum absolute restricted Walsh–Hadamard spectrum value is indeed much higher, which is $\geq \frac{1}{2^{18.49}}$. Following the work of [2] it can be calculated that this is also less than $\frac{1}{2^{13.59}}$. Thus, the bound obtained by simple Walsh–Hadamard transform does not provide the actual picture and it is indeed much higher in FLIP stream cipher when restricted inputs are considered.

2.1   Our Idea: Frequency Distribution of Concatenated Sub-strings of a Fixed Weight Bit String

Recall that each element $\mathbf{x} = (x_1, x_2, \ldots, x_n) \in E_{n,k}$ is an n bit binary string with weight $wt(\mathbf{x}) = k$. In $\mathbf{x}$, if we consider the first n_1 components x_i's, then the weight distributions may not be uniform. This may affect different cryptographic properties of a Boolean function defined over the first n_1 number of variables of $\mathbf{x} \in E_{n,k}$. Carlet et al. [2] did not consider this issue, although they studied several properties of the complete function $F = f_1 + f_2 + f_3$, when the input is restricted to a set. As the nonlinear filter function in the FLIP stream cipher is $F = f_1 + f_2 + f_3$, we need to study the individual functions f_1, f_2 and f_3 by considering the weight distribution of inputs for each of these functions.

For $n = n_1 + n_2 + n_3$ and $\mathbf{x} \in \mathbb{F}_2^n$, we write $\mathbf{x} = \mathbf{x}'||\mathbf{x}''||\mathbf{x}'''$, where $\mathbf{x}' \in \mathbb{F}_2^{n_1}$, $\mathbf{x}'' \in \mathbb{F}_2^{n_2}$ and $\mathbf{x}''' \in \mathbb{F}_2^{n_3}$ with $Pr(\mathbf{x}) = \frac{1}{2^n}$, representing the probability of picking any element $\mathbf{x} \in \mathbb{F}_2^n$. The cardinality of $E_{n,k}$ is equal to $\binom{n}{k}$, $0 \leq k \leq n$, which follows the normal distribution (when n approaches to ∞). Also, if we consider the first n_1 bits of $\mathbb{F}_2^n$, then all elements belonging to $\mathbb{F}_2^{n_1}$ of whatever weight distribution will follow the normal distribution. However, fixing k, that is, by considering only one set $E_{n,k}$, the cardinality of $E_{n_1,i}^{n=n_1+n_2+n_3,k}$, $0 \leq i \leq n_1$, does not follow the normal distribution.

For example, let $n = 4$ and $n_1 = 2$. Then $E_{4,i}$, $0 \leq i \leq 4$, satisfy the normal distribution, as well as $E_{2,j}^{4=2+2,i}$, $0 \leq j \leq 2$, as $|E_{2,0}^{4=2+2,i}| = 4 = |E_{2,2}^{4=2+2,i}| = |E_{2,1}^{4=2+2,i}|/2$. ($|E_{2,1}^{4=2+2,i}| = 8$, as for each one weight element of length two there are four possibilities in the last two bits for $0 < i < 4$.) Let us consider only the set $E_{4,2}$ of cardinality 6 and, if we consider the first two bits then $|E_{2,0}^{4=2+2,2}| = 1$, $|E_{2,1}^{4=2+2,2}| = 4$ and $|E_{2,2}^{4=2+2,2}| = 1$, then $Pr(\mathbf{x}' = 00) = \frac{1}{6} = Pr(\mathbf{x}' = 11)$, $Pr(\mathbf{x}' = 01) = \frac{1}{3} = Pr(\mathbf{x}' = 10)$. The probability distribution is provided in Fig. 2.

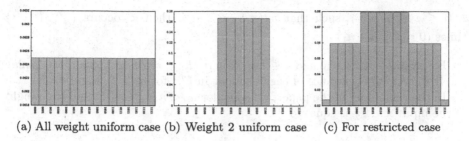

(a) All weight uniform case (b) Weight 2 uniform case (c) For restricted case

Fig. 2. Probability distributions

Let P be any permutation of the set $\{x_1, x_2, \ldots, x_n\}$. Then $P(\mathbf{x}) \in E_{n,k}$, for all $\mathbf{x} \in E_{n,k}$. In the rest of the paper, we consider $\mathbf{x} = (x_1, x_2, \ldots, x_n) \in E_{n,\frac{n}{2}}$ and $\mathbf{x}' = (y_1, y_2, \ldots, y_{n_1}) \in \mathbb{F}_2^{n_1}$ where $y_i = x_i$, $1 \le i \le n_1$. We now calculate the frequency distributions of $\mathbf{x}'$'s with respect to their weights.

Case 1: Let $n_1 = \frac{n}{2}$. Then all possible elements $\mathbf{x}'$ of $\mathbb{F}_2^{n_1}$ exist. We observe that there are $\binom{\frac{n}{2}}{i}\binom{\frac{n}{2}}{\frac{n}{2}-i}$ elements $\mathbf{x}'$ such that $wt(\mathbf{x}') = i$, $0 \le i \le \frac{n}{2}$ and each bit pattern of the same weight will occur an equal number of times.

Case 2: Let $n_1 < \frac{n}{2}$. Again all possible elements $\mathbf{x}'$ of $\mathbb{F}_2^{n_1}$ exist. We observe that there are $\binom{n_1}{i}\binom{n-n_1}{\frac{n}{2}-i}$ elements $\mathbf{x}' \in \mathbb{F}_2^{n_1}$ such that $wt(\mathbf{x}') = i$; $0 \le i \le n_1$.

Case 3: Let $n_1 > \frac{n}{2}$. Now we find the number of possible $\mathbf{x}'$ with weight $wt(\mathbf{x}') = i$, $n_1 - \frac{n}{2} \le i \le \frac{n}{2}$. We observe that the cardinality $\left|\{\mathbf{x}' \in \mathbb{F}_2^{n_1} \mid wt(\mathbf{x}') = \frac{n}{2}\}\right| = \binom{n_1}{\frac{n}{2}}$, where every such element occurs exactly once. In general, for each

Table 1. Frequency distribution of $\mathbb{F}_2^{n_1}$ for $n_1 = 2, 3$ and 4

x_2x_1	Frequency
00	4
01	6
10	6
11	4

$x_3x_2x_1$	Frequency
000	1
001	3
010	3
011	3
100	3
101	3
110	3
111	1

$x_4x_3x_2x_1$	frequency
0000	0
0001	1
0010	1
0011	2
0100	1
0101	2
0110	2
0111	1
1000	1
1001	2
1010	2
1011	1
1100	2
1101	1
1110	1
1111	0

i, $0 \leq i \leq n - n_1$, such that $wt(\mathbf{x}') = \frac{n}{2} - i$, then $\mathbf{x}'$ occurs $\binom{n_1}{\frac{n}{2} - i}\binom{n - n_1}{i}$ times $(0 \leq i \leq n - n_1)$.

For example, let $n = 6$ and $\mathbf{x} \in E_{6,3}$, where $|E_{6,3}| = 20$. In Table 1, we display the frequency of occurrence of each element in $\mathbb{F}_2^{n_1}$ for $n_1 = 2, 3$ and 4.

In the remainder of the paper, we consider $n_i < n$, $1 \leq i \leq 3$, and the weight of the input $\mathbf{x} \in \mathbb{F}_2^n$ is equal to $\frac{n}{2}$.

3   Biased Walsh–Hadamard Transform

We define the Walsh–Hadamard transform of a Boolean function when the input elements have different probabilities, not necessarily uniform. We shall call this transform, a biased Walsh–Hadamard transform of a Boolean function (see [4], for yet another definition), which is the same as the bias between a Boolean function and a linear function over a non-uniform domain. If the input to a Boolean function does not follow the uniform distribution, several properties of the function change significantly.

Let $p(\mathbf{a})$ be the probability of an input element $\mathbf{a} \in \mathbb{F}_2^n$ in $f \in \mathcal{B}_n$. Recall that $0 \leq p(\mathbf{a}) \leq 1$, for all $\mathbf{a} \in \mathbb{F}_2^n$, and $\sum_{\mathbf{a} \in \mathbb{F}_2^n} p(\mathbf{a}) = 1$. For any $f, g \in \mathcal{B}_n$, we let $\mathcal{S}(f,g) = \{\mathbf{x} \in \mathbb{F}_2^n : f(\mathbf{x}) \neq g(\mathbf{x})\}$ and $\bar{\mathcal{S}}(f,g) = \mathbb{F}_2^n \setminus \mathcal{S}(f,g) = \{\mathbf{x} \in \mathbb{F}_2^n : f(\mathbf{x}) = g(\mathbf{x})\}$.

The *biased Hamming distance*, $d_H^B(f,g)$, between two Boolean functions $f, g \in \mathcal{B}_n$, when the inputs are not uniformly distributed, is defined by $d_H^B(f,g) = \sum_{\mathbf{x} \in \mathcal{S}(f,g)} p(\mathbf{x})$. Further, the *biased Hamming distance* between two Boolean functions $f, g \in \mathcal{B}_n$ is

$$
d_H^B(f,g) = \frac{1}{2}\left\{ \sum_{\mathbf{x} \in \bar{\mathcal{S}}(f,g)} p(\mathbf{x}) + \sum_{\mathbf{x} \in \mathcal{S}(f,g)} p(\mathbf{x}) \right\} - \frac{1}{2}\left\{ \sum_{\mathbf{x} \in \bar{\mathcal{S}}(f,g)} p(\mathbf{x}) - \sum_{\mathbf{x} \in \mathcal{S}(f,g)} p(\mathbf{x}) \right\}
$$
$$
= \frac{1}{2} - \frac{1}{2} \sum_{\mathbf{x} \in \mathbb{F}_2^n} p(\mathbf{x})(-1)^{f(\mathbf{x}) + g(\mathbf{x})}.
$$

In particular, $d_H^B(f, l_{\mathbf{a},\varepsilon}) = \frac{1}{2} - \frac{(-1)^\varepsilon}{2} \sum_{\mathbf{x} \in \mathbb{F}_2^n} p(\mathbf{x})(-1)^{f(\mathbf{x}) + \mathbf{a} \cdot \mathbf{x}} = \frac{1}{2} - \frac{(-1)^\varepsilon}{2} \mathcal{W}_f^B(\mathbf{a})$, where $\mathcal{W}_f^B(\mathbf{a}) = \sum_{\mathbf{x} \in \mathbb{F}_2^n} p(\mathbf{x})(-1)^{f(\mathbf{x}) + \mathbf{a} \cdot \mathbf{x}}$ is the *biased Walsh–Hadamard transform* of $f \in \mathcal{B}_n$ at $\mathbf{a} \in \mathbb{F}_2^n$. The multiset $\left[\mathcal{W}_f^B(\mathbf{a}) : \mathbf{a} \in \mathbb{F}_2^n \right]$ is the *biased Walsh–Hadamard spectrum* of $f \in \mathcal{B}_n$. Note that $\mathcal{W}_{l_{\mathbf{a},0}}^B(\mathbf{a}) = 1$ and for any other point, the value of this biased Walsh–Hadamard transform may or may not be zero, which is not the case for the uniform domain.

Further, for the non-uniform case, we define the $\mathsf{corr}^B(f,g)$ between $f, g \in \mathcal{B}_n$, by

$$\mathrm{corr}^B(f,g) = \left| \sum_{\mathbf{x}\in \bar{S}(f,g)} p(\mathbf{x}) - \sum_{\mathbf{x}\in S(f,g)} \hat{p}(\mathbf{x}) \right|.$$

Note that $\mathrm{corr}^B(f, l_{\mathbf{a},0}) = \left| \mathcal{W}_f^B(\mathbf{a}) \right|$.

3.1 The Biased Walsh–Hadamard Transform of a Direct Sum of Boolean Functions

This section presents a convolution theorem in the biased domain and several bounds related to a direct sum of Boolean functions. Let $n = n_1 + n_2$, and $\mathbf{x} = \mathbf{x}'||\mathbf{x}'' \in \mathbb{F}_2^n$, where $\mathbf{x}' \in \mathbb{F}_2^{n_1}$ and $\mathbf{x}'' \in \mathbb{F}_2^{n_2}$. Then, $Pr[\mathbf{x}] = Pr[\mathbf{x}', \mathbf{x}''] = Pr[\mathbf{x}'/\mathbf{x}'']Pr[\mathbf{x}''] = Pr[\mathbf{x}''/\mathbf{x}']Pr[\mathbf{x}']$, for any $\mathbf{x} \in \mathbb{F}_2^n$. The biased Walsh–Hadamard transform of $f(\mathbf{x}) = f_1(\mathbf{x}') + f_2(\mathbf{x}'')$ at $\mathbf{a} = \mathbf{a}'||\mathbf{a}''$ is equal to

$$\begin{aligned}
\mathcal{W}_f^B(\mathbf{a}) &= \sum_{\mathbf{x}\in\mathbb{F}_2^n} p(\mathbf{x})(-1)^{f(\mathbf{x})+\mathbf{a}\cdot\mathbf{x}} \\
&= \sum_{\mathbf{x}''\in\mathbb{F}_2^{n_2}} p(\mathbf{x}'')(-1)^{f_2(\mathbf{x}'')+\mathbf{a}''\cdot\mathbf{x}''} \sum_{\mathbf{x}'\in\mathbb{F}_2^{n_1}} p(\mathbf{x}'/\mathbf{x}'')(-1)^{f_1(\mathbf{x}')+\mathbf{a}'\cdot\mathbf{x}'},
\end{aligned} \tag{1}$$

where $p(\mathbf{x}'/\mathbf{x}'') = Pr[\mathbf{x}'/\mathbf{x}'']$. From Eq. (1), it is clear that we are unable to directly calculate the biased Walsh–Hadamard transform of $f = f_1 + f_2$ even though we may know the biased Walsh–Hadamard transform of two component functions f_1 and f_2, as $Pr[\mathbf{x}'/\mathbf{x}''] \neq Pr[\mathbf{x}']$, in general.

Let now $f = f_1 + f_2$ on $\mathbb{F}_2^n$, where f_1, f_2 depend upon independent sets of variables. If the domain is uniform, to calculate the Walsh–Hadamard transform of f at any point $\mathbf{a} \in \mathbb{F}_2^n$, we only need to calculate the Walsh–Hadamard transform of the component functions f_1 and f_2 at the points $\mathbf{a}'$ and $\mathbf{a}''$, respectively. Thus, we only need two tables of sizes 1×2^{n_1} and 1×2^{n_2} corresponding to the Walsh–Hadamard values of f_1 and f_2, respectively. However, for the biased domain, we need more data to calculate the biased Walsh–Hadamard value at any point $\mathbf{a}$, as it can be seen from Theorem 1 and Corollary 1 (under the assumption that if $wt(\mathbf{x}) = wt(\mathbf{y})$, then $Pr[\mathbf{x}] = Pr[\mathbf{y}]$). To compute the biased Walsh–Hadamard transform values of f at any point, we need three probability tables P_1, P_2 and P_3 of sizes $1\times(n+1)$, $1\times(n_1+1)$ and $1\times(n_2+1)$ corresponding to the probabilities $\mathbf{x} \in \mathbb{F}_2^n$, $\mathbf{x}' \in \mathbb{F}_2^{n_1}$ and $\mathbf{x}'' \in \mathbb{F}_2^{n_2}$, respectively. We also need two tables T_{f_1} and T_{f_2} of sizes $2^{n_1} \times (n_1 + 1)$ and $2^{n_2} \times (n_2 + 1)$ (worst case) corresponding to the restricted biased Walsh–Hadamard values of f_1 and f_2, respectively. Certainly, the complexity increases when the size of the partition for n gets larger.

Further, we show a convolution theorem, which will depend on the Walsh–Hadamard transform over the inputs of fixed weight. We want to compute $\mathcal{W}_f^{B(k)}$ (defined as in Eq. (1) but summing for $\mathbf{x} \in E_{n,k}$), where $f = f_1 + f_2$ and $0 \leq k \leq n$. Here we use the fact that if $\mathbf{x}, \mathbf{y} \in \mathbb{F}_2^m$ with $wt(\mathbf{x}) = wt(\mathbf{y})$, then $p(\mathbf{x}) = p(\mathbf{y})$, and $p_{m,i} = Pr[\mathbf{x}]$, for all $\mathbf{x} \in E_{m,i}$, $0 \leq i \leq m$. From the definition of the Walsh–Hadamard transform of a Boolean function in both uniform and non-uniform cases, we infer the following relation

$$\mathcal{W}_f^{B(k)}(\mathbf{a}) = p\binom{n}{k}\mathcal{W}_f^{(k)}(\mathbf{a}), \quad \forall \mathbf{a} \in \mathbb{F}_2^n, \tag{2}$$

where $0 \le k \le n$ and $p = Pr[\mathbf{x} : wt(\mathbf{x}) = k]$.

Theorem 1 (Restricted Domain Convolution). *Let $n = n_1 + n_2$ and $f = f_1 + f_2$, where $f_i \in \mathcal{B}_{n_i}$, $i \in \{1, 2\}$. Then, for any $\mathbf{a} = \mathbf{a}' || \mathbf{a}'' \in \mathbb{F}_2^n$ and $0 \le k \le n$,*

$$\mathcal{W}_f^{B(k)}(\mathbf{a}) = p_{n,k} \sum_{i=0}^{k} \binom{n_1}{i}\binom{n_2}{k-i}\mathcal{W}_{f_1}^{(i)}(\mathbf{a}')\mathcal{W}_{f_2}^{(k-i)}(\mathbf{a}'')$$

$$= \sum_{i=0}^{k} \frac{p_{n,k}}{q_{n_1,i}q_{n_2,k-i}}\mathcal{W}_{f_1}^{B(i)}(\mathbf{a}')\mathcal{W}_{f_2}^{B(k-i)}(\mathbf{a}''),$$

where $q_{n_1,i} = \frac{\binom{n_2}{k-i}}{\binom{n}{k}}$, $q_{n_2,k-i} = \frac{\binom{n_1}{i}}{\binom{n}{k}}$.

Proof. For any $\mathbf{a} = \mathbf{a}' || \mathbf{a}'' \in \mathbb{F}_2^n$ and $0 \le k \le n$, we have

$$\mathcal{W}_f^{B(k)}(\mathbf{a}) = \sum_{\mathbf{x} \in E_{n,k}} Pr[\mathbf{x}](-1)^{f(\mathbf{x})+\mathbf{a}\cdot\mathbf{x}} = p_{n,k} \sum_{\mathbf{x} \in E_{n,k}} (-1)^{f(\mathbf{x})+\mathbf{a}\cdot\mathbf{x}}$$

$$= p_{n,k} \sum_{i=0}^{k} \sum_{\mathbf{x}' \in E_{n_1,i}^{n=n_1+n_2,k}} \sum_{\mathbf{x}'' \in E_{n_2,k-i}^{n=n_1+n_2,k}} (-1)^{f_1(\mathbf{x}')+\mathbf{a}'\cdot\mathbf{x}'}(-1)^{f_2(\mathbf{x}'')+\mathbf{a}''\cdot\mathbf{x}''}$$

$$= p_{n,k} \sum_{i=0}^{k} \sum_{\mathbf{x}' \in E_{n_1,i}^{n=n_1+n_2,k}} (-1)^{f_1(\mathbf{x}')+\mathbf{a}'\cdot\mathbf{x}'} \sum_{\mathbf{x}'' \in E_{n_2,k-i}^{n=n_1+n_2,k}} (-1)^{f_2(\mathbf{x}'')+\mathbf{a}''\cdot\mathbf{x}''}$$

$$= p_{n,k} \sum_{i=0}^{k} \binom{n_1}{i}\binom{n_2}{k-i}\mathcal{W}_{f_1}^{(i)}(\mathbf{a}')\mathcal{W}_{f_2}^{(k-i)}(\mathbf{a}'').$$

We can also rewrite the above in terms of the biased Walsh–Hadamard transform by using Eq. (2), obtaining

$$\mathcal{W}_f^{B(k)}(\mathbf{a}) = \sum_{i=0}^{k} \frac{p_{n,k}}{q_{n_1,i}q_{n_2,k-i}}\mathcal{W}_{f_1}^{B(i)}(\mathbf{a}')\mathcal{W}_{f_2}^{B(k-i)}(\mathbf{a}''),$$

Hence we get both equalities in terms of the Walsh–Hadamard transform in the uniform and biased domains. □

From Theorem 1, we obtain the next corollary.

Corollary 1. *Let* $n = n_1 + n_2$ *and* $f = f_1 + f_2$, *where* $f_i \in \mathcal{B}_{n_i}$, $i \in \{1,2\}$. *For any* $\mathbf{a} = \mathbf{a}'\|\mathbf{a}'' \in \mathbb{F}_2^n$,

$$\mathcal{W}_f^B(\mathbf{a}) = \sum_{k=0}^{n} \mathcal{W}_f^{B(k)}(\mathbf{a}) = \sum_{k=0}^{n} p_{n,k} \sum_{i=0}^{k} \binom{n_1}{i}\binom{n_2}{k-i} \mathcal{W}_{f_1}^{(i)}(\mathbf{a}') \mathcal{W}_{f_2}^{(k-i)}(\mathbf{a}'')$$

$$= \sum_{k=0}^{n} \sum_{i=0}^{k} \frac{p_{n,k}}{q_{n_1,i} q_{n_2,k-i}} \mathcal{W}_{f_1}^{B(i)}(\mathbf{a}') \mathcal{W}_{f_2}^{B(k-i)}(\mathbf{a}'').$$

We observe that it is very difficult to compute the biased Walsh–Hadamard transform for a direct sum of two Boolean functions, in arbitrary (large) number of variables. So we have to find an appropriate bound for the biased Walsh–Hadamard transform of $f \in \mathcal{B}_n$, where $f = f_1 + f_2 \in \mathcal{B}_n$, with $f_i \in \mathcal{B}_{n_i}$, $i = 1, 2$.

In the following theorem, we show that the biased Walsh–Hadamard transform may help us obtain a better bound.

Theorem 2. *For all* $0 \le k \le n$, *the following inequality holds*

$$\sum_{i=0}^{k} p_{n,k} \max_{\mathbf{a}_1 \in \mathbb{F}_2^{n_1}} \left| \sum_{\mathbf{x}_1 \in E_{n_1,i}} (-1)^{f_1(\mathbf{x}_1)+\mathbf{a}_1 \cdot \mathbf{x}_1} \right| \max_{\mathbf{a}_2 \in \mathbb{F}_2^{n_2}} \left| \sum_{\mathbf{x}_2 \in E_{n_2,k-i}} (-1)^{f_1(\mathbf{x}_2)+\mathbf{a}_2 \cdot \mathbf{x}_2} \right|$$

$$\ge \sum_{i=0}^{k} \max_{\mathbf{a}_1 \in \mathbb{F}_2^{n_1}} \left| \mathcal{W}_{f_1}^{B(i)}(\mathbf{a}_1) \right| \max_{\mathbf{a}_2 \in \mathbb{F}_2^{n_2}} \left| \mathcal{W}_{f_2}^{B(k-i)}(\mathbf{a}_2) \right|.$$

Proof. Using Vandermonde's identity, $\binom{n}{k} = \sum_{i=0}^{k} \binom{n_1}{i}\binom{n_2}{k-i}$, or directly using Stirling's formula, we infer that $\dfrac{\binom{n}{k}}{\binom{n_1}{i}\binom{n_2}{k-i}} \ge 1$, for all $0 \le i \le k$. Further,

$$\sum_{i=0}^{k} p_{n,k} \max_{\mathbf{a}_1 \in \mathbb{F}_2^{n_1}} \left| \sum_{\mathbf{x}_1 \in E_{n_1,i}} (-1)^{f_1(\mathbf{x}_1)+\mathbf{a}_1 \cdot \mathbf{x}_1} \right| \max_{\mathbf{a}_2 \in \mathbb{F}_2^{n_2}} \left| \sum_{\mathbf{x}_2 \in E_{n_2,k-i}} (-1)^{f_1(\mathbf{x}_2)+\mathbf{a}_2 \cdot \mathbf{x}_2} \right|$$

$$= \sum_{i=0}^{k} \frac{p_{n,k}}{q_{n_1,i} q_{n_2,k-i}} \max_{\mathbf{a}_1 \in \mathbb{F}_2^{n_1}} \left| \mathcal{W}_{f_1}^{B(i)}(\mathbf{a}_1) \right| \max_{\mathbf{a}_2 \in \mathbb{F}_2^{n_2}} \left| \mathcal{W}_{f_2}^{B(k-i)}(\mathbf{a}_2) \right|$$

$$= \sum_{i=0}^{k} \frac{1}{p_{n,k}\binom{n_1}{i}\binom{n_2}{k-i}} \max_{\mathbf{a}_1 \in \mathbb{F}_2^{n_1}} \left| \mathcal{W}_{f_1}^{B(i)}(\mathbf{a}_1) \right| \max_{\mathbf{a}_2 \in \mathbb{F}_2^{n_2}} \left| \mathcal{W}_{f_2}^{B(k-i)}(\mathbf{a}_2) \right|$$

$$= \sum_{i=0}^{k} \frac{\binom{n}{k}}{\binom{n_1}{i}\binom{n_2}{k-i}} \max_{\mathbf{a}_1 \in \mathbb{F}_2^{n_1}} \left| \mathcal{W}_{f_1}^{B(i)}(\mathbf{a}_1) \right| \max_{\mathbf{a}_2 \in \mathbb{F}_2^{n_2}} \left| \mathcal{W}_{f_2}^{B(k-i)}(\mathbf{a}_2) \right|$$

$$\ge \sum_{i=0}^{k} \max_{\mathbf{a}_1 \in \mathbb{F}_2^{n_1}} \left| \mathcal{W}_{f_1}^{B(i)}(\mathbf{a}_1) \right| \max_{\mathbf{a}_2 \in \mathbb{F}_2^{n_2}} \left| \mathcal{W}_{f_2}^{B(k-i)}(\mathbf{a}_2) \right|,$$

and the result is shown. $\square$

From the above inequality we can theoretically claim that the bound provided by Carlet et al. [2] is much higher than our bound

$$G = \sum_{i=0}^{k} \max_{\mathbf{a}_1 \in \mathbb{F}_2^{n_1}} |\mathcal{W}_{f_1}^{B(i)}(\mathbf{a}_1)| \max_{\mathbf{a}_2 \in \mathbb{F}_2^{n_2}} |\mathcal{W}_{f_2}^{B(k-i)}(\mathbf{a}_2)|.$$

It might be tempting to conjecture that G is smaller than $\max_{\mathbf{a} \in \mathbb{F}_2^n} |\mathcal{W}_f^{(k)}(\mathbf{a})|$. Experimentally, using some small functions, we found that in many cases $G \leq \max_{\mathbf{a} \in \mathbb{F}_2^n} |\mathcal{W}_f^{(k)}(\mathbf{a})|$, but we also observed that under some conditions this inequality will change its direction. Now we are interested to find such conditions for which $G \geq \max_{\mathbf{a} \in \mathbb{F}_2^n} |\mathcal{W}_f^{(k)}(\mathbf{a})|$. We start with the following lemma.

Lemma 1. *Let a_i be positive numbers and b_i be any integer numbers (positive or negative), where $i = 0, 1, \ldots, k$. If* $\left| \left| \sum_{i=0}^{k} a_i b_i \right| - \left| \sum_{i,j=0; i \neq j=0}^{k} a_i b_j \right| \right| \leq \left| \sum_{i=0}^{k} a_i b_i \right|,$ *and the sums $\sum_{i=0}^{k} a_i b_i$, $\sum_{i,j=0; i \neq j}^{k} a_i b_j$ have opposite signs, then*

$$\left| \sum_{i=0}^{k} a_i b_i \right| \geq \left(\sum_{i=0}^{k} a_i \right) \left| \sum_{j=0}^{k} b_j \right|.$$

Proof. We start with the following simple observation $\left(\sum_{i=0}^{k} a_i \right) \left(\sum_{j=0}^{k} b_j \right) = \sum_{i=0}^{k} a_i b_i + \sum_{i,j=0; i \neq j}^{k} a_i b_j$. By our assumption, $\left| \left| \sum_{i=0}^{k} a_i b_i \right| - \left| \sum_{i,j=0; i \neq j}^{k} a_i b_j \right| \right| \leq \left| \sum_{i=0}^{k} a_i b_i \right|$, and $\sum_{i=0}^{k} a_i b_i$, $\sum_{i,j=0; i \neq j}^{k} a_i b_j$ have opposite signs, so,

$$\left| \sum_{i=0}^{k} a_i b_i \right| \geq \left| \left| \sum_{i=0}^{k} a_i b_i \right| - \left| \sum_{i,j=0; i \neq j}^{k} a_i b_j \right| \right| = \left| \sum_{i=0}^{k} a_i b_i + \sum_{i,j=0; i \neq j}^{k} a_i b_j \right|$$

$$= \left| \left(\sum_{i=0}^{k} a_i \right) \left(\sum_{j=0}^{k} b_j \right) \right| = \left(\sum_{i=0}^{k} a_i \right) \left| \sum_{j=0}^{k} b_j \right|,$$

and the lemma is shown. $\square$

With the help of the above lemma we will prove that $G \geq \max_{\mathbf{a} \in \mathbb{F}_2^n} |\mathcal{W}_f^{(k)}(\mathbf{a})|$ holds under some conditions.

Theorem 3. *Let* $f = f_1 + f_2 \in \mathcal{B}_n$, $f_i \in \mathcal{B}_{n_i}$, $i = 1, 2$, $A_i := q_{n_1,i} q_{n_2,k-i}$ *and*
$B_i := \sum_{\mathbf{x}_1 \in E_{n_1,i}} (-1)^{f_1(\mathbf{x}_1) + \mathbf{a}_1 \cdot \mathbf{x}_1} \sum_{\mathbf{x}_2 \in E_{n_2,k-i}} (-1)^{f_2(\mathbf{x}_2) + \mathbf{a}_2 \cdot \mathbf{x}_2}$, *for all* $0 \leq i \leq k$
(here $q_{n_1,i} = \frac{\binom{n_2}{k-i}}{\binom{n}{k}}$, $q_{n_2,k-i} = \frac{\binom{n_1}{i}}{\binom{n}{k}}$). *Then*

$$\max_{\mathbf{a} \in \mathbb{F}_2^n} \left| \mathcal{W}_f^{(k)}(\mathbf{a}) \right| \leq \sum_{i=0}^{k} \max_{\mathbf{a}_1 \in \mathbb{F}_2^{n_1}} \left| \mathcal{W}_{f_1}^{B(i)}(\mathbf{a}_1) \right| \max_{\mathbf{a}_2 \in \mathbb{F}_2^{n_2}} \left| \mathcal{W}_{f_2}^{B(k-i)}(\mathbf{a}_2) \right|,$$

if $\left| \left| \sum_{i=0}^{k} A_i B_i \right| - \left| \sum_{i=0}^{k} A_i B_i - p_{n,k} \sum_{j=0}^{k} B_j \right| \right| \leq \left| \sum_{i=0}^{k} A_i B_i \right|$, *where* $p_{n,k} = \frac{1}{\binom{n}{k}}$, *and,*

the expressions $\sum_{i=0}^{k} A_i B_i$, $p_{n,k} \sum_{j=0}^{k} B_j - \sum_{i=0}^{k} A_i B_i$ *have opposite signs.*

Proof. We compute

$$\sum_{i=0}^{k} \max_{\mathbf{a}_1 \in \mathbb{F}_2^{n_1}} \left| \mathcal{W}_{f_1}^{B(i)}(\mathbf{a}_1) \right| \max_{\mathbf{a}_2 \in \mathbb{F}_2^{n_2}} \left| \mathcal{W}_{f_2}^{B(k-i)}(\mathbf{a}_2) \right| \geq \max_{\mathbf{a}_1 || \mathbf{a}_2} \sum_{i=0}^{k} \left| \mathcal{W}_{f_1}^{B(i)}(\mathbf{a}_1) \right| \left| \mathcal{W}_{f_2}^{B(k-i)}(\mathbf{a}_2) \right|$$

$$= \max_{\mathbf{a}_1, \mathbf{a}_2} \sum_{i=0}^{k} \left| q_{n_1,i} q_{n_2,k-i} \sum_{\mathbf{x}_1 \in E_{n_1,i}} (-1)^{f_1(\mathbf{x}_1) + \mathbf{a}_1 \cdot \mathbf{x}_1} \sum_{\mathbf{x}_2 \in E_{n_2,k-i}} (-1)^{f_2(\mathbf{x}_2) + \mathbf{a}_2 \cdot \mathbf{x}_2} \right|$$

$$\geq \max_{\mathbf{a}_1, \mathbf{a}_2} \left| \sum_{i=0}^{k} q_{n_1,i} q_{n_2,k-i} \sum_{\mathbf{x}_1 \in E_{n_1,i}} (-1)^{f_1(\mathbf{x}_1) + \mathbf{a}_1 \cdot \mathbf{x}_1} \sum_{\mathbf{x}_2 \in E_{n_2,k-i}} (-1)^{f_2(\mathbf{x}_2) + \mathbf{a}_2 \cdot \mathbf{x}_2} \right|$$

$$\geq \left(\sum_{i=0}^{k} q_{n_1,i} q_{n_2,k-i} \right) \max_{\mathbf{a}_1, \mathbf{a}_2} \left| \sum_{i=0}^{k} \sum_{\mathbf{x}_1 \in E_{n_1,i}} (-1)^{f_1(\mathbf{x}_1) + \mathbf{a}_1 \cdot \mathbf{x}_1} \sum_{\mathbf{x}_2 \in E_{n_2,k-i}} (-1)^{f_2(\mathbf{x}_2) + \mathbf{a}_2 \cdot \mathbf{x}_2} \right|.$$

By Lemma 1, the last inequality holds if $\sum_{i=0}^{k} A_i B_i$ and $\sum_{i,j=0; i \neq j}^{k} A_i B_j$ have

opposite signs, and $\left| \left| \sum_{i=0}^{k} A_i B_i \right| - \left| \sum_{i,j=0; i \neq j}^{k} A_i B_j \right| \right| \leq \left| \sum_{i=0}^{k} A_i B_i \right|$. We argue that

this last condition is equivalent to $\left| \left| \sum_{i=0}^{k} A_i B_i \right| - \left| \sum_{i=0}^{k} A_i B_i - p_{n,k} \sum_{j=0}^{k} B_j \right| \right| \leq$

$\left| \sum_{i=0}^{k} A_i B_i \right|$, where $p_{n,k} = \frac{1}{\binom{n}{k}}$. That follows from the observation that for any

j, $0 \leq j \leq k$, we have $\displaystyle\sum_{i=0;i\neq j}^{k} A_i B_j = \sum_{i=0}^{k} A_i B_j - A_j B_j = p_{n,k} B_j - A_j B_j$.

Further, using Vandermonde's identity, $\displaystyle\sum_{i=0}^{k} q_{n_1,i} q_{n_2,k-i} = \sum_{i=0}^{k} \frac{\binom{n_2}{k-i}\binom{n_1}{i}}{\binom{n}{k}\binom{n}{k}} = \frac{1}{\binom{n}{k}}$,

therefore,

$$\sum_{i=0}^{k} \max_{\mathbf{a}_1 \in \mathbb{F}_2^{n_1}} \left| \mathcal{W}_{f_1}^{B(i)}(\mathbf{a}_1) \right| \cdot \max_{\mathbf{a}_2 \in \mathbb{F}_2^{n_2}} \left| \mathcal{W}_{f_2}^{B(k-i)}(\mathbf{a}_2) \right|$$

$$\geq \max_{\mathbf{a}_1,\mathbf{a}_2} \frac{1}{\binom{n}{k}} \left| \sum_{i=0}^{k} \sum_{\mathbf{x}_1 \in E_{n_1,i}} (-1)^{f_1(\mathbf{x}_1)+\mathbf{a}_1\cdot\mathbf{x}_1} \sum_{\mathbf{x}_2 \in E_{n_2,k-i}} (-1)^{f_2(\mathbf{x}_2)+\mathbf{a}_2\cdot\mathbf{x}_2} \right|$$

$$= \max_{\mathbf{a}} \left| \mathcal{W}_f^{(k)}(\mathbf{a}) \right|,$$

and the claim is shown. $\qquad\qquad\qquad\qquad\qquad\qquad\qquad\qquad\qquad\qquad\qquad$ $\square$

In our next result we show that under some conditions we could achieve the lower bound of $\max_{\mathbf{a}\in\mathbb{F}_2^n} |\mathcal{W}_f^{(k)}(\mathbf{a})|$ in terms of the biased Walsh–Hadamard transform.

Theorem 4. *Let* $0 \leq i \leq k$, $\mathbf{c}_i \in \mathbb{F}_2^{n_1}$, $\mathbf{d}_i \in \mathbb{F}_2^{n_2}$, $q_{n_1,i} = \dfrac{\binom{n_2}{k-i}}{\binom{n}{k}}$, $q_{n_2,k-i} = \dfrac{\binom{n_1}{i}}{\binom{n}{k}}$, *and*

$$\max_{\mathbf{a}_1 \in \mathbb{F}_2^{n_1}} \left| \mathcal{W}_{f_1}^{B(i)}(\mathbf{a}_1) \right| = q_{n_1,i} \left| \sum_{\mathbf{x}_1 \in E_{n_1,i}} (-1)^{f_1(\mathbf{x}_1)+\mathbf{c}_i\cdot\mathbf{x}_1} \right|,$$

$$\max_{\mathbf{a}_2 \in \mathbb{F}_2^{n_2}} \left| \mathcal{W}_{f_2}^{B(k-i)}(\mathbf{a}_2) \right| = q_{n_2,k-i} \left| \sum_{\mathbf{x}_2 \in E_{n_2,k-i}} (-1)^{f_2(\mathbf{x}_2)+\mathbf{d}_i\cdot\mathbf{x}_2} \right|.$$

If $\displaystyle\sum_{\mathbf{x}_1 \in E_{n_1,i}} (-1)^{f_1(\mathbf{x}_1)+\mathbf{c}_i\cdot\mathbf{x}_1} \sum_{\mathbf{x}_2 \in E_{n_2,k-i}} (-1)^{f_2(\mathbf{x}_2)+\mathbf{d}_i\cdot\mathbf{x}_2}$ *has constant sign, for all* $0 \leq i \leq k$, *then,*

$$\sum_{i=0}^{k} \max_{\mathbf{a}_1 \in \mathbb{F}_2^{n_1}} \left| \mathcal{W}_{f_1}^{B(i)}(\mathbf{a}_1) \right| \max_{\mathbf{a}_2 \in \mathbb{F}_2^{n_2}} \left| \mathcal{W}_{f_2}^{B(k-i)}(\mathbf{a}_2) \right| \leq \max_{\mathbf{a}\in\mathbb{F}_2^n} \left| \mathcal{W}_f^{(k)}(\mathbf{a}) \right|.$$

Proof. We compute

$$\sum_{i=0}^{k} \max_{\mathbf{a}_1 \in \mathbb{F}_2^{n_1}} \left| \mathcal{W}_{f_1}^{B(i)}(\mathbf{a}_1) \right| \max_{\mathbf{a}_2 \in \mathbb{F}_2^{n_2}} \left| \mathcal{W}_{f_2}^{B(k-i)}(\mathbf{a}_2) \right|$$

$$= \sum_{i=0}^{k} \left| \mathcal{W}_{f_1}^{B(i)}(\mathbf{c}_i) \right| \left| \mathcal{W}_{f_2}^{B(k-i)}(\mathbf{d}_i) \right| = \sum_{i=0}^{k} \left| \mathcal{W}_{f_1}^{B(i)}(\mathbf{c}_i) \mathcal{W}_{f_2}^{B(k-i)}(\mathbf{d}_i) \right|$$

$$= \sum_{i=0}^{k} \left| q_{n_1,i} q_{n_2,k-i} \sum_{\mathbf{x}_1 \in E_{n_1,i}} (-1)^{f_1(\mathbf{x}_1) + \mathbf{c}_i \cdot \mathbf{x}_1} \sum_{\mathbf{x}_2 \in E_{n_2,k-i}} (-1)^{f_2(\mathbf{x}_2) + \mathbf{d}_i \cdot \mathbf{x}_2} \right|$$

$$= \sum_{i=0}^{k} q_{n_1,i} q_{n_2,k-i} \left| \sum_{\mathbf{x}_1 \in E_{n_1,i}} (-1)^{f_1(\mathbf{x}_1) + \mathbf{c}_i \cdot \mathbf{x}_1} \sum_{\mathbf{x}_2 \in E_{n_2,k-i}} (-1)^{f_2(\mathbf{x}_2) + \mathbf{d}_i \cdot \mathbf{x}_2} \right| \quad (3)$$

$$\leq \sum_{i=0}^{k} q_{n_1,i} q_{n_2,k-i} \sum_{i=0}^{k} \left| \sum_{\mathbf{x}_1 \in E_{n_1,i}} (-1)^{f_1(\mathbf{x}_1) + \mathbf{c}_i \cdot \mathbf{x}_1} \sum_{\mathbf{x}_2 \in E_{n_2,k-i}} (-1)^{f_2(\mathbf{x}_2) + \mathbf{d}_i \cdot \mathbf{x}_2} \right|$$

$$= \frac{1}{\binom{n}{k}} \sum_{i=0}^{k} \left| \sum_{\mathbf{x}_1 \in E_{n_1,i}} (-1)^{f_1(\mathbf{x}_1) + \mathbf{c}_i \cdot \mathbf{x}_1} \sum_{\mathbf{x}_2 \in E_{n_2,k-i}} (-1)^{f_2(\mathbf{x}_2) + \mathbf{d}_i \cdot \mathbf{x}_2} \right| \quad (4)$$

$$= \frac{1}{\binom{n}{k}} \left| \sum_{i=0}^{k} \sum_{\mathbf{x}_1 \in E_{n_1,i}} (-1)^{f_1(\mathbf{x}_1) + \mathbf{c}_i \cdot \mathbf{x}_1} \sum_{\mathbf{x}_2 \in E_{n_2,k-i}} (-1)^{f_2(\mathbf{x}_2) + \mathbf{d}_i \cdot \mathbf{x}_2} \right|,$$

$$\leq \frac{1}{\binom{n}{k}} \max_{\mathbf{a} = \mathbf{b}_1 \| \mathbf{b}_2} \left| \sum_{i=0}^{k} \sum_{\mathbf{x}_1 \in E_{n_1,i}} (-1)^{f_1(\mathbf{x}_1) + \mathbf{b}_1 \cdot \mathbf{x}_1} \sum_{\mathbf{x}_2 \in E_{n_2,k-i}} (-1)^{f_2(\mathbf{x}_2) + \mathbf{b}_2 \cdot \mathbf{x}_2} \right|$$

$$\leq \max_{\mathbf{a} \in \mathbb{F}_2^n} \left| \mathcal{W}_f^{(k)}(\mathbf{a}) \right|,$$

and the theorem is shown.   □

4   More Accurate Calculations of Biases by Our Technique and Comparisons with Previous Work

In this section we compare our bound with the one proposed by Carlet et al. [2]. We first consider a small Boolean function, then we further do the same for the nonlinear filter function used in the FLIP stream cipher.

4.1   Comparison for a Small Boolean Function

Here, we consider a small Boolean function (of FLIP type), involving 12 variables to compare our result with the one obtained from Carlet et al.'s technique [2]. The function $f = f_1 + f_2 + f_3$ is a direct sum of three Boolean functions f_1, f_2 and f_3, where $f_1(x_0, x_1) = x_0 + x_1$ is linear, $f_2(x_0, \ldots, x_3) = x_0 x_1 + x_2 x_3$ is quadratic bent, and $f_3(x_0, \ldots, x_5) = x_0 + x_1 x_2 + x_3 x_4 x_5$ is triangular, respectively. Note that out of 12 variables of f, f_1 depends on the first 2 variables, f_2 depends on the next 4 variables and f_3 depends on the last 6 variables. As before, we let $E_{n,k} = \{\mathbf{x} \mid wt(\mathbf{x}) = k\}$. We here assume that f takes inputs from the set $E_{12,6}$.

To obtain a bound for the bias of f, we consider the two types of Walsh–Hadamard transforms. First, the classical Walsh–Hadamard transform $\mathcal{W}_f$, which is also used in the paper of Carlet et al. [2], and the second is our newly defined biased Walsh–Hadamard transform $\mathcal{W}_f^B$. We compute both types of Walsh–Hadamard transform values for f_1, f_2, f_3 for all possible weights of $\mathbf{x}_1$, $\mathbf{x}_2$, $\mathbf{x}_3$. Here the functions depend on the variables $\mathbf{x}_1$, $\mathbf{x}_2$, $\mathbf{x}_3$ and $\mathbf{x} = \mathbf{x}_1 \| \mathbf{x}_2 \| \mathbf{x}_3$. For each weight of $\mathbf{x}_1$, $\mathbf{x}_2$ and $\mathbf{x}_3$ we find the maximum absolute Walsh–Hadamard transform values of f_1, f_2, f_3. From these maximum absolute Walsh–Hadamard transform values we can compute the maximum absolute value of Walsh–Hadamard transform of $f = f_1 + f_2 + f_3$, when $wt(\mathbf{x}) = 6$ is fixed. We multiply those maximum absolute Walsh–Hadamard transform values (corresponding to weights) when $wt(\mathbf{x}_1) + wt(\mathbf{x}_2) + wt(\mathbf{x}_3) = 6$ and add them. Finally, the bias bound of the function f in the classical Walsh–Hadamard transform set up can be found by dividing the maximum absolute value by $\binom{12}{6}$. We provide the bias comparison for the classical and biased Walsh–Hadamard transforms with the original bias in Table 2.

Table 2. Correlation bound comparison

Original bias	≈ 0.264069
Carlet et al. [2]	≤ 0.772727
This paper	≥ 0.20857

The comparison of Table 2 clearly shows that our correlation bound is much tighter than Carlet et al. [2]. For better understanding, we refer to Appendix A.

4.2   Comparison for the Actual Nonlinear Filter Function of FLIP

Here we compare the bound for the bias of the nonlinear filter function of the FLIP stream cipher, by extending the ideas explained in Sect. 4.1. The nonlinear filter function of the $\text{FLIP}_{530}(42, 128, 360)$ stream cipher is a direct sum of a linear function of 42 variables, a quadratic bent function of 128 variables and a direct sum of 8 triangular functions each of 45 variables. Since in the triangular part there are 8 terms of degree 1, the final linear function is of 50 variables. Also, since there are 8 terms of degree 2 in the triangular part, the complete quadratic function will be of 144 variables.

As in the toy example, we compute the bias by using the classical Walsh–Hadamard transform and our biased Walsh–Hadamard transform. To compute the bias of the complete function (in classical and biased domain) we break the function in the following form: 5 linear Boolean function involving 10 variables, 18 quadratic function involving 8 variables and 8 degree $3, 4, \ldots, 9$ terms.

By following the same process, as described in Sect. 4.1 we compute the bias bound value for the normal and biased domain for the Carlet et al.'s study [2] and our study. For Carlet et al.'s case we get the bias bound $G_c = \frac{1}{2^{13.59}}$ and for our biased case the bias bound is $G_o = \frac{1}{2^{18.49}}$. Now, this shows that the computed bias value for Carlet et al.'s study [2], that is, G_c will be an upper bound of the original bias (which can be found in Lemma 3 of [2]).

Table 3. Correlation comparison

Carlet et al. [2]	$\leq \frac{1}{2^{13.59}}$
This paper	$\geq \frac{1}{2^{18.49}}$

Next, we show that G_o is a lower bound of the original bias. To show this we use our Theorem 4. In our computation, the product of the probabilities $(q_{n_1,i}, q_{n_2,k-i}$ of Theorem 4) will be the product of the probabilities corresponding to the 5 linear Boolean function involving 10 variables, probabilities corresponding to 18 quadratic function involving 8 variables, and probabilities corresponding to the 8 degree $3, 4, \ldots, 9$ terms. We have observed that the maximum of this for all product terms is much smaller than $\frac{1}{\binom{530}{265}}$. So we replace all these products of probabilities by $\frac{1}{\binom{530}{265}}$ to get the inequality between Eqs. (3) and (4) from the proof of Theorem 4. Computationally, we found that all these functions $f_1 = x_0 + x_1 + x_2 + x_3 + x_4 + x_5 + x_6 + x_7 + x_8 + x_9$, $f_2 = x_0 x_1 + x_2 x_3 + x_4 x_5 + x_6 x_7$, $f_3 = x_0 x_1 x_2$, $f_4 = x_0 x_1 x_2 x_3$, $f_5 = x_0 x_1 x_2 x_3 x_4$, $f_6 = x_0 x_1 x_2 x_3 x_4 x_5$, $f_7 = x_0 x_1 x_2 x_3 x_4 x_5 x_6$, $f_8 = x_0 x_1 x_2 x_3 x_4 x_5 x_6 x_7$, $f_9 = x_0 x_1 x_2 x_3 x_4 x_5 x_6 x_7 x_8$ satisfy the required condition of Theorem 4. More specifically, in the case of all these functions f_j, there exists at least one point **b** for which $\sum_{\mathbf{x} \in E_{n,i}} (-1)^{f_j(\mathbf{x}) + \mathbf{b} \cdot \mathbf{x}}$ attains $\max_{\mathbf{a}} \left| \sum_{\mathbf{x} \in E_{n,i}} (-1)^{f_j(\mathbf{x}) + \mathbf{a} \cdot \mathbf{x}} \right|$ for all weights i (See Appendix B). Thus, by Theorem 4 we infer that $G_o = \frac{1}{2^{18.49}}$ is the lower bound of the original bias of the nonlinear filter function of FLIP. Table 3 summarizes the comparison of the biases.

4.3   Computation Process

To compute the bias by using the normal and biased Walsh–Hadamard transforms, we split each component function into very small functions and the choices we made are simply dependent upon the power of the machine we ran the code on. We first divide the 50 variables linear function into 5 functions each involving 10 variables. We divide the second function (which is quadratic) in 144 variables into 18 Boolean functions involving 8 variables and similarly, for the other degree terms. We compute the Walsh–Hadamard transform for each component function of the linear, bent, and the combination of the other degree terms, separately. We find the Walsh–Hadamard transform values (corresponding to each weight of the input) for all functions and save them in separate files. Now we need to combine all these Walsh–Hadamard transform values to calculate a bound of the correlation value. We do the following to compute that bound.

1. For the linear function involving 50 variables, we do the following: from the Walsh–Hadamard transform value corresponding to each weight of the 10 variable linear function we compute the maximum absolute Walsh–Hadamard transform value corresponding to each weight.
2. Now we compute the maximum Walsh–Hadamard transform values corresponding to each weight of the quadratic function involving 144 variables. We first compute the maximum absolute Walsh–Hadamard transform values corresponding to each weight of the quadratic function involving 8 variables. Then we evaluate the maximum absolute Walsh–Hadamard coefficients of the 16 variable function by using the data for the 8 variable function. By doing this, we go up to a quadratic function involving 128 variables. After that we merge 128 variables with a 16 variable function to obtain a similar type of data for 144 variables.
3. Further, we need to do a similar analysis for the combination of degree 3 to degree 9 of the 8 triangular functions, each involving 42 variables. We first compute the maximum absolute Walsh–Hadamard transform values corresponding to each weight of the 42 variable function by considering each monomial as a separate Boolean function. From the maximum absolute Walsh–Hadamard transform values corresponding to each weight of the 42 variable function, we compute the maximum absolute Walsh–Hadamard transform values corresponding to each weight of the 84 variable function. By following the same technique we can compute the maximum absolute Walsh–Hadamard transform values corresponding to each weight of the combination of the degree 3 to degree 9 of 8 triangular functions involving 336 variables.
4. Finally we combine all these absolute Walsh–Hadamard transform values to compute the bias of the complete function F involving 530 variables which takes input of weight 265, only.

Finally, let us summarize the theoretical formulae of our work as well as those provided in [2,7]. Carlet et al. [2] showed the lower bound of the bias in restricted domain is $\max_{\mathbf{a} \in \mathbb{F}_2^n} |\mathcal{W}_f^{(k)}(\mathbf{a})| \geq \frac{1}{\binom{n}{k}} \sqrt{\binom{n}{k} + \lambda}$ (for a parameter λ defined in [2, Prop. 8, p. 207]). Later, Mesnager et al. [7] improved the lower bound of the bias to $\max_{\mathbf{a} \in \mathbb{F}_2^n} |\mathcal{W}_f^{(k)}(\mathbf{a})| \geq \frac{1}{\binom{n}{k}} \sqrt{\binom{n}{k} + \lambda + \max\left(\theta, \frac{1}{\binom{n}{k}} \gamma - \lambda\right)}$ (where λ, γ, θ are defined in [7, Thm. 16]). These two bounds are not related to the direct sum of functions in restricted domain. In this paper we have shown that the bias of direct sum of two functions $f_1 + f_2$ in a restricted domain can be expressed in terms of biased

Walsh–Hadamard transform of f_1, f_2. The lower bound of the bias under some constraints is $\max_{\mathbf{a}\in\mathbb{F}_2^n}\left|\mathcal{W}^{(k)}_{f_1+f_2}(\mathbf{a})\right| \geq \sum_{i=0}^{k}\max_{\mathbf{a}_1\in\mathbb{F}_2^{n_1}}\left|\mathcal{W}^{B(i)}_{f_1}(\mathbf{a}_1)\right|\max_{\mathbf{a}_2\in\mathbb{F}_2^{n_2}}\left|\mathcal{W}^{B(k-i)}_{f_2}(\mathbf{a}_2)\right|$.

Carlet et al. [2] found an upper bound of the bias of a direct sum of two functions in a restricted domain. The expression of the bound is $\max_{\mathbf{a}\in\mathbb{F}_2^n}\left|\mathcal{W}^{(k)}_{f_1+f_2}(\mathbf{a})\right| \leq$

$$\frac{1}{\binom{n}{k}}\sum_{i=0}^{k}\left(\max_{\mathbf{a}\in\mathbb{F}_2^{n_1}}\left|\sum_{\mathbf{x}\in E_{n_1,i}}(-1)^{f_1(\mathbf{x})+\mathbf{a}\cdot\mathbf{x}}\right|\max_{\mathbf{b}\in\mathbb{F}_2^{n_2}}\left|\sum_{\mathbf{y}\in E_{n_2,k-i}}(-1)^{f_2(\mathbf{y})+\mathbf{b}\cdot\mathbf{y}}\right|\right).$$ We note that

the paper [7] of Mesnager et al. does not contain any result related to the direct sum of Boolean functions in a restricted domain. Here, we found (under some technical conditions) an upper bound of the bias of a direct sum of two functions $f_1 + f_2$ in a restricted domain in terms of the biased Walsh–Hadamard transform of f_1 and f_2, namely, $\max_{\mathbf{a}\in\mathbb{F}_2^n}|\mathcal{W}^{(k)}_{f_1+f_2}(\mathbf{a})| \leq$

$$\sum_{i=0}^{k}\max_{\mathbf{a}_1\in\mathbb{F}_2^{n_1}}\left|\mathcal{W}^{B(i)}_{f_1}(\mathbf{a}_1)\right|\max_{\mathbf{a}_2\in\mathbb{F}_2^{n_2}}\left|\mathcal{W}^{B(k-i)}_{f_2}(\mathbf{a}_2)\right|.$$

5 Conclusion

In this paper we have proposed a *non-uniform (biased)* way to investigate the cryptographic properties of a Boolean function, when the inputs to the Boolean function do not follow a uniform distribution. To study this we first define the notion of correlation (biased Walsh–Hadamard transform) for a non-uniform domain, along with the necessary tools. Further, we show how this correlation is related with our newly defined biased Walsh–Hadamard transform, which is used to study several cryptographic properties of a Boolean function in a non-uniform domain. As the computation using our theoretical convolution theorem for the biased Walsh–Hadamard transform cannot be done in an efficient way for Boolean functions with a large number of variables, we use several inequalities for these coefficients. Consequently, we find a lower bound for the bias of the nonlinear filter function of the FLIP stream cipher by exploiting the biased Walsh–Hadamard transform, and compare that with previous work. Certainly, the properties when the domain of the Boolean function does not follow a uniform distribution is worthy of investigation. In this context, our results provide a more accurate calculation of biases related to Boolean functions. This is important in the security evaluation of the stream ciphers, in particular, the ones used in efficient homomorphic encryption schemes.

Acknowledgments. We would like to thank the anonymous reviewers of Indocrypt 2018 for their valuable suggestions and comments, which considerably improved the quality of our paper. The work of T.M. and P.S. started during an enjoyable visit to ISI-Kolkata in March 2018. They would like to thank the hosts and the institution for the excellent working conditions. T.M. also acknowledges support from the Omar Nelson Bradley foundation officer research fellowship in mathematics.

A   Biases for 12-variable Function

In our example, the function $F = x_0 + x_1 + x_2x_3 + x_4x_5 + x_6 + x_7x_8 + x_9x_{10}x_{11}$ takes input from $E_{12,6}$. The bias of the function F in this restricted domain is ≈ 0.264069. It is worth noticing that in the uniform domain (i.e., the function takes input from $\mathbb{F}_2^{12}$ instead of $E_{12,6}$) the bias between the original function F and the linear function $l_1 = l_{\mathbf{a}_1,0} = x_0 + x_1 + x_6$ is high, as the monomial of the form x_ix_j or $x_ix_jx_k$ is always 0 unless all variables involved in the monomials are 1. It can be observed that, the bias between F and l_1 in the domain $\mathbb{F}_2^{12}$ and $E_{12,6}$ are $|\mathcal{W}_F(\mathbf{a}_1)| = 0.09375$ and $|\mathcal{W}_F^{(6)}(\mathbf{a}_1)| = 0.099567$, respectively.

The situation is different when the domain of the function F is $E_{12,6}$ (restricted domain). In this domain, the bias between the original function F and a linear function is highest for $l_2 = l_{\mathbf{a}_2,0} = x_0 + x_1 + x_2 + x_3 + x_4 + x_5 + x_6$ instead of $l_1 = x_0 + x_1 + x_6$. The bias between F and l_2 in restricted domain $E_{12,6}$ is $|\mathcal{W}_F^{(6)}(\mathbf{a}_2)| = 0.264069$, but the bias between F and l_1 in the restricted domain $E_{12,6}$ is $|\mathcal{W}_F^{(6)}(\mathbf{a}_1)| = 0.099567$. All the linear function for which the bias is high in the restricted domain $E_{12,6}$ are provided below:

1. $l_{\mathbf{a}_2,0} = l_2 = x_0 + x_1 + x_2 + x_3 + x_4 + x_5 + x_6$: $|\mathcal{W}_F^{(6)}(\mathbf{a}_2)| = 0.264069$, $|\mathcal{W}_F(\mathbf{a}_2)| = 0.09375$.
2. $l_{\mathbf{a}_3,0} = l_3 = x_0 + x_1 + x_2 + x_3 + x_6 + x_7 + x_8$: $|\mathcal{W}_F^{(6)}(\mathbf{a}_3)| = 0.264069$, $|\mathcal{W}_F(\mathbf{a}_3)| = 0.09375$.
3. $l_{\mathbf{a}_4,0} = l_4 = x_0 + x_1 + x_4 + x_5 + x_6 + x_7 + x_8$: $|\mathcal{W}_F^{(6)}(\mathbf{a}_4)| = 0.264069$, $|\mathcal{W}_F(\mathbf{a}_4)| = 0.09375$.
4. $l_{\mathbf{a}_5,0} = l_5 = x_2 + x_3 + x_9 + x_{10} + x_{11}$: $|\mathcal{W}_F^{(6)}(\mathbf{a}_5)| = 0.264069$, $|\mathcal{W}_F(\mathbf{a}_5)| = 0$.
5. $l_{\mathbf{a}_6,0} = l_6 = x_4 + x_5 + x_9 + x_{10} + x_{11}$: $|\mathcal{W}_F^{(6)}(\mathbf{a}_6)| = 0.264069$, $|\mathcal{W}_F(\mathbf{a}_6)| = 0$.
6. $l_{\mathbf{a}_7,0} = l_7 = x_7 + x_8 + x_9 + x_{10} + x_{11}$: $|\mathcal{W}_F^{(6)}(\mathbf{a}_7)| = 0.264069$, $|\mathcal{W}_F(\mathbf{a}_7)| = 0$.

B   Existence of a Point b Referred to in Sect. 4.2

This appendix describes the existence of a point $\mathbf{b}$ for each function f_j at which

$$\sum_{\mathbf{x} \in E_{n,i}} (-1)^{f_j(\mathbf{x}) + \mathbf{b} \cdot \mathbf{x}} \text{ attains } \max_{\mathbf{a}} \left| \sum_{\mathbf{x} \in E_{n,i}} (-1)^{f_j(\mathbf{x}) + \mathbf{a} \cdot \mathbf{x}} \right| \text{ for all weight } i.$$

1. First, let $f_1 = x_0 + x_1 + x_2 + x_3 + x_4 + x_5 + x_6 + x_7 + x_8 + x_9$. The existence of a point $\mathbf{b}$ corresponding to each weight starting from weight zero to weight ten is given below (points are provided in integer form): $0, 1023, 0, 1023, 0, 1023, 0, 1023, 0, 1023, 0$.
2. For $f_2 = x_0x_1 + x_2x_3 + x_4x_5 + x_6x_7$, the existence of a point $\mathbf{b}$ corresponding to each weight starting from weight zero to weight eight is mentioned below (points are provided in integer form): $0, 0, 0, 63, 15, 3, 0, 255, 0$.
3. For $f_3 = x_0x_1x_2$, the existence of a point $\mathbf{b}$ corresponding to each weight starting from weight zero to weight three is provided below (points are provided in integer form): $0, 0, 0, 1$.

4. For $f_4 = x_0 x_1 x_2 x_3$, the existence of a point $\mathbf{b}$ corresponding to each weight starting from weight zero to weight four is mentioned below (points are provided in integer form): $0, 0, 0, 0, 1$.

5. For $f_5 = x_0 x_1 x_2 x_3 x_4$, the existence of a point $\mathbf{b}$ corresponding to each weight starting from weight zero to weight five is given below (points are provided in integer form): $0, 0, 0, 0, 0, 1$.

6. For $f_6 = x_0 x_1 x_2 x_3 x_4 x_5$, the existence of a point $\mathbf{b}$ corresponding to each weight starting from weight zero to weight six is provided below (points are provided in integer form): $0, 0, 0, 0, 0, 0, 1$.

7. For $f_7 = x_0 x_1 x_2 x_3 x_4 x_5 x_6$, the existence of a point $\mathbf{b}$ corresponding to each weight starting from weight zero to weight seven is mentioned below (points are provided in integer form): $0, 0, 0, 0, 0, 0, 0, 1$.

8. For $f_8 = x_0 x_1 x_2 x_3 x_4 x_5 x_6 x_7$, the existence of a point $\mathbf{b}$ corresponding to each weight starting from weight zero to weight eight is given below (points are provided in integer form): $0, 0, 0, 0, 0, 0, 0, 0, 1$.

9. For $f_9 = x_0 x_1 x_2 x_3 x_4 x_5 x_6 x_7 x_8$, the existence of a point $\mathbf{b}$ corresponding to each weight starting from weight zero to weight nine is mentioned below (points are provided in integer form): $0, 0, 0, 0, 0, 0, 0, 0, 0, 1$.

References

1. Canteaut, A., et al.: Stream ciphers: a practical solution for efficient homomorphic-ciphertext compression. In: Peyrin, T. (ed.) FSE 2016. LNCS, vol. 9783, pp. 313–333. Springer, Heidelberg (2016). https://doi.org/10.1007/978-3-662-52993-5_16

2. Carlet, C., Méaux, P., Rotella, Y.: Boolean functions with restricted input and their robustness, application to the FLIP cipher. IACR Trans. Symmetric Cryptology **3**, 192–227 (2017). (presented at FSE 2018)

3. Duval, S., Lallemand, V., Rotella, Y.: Cryptanalysis of the FLIP family of stream ciphers. In: Robshaw, M., Katz, J. (eds.) CRYPTO 2016. LNCS, vol. 9814, pp. 457–475. Springer, Heidelberg (2016). https://doi.org/10.1007/978-3-662-53018-4_17

4. Gangopadhyay, S., Gangopadhyay, A.K., Pollatos, S., Stănică, P.: Cryptographic Boolean functions with biased inputs. Crypt. Commun. **9**(2), 301–314 (2017)

5. Méaux, P.: Symmetric Encryption Scheme adapted to Fully Homomorphic Encryption Scheme. In: Journées Codage et Cryptographie - JC2 2015–12éme édition des Journées Codage et Cryptographie du GT C2, 5 au 9 octobre 2015, La Londeles-Maures, France (2015). http://imath.univ-tln.fr/C2/

6. Méaux, P., Journault, A., Standaert, F.-X., Carlet, C.: Towards stream ciphers for efficient FHE with low-noise ciphertexts. In: Fischlin, M., Coron, J.-S. (eds.) EUROCRYPT 2016. LNCS, vol. 9665, pp. 311–343. Springer, Heidelberg (2016). https://doi.org/10.1007/978-3-662-49890-3_13

7. Mesnager, S., Zhou, Z., Ding, C.: On the nonlinearity of Boolean functions with restricted input. Crypt. Commun. (2018). https://doi.org/10.1007/s12095-018-0293-6

Theory

Non-malleable Codes Against Lookahead Tampering

Divya Gupta[1]([⊠]), Hemanta K. Maji[2]([⊠]), and Mingyuan Wang[2]([⊠])

[1] Microsoft Research, Bangalore, India
divya.gupta@microsoft.com
[2] Department of Computer Science, Purdue University, West Lafayette, USA
{hmaji,wang1929}@purdue.edu

Abstract. There are natural cryptographic applications where an adversary only gets to tamper a high-speed data stream on the fly based on her view so far, namely, the lookahead tampering model. Since the adversary can easily substitute transmitted messages with her messages, it is farfetched to insist on strong guarantees like error-correction or, even, manipulation detection. Dziembowski, Pietrzak, and Wichs (ICS–2010) introduced the notion of non-malleable codes that provide a useful message integrity for such scenarios. Intuitively, a non-malleable code ensures that the tampered codeword encodes the original message or a message that is entirely independent of the original message.

Our work studies the following tampering model. We encode a message into $k \geqslant 1$ secret shares, and we transmit each share as a separate stream of data. Adversaries can perform lookahead tampering on each share, albeit, independently. We call this *k-lookahead model*.

First, we show a hardness result for the k-lookahead model. To transmit an ℓ-bit message, the cumulative length of the secret shares must be at least $\frac{k}{k-1}\ell$. This result immediately rules out the possibility of a solution with $k = 1$. Next, we construct a solution for 2-lookahead model such that the total length of the shares is 3ℓ, which is only 1.5x of the optimal encoding as indicated by our hardness result.

Prior work considers stronger model of split-state encoding that creates $k \geqslant 2$ secret shares, but protects against adversaries who perform arbitrary (but independent) tampering on each secret share. The size of the secret shares of the most efficient 2-split-state encoding is $\ell \log \ell / \log \log \ell$ (Li, ECCC–2018). Even though k-lookahead is a weaker tampering class, our hardness result matches that of k-split-state tampering by Cheraghchi and Guruswami (TCC–2014). However, our explicit constructions above achieve much higher efficiency in encoding.

Keywords: Non-malleable codes · Lookahead tampering · Split-state Constant-rate

H. K. Maji—The research effort is supported in part by an NSF CRII Award CNS–1566499, an NSF SMALL Award CNS–1618822, and an REU CNS–1724673.
H. K. Maji and M. Wang—The research effort is supported in part by a Purdue Research Foundation grant.

D. Chakraborty and T. Iwata (Eds.): INDOCRYPT 2018, LNCS 11356, pp. 307–328, 2018.
https://doi.org/10.1007/978-3-030-05378-9_17

1   Introduction

Dziembowski, Pietrzak, and Wichs [15] introduced the powerful notion of *non-malleable codes* for message integrity for scenarios where error-correction or, even, error-detection is impossible. Some of the main applications of non-malleable codes are tamper resilient storage and computation [15], and *non-malleable message transmission* between two parties [17]. In this work, we focus on the application of non-malleable message transmission. Intuitively, non-malleable coding scheme guarantees that the decoding of the *tampered* codeword is either the original message or an unrelated message and the probability of either of these events happening is independent of the original message. To build such a scheme against some of the simpler tampering functions such as adding an arbitrary low Hamming weight error, the sender can encode the message using appropriate error-correcting codes, and the receiver would always recover the original message (by error correcting). Moreover, against the family of tampering functions that add an arbitrary constant, the sender can use Algebraic Manipulation Detection codes to help the receiver detect the tampering with high probability [12]. However, against more complex tampering functions, where error correction or detection are impossible, non-malleable codes can still give the following meaningful guarantee: Let $(\mathsf{Enc}, \mathsf{Dec})$ be the encoding and decoding algorithms for messages in $\{0,1\}^\ell$ against the tampering family $\mathcal{F}$. Then, for any message $m \in \{0,1\}^\ell$, $f \in \mathcal{F}$, the decoding of the tampered codeword, i.e., the message $\mathsf{Dec}(f(\mathsf{Enc}(m)))$, is either the original message m or a simulator Sim_f, which is entirely independent of the original message, can simulate its distribution. Ensuring this weak message integrity turns out to be extremely useful for cryptography. For example, tampering the secret-key of a signature scheme either yields the original secret-key (in which case the signature's security already holds) or yields an unrelated secret-key (which, again, is useless for forging signatures using the original secret-key).

However, it is impossible to construct non-malleable codes that are secure against class of all tampering functions. For instance, the adversary can intercept the entire encoding, decode the transmitted codeword c to retrieve the original message m and then write a particular encoding of the related message m_1^ℓ, where m_1 is the first bit of m. So, it is necessary to ensure that the decoding algorithm Dec (or any of its approximations) does not lie in the tampering function family itself. Therefore, non-malleable codes are typically constructed against a restricted class of tampering functions. Next, we discuss some tampering families considered in this work.

Lookahead Tampering and Non-malleable Messaging. Consider the motivating application of non-malleable message transmission, where the high-speed network switches routing the communication between parties shall forward their data packets at several gigabits per second. An adversary, who is monitoring the communication at a network switch, cannot block or slow the information stream, which would outrightly signal her intrusion. So, the adversary is naturally left to innocuously substituting data packets based on all the information that she has seen so far, namely, the *lookahead tampering* model [2,6]. This restricts the tampering power of the adversary as she cannot tamper the encoding arbitrarily.

Split-State Tampering. A widely studied setting is k-split-state tampering [2,3, 8,14,20,22]. Here, message is encoded into k states and the adversary can only tamper each of the states independently (and arbitrarily). More formally, the message $m \in \{0,1\}^\ell$ is encoded as $c = (c_1, c_2, \ldots, c_k) \in \{0,1\}^{n_1} \times \{0,1\}^{n_2} \times \cdots \times \{0,1\}^{n_k}$. A tampering function is a k-tuple of functions $f = (f_1, f_2, \ldots, f_k)$ s.t. the function $f_i : \{0,1\}^{n_i} \to \{0,1\}^{n_i}$ is an arbitrary function. Note that the tampering function only sees single states locally, and decoding requires aggregating information across all states.

Our Objective. Motivated by applications like non-malleable message transmission over high-speed networks, our work studies the limits of the efficiency of constructing non-malleable codes in the k-split-state model where a lookahead adversary tampers each state independently, i.e., the *k-lookahead model*. We know that constructing non-malleable codes against single state, i.e., $k = 1$, lookahead adversary is impossible [6]. So, we consider the next best setting of 2-split-state lookahead tampering, where the message is encoded into 2 states and transmitted using 2 independent paths. Each of these states is tampered independently using lookahead tampering. Since split-state lookahead tampering is a sub-class of split-state tampering, a conservative approach is to use generic non-malleable codes in the k-split-state, which protect against arbitrary split-state tamperings. Prior to our work, the most efficient non-malleable codes achieved rate $R = \log\log \ell / \log \ell$ for $k = 2$ [23], and rate $R = 1/3$ for $k = 4$ [20]. In a concurrent and independent work, [21] achieves rate $R = 1/3$ for $k = 3$.

As illustrated above, there are natural cryptographic applications where lookahead attacks appropriately model the adversarial threat. We ask the following question: Can we leverage the structure of the lookahead tampering to construct a constant rate non-malleable code that requires establishing least number of, i.e., only 2, independent communication routes between the sender and the receiver?

Our Results. We first prove an upper-bound that the rate of any non-malleable code in the 2-split-state lookahead model is at most $1/2$. Next, we construct a non-malleable code for the 2-lookahead model with rate $R = 1/3$, which is $2/3$-close to the above mentioned optimal upper-bound. En route, we also independently construct a 3-split-state non-malleable code that achieves rate $R = 1/3$. The starting point of all our non-malleable code constructions is the recent construction of [20] in the 4-split-state model.

Finally, we interpret our results in the context of the original motivating example of non-malleable message transmission. It is necessary to establish at least two independent routes of communication to facilitate non-malleable message transmission between two parties. We show that the cumulative size of the encoding of the message sent by the sender must be at least twice the message length when the sender transmits the shares of the encoded message over two independent routes. For this setting, we provide a construction where the encoding of the message is (roughly) three-times the size of the message (1.5x the optimal solution).

1.1 Our Contribution

Let $\mathcal{S}_n$ represent the set of all functions from $\{0,1\}^n$ to $\{0,1\}^n$. We call any subset $\mathcal{F} \subseteq \mathcal{S}_n$ a *tampering family* on $\{0,1\}^n$. We denote k-split-state tampering families on $\{0,1\}^{n_1+n_2+\cdots+n_k}$ by $\mathcal{F}_1 \times \mathcal{F}_2 \times \cdots \times \mathcal{F}_k$, where $\mathcal{F}_1, \mathcal{F}_2, \ldots, \mathcal{F}_k$ are tampering families on $\{0,1\}^{n_1}, \{0,1\}^{n_2}, \ldots, \{0,1\}^{n_k}$. Here, the codeword is distributed over k states of size $n_1, n_2, \ldots, n_k$.

(Split-State) Lookahead Tampering. Motivated by the example in the introduction, instead of considering an arbitrary tampering function for each state, we consider tampering functions that encounter the information as a stream. Let $\mathcal{LA}_{n_1,n_2,\ldots,n_B}$ be the set of all functions $f \colon \{0,1\}^{n_1+n_2+\cdots+n_B} \rightarrow \{0,1\}^{n_1+n_2+\cdots+n_B}$ such that there exists functions $f^{(1)}, f^{(2)}, \ldots, f^{(B)}$ with the following properties.

1. For each $1 \le i \le B$, we have $f^{(i)} \colon \{0,1\}^{n_1+n_2+\cdots+n_i} \rightarrow \{0,1\}^{n_i}$, and
2. The function $f(x_1, x_2, \ldots, x_B)$ is the concatenation of $f^{(i)}(x_1, x_2, \ldots, x_i)$, i.e.,
$$f(x_1, x_2, \ldots, x_B) = f^{(1)}(x_1) \| f^{(2)}(x_1, x_2) \| \cdots \| f^{(B)}(x_1, x_2, \ldots, x_B)$$

Intuitively, the codeword arrives as B blocks of information, and the i-th block is tampered based on all the blocks so far $\{1, 2, \ldots, i\}$. In the k-split-state lookahead tampering, denoted by k-*lookahead*, the tampering function for each state is a lookahead function. The k-lookahead tampering family was introduced in [2] for the purpose of constructing non-malleable codes in the 2-split-state model. A similar notion called block-wise tampering function was introduced by [6]. Our first result is the hardness result. We give a more precise statement of this result in Theorem 5.

Theorem 1. *For k-lookahead tampering family, the best achievable rate is $1 - 1/k$.*

In fact, we prove the above upper bound for the weakest tampering family in this class where each block in lookahead tampering is a single bit, i.e., $\mathcal{LA}_{n_1,n_2,\ldots,n_B}$ s.t. $B = n$ and $n_i = 1$. For brevity, we represent this function by $\mathcal{LA}_{1 \otimes n}$. Surprisingly, analogous to the result of Cheraghchi and Guruswami [10] for the k-split-state model, we prove that even against significantly more restricted k-lookahead tampering $\mathcal{LA}_{1 \otimes n_1} \times \cdots \times \mathcal{LA}_{1 \otimes n_k}$, the rate of any non-malleable code is at most $1 - 1/k$ (see Subsect. 3.1).

We use Fig. 1 to summarize our positive results in k-lookahead and k-split-state model and position our results relative to relevant prior works. Intuitively, lower the k, the more powerful is the tampering family, and the harder it is to construct the non-malleable codes. The state-of-the-art in non-malleable code construction against k-lookahead coincides with the general k-split-state model. In particular, no constant-rate non-malleable codes are known even against the restricted 2-lookahead model. We resolve this open question in the positive (with $2/3$ the optimal rate).

Theorem 2 (Rate-1/3 NMC against 2-Lookahead). *There exists a computationally efficient non-malleable code, with negligible simulation error, against the 2-lookahead tampering $\mathcal{LA}_{n_1,n_2} \times \mathcal{LA}_{n_3,n_4}$, where $n_1 = (2+o(1))\ell$, $n_2 = o(\ell)$, $n_3 = o(\ell)$, $n_4 = \ell$, where ℓ is the length of the message.*

We start from the construction of 4-split-state non-malleable codes by Kanukruthi et al. [20] and leverage a unique characteristic of the (rate-0) 2-split-state code of Aggarwal, Dodis, and Lovett [3], namely *augmented non-malleability* that was identified by [1].

By manipulating the way we store information in the construction of Theorem 2, we also obtain the first constant-rate non-malleable codes in 3-split-state[1].

Theorem 3 (Rate-1/3 NMC in 3-Split-State). *There exists a computationally efficient non-malleable code, with negligible simulation error, in the 3-split-state model $\mathcal{S}_{n_1} \times \mathcal{S}_{n_2} \times \mathcal{S}_{n_3}$, where $n_1 = \ell$, $n_2 = (2+o(1))\ell$, $n_3 = o(\ell)$, where ℓ is the length of the message.*

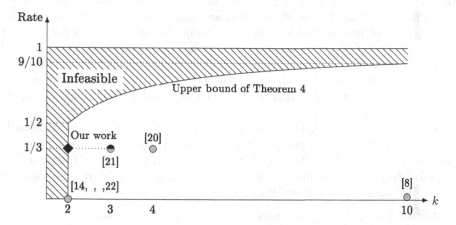

Fig. 1. A comparison of the efficiency of our 2-lookahead non-malleable code with the efficiency of generic k-split-state non-malleable codes in the information-theoretic setting. The diamond represents a k-lookahead result, and the circles represent k-split-state results. Black color represents our results, and gray color represents other known results (includes both prior and concurrent works).

Lastly, [2] motivated constant-rate construction achieving non-malleability against 2-lookahead tampering along with another particular family of functions (namely, *forgetful functions*) as an intermediate step to constructing constant-rate non-malleable codes in the 2-split-state model. We achieve partial progress towards this goal, and Theorem 6 summarizes this result.

---

[1] Concurrent and independent work of [21] obtained similar result.

1.2   Prior Relevant Works

As explained earlier, it is impossible to construct non-malleable codes against the set of all tampering functions. If the size of the tampering family $\mathcal{F}$ is bounded then Monte-Carlo constructions of non-malleable codes exist [10,16]. However, for a single state, explicit constructions are known only for a few tampering families. For example, (1) bit-level perturbation and permutations [4,11,15], and (2) local or AC^0 tampering functions [5,7] are a few representative families of tampering functions.

Another well-studied restriction on the tampering class is the k-split-state, for $k \geqslant 2$, where the tampering function tampers each state independently. Cheraghchi and Guruswami [10] proved an upper bound of $1 - 1/k$ on the rate of any non-malleable code in the k-split-state model. Decreasing the number of states k escalates the complexity of constructing non-malleable codes significantly. For $k = 2$, technically the most challenging problem and most reliable for cryptographic applications, [14] constructed the first explicit non-malleable code for one-bit messages. In a breakthrough result, Aggarwal, Dodis, and Lovett [3] presented the first multi-bit non-malleable code with rate $O(\ell^{-\rho})$, for a suitable constant $\rho > 1$. The subsequent work of [2] introduced the general notions of non-malleable reductions and transformations and exhibited their utility for modular constructions of non-malleable codes. Currently, the best rate of $\log \log \ell / \log \ell$ is achieved by [23]. For higher values of k, Chattopadhyay, and Zuckerman [8] constructed the first constant-rate non-malleable code when $k = 10$. Recently, [20] constructed a rate-$1/3$ non-malleable code in the 4-split-state model. The construction of constant rate non-malleable codes in the 2-split-state and 3-split-state models was open.

The computational version of this problem restricts to only computationally efficient tampering, and [1] provided the qualitatively and quantitatively optimal solution. In the 2-split-state model, they showed that one-way functions are necessary to surpass the upper bound of rate-$1/2$ in the information-theoretic setting [10], and one-way functions suffice to achieve rate-1.

Lookahead Model. In the lookahead model, tampering functions encounter the state as a stream, and the tampering functions tampers a block of the state based solely on the blocks of the state it has seen thus far. [6] first considered this family of tampering functions (referred to as block-wise tampering). In fact, they focused on the 1-lookahead family and showed the impossibility even in the computational setting. Thus, they relaxed the non-malleability guarantee and gave a construction using computational assumptions.

This family of tampering has also been considered by [2] as an interesting pit stop on the route to constructing non-malleable codes in the 2-split-state model. Specifically, they showed that given any non-malleable codes against k-lookahead tampering family together with another so-called forgetful tampering, they can transform it to get a 2-split-state non-malleable codes with only constant overhead on the rate.

Observe that a non-malleable code in the k-split-state model is also a non-malleable code in the k-lookahead version. Currently, the state-of-the-art in the

k-lookahead model coincides with the general k-split-state model.[2] In particular, there are no known constant-rate non-malleable codes in the information-theoretic setting for $k = 2$ and $k = 3$.

Concurrent and Independent Work. In a recent, concurrent and independent work, Kanukurthi et al. [21] obtain a similar construction for non-malleable codes in the 3-split-state setting that achieves an identical rate as our construction. In their work, they study the problem of storing random secrets. While, the primary focus of our work is to study the family of tampering function in the lookahead model and explore the hardness of achieving non-malleability against this family of tampering functions.

2   Preliminaries

For any natural number n, the symbol $[n]$ denotes the set $\{1, 2, \ldots, n\}$. For a probability distribution A over a finite sample space Ω, $A(x)$ denotes the probability of sampling $x \in \Omega$ according to the distribution A and $x \sim A$ denotes that x is sampled from Ω according to A. For any $n \in \mathbb{N}$, U_n denotes the uniform distribution over $\{0,1\}^n$. Similarly, for a set S, U_S denotes the uniform distribution over S. For two probability distributions A and B over the same sample space Ω, the *statistical distance* between A and B, represented by $\mathrm{SD}(A, B)$, is defined to be $\frac{1}{2} \sum_{x \in \Omega} |A(x) - B(x)|$.

Let $f : \{0,1\}^p \times \{0,1\}^q \longrightarrow \{0,1\}^p \times \{0,1\}^q$. For any $x \in \{0,1\}^p, y \in \{0,1\}^q$, let $(\widetilde{x}, \widetilde{y}) = f(x, y)$. Then, we define $f_x(y) = \widetilde{y}$ and $f_y(x) = \widetilde{x}$. Note that $f_x : \{0,1\}^q \to \{0,1\}^q$ and $f_y : \{0,1\}^p \to \{0,1\}^p$.

2.1   Non-malleable Codes

We follow the presentation in previous works and define non-malleable codes below.

Definition 1 (Coding Schemes). *Let* Enc: $\{0,1\}^\ell \to \{0,1\}^n$ *and* Dec: $\{0,1\}^n \to \{0,1\}^\ell \cup \{\bot\}$ *be functions such that* Enc *is a randomized function (that is, it has access to private randomness) and* Dec *is a deterministic function. The pair* (Enc, Dec) *is called a coding scheme with block length n and message length ℓ if it satisfies perfect correctness, i.e., for all $m \in \{0,1\}^\ell$, over the randomness of* Enc, $\Pr[\mathrm{Dec}(\mathrm{Enc}(m)) = m] = 1$.

A non-malleable code is defined w.r.t. a family of tampering functions. For an encoding scheme with block length n, let $\mathcal{F}_n$ denote the set of all functions $f : \{0,1\}^n \to \{0,1\}^n$. Any subset $\mathcal{F} \subseteq \mathcal{F}_n$ is considered to a family of tampering functions. Please refer to Sect. 1.1 for definition of k-split-state tampering function family $\mathcal{S}_{n_1} \times \mathcal{S}_{n_2} \times \cdots \times \mathcal{S}_{n_k}$ and the lookahead version of the k-split-state tampering function family $\mathcal{LA}_{1 \otimes n_1} \times \cdots \times \mathcal{LA}_{1 \otimes n_k}$.

---

[2] In light of the objection raised by [22] in the argument of [2], their constructions against lookahead tampering are flawed.

Next, we define the non-malleable codes against a family $\mathcal{F}$ of tampering functions. We need the following $\text{copy}(x, y)$ function defined as follows:

$$\text{copy}(x, y) = \begin{cases} y, & \text{if } x = \text{same}^*; \\ x, & \text{otherwise.} \end{cases}$$

Definition 2 ((n, ℓ, ε)-Non-malleable Codes). *A coding scheme* (Enc, Dec) *with block length n and message length ℓ is said to be non-malleable against tampering family $\mathcal{F} \subseteq \mathcal{F}_n$ with error ε if for all function $f \in \mathcal{F}$, there exists a distribution* Sim_f *over $\{0, 1\}^\ell \cup \{\bot\} \cup \{\text{same}^*\}$ such that for all $m \in \{0, 1\}^\ell$,*

$$\text{Tamper}_f^m \approx_\varepsilon \text{copy}(\text{Sim}_f, m)$$

where Tamper_f^m stands for the following tampering distribution

$$\text{Tamper}_f^m := \left\{ \begin{array}{c} c \sim \text{Enc}(m), \ \tilde{c} = f(c), \ \tilde{m} = \text{Dec}(\tilde{c}) \\ \textit{Output: } \tilde{m}. \end{array} \right\}$$

The rate of a non-malleable code is defined as ℓ/n.

Our constructions rely on leveraging a unique characteristic of the non-malleable code in 2-split-state ($\mathcal{S}_{n_1} \times \mathcal{S}_{n_2}$ s.t. $n_1 + n_2 = n$) provided by Aggarwal, Dodis, and Lovett [3], namely *augmented non-malleability*, which was identified by [1]. We formally define this notion next. Below, we denote the two states of the codeword as $(L, R) \in \{0, 1\}^{n_1} \times \{0, 1\}^{n_2}$.

Definition 3 ($(n_1, n_2, \ell, \varepsilon)$-Augmented Non-malleable Codes against 2-split-state tampering family). *A coding scheme* (Enc, Dec) *with message length ℓ is said to be an augmented non-malleable coding scheme against tampering family $\mathcal{S}_{n_1} \times \mathcal{S}_{n_2}$ with $n_1 + n_2 = n$ and error ε if for all functions $(f, g) \in \mathcal{S}_{n_1} \times \mathcal{S}_{n_2}$, there exists a distribution $\text{SimPlus}_{f,g}$ over $\{0, 1\}^{n_1} \times (\{0, 1\}^\ell \cup \{\bot\} \cup \{\text{same}^*\})$ such that for all $m \in \{0, 1\}^\ell$,*

$$\text{TamperPlus}_{f,g}^m \approx_\varepsilon \text{copy}(\text{SimPlus}_{f,g}, m)$$

where $\text{TamperPlus}_{f,g}^m$ stands for the following augmented tampering distribution

$$\text{TamperPlus}_{f,g}^m := \left\{ \begin{array}{c} (L, R) \sim \text{Enc}(m), \ \tilde{L} = f(L), \ \tilde{R} = g(R) \\ \textit{Output } \left(L, \text{Dec}(\tilde{L}, \tilde{R}) \right) \end{array} \right\}$$

Note that above we abuse notation for $\text{copy}(\text{SimPlus}_{f,g}, m)$. Formally, it is defined as follows: $\text{copy}(\text{SimPlus}_{f,g}, m) = (L, m)$ when $\text{SimPlus}_{f,g} = (L, \text{same}^*)$ and $\text{SimPlus}_{f,g}$ otherwise.

It was shown in [1] that the construction of Aggarwal et al. [3] satisfies this stronger definition of augmented non-malleability with rate $1/\text{poly}(\ell)$ and negligible error ε. More formally, the following holds.

Imported Theorem 1 ([1]). *For any message length ℓ, there is a coding scheme* $(\text{Enc}^+, \text{Dec}^+)$ *of block length $n = p(\ell)$ (where p is a polynomial) that satisfies augmented non-malleability against 2-split-state tampering functions with error that is negligible in ℓ.*

2.2   Building Blocks

Next, we describe average min-entropy seeded extractors with small seed and one-time message authentication codes that we use in our construction.

Definition 4 (Average conditional min-entropy). *The average conditional min-entropy of a distribution A conditioned on distribution L is defined to be*

$$\tilde{H}_\infty(A|L) = -\log\left(\mathbb{E}_{\ell \sim L}\left[2^{-H_\infty(A|L=\ell)}\right]\right)$$

Following lemma holds for average conditional min-entropy in the presence of leakage.

Lemma 1 ([13]). *Let L be an arbitrary κ-bit leakage on A, then $\tilde{H}_\infty(A|L) \geqslant H_\infty(A) - \kappa$.*

Definition 5 (Seeded Average Min-entropy Extractor). *We say* Ext : $\{0,1\}^n \times \{0,1\}^d \longrightarrow \{0,1\}^\ell$ *is a (k, ε)-average min-entropy strong extractor if for every joint distribution (A, L) such that $\tilde{H}_\infty(A|L) \geqslant k$, we have that* $(\mathrm{Ext}(A, U_d), U_d, L) \approx_\varepsilon (U_\ell, U_d, L)$.

It is proved in [24] that any extractor is also a average min-entropy extractor with only a loss of constant factor on error. Also, [19] gave strong extractors with small seed length that extract arbitrarily close to k uniform bits. We summarize these in the following lemma.

Combining these results with the following known construction for extractors, we have that there exists average min-entropy extractor that require seed length $O(\log n + \log(1/\varepsilon))$ and extracts uniform random strings of length arbitrarily close to the conditional min-entropy of the source.

Lemma 2 ([19,24]). *For all constants $\alpha > 0$ and all integers $n \geqslant k$, there exists an efficient (k, ε)-average min-entropy strong extractor* Ext : $\{0,1\}^n \times \{0,1\}^d \longrightarrow \{0,1\}^\ell$ *with seed length $d = O(\log n + \log(1/\varepsilon))$ and $\ell = (1 - \alpha)k - O(\log(n) + \log(1/\varepsilon))$.*

Next, we define one-time message authentication codes.

Definition 6 (Message authentication code). *A μ-secure one-time message authentication code (MAC) is a family of pairs of function*

$$\{\mathrm{Tag}_k : \{0,1\}^\alpha \longrightarrow \{0,1\}^\beta, \ \mathrm{Verify}_k : \{0,1\}^\alpha \times \{0,1\}^\beta \longrightarrow \{0,1\}\}_{k \in K}$$

such that

(1) For all m, k, $\mathrm{Verify}_k(m, \mathrm{Tag}_k(m)) = 1$.

(2) For all $m \neq m'$ and t, t', $\Pr_{k \sim U_K}[\mathrm{Tag}_k(m) = t \mid \mathrm{Tag}_k(m') = t'] \leq \mu$.

Remark 1. Message authentication code can be constructed from μ-almost pairwise hash function family with the key length $2\log(1/\mu)$. An example construction can be found in the full version [18] of this paper.

3   Non-malleable Codes Against k-Lookahead

In this section, we study the k-lookahead tampering family. We first prove an upper-bound on the maximum rate that can be achieved for any non-malleable code against k-lookahead tampering family. For this, Theorem 5 states that the maximum rate that can be achieved is roughly $1 - 1/k$. Surprisingly, this matches the upper bound on the rate non-malleable codes against much stronger tampering family of k-split-state by [10]. Our upper bound as well as the impossibility result by [6] rules the information theoretic construction against single state lookahead tampering. On the constructive side, for 2-lookahead model, the technically most challenging setting among k-lookahead tampering families, we construct a non-malleable code that achieves rate $1/3$.

Notation. Recall that $\mathcal{LA}_{n_1,n_2,...,n_B} \subseteq (\{0,1\}^n)^{\{0,1\}^n}$, where $n = \sum_{i\in[B]} n_i$, denotes the family of lookahead tampering functions $f = (f^{(1)}, f^{(2)}, \ldots, f^{(B)})$ for $f^{(i)} : \{0,1\}^{\sum_{j\in[i]} n_j} \to \{0,1\}^{n_i}$ such that

$$\tilde{c} := f(c) = f^{(1)}(c_1)\|f^{(2)}(c_1,c_2)\| \ldots \|f^{(i)}(c_1,\ldots,c_i)\| \ldots \|f^{(B)}(c_1,\ldots,c_B)$$

for $c = c_1\|c_2\| \ldots \|c_B$ and for all $i \in [B]$, $c_i \in \{0,1\}^{n_i}$. That is, if c consists of B parts such that i^{th} part has length n_i, then i^{th} tampered part depends on first i parts of c. We also use $\mathcal{LA}_{m^{\otimes B}}$ to denote the family of lookahead tampering functions $\underbrace{\mathcal{LA}_{m, m, \ldots, m}}_{B\text{-times}}$, i.e., the codeword has B parts of length m each.

3.1   Impossibility Results for the Split-State Lookahead Model

In this section, we first prove an upper-bound on the rate of any non-malleable encoding against 2-lookahead tampering function, where each bit is treated as a block, i.e., $\mathcal{LA}_{1^{\otimes n/2}} \times \mathcal{LA}_{1^{\otimes n/2}}$. In our proof, we use ideas similar to [10] and the following imported lemma is used in their proof of theorem 5.3 (see [9]).[3]

Imported Lemma 1. *For any constant $0 < \delta < \alpha$ and any encoding scheme* (Enc, Dec) *with block length n and rate $1 - \alpha + \delta$, the following holds. Let the codeword c be written as $(c_1, c_2) \in \{0,1\}^{\alpha n} \times \{0,1\}^{(1-\alpha)n}$. Let $\eta = \frac{\delta}{4\alpha}$. Then, there exists a set $X_\eta \subseteq \{0,1\}^{\alpha n}$ and two messages m_0, m_1 such that*

$$\Pr[c_1 \in X_\eta \,|\, \mathrm{Dec}(c) = m_0] \geqslant \eta$$
$$\Pr[c_1 \in X_\eta \,|\, \mathrm{Dec}(c) = m_1] \leq \eta/2$$

Theorem 4. *Let* (Enc, Dec) *be any encoding scheme that is non-malleable against the family of tampering functions $\mathcal{LA}_{1^{\otimes n/2}} \times \mathcal{LA}_{1^{\otimes n/2}}$ and achieves rate $1/2 + \delta$, for any constant $\delta > 0$ and simulation error ε. Then, $\varepsilon > \delta/8$.*

---

[3] Specifically, in their proof of Theorem 5.3, they picked two messages s_0, s_1 along with X_η that satisfy the property we require for m_0, m_1 in the imported lemma. Also, we stress that their proof not only showed s_0 and s_1 exist, but there are *multiple* choices for the pair. This gives us the freedom when we pick our m_0 and m_1. We make use of this in our proof.

Proof. Note that any codeword c in support of Enc consists of two states c_1 and c_2, each of length $n/2$. We use $c_{i,j}$ for $i \in \{1,2\}$ and $j \in \{1,\dots,n/2\}$ to denote the jth bit in state i. Any tampering function $f = (f_1, f_2)$ generates a tampered codeword $\tilde{c} = (\tilde{c}_1, \tilde{c}_2) = (f_1(c_1), f_2(c_2))$. Below, we will construct a tampering function f^* such that any simulated distribution Sim_{f^*} will be ε far from tampering distribution Tamper_{f^*}.

Next, we fix a message $\widehat{m}$ and its codeword $\widehat{c}^{(0)} = (\widehat{c}_1^{(0)}, \widehat{c}_2^{(0)}) \in \mathrm{Enc}(\widehat{m})$ such that the following holds. Let $\widehat{c}^{(1)} \in \{0,1\}^n$ be such that for all $j \in \{1,\dots,n/2-1\}$, $\widehat{c}_{1,j}^{(0)} = \widehat{c}_{1,j}^{(1)}$, $\widehat{c}_{1,n/2}^{(0)} \neq \widehat{c}_{1,n/2}^{(1)}$ and $\widehat{c}_2^{(0)} = \widehat{c}_2^{(1)}$. Moreover, we require that $\mathrm{Dec}(\widehat{c}^{(1)}) \neq \widehat{m}$. That is, the two codewords are identical except the last bit of first block and the second codeword does not encode the same message[4] $\widehat{m}$. Above condition is still satisfied if $\mathrm{Dec}(\widehat{c}^{(1)}) = \bot$.

Since the rate of the given scheme (Enc, Dec) is $1 - 1/2 + \delta$ (with a constant δ), by Imported Lemma 1, we have that there exist special messages m_0, m_1 and set X_η with the above guarantees where c_1 corresponds to the first state. In fact, Imported Lemma 1 gives many such pair of messages and we will pick such that $\widehat{m}, m_0, m_1$ are all unique.

Now, our tampering function $f^* = (f^*_1, f^*_2)$ is as follows: f^* tampers a codeword $c = (c_1, c_2)$ to $\tilde{c} = (\tilde{c}_1, \tilde{c}_2)$ such that for all $j \in \{1,\dots,n/2-1\}$, $\tilde{c}_{1,j} = \widehat{c}_{1,j}^{(0)}$, $\tilde{c}_{1,n/2} = \widehat{c}_{1,n/2}^{(0)}$ if $c_1 \in X_\eta$, else $\widehat{c}_{1,n/2}^{(1)}$ and $\tilde{c}_2 = \widehat{c}_2^{(0)}$. That is, if $c_1 \in X_\eta$, the resulting codeword is $\widehat{c}^{(0)}$, else it is $\widehat{c}^{(1)}$. Note that the above tampering attack can be done using a split-state lookahead tampering function.

Finally, it is evident that for message m_0, the tampering experiment results in $\widehat{m}$ with probability at least η. On the other hand, for message m_1, the tampering experiment results in $\widehat{m}$ with probability at most $\eta/2$. Hence, probability assigned by $\mathrm{Tamper}_{f^*}^{m_0}$ and $\mathrm{Tamper}_{f^*}^{m_1}$ to message $\widehat{m}$ differs by at least $\eta/2$. Since $\widehat{m}$ is different from m_0, m_1, it holds that ε, the simulation error of non-malleable code, is at least $\eta/4$ by triangle inequality.

The above result can be extended to k-lookahead tampering as follows:

Theorem 5. *Let* (Enc, Dec) *be any encoding scheme that is non-malleable against the family of tampering functions* $\mathcal{LA}_{1^{\otimes n_1}} \dots \times \dots \mathcal{LA}_{1^{\otimes n_k}}$ *and achieves rate* $1-1/k+\delta$, *for any constant* $\delta > 0$ *and simulation error* ε. *Then,* $\varepsilon > k\delta/16$.

Proof Outline. The proof follows by doing a similar analysis as above for the largest state. Without loss of generality, let the first state be the largest state, i.e., $n_1 \geqslant n_i$ for all $i \in \{2,\dots,k\}$. By averaging argument it holds that $n_1 \geqslant n/k$, where n is the block length. Now, the theorem follows along the same lines as the proof of 2-lookahead tampering above when we consider the code for the first state as c_1 and rest of the code as c_2. We note that the above proof does not require c_1 and c_2 to have the same size.

---

[4] We note that such codewords would exist otherwise we can show that the last bit of the first state is redundant for decoding. This way we can obtain a smaller encoding. Then, w.l.o.g., we can apply our argument on this new encoding.

3.2 Rate-1/3 Non-malleable Code in 2-Lookahead Model

In this section, we present our construction for non-malleable codes against 2-lookahead tampering functions. Our construction relies on the following tools. Let (Tag, Verify) (resp., (Tag$'$, Verify$'$)) be a μ (resp., μ') secure message authentication code with message length ℓ (resp., n), tag length β (resp., β') and key length γ (resp., γ'). Let Ext : $\{0,1\}^n \times \{0,1\}^d \rightarrow \{0,1\}^\ell$ be a (k, ε_1) average min-entropy strong extractor. We define k later during parameter setting. Finally, let $(\mathrm{Enc}^+, \mathrm{Dec}^+)$ be $(n_1^+, n_2^+, \ell^+, \varepsilon^+)$-augmented 2-split-state non-malleable code (see Definition 3), where $\ell^+ = \gamma + \gamma' + \beta + \beta' + d$. We denote the codewords of this scheme as (L, R) and given a tampering function, we denote the output of the simulator SimPlus as (L, Ans).

Construction Overview. We define our encoding and decoding functions formally in Fig. 2. In our encoding procedure, we first sample a uniform source w of n bits and a uniform seed s of d bits. Next, we extract a randomness r from (w, s) using the strong extractor Ext. We hide the message m using r as the one-time pad to obtain a ciphertext c. Next, we sample random keys k_1, k_2 and authenticate the ciphertext c using Tag_{k_1} and the source w using Tag'_{k_2} to obtain tags t_1 and t_2, respectively. Now, we think of (k_1, k_2, t_1, t_2, s) as the digest and protect it using an augmented 2-state non-malleable encoding Enc^+ to obtain (L, R). Finally, our codeword is $((c_1, c_2), (c_3, c_4))$ where $c_1 = w$, $c_2 = R$, $c_3 = L$ and $c_4 = c$.

We also note that $n_1 := |c_1| = |w| = n$, $n_2 := |c_2| = |R| = n_2^+$, $n_3 := |c_3| = |L| = n_1^+$ and $n_4 := |c_4| = |c| = \ell$. From Fig. 2, it is evident that our construction satisfies perfect correctness.

Enc(m):	Dec$\Big((c_1, c_2), (c_3, c_4)\Big)$:
1. Sample $w \sim U_n$, $s \sim U_d$, $k_1 \sim U_\gamma$, $k_2 \sim U_{\gamma'}$	1. Let the tampered states be $\widetilde{w} := c_1$, $\widetilde{R} := c_2$, $\widetilde{L} := c_3$, $\widetilde{c} := c_4$
2. Compute $r = \mathrm{Ext}(w, s)$, $c = m \oplus r$	2. Decrypt $\quad (\widetilde{k}_1, \widetilde{k}_2, \widetilde{t}_1, \widetilde{t}_2, \widetilde{s}) =$ $\mathrm{Dec}^+(\widetilde{L}, \widetilde{R})$
3. Compute the tags $t_1 = \mathrm{Tag}_{k_1}(c)$, $t_2 = \mathrm{Tag}'_{k_2}(w)$	3. If $(\widetilde{k}_1, \widetilde{k}_2, \widetilde{t}_1, \widetilde{t}_2, \widetilde{s}) = \bot$, output $\bot$
4. Compute the 2-state non-malleable encoding $(L, R) \sim \mathrm{Enc}^+(k_1, k_2, t_1, t_2, s)$	4. (Else) If $\mathrm{Verify}_{\widetilde{k}_1}(\widetilde{c}, \widetilde{t}_1) = 0$ or $\mathrm{Verify}'_{\widetilde{k}_2}(\widetilde{w}, \widetilde{t}_2) = 0$, output $\bot$
5. Output the states $\Big((w, R), (L, c)\Big)$	5. (Else) Output $\widetilde{c} \oplus \mathrm{Ext}(\widetilde{w}, \widetilde{s})$

Fig. 2. Non-malleable coding scheme against $\mathcal{LA}_{n_1, n_2} \times \mathcal{LA}_{n_3, n_4}$, where $n_1 = |w|$, $n_2 = |R|$, $n_3 = |L|$, and $n_4 = |c|$.

Proof of Non-malleability against 2-lookahead tampering. Given a tampering function $(f, g) \in \mathcal{LA}_{n_1, n_2} \times \mathcal{LA}_{n_3, n_4}$, where $f = (f^{(1)}, f^{(2)})$ and $g = (g^{(1)}, g^{(2)})$, we formally describe our simulator in Fig. 3.

Our simulator describes a leakage function $\mathcal{L}(w)$ that captures the leakage required on the source w in order to simulate the tampering experiment. This leakage has five parts $(L, \text{Ans}, \text{flag}_1, \text{flag}_2, \text{mask})$. The values L and Ans are the outputs of simulator SimPlus on tampering function $(g^{(1)}, f_w^{(2)})$, where $f_w^{(2)}$ represents the tampering function on R given w. Next, for the case when Ans $=$ same*, flag_1 denotes the bit $\widetilde{w} = w$. When Ans $= (\widetilde{k}_1, \widetilde{k}_2, \widetilde{t}_1, \widetilde{t}_2, \widetilde{s})$, flag_2 captures the bit $\text{Verify}'_{\widetilde{k}_2}(\widetilde{w}, \widetilde{t}_2)$, i.e., whether the new key $\widetilde{k}_2$ and tag $\widetilde{t}_2$ are valid authentication on new source $\widetilde{w}$. In this case, the value mask is the extracted output of tampered source $\widetilde{w}$ using tampered seed $\widetilde{s}$.

---

1. $w \sim U_n$
2. Define leakage function $\mathcal{L}(w) : \{0,1\}^n \longrightarrow \{0,1\}^{n_1^+} \times \{0,1\}^{\beta+\beta'+\gamma+\gamma'+d+1} \times \{0,1\} \times \{0,1\} \times \{0,1\}^\ell$ as the following function:
   (a) $(L, \text{Ans}) \sim \text{SimPlus}_{g^{(1)}, f_w^{(2)}}$, $\widetilde{w} = f^{(1)}(w)$
   (b) If Ans $=$
      − Case $\bot$: $\text{flag}_1 = 0$, $\text{flag}_2 = 0$, $\text{mask} = 0^\ell$
      − Case same*: If $(\widetilde{w} = w)$, $\text{flag}_1 = 1$; Else $\text{flag}_1 = 0$
         $\text{flag}_2 = 0$, $\text{mask} = 0^\ell$
      − Case $(\widetilde{k}_1, \widetilde{k}_2, \widetilde{t}_1, \widetilde{t}_2, \widetilde{s})$: $\text{flag}_1 = 0$, Let $\text{mask} = \text{Ext}(\widetilde{w}, \widetilde{s})$
         If $\left(\text{Verify}'_{\widetilde{k}_2}(\widetilde{w}, \widetilde{t}_2) \right) = 1$, $\text{flag}_2 = 1$; Else $\text{flag}_2 = 0$
   (c) $\mathcal{L}(w) := (L, \text{Ans}, \text{flag}_1, \text{flag}_2, \text{mask})$
3. $r \sim U_\ell$, $c = 0^\ell \oplus r$, $\widetilde{c} = g_L^{(2)}(c)$
4. If Ans $=$
      − Case $\bot$: Output $\bot$
      − Case same*: If $\left(\widetilde{c} = c \text{ and } \text{flag}_1 \right) = 1$, output same*
         Else output $\bot$
      − Case $(\widetilde{k}_1, \widetilde{k}_2, \widetilde{t}_1, \widetilde{t}_2, \widetilde{s})$: If $\left(\text{Verify}_{\widetilde{k}_1}(\widetilde{c}, \widetilde{t}_1) = 0 \text{ or } \text{flag}_2 = 0 \right)$, output $\bot$
         Else output $\widetilde{c} \oplus \text{mask}$

Fig. 3. The simulator $\text{Sim}_{f,g}$ for the non-malleable code against 2-lookahead tampering family.

We give the formal proof on indistinguishability between simulated and tampering distributions in Subsect. 3.3 using a series of statistically close hybrids.

Rate analysis. We will use λ as our security parameter. By Remark 1, we will let k_1, k_2 be of length 2λ, i.e. $\gamma = \gamma' = 2\lambda$ and t_1, t_2 will have length λ, i.e. $\beta = \beta' = \lambda$ and both $(\text{Tag}, \text{Verify})$ and $(\text{Tag}', \text{Verify}')$ will have error $2^{-\lambda}$.

Since we will need to extract ℓ bits as a one-time pad to mask the message, by Lemma 2, we will set min-entropy k to be $(1+\alpha')\ell$ for some constant α' and let Ext be a $((1+\alpha')\ell, 2^{-\lambda})$-strong average min-entropy extractor that extract ℓ-bit randomness with seed length $O(\log n + \lambda)$. By our analysis in Subsect. 3.3, it suffices to have $n - (\ell + n_1^+ + \ell^+ + 3) = n - \ell - p(\log n + \lambda) \geqslant (1+\alpha')\ell$. Hence, we will set $n = (2+\alpha)\ell$ for some constant $\alpha > \alpha'$.

Now the message length for our augmented 2-state non-malleable code will be $2\lambda + 2\lambda + \lambda + \lambda + O(\log n + \lambda) = O(\log n + \lambda)$. Now by Theorem 1, we will let ζ be the constant such that $p(n^\zeta) = o(n)$ and set $\lambda = O(n^\zeta)$. Hence, the length of (L, R) will be $o(n)$. Therefore, the total length of our coding scheme will be $\ell + (2 + \alpha)\ell + o(n)$ and the rate is $\frac{1}{3+\alpha}$ with error $O(2^{-n^\zeta})$. This completes the proof for Theorem 2.

3.3   Proof of Non-malleability Against 2-Lookahead (Theorem 2)

In this section, we prove that our code scheme Fig. 2 is secure against the tampering family $\mathcal{LA}_{n_1,n_2} \times \mathcal{LA}_{n_3,n_4}$. In order to prove the non-malleability, we need to show that for all tampering functions $(f, g) \in \mathcal{LA}_{n_1,n_2} \times \mathcal{LA}_{n_3,n_4}$, where $f = (f^{(1)}, f^{(2)})$ and $g = (g^{(1)}, g^{(2)})$, our simulator as defined in Fig. 3 satisfies that, for all m, we have

$$
\left\{
\begin{array}{l}
\big((w, R), (L, c)\big) \sim \mathrm{Enc}(m) \\
\widetilde{w} = f^{(1)}(w), \ \widetilde{R} = f^{(2)}(w, R) \\
\widetilde{L} = g^{(1)}(L), \ \widetilde{c} = g^{(2)}(L, c) \\
\text{Output: } \widetilde{m} = \mathrm{Dec}\big((\widetilde{w}, \widetilde{R}), (\widetilde{L}, \widetilde{c})\big)
\end{array}
\right\} = \mathrm{Tamper}^m_{f,g} \approx \mathrm{copy}\left(\mathrm{Sim}_{f,g}, m\right)
$$

The following sequence of hybrids will lead us from tampering experiment to the simulator. Throughout this section, we use the following color/highlight notation. In a current hybrid, the text in red denotes the changes from the previous hybrid. The text in shaded part represents the steps that will be replaced by red part of the next hybrid.

The initial hybrid represents the tampering experiment $\mathrm{Tamper}^m_{f,g}$ and the last hybrid represents $\mathrm{copy}(\mathrm{Sim}_{f,g}, m)$.

$H_0(f, g, m)$:

1. $w \sim U_n$, $s \sim U_d$, $k_1 \sim U_\gamma$, $k_2 \sim U_{\gamma'}$
2. $r = \mathrm{Ext}(w, s)$, $c = m \oplus r$, $t_1 = \mathrm{Tag}_{k_1}(c)$, $t_2 = \mathrm{Tag}'_{k_2}(w)$
3. $(L, R) \sim \mathrm{Enc}^+(k_1, k_2, t_1, t_2, s)$
4. $\widetilde{w} = f^{(1)}(w)$, $\widetilde{R} = f^{(2)}(w, R)$, $\widetilde{L} = g^{(1)}(L)$, $\widetilde{c} = g^{(2)}(L, c)$
5. $(\widetilde{k}_1, \widetilde{k}_2, \widetilde{t}_1, \widetilde{t}_2, \widetilde{s}) = \mathrm{Dec}^+(\widetilde{L}, \widetilde{R})$
6. If $(\widetilde{k}_1, \widetilde{k}_2, \widetilde{t}_1, \widetilde{t}_2, \widetilde{s}) = \bot$, output $\bot$
7. Else If $\left(\mathrm{Verify}_{\widetilde{k}_1}(\widetilde{c}, \widetilde{t}_1) = 0 \text{ or } \mathrm{Verify}'_{\widetilde{k}_2}(\widetilde{w}, \widetilde{t}_2) = 0 \right)$, output $\bot$
8. Else Output $\widetilde{c} \oplus \mathrm{Ext}(\widetilde{w}, \widetilde{s})$

Next, we rewrite $\widetilde{R} = f^{(2)}(w, R)$ and $\widetilde{c} = g^{(2)}(L, c)$ as $\widetilde{R} = f_w^{(2)}(R)$ and $\widetilde{c} = g_L^{(2)}(c)$. Now, rearrange the steps leads us to the next hybrid.

$H_1(f, g, m)$:

1. $w \sim U_n$, $s \sim U_d$, $k_1 \sim U_\gamma$, $k_2 \sim U_{\gamma'}$
2. $r = \mathrm{Ext}(w, s)$, $c = m \oplus r$, $t_1 = \mathrm{Tag}_{k_1}(c)$, $t_2 = \mathrm{Tag}'_{k_2}(w)$
3. $\widetilde{w} = f^{(1)}(w)$
4. $(L, R) \sim \mathrm{Enc}^+(k_1, k_2, t_1, t_2, s)$
5. $\widetilde{L} = g^{(1)}(L)$, $\widetilde{R} = f_w^{(2)}(R)$
6. $(\widetilde{k_1}, \widetilde{k_2}, \widetilde{t_1}, \widetilde{t_2}, \widetilde{s}) = \mathrm{Dec}^+(\widetilde{L}, \widetilde{R})$
7. $\widetilde{c} = g_{\widetilde{L}}^{(2)}(c)$
8. If $(\widetilde{k_1}, \widetilde{k_2}, \widetilde{t_1}, \widetilde{t_2}, \widetilde{s}) = \bot$, output $\bot$
9. Else If $\left(\mathrm{Verify}_{\widetilde{k_1}}(\widetilde{c}, \widetilde{t_1}) = 0 \text{ or } \mathrm{Verify}'_{\widetilde{k_2}}(\widetilde{w}, \widetilde{t_2}) = 0 \right)$, output $\bot$
10. Else Output $\widetilde{c} \oplus \mathrm{Ext}(\widetilde{w}, \widetilde{s})$

Note that shaded steps in the previous hybrid formulate a 2-state tampering experiment onto (L, R). Therefore, we could use the augmented simulator to replace the tampering experiment of augmented two-state non-malleable codes.

$H_2(f, g, m)$:

1. $w \sim U_n$, $s \sim U_d$, $k_1 \sim U_\gamma$, $k_2 \sim U_{\gamma'}$
2. $r = \mathrm{Ext}(w, s)$, $c = m \oplus r$, $t_1 = \mathrm{Tag}_{k_1}(c)$, $t_2 = \mathrm{Tag}'_{k_2}(w)$
3. $\widetilde{w} = f^{(1)}(w)$
4. $(L, \mathrm{Ans}) \sim \mathrm{SimPlus}_{g^{(1)}, f_w^{(2)}}$
5. $(\widetilde{k_1}, \widetilde{k_2}, \widetilde{t_1}, \widetilde{t_2}, \widetilde{s}) = \mathrm{copy}\left(\mathrm{Ans}, (k_1, k_2, t_1, t_2, s) \right)$.
6. $\widetilde{c} = g_L^{(2)}(c)$
7. If $(\widetilde{k_1}, \widetilde{k_2}, \widetilde{t_1}, \widetilde{t_2}, \widetilde{s}) = \bot$, output $\bot$
8. Else If $\left(\mathrm{Verify}_{\widetilde{k_1}}(\widetilde{c}, \widetilde{t_1}) = 0 \text{ or } \mathrm{Verify}'_{\widetilde{k_2}}(\widetilde{w}, \widetilde{t_2}) = 0 \right)$, output $\bot$
9. Else Output $\widetilde{c} \oplus \mathrm{Ext}(\widetilde{w}, \widetilde{s})$

Now in hybrid $H_3(f, g, m)$, instead of doing copy(), we do a case analysis on Ans. We note that the hybrids $H_2(f, g, m)$ and $H_3(f, g, m)$ are identical.

$H_3(f, g, m)$:

1. $w \sim U_n$, $s \sim U_d$, $k_1 \sim U_\gamma$, $k_2 \sim U_{\gamma'}$
2. $r = \text{Ext}(w, s)$, $c = m \oplus r$, $t_1 = \text{Tag}_{k_1}(c)$, $t_2 = \text{Tag}'_{k_2}(w)$
3. $\widetilde{w} = f^{(1)}(w)$
4. $(L, \text{Ans}) \sim \text{SimPlus}_{g^{(1)}, f_w^{(2)}}$
5. $\widetilde{c} = g_L^{(2)}(c)$
6. If Ans =
   - Case $\perp$: Output $\perp$
   - Case same*: If $\left(\text{Verify}_{k_1}(\widetilde{c}, t_1) = 0 \text{ or } \text{Verify}'_{k_2}(\widetilde{w}, t_2) = 0 \right)$,

     output $\perp$; Else output $\widetilde{c} \oplus \text{Ext}(\widetilde{w}, s)$
   - Case $(\widetilde{k_1}, \widetilde{k_2}, \widetilde{t_1}, \widetilde{t_2}, \widetilde{s})$: If $\left(\text{Verify}_{\widetilde{k_1}}(\widetilde{c}, \widetilde{t_1}) = 0 \text{ or } \text{Verify}'_{\widetilde{k_2}}(\widetilde{w}, \widetilde{t_2}) = \right.$

     $\left. 0 \right)$, output $\perp$

     Else output $\widetilde{c} \oplus \text{Ext}(\widetilde{w}, \widetilde{s})$

Next, in hybrid $H_3(f, g, m)$ we change the case when Ans = same*. Note that Ans = same* says that the both the authentication keys k_1, k_2 as well as the tags are unchanged. Hence, with probability at least $(1 - \mu - \mu')$, both authentications would verify only if w and c are unchanged. Hence, in $H_4(f, g, m)$, we check if the ciphertext c and source w are the same.

Given that (Tag, Verify) and (Tag', Verify') are μ and μ'-secure message authentication codes, $H_3(f, g, m) \approx_{\mu + \mu'} H_4(f, g, m)$.

$H_4(f, g, m)$:

copy

$\left(\vphantom{\begin{array}{c}1\\2\\3\\4\\5\\6\end{array}}\right.$

1. $w \sim U_n$, $s \sim U_d$, $k_1 \sim U_\gamma$, $k_2 \sim U_{\gamma'}$
2. $r = \text{Ext}(w, s)$, $c = m \oplus r$, $t_1 = \text{Tag}_{k_1}(c)$, $t_2 = \text{Tag}'_{k_2}(w)$
3. $\widetilde{w} = f^{(1)}(w)$
4. $(L, \text{Ans}) \sim \text{SimPlus}_{g^{(1)}, f_w^{(2)}}$
5. $\widetilde{c} = g_L^{(2)}(c)$
6. If Ans =
   - Case $\perp$: Output $\perp$
   - Case same*: If $\left(\widetilde{c} = c \text{ and } \widetilde{w} = w \right) = 1$, output same*

     Else output $\perp$
   - Case $(\widetilde{k_1}, \widetilde{k_2}, \widetilde{t_1}, \widetilde{t_2}, \widetilde{s})$: If $\left(\text{Verify}_{\widetilde{k_1}}(\widetilde{c}, \widetilde{t_1}) = 0 \text{ or} \right.$

     $\left. \text{Verify}'_{\widetilde{k_2}}(\widetilde{w}, \widetilde{t_2}) = 0 \right)$, output $\perp$

     Else output $\widetilde{c} \oplus \text{Ext}(\widetilde{w}, \widetilde{s})$

$\left.\vphantom{\begin{array}{c}1\\2\\3\\4\\5\\6\end{array}}, m\right)$

We note that the variables k_1, k_2, t_1, t_2 are no longer used in the hybrid. Hence, we remove the sampling of these in the next hybrid. It is clear that the two hybrids $H_4(f, g, m)$ and $H_5(f, g, m)$ are identical.

$H_5(f, g, m)$:
copy

$$\left(\begin{array}{l}
\begin{array}{l}
\text{1. } w \sim U_n, \; s \sim U_d, \; r = \text{Ext}(w, s), \; c = m \oplus r \\
\text{2. } (L, \text{Ans}) \sim \text{SimPlus}_{g^{(1)}, f_w^{(2)}} \\
\text{3. } \widetilde{w} = f^{(1)}(w), \; \widetilde{c} = g_L^{(2)}(c) \\
\text{4. If Ans } = \\
\quad - \text{ Case } \perp: \text{ Output } \perp \\
\quad - \text{ Case same*: If } \left(\widetilde{c} = c \text{ and } \boxed{\widetilde{w} = w}\right) = 1, \text{ output same*} \\
\qquad\qquad \text{Else output } \perp \\
\quad - \text{ Case } (\widetilde{k}_1, \widetilde{k}_2, \widetilde{t}_1, \widetilde{t}_2, \widetilde{s}): \text{ If } \Big(\text{Verify}_{\widetilde{k}_1}(\widetilde{c}, \widetilde{t}_1) = 0 \text{ or} \\
\qquad\qquad \text{Verify}'_{\widetilde{k}_2}(\widetilde{w}, \widetilde{t}_2) = 0 \Big), \text{ output } \perp \\
\qquad\qquad \text{Else output } \widetilde{c} \oplus \text{Ext}(\widetilde{w}, \widetilde{s})
\end{array}
\end{array}\right. , m \right)$$

Now, we wish to use the property of average min-entropy extractor to remove the dependence between c and w. Before we do the trick, we shall first rearrange the steps in $H_5(f, g, m)$ to get $H_6(f, g, m)$. We process all the leakage we need at the first part of our hybrid and use only the leakage of w in the remaining. Intuitively, when Ans = same*, flag_1 records whether $\widetilde{w} = w$ and when Ans = $(\widetilde{k}_1, \widetilde{k}_2, \widetilde{t}_1, \widetilde{t}_2, \widetilde{s})$, flag_2 records whether $\widetilde{w}$ can pass the MAC verification under new key and tag and mask is the new one-time pad we need for decoding the tampered message. We note that the hybrids $H_5(f, g, m)$ and $H_6(f, g, m)$ are identical.

$H_6(f, g, m)$:
copy

$$\left(\begin{array}{l}
\begin{array}{l}
\text{1. } w \sim U_n \\
\text{2. } (L, \text{Ans}) \sim \text{SimPlus}_{g^{(1)}, f_w^{(2)}}, \; \widetilde{w} = f^{(1)}(w) \\
\text{3. If Ans } = \\
\quad - \text{ Case same*: If } (\widetilde{w} = w), \text{ flag}_1 = 1; \text{ Else flag}_1 = 0 \\
\quad - \text{ Case } (\widetilde{k}_1, \widetilde{k}_2, \widetilde{t}_1, \widetilde{t}_2, \widetilde{s}): \text{ If } \left(\text{Verify}'_{\widetilde{k}_2}(\widetilde{w}, \widetilde{t}_2) \right) = 1, \text{ flag}_2 = 1; \\
\qquad\qquad \text{Else flag}_2 = 0. \\
\qquad\qquad \text{Let mask} = \text{Ext}(\widetilde{w}, \widetilde{s}) \\
\text{4. } s \sim U_d, \; r = \text{Ext}(w, s), \; c = m \oplus r, \; \widetilde{c} = g_L^{(2)}(c) \\
\text{5. If Ans } = \\
\quad - \text{ Case } \perp: \text{ Output } \perp \\
\quad - \text{ Case same*: If } \left(\widetilde{c} = c \text{ and flag}_1\right) = 1, \text{ output same*} \\
\qquad\qquad \text{Else output } \perp \\
\quad - \text{ Case } (\widetilde{k}_1, \widetilde{k}_2, \widetilde{t}_1, \widetilde{t}_2, \widetilde{s}): \text{ If } \left(\text{Verify}_{\widetilde{k}_1}(\widetilde{c}, \widetilde{t}_1) = 0 \text{ or flag}_2 = 0 \right), \\
\qquad\qquad \text{output } \perp \\
\qquad\qquad \text{Else output } \widetilde{c} \oplus \text{mask}
\end{array}
\end{array}\right. , m \right)$$

In the next hybrid, we formalize $(L, \text{Ans}, \text{flag}_1, \text{flag}_2, \text{mask})$ as the leakage on source w. Note that the hybrids $H_6(f, g, m)$ and $H_7(f, g, m)$ are identical.

$H_7(f, g, m)$:
copy

$$\left(\begin{array}{l}
\text{1. } w \sim U_n \\
\text{2. Leakage function } \mathcal{L}(w) : \{0,1\}^n \longrightarrow \{0,1\}^{n_1^+} \times \{0,1\}^{\beta+\beta'+\gamma+\gamma'+d+1} \times \\
\quad \{0,1\} \times \{0,1\} \times \{0,1\}^\ell \text{ be the following function:} \\
\quad \text{(a) } (L, \text{Ans}) \sim \text{SimPlus}_{g^{(1)}, f_w^{(2)}}, \; \widetilde{w} = f^{(1)}(w) \\
\quad \text{(b) If Ans =} \\
\qquad - \text{ Case } \bot: \text{flag}_1 = 0, \; \text{flag}_2 = 0, \; \text{mask} = 0^\ell \\
\qquad - \text{ Case same*: If } (\widetilde{w} = w), \text{ flag}_1 = 1; \text{ Else flag}_1 = 0 \\
\qquad\qquad\qquad \text{flag}_2 = 0, \; \text{mask} = 0^\ell \\
\qquad - \text{ Case } (\widetilde{k_1}, \widetilde{k_2}, \widetilde{t_1}, \widetilde{t_2}, \widetilde{s}): \text{flag}_1 = 0, \text{ Let mask} = \text{Ext}(\widetilde{w}, \widetilde{s}) \\
\qquad\qquad\qquad \text{If } \left(\text{Verify}'_{\widetilde{k_2}}(\widetilde{w}, \widetilde{t_2})\right) = 1, \text{flag}_2 = 1; \text{ Else flag}_2 = 0 \\
\quad \text{(c) } \mathcal{L}(w) := (L, \text{Ans}, \text{flag}_1, \text{flag}_2, \text{mask}) \\
\text{3. } s \sim U_d, \; r = \text{Ext}(w, s), \; c = m \oplus r, \; \widetilde{c} = g_L^{(2)}(c) \\
\text{4. If Ans =} \\
\qquad - \text{ Case } \bot: \text{Output } \bot \\
\qquad - \text{ Case same*: If } \left(\widetilde{c} = c \text{ and flag}_1\right) = 1, \text{ output same*; Else output } \bot \\
\qquad - \text{ Case } (\widetilde{k_1}, \widetilde{k_2}, \widetilde{t_1}, \widetilde{t_2}, \widetilde{s}): \text{If } \left(\text{Verify}_{\widetilde{k_1}}(\widetilde{c}, \widetilde{t_1}) = 0 \text{ or flag}_2 = 0\right), \text{ output } \bot \\
\qquad\qquad\qquad\qquad \text{Else output } \widetilde{c} \oplus \text{mask}
\end{array}\right., m\right)$$

In the next hybrid, we replace the extracted output r with a uniform random ℓ bit string. We argue that the hybrids $H_7(f, g, m)$ and $H_8(f, g, m)$ are ε_1 close for appropriate length n of source w.

Since $\mathcal{L}(w)$ outputs a $\ell + n_1^+ + \ell^+ + 3$ bits of leakage, by Lemma 1, $H_\infty(W|\mathcal{L}(W)) \geqslant n - (\ell + n_1^+ \ell^+ + 3)$. Here, W denotes the random variable corresponding to w. We will pick n such that $n - (\ell + n_1^+ + \ell^+ + 3) > \ell$ for the min-entropy extraction to give a uniform string (see Lemma 2).

$H_8(f, g, m)$:
copy

$$\left(\begin{array}{l}
\text{1. } w \sim U_n \\
\text{2. Leakage function } \mathcal{L}(w) : \{0,1\}^n \longrightarrow \{0,1\}^{n_1^+} \times \{0,1\}^{\beta+\beta'+\gamma+\gamma'+d+1} \times \\
\quad \{0,1\} \times \{0,1\} \times \{0,1\}^\ell \text{ be the following function:} \\
\quad \text{(a) } (L, \text{Ans}) \sim \text{SimPlus}_{g^{(1)}, f_w^{(2)}}, \; \widetilde{w} = f^{(1)}(w) \\
\quad \text{(b) If Ans =} \\
\qquad - \text{ Case } \bot: \text{flag}_1 = 0, \; \text{flag}_2 = 0, \; \text{mask} = 0^\ell \\
\qquad - \text{ Case same*: If } (\widetilde{w} = w), \text{ flag}_1 = 1; \text{ Else flag}_1 = 0 \\
\qquad\qquad\qquad \text{flag}_2 = 0, \; \text{mask} = 0^\ell \\
\qquad - \text{ Case } (\widetilde{k_1}, \widetilde{k_2}, \widetilde{t_1}, \widetilde{t_2}, \widetilde{s}): \text{flag}_1 = 0, \text{ Let mask} = \text{Ext}(\widetilde{w}, \widetilde{s}) \\
\qquad\qquad\qquad \text{If } \left(\text{Verify}'_{\widetilde{k}}(\widetilde{w}, \widetilde{t_2})\right) = 1, \text{flag}_2 = 1; \text{ Else flag}_2 = 0 \\
\quad \text{(c) } \mathcal{L}(w) := (L, \text{Ans}, \text{flag}_1, \text{flag}_2, \text{mask}) \\
\text{3. } r \sim U_\ell, \; c = m \oplus r, \; \widetilde{c} = g_L^{(2)}(c) \\
\text{4. If Ans =} \\
\qquad - \text{ Case } \bot: \text{Output } \bot \\
\qquad - \text{ Case same*: If } \left(\widetilde{c} = c \text{ and flag}_1\right) = 1, \text{ output same*; Else output } \bot \\
\qquad - \text{ Case } (\widetilde{k_1}, \widetilde{k_2}, \widetilde{t_1}, \widetilde{t_2}, \widetilde{s}): \text{If } \left(\text{Verify}_{\widetilde{k_1}}(\widetilde{c}, \widetilde{t_1}) = 0 \text{ or flag}_2 = 0\right), \text{ output } \bot \\
\qquad\qquad\qquad\qquad \text{Else output } \widetilde{c} \oplus \text{mask}
\end{array}\right., m\right)$$

Finally, notice that the distribution of c is independent of m and we can use the message 0^ℓ. This gives us our simulator. Clearly $H_8(f, g, m) = H_9(f, g, m)$. Notice that $H_9(f, g, m) = \text{copy}\left(\text{Sim}_{f,g}, m\right)$.

$H_9(f, g, m)$:
copy

$\left(\begin{array}{l} \text{1. } w \sim U_n \\ \text{2. Leakage function } \mathcal{L}(w) : \{0,1\}^n \longrightarrow \{0,1\}^{n_1^+} \times \{0,1\}^{\beta+\beta'+\gamma+\gamma'+d+1} \times \\ \quad \{0,1\} \times \{0,1\} \times \{0,1\}^\ell \text{ be the following function:} \\ \quad \text{(a) } (L, \text{Ans}) \sim \text{SimPlus}_{g^{(1)}, f_w^{(2)}}, \ \widetilde{w} = f^{(1)}(w) \\ \quad \text{(b) If Ans} = \\ \qquad - \text{ Case } \bot: \text{flag}_1 = 0, \ \text{flag}_2 = 0, \ \text{mask} = 0^\ell \\ \qquad - \text{ Case same*: If } (\widetilde{w} = w), \text{flag}_1 = 1; \text{ Else flag}_1 = 0 \\ \qquad\qquad \text{flag}_2 = 0, \ \text{mask} = 0^\ell \\ \qquad - \text{ Case } (\widetilde{k_1}, \widetilde{k_2}, \widetilde{t_1}, \widetilde{t_2}, \widetilde{s}): \text{flag}_1 = 0, \text{ Let mask} = \text{Ext}(\widetilde{w}, \widetilde{s}) \\ \qquad\qquad \text{If } \left(\text{Verify}'_{\widetilde{k_2}}(\widetilde{w}, \widetilde{t_2})\right) = 1, \text{flag}_2 = 1; \text{ Else flag}_2 = 0 \\ \quad \text{(c) } \mathcal{L}(w) := (L, \text{Ans}, \text{flag}_1, \text{flag}_2, \text{mask}) \\ \text{3. } r \sim U_\ell, \ c = 0^\ell \oplus r, \ \widetilde{c} = g_L^{(2)}(c) \\ \text{4. If Ans} = \\ \quad - \text{ Case } \bot: \text{Output } \bot \\ \quad - \text{ Case same*: If } \left(\widetilde{c} = c \text{ and flag}_1\right) = 1, \text{ output same*} \\ \qquad\qquad \text{Else output } \bot \\ \quad - \text{ Case } (\widetilde{k_1}, \widetilde{k_2}, \widetilde{t_1}, \widetilde{t_2}, \widetilde{s}): \text{If } \left(\text{Verify}_{\widetilde{k_1}}(\widetilde{c}, \widetilde{t_1}) = 0 \text{ or flag}_2 = 0\right), \text{ output } \bot \\ \qquad\qquad \text{Else output } \widetilde{c} \oplus \text{mask} \end{array}\right), m$

---

4   Construction for 3-Split-State Non-malleable Code

By re-organizing the information between states, we also obtain a rate-1/3 3-split-state non-malleable codes. Our coding scheme is defined in Fig. 4. Specifically, instead of storing w with R and L with c, we merge w and L into one

---

$\text{Enc}(m)$:	$\text{Dec}(c_1, c_2, c_3)$:
1. Sample $w \sim U_n$, $s \sim U_d$, $k_1 \sim U_\gamma$, $k_2 \sim U_{\gamma'}$	1. Let the tampered states be $\widetilde{c} := c_1$, $(\widetilde{w}, \widetilde{L}) := c_2$, $\widetilde{R} := c_3$
2. Compute $r = \text{Ext}(w, s)$, $c = m \oplus r$	2. Decrypt $(\widetilde{k_1}, \widetilde{k_2}, \widetilde{t_1}, \widetilde{t_2}, \widetilde{s}) = \text{Dec}^+(\widetilde{L}, \widetilde{R})$
3. Compute the tags $t_1 = \text{Tag}_{k_1}(c)$ and $t_2 = \text{Tag}'_{k_2}(w)$	3. If $(\widetilde{k_1}, \widetilde{k_2}, \widetilde{t_1}, \widetilde{t_2}, \widetilde{s}) = \bot$, output $\bot$
4. Compute the 2-state non-malleable encoding: $(L, R) \sim \text{Enc}^+(k_1, k_2, t_1, t_2, s)$	4. (Else) If $\text{Verify}_{\widetilde{k_1}}(\widetilde{c}, \widetilde{t_1}) = 0$ or $\text{Verify}'_{\widetilde{k_2}}(\widetilde{w}, \widetilde{t_2}) = 0$, then output $\bot$
5. Output the three states $\left(c, (w, L), R\right)$	5. (Else) Output $\widetilde{c} \oplus \text{Ext}(\widetilde{w}, \widetilde{s})$

Fig. 4. Non-malleable coding scheme against 3-split-state tampering.

state and store c, (w, L) and R independently. We defer the proof of non-malleability for this coding scheme to the full version [18]. By similar analysis as in 2-lookahead case, it is easy to see our non-malleable codes in 3-split-state scheme also has rate-$1/3$.

5   Forgetful Tampering in the 2-Lookahead Model

In this section we restrict ourselves to the 2-lookahead model. Let us define an additional family of tampering functions. Consider a tampering function $f\colon \{0,1\}^{n_1+n_2+n_3+n_4} \to \{0,1\}^{n_1+n_2+n_3+n_4}$. The function f is *1-forgetful*, if there exists a function $g\colon \{0,1\}^{n_2+n_3+n_4} \to \{0,1\}^{n_1+n_2+n_3+n_4}$ such that $f(x_1, x_2, x_3, x_4) = g(x_2, x_3, x_4)$ for all $x_1 \in \{0,1\}^{n_1}$, $x_2 \in \{0,1\}^{n_2}$, $x_3 \in \{0,1\}^{n_3}$, and $x_4 \in \{0,1\}^{n_4}$. Intuitively, the tampering function f forgets its first n_1-bits of the codeword and do the entire tampering using only x_2, x_3, x_4. The set of all functions that are 1-forgetful are represented by $\mathcal{FOR}_{n_1,n_2,n_3,n_4-\{1\}}$. Analogously, we define $\mathcal{FOR}_{n_1,n_2,n_3,n_4-\{i\}}$, for each $i \in \{2,3,4\}$.

Aggarwal et al. [2] proved that we can construct constant-rate non-malleable code in the 2-split-state from a constant-rate non-malleable code that protects against the following tampering family[5]

$$\left(\mathcal{LA}_{n_1,n_2} \times \mathcal{LA}_{n_3,n_4}\right) \bigcup_{i=1}^{4} \mathcal{FOR}_{n_1,n_2,n_3,n_4-\{i\}}$$

We make partial progress towards the goal of constructing non-malleable codes secure against above tampering family (and hence, constant rate codes against 2-split-state family), and prove the following theorem.

Theorem 6. *For all constants α, there exists a constant ζ and a computationally efficient non-malleable coding scheme against $\left(\mathcal{LA}_{n_1,n_2} \times \mathcal{LA}_{n_3,n_4}\right) \cup \mathcal{FOR}_{n_1,n_2,n_3,n_4-\{1\}} \cup \mathcal{FOR}_{n_1,n_2,n_3,n_4-\{3\}}$ with rate $\frac{1}{4+\alpha}$ and error $2^{-n^{\zeta}}$.*

We defer the proof of Theorem 6 to the full version [18].

References

1. Aggarwal, D., Agrawal, S., Gupta, D., Maji, H.K., Pandey, O., Prabhakaran, M.: Optimal computational split-state non-malleable codes. In: Kushilevitz, E., Malkin, T. (eds.) TCC 2016. LNCS, vol. 9563, pp. 393–417. Springer, Heidelberg (2016). https://doi.org/10.1007/978-3-662-49099-0_15
2. Aggarwal, D., Dodis, Y., Kazana, T., Obremski, M.: Non-malleable reductions and applications. In: Servedio, R.A., Rubinfeld, R. (eds.) 47th Annual ACM Symposium on Theory of Computing, Portland, OR, USA, 14–17 June 2015, pp. 459–468. ACM Press (2015)

---

[5] Specifically, Theorem 30 in [2] states that there exists a constant-rate non-malleable reduction from 2-split-state tampering family to the following tampering function family consisting of union of split-state lookahead and forgetful tampering functions.

3. Aggarwal, D., Dodis, Y., Lovett, S.: Non-malleable codes from additive combinatorics. In: Shmoys, D.B. (ed.) 46th Annual ACM Symposium on Theory of Computing, New York, NY, USA, 31 May–3 June 2014, pp. 774–783. ACM Press (2014)

4. Agrawal, S., Gupta, D., Maji, H.K., Pandey, O., Prabhakaran, M.: A rate-optimizing compiler for non-malleable codes against bit-wise tampering and permutations. In: Dodis, Y., Nielsen, J.B. (eds.) TCC 2015. LNCS, vol. 9014, pp. 375–397. Springer, Heidelberg (2015). https://doi.org/10.1007/978-3-662-46494-6_16

5. Ball, M., Dachman-Soled, D., Kulkarni, M., Malkin, T.: Non-malleable codes for bounded depth, bounded fan-in circuits. In: Fischlin, M., Coron, J.-S. (eds.) EUROCRYPT 2016. LNCS, vol. 9666, pp. 881–908. Springer, Heidelberg (2016). https://doi.org/10.1007/978-3-662-49896-5_31

6. Chandran, N., Goyal, V., Mukherjee, P., Pandey, O., Upadhyay, J.: Block-wise non-malleable codes. In: Chatzigiannakis, I., Mitzenmacher, M., Rabani, Y., Sangiorgi, D. (eds.) 43rd International Colloquium on Automata, Languages and Programming, ICALP 2016. LIPIcs, Rome, Italy, 11–15 July 2016, vol. 55, pp. 31:1–31:14. Schloss Dagstuhl - Leibniz-Zentrum fuer Informatik (2016)

7. Chattopadhyay, E., Li, X.: Explicit non-malleable extractors, multi-source extractors, and almost optimal privacy amplification protocols. In: Dinur, I. (ed.) 57th Annual Symposium on Foundations of Computer Science, New Brunswick, NJ, USA, 9–11 October 2016, pp. 158–167. IEEE Computer Society Press (2016)

8. Chattopadhyay, E., Zuckerman, D.: Non-malleable codes against constant split-state tampering. In: 55th Annual Symposium on Foundations of Computer Science, Philadelphia, PA, USA, 18–21 October 2014, pp. 306–315. IEEE Computer Society Press (2014)

9. Cheraghchi, M., Guruswami, V.: Capacity of non-malleable codes. CoRR, abs/1309.0458 (2013)

10. Cheraghchi, M., Guruswami, V.: Capacity of non-malleable codes. In: Naor, M. (ed.) 5th Innovations in Theoretical Computer Science, ITCS 2014, Princeton, NJ, USA, 12–14 January 2014, pp. 155–168. Association for Computing Machinery (2014)

11. Cheraghchi, M., Guruswami, V.: Non-malleable coding against bit-wise and split-state tampering. In: Lindell, Y. (ed.) TCC 2014. LNCS, vol. 8349, pp. 440–464. Springer, Heidelberg (2014). https://doi.org/10.1007/978-3-642-54242-8_19

12. Cramer, R., Dodis, Y., Fehr, S., Padró, C., Wichs, D.: Detection of algebraic manipulation with applications to robust secret sharing and fuzzy extractors. In: Smart, N. (ed.) EUROCRYPT 2008. LNCS, vol. 4965, pp. 471–488. Springer, Heidelberg (2008). https://doi.org/10.1007/978-3-540-78967-3_27

13. Dodis, Y., Ostrovsky, R., Reyzin, L., Smith, A.: Fuzzy extractors: How to generate strong keys from biometrics and other noisy data. SIAM J. Comput. 38(1), 97–139 (2008)

14. Dziembowski, S., Kazana, T., Obremski, M.: Non-malleable codes from two-source extractors. In: Canetti, R., Garay, J.A. (eds.) CRYPTO 2013. LNCS, vol. 8043, pp. 239–257. Springer, Heidelberg (2013). https://doi.org/10.1007/978-3-642-40084-1_14

15. Dziembowski, S., Pietrzak, K., Wichs, D.: Non-malleable codes. In: Yao, A.-C.-C. (ed.) 1st Innovations in Computer Science, ICS 2010, Tsinghua University, Beijing, China, 5–7 January 2010, pp. 434–452. Tsinghua University Press (2010)

16. Faust, S., Mukherjee, P., Venturi, D., Wichs, D.: Efficient non-malleable codes and key-derivation for poly-size tampering circuits. In: Nguyen, P.Q., Oswald, E. (eds.) EUROCRYPT 2014. LNCS, vol. 8441, pp. 111–128. Springer, Heidelberg (2014). https://doi.org/10.1007/978-3-642-55220-5_7

17. Goyal, V., Kumar, A.: Non-malleable secret sharing. In: Proceedings of the 50th Annual ACM SIGACT Symposium on Theory of Computing, STOC 2018, Los Angeles, CA, USA, 25–29 June 2018 (2018)

18. Gupta, D., Maji, H.K., Wang, M.: Non-malleable codes against lookahead tampering. Cryptology ePrint Archive, report 2017/1048 (2017). https://eprint.iacr.org/2017/1048

19. Guruswami, V., Umans, C., Vadhan, S.P.: Unbalanced expanders and randomness extractors from Parvaresh-Vardy codes. In: 22nd Annual IEEE Conference on Computational Complexity (CCC 2007), 13–16 June 2007, San Diego, California, USA, pp. 96–108 (2007)

20. Kanukurthi, B., Obbattu, S.L.B., Sekar, S.: Four-state non-malleable codes with explicit constant rate. In: Kalai, Y., Reyzin, L. (eds.) TCC 2017. LNCS, vol. 10678, pp. 344–375. Springer, Cham (2017). https://doi.org/10.1007/978-3-319-70503-3_11

21. Kanukurthi, B., Obbattu, S.L.B., Sekar, S.: Non-malleable randomness encoders and their applications. In: Nielsen, J.B., Rijmen, V. (eds.) EUROCRYPT 2018. LNCS, vol. 10822, pp. 589–617. Springer, Cham (2018). https://doi.org/10.1007/978-3-319-78372-7_19

22. Li, X.: Improved non-malleable extractors, non-malleable codes and independent source extractors. In: Hatami, H., McKenzie, P., King, V. (eds.) 49th Annual ACM Symposium on Theory of Computing, Montreal, QC, Canada, 19–23 June 2017, pp. 1144–1156. ACM Press (2017)

23. Li, X.: Pseudorandom correlation breakers, independence preserving mergers and their applications. In: Electronic Colloquium on Computational Complexity (ECCC), vol. 25, p. 28 (2018)

24. Vadhan, S.P.: Pseudorandomness. Foundations and Trends in Theoretical Computer Science. Now Publishers (2012)

Obfuscation from Low Noise Multilinear Maps

Nico Döttling[1], Sanjam Garg[2], Divya Gupta[3], Peihan Miao[2],
and Pratyay Mukherjee[4(✉)]

[1] Center of IT-Security, Privacy and Accountability, Saarbrücken, Germany
[2] University of California, Berkeley, USA
[3] Microsoft Research India, Bengaluru, India
[4] Visa Research, Palo Alto, USA
pratyay85@gmail.com

Abstract. Multilinear maps enable homomorphic computation on encoded values and a public procedure to check if the computation on the encoded values results in a zero. Encodings in known candidate constructions of multilinear maps have a (growing) noise component, which is crucial for security. For example, noise in GGH13 multilinear maps grows with the number of levels that need to be supported and must remain below the maximal noise supported by the multilinear map for correctness. A smaller maximal noise, which must be supported, is desirable both for reasons of security and efficiency.

In this work, we put forward new candidate constructions of obfuscation for which the maximal supported noise is polynomial (in the security parameter). Our constructions are obtained by instantiating a modification of Lin's obfuscation construction (EUROCRYPT 2016) with composite order variants of the GGH13 multilinear maps. For these schemes, we show that the maximal supported noise only needs to grow polynomially in the security parameter. We prove the security of these constructions in the weak multilinear map model that captures *all known* vulnerabilities of GGH13 maps. Finally, we investigate the security of the considered composite order variants of GGH13 multilinear maps from a cryptanalytic standpoint.

1 Introduction

Program obfuscation aims to make computer programs "unintelligible" while keeping their functionalities intact. The known obfuscation constructions [5–8,11,12,26,27,30,39,46,51] are all based on new candidate constructions

Research supported in part from DARPA/ARL SAFEWARE Award W911NF15C0210, AFOSR Award FA9550-15-1-0274, AFOSR YIP Award, DARPA and SPAWAR under contract N66001-15-C-4065, a Hellman Award and research grants by the Okawa Foundation, Visa Inc., and Center for Long-Term Cybersecurity (CLTC, UC Berkeley). The views expressed are those of the author and do not reflect the official policy or position of the funding agencies.

© Springer Nature Switzerland AG 2018
D. Chakraborty and T. Iwata (Eds.): INDOCRYPT 2018, LNCS 11356, pp. 329–352, 2018.
https://doi.org/10.1007/978-3-030-05378-9_18

[17, 18, 25, 28] of multilinear maps [10], security of which is poorly understood [14–16, 25, 34, 44].

Briefly, multilinear maps (a.k.a. graded encodings) allow "leveled" homomorphic computations of a-priori bounded degree (say κ) polynomials on "encoded" values. Furthermore, they provide a mechanism to publicly check if the result of a polynomial computation is a zero or not. At a high level, known obfuscation methods map the program to a sequence of encodings. These encodings are such that the output of the program on a specific input is zero if and only if the output of a corresponding input dependent polynomial (of degree κ) on the encoded values yields a zero.

Noise in GGH-Based Obfuscations. Encodings in the known candidate multilinear map[1] constructions are generated to have a noise component (referred to as "fresh" encoding/noise[2]) that is necessary for security. Homomorphic computations on these fresh encodings yield encodings with increased noise due to accumulation of the fresh noise (hence called "accumulated" noise). In the candidate construction by Garg, Gentry and Halevi [25] (a.k.a. GGH), the noise level in the fresh level-1 encodings can be set to be as low as a polynomial in the security parameter, without hurting the security. However, the noise level in the fresh level-i encodings grows exponentially in i.[3] Furthermore, the accumulated noise also grows with the number of homomorphic multiplications. The GGH construction is parameterized by a modulus q that needs to be greater than the maximum supported noise (referred to as "noise bound") of any encoding in the system in order to preserve functionality. Most obfuscation constructions involve homomorphic multiplication of polynomially many "fresh" encodings. Therefore, these constructions need to support exponentially large noise. An exception[4] to this is the recent construction of Lin [39] that only needs a constant number of multiplications on composite-order multilinear maps. However, this construction still needs to give out "fresh" encodings at polynomially high levels. Thus it would still need exponential noise if one was to use a composite order variant of GGH multilinear maps (e.g. the one eluded to in the first EPRINT draft of GGH [24] or the one from [31, Appendix B.3]). Another alternative is to use a composite order variant of the [17, 18] multilinear maps, e.g. the one by Gentry et al. [31, Appendix B.3 and B.4]. Note that in the CLT based constructions the number of primes needed is always polynomial in the security parameter. This

---

[1] Throughout this paper, we use multilinear maps to refer to private encoding multilinear maps. In other words, no public low-level encodings of zero are provided in our constructions.

[2] By "fresh" encodings we mean that it is generated via the encoding procedure using the secret parameters and *not* produced as a result of homomorphic computations.

[3] The reported noise is for the recommended version of the GGH multilinear maps [24, Sect. 6.4]. This recommendation was made in [25] with the goal of avoiding averaging attacks [20, 33, 45]. Similar recommendation is made in [3, Sect. 4.2].

[4] Another exception is [37] which uses Reniy divergence to construct a map called GGHLite that supports more efficient concrete parameters than GGH.

is the case even if the construction itself uses a constant number of slots, as is the case in Lin's scheme. This use of polynomially many primes is essential for security — specifically, in order to avoid lattice attacks. Consequently, the noise growth in this case is also exponential (as elaborated on in [31, Appendix B.5]).

In the context of GGH multilinear maps, the use of an exponential "noise bound," and hence the modulus q, is not desirable in light of the recent NTRU attacks[5] [3,35]. It is desirable to have a much smaller value of q (say $\mathsf{poly}(\lambda)$). Furthermore, having a smaller modulus offers asymptotic efficiency improvements.

Weak Multilinear Map Model for GGH. Typically, candidate obfuscation schemes (including the above constructions) are proven secure in so-called *ideal graded encoding model*, that does not capture all the known vulnerabilities of the GGH multilinear maps [25,34,44]. In particular, Miles, Sahai and Zhandry [44] exploit these vulnerabilities of GGH to show attacks against obfuscation constructions. In light of these attacks, [44] proposed the *weak multilinear map model* that better captures the known vulnerabilities of the GGH multilinear maps. Subsequently, Garg et al. [27] gave an obfuscation scheme provable secure in this model.

In this work, we ask the following question.

Can we construct an obfuscation scheme using low noise multilinear maps and prove its security in the weak multilinear map model?

1.1 Our Result

In this work, we resolve the above question affirmatively providing new candidate indistinguishability obfuscation constructions such that: (i) they only require a modulus q which grows polynomially in the security parameter, and (ii) they are provably secure in the weak multilinear map model.

Our construction is instantiated using composite order GGH multilinear maps[6] that are the same as the composite order proposal of [24] except that

---

[5] Specifically, the subfield lattice attack is sub-exponential as soon as q is super-polynomial. Furthermore, using attack of [35] becomes polynomial for power-of-two cyclotomic fields when $q = 2^{\Omega(\sqrt{n \log \log n})}$. We note that the attack of [35] is much more general, but we are only concerned with these parameter choices.

[6] In the first draft of [24], authors suggested a composite order variant of multilinear maps. However, in later versions they restricted their claims to the prime order construction. This was in light of the weak-discrete log attacks they found against their construction. However, these attacks worked only when public encodings of zero are provided and rendered assumptions such as subgroup hiding easy. In particular, all known attacks against composite order GGH maps use low level encodings of zero [25] or some specific high-level encodings of zero [15]. In light of the Miles et al. attacks [44] we envision more (potential) attacks but they are all captured by the weak multilinear map model.

we use a specific choice of the Lagrange Coefficients used in Chinese Remaindering in our construction.[7] This specific choice of Lagrange Coefficients is done in order to strengthen security — specifically, in order to obtain a proof of security in the weak multilinear map model. We evaluate the security of the GGH composite order multilinear maps (with our choice of Lagrange Coefficients) in light of known attack strategies (see Sect. C.3 of the full version [19]).

Next, in order to enable constructions with low noise, we suggest two ways to modify the GGH sampling procedure [24, Sect. 6.4] such that: (i) The first one is a variation of the original GGH sampling procedure. (ii) Our second variant departs more from the GGH sampling procedure but obtains better efficiency in terms of the dimension of the lattice necessary. From a cryptanalytic standpoint (see Sect. C.3 of the full version [19]), we do not know of any attacks against this more aggressive variation.

Additional Contribution. As mentioned earlier, recent work by Garg et al. [27] provides the first construction of obfuscation in the weak multilinear map model. However, this construction works by converting a circuit into a branching program. Our work also provides a direct construction (obtained by slightly modifying our main construction) for circuits, for which security can be argued in the weak multilinear map model. Previous works [6,51] in this direction proved security only in the ideal graded encoding model.

Independent and Follow-Up Work. In a concurrent and independent work, Lin and Vaikuntanathan [42] obtain a construction which when instantiated with GGH multilinear maps would yield a construction that supports low noise. However, a bonus of our scheme is that it is proved secure in the weak multilinear map model. Furthermore, the techniques introduced in this work are orthogonal to the work of Lin and Vaikuntanathan [42] and are of independent interest. Following [42], Lin [40] and Ananth and Sahai [4] provided constructions of iO from degree 5 multilinear maps, which were subsequently improved to degree 3 by Lin and Tessaro [41]. Both these results rely on yet unrealized notions of noise-free multilinear maps and do not deal with imperfections of candidate constructions of multilinear maps. In particular, they have no mechanism to protect against zeroizing attacks.

Ducas and Pellet-Mary [21] consider the security of the modified straddling set systems proposed in this work and identify a statistical leak in one of our two candidates. This statistical leak weakly depends on the (secret) ideal generator g of the ideal lattice $\mathcal{I}$. While [21] shows that this leak can be used to attack a simplified instantiation of our multilinear maps, it falls short of a full attack.

---

[7] We do not provide public encodings of zero in our constructions. Therefore, they are insufficient to instantiate the assumptions made by Gentry et al. [29,30].

1.2   Technical Overview

We start from a brief overview of Lin's construction [39].

Overview of Lin's Construction: $i\mathcal{O}$ **from constant-degree multilinear map.** It has two main steps.

Step-1: Stronger bootstrapping. All existing candidates of indistinguishability obfuscation ($i\mathcal{O}$ for short) for all circuits (i.e., P/Poly) rely on "bootstrapping" $i\mathcal{O}$ for weaker class of circuits. Known techniques [13,26] require $i\mathcal{O}$ for NC^1 to start with: the idea is to first construct a scheme only for NC^1 circuits and then use cryptographic primitives (e.g., fully homomorphic encryption) to "bootstrap" this into a construction for P/Poly. In contrast, [39] uses a much stronger bootstrapping technique that only requires $i\mathcal{O}$ (with some necessary efficiency requirements) for *specific* constant-degree circuits (as opposed to *general* NC^1 circuits in the earlier constructions). To realize that, only multilinear maps supporting a constant number of multiplications suffice. Such specific circuit class is referred to as the "seed class" and denoted by $\mathcal{C}_{seed}$ in the following.

Step-2: Special purpose $i\mathcal{O}$ **for** $\mathcal{C}_{seed}$**.** In the second step, [39] gives a candidate $i\mathcal{O}$-construction for this seed class. The construction builds on the techniques from [6,51] for obfuscating NC^1 circuits directly while ensuring constant-degree computation. Lin then proves that her construction is secure in the ideal graded encoding model.

Our Techniques: Main Steps. To achieve our result, we build on the bootstraping result of [39] and focus on building the $i\mathcal{O}$-candidate (Step-2) for $\mathcal{C}_{seed}$ such that it only requires a polynomial sized modulus and is secure in weak multilinear map model. Our main steps of construction are as follows:

1. **Composite-order GGH multilinear map.** We propose a composite-order extension of the GGH multilinear map candidate. Our candidate is the same as the first proposal of GGH maps (as in the first EPRINT version of [24]) except that we use specific Lagrange Coefficients in Chinese Remaindering in the construction. This choice allows us to argue security in the weak multilinear map model.

2. **Security in the weak multilinear map model.** We strengthen the security of the basic $i\mathcal{O}$ construction of [39] via the so-called *self-fortification* technique, similar to [27]. As a result we are able to prove that our construction is secure in the (GGH-based) weak multilinear map model (see Appendix F of the full version [19] for details on the model).

3. **GGH with low-noise.** We propose two modifications of composite-order GGH multilinear maps such that all "fresh" encodings that need to be provided in our construction can be provided with noise of size $\mathsf{poly}(\lambda)$. Moreover, any κ degree computation results in final encodings with noise of size $O(\exp(\kappa)\mathsf{poly}(\lambda))$. Using the fact that $\mathcal{C}_{seed}$ has constant degree (i.e., κ is constant), we obtain polynomial sized modulus q.

Overview of Composite-Order GGH Multilinear Maps. An instance of the GGH scheme is parameterized by the security parameter λ and the required multilinearity level $\kappa \leq \text{poly}(\lambda)$. Based on these parameters, consider the $2n$-th cyclotomic ring $R = \mathbb{Z}[X]/(X^n + 1)$, where n is a power of 2 (n is set large enough to ensure security), and a modulus q that defines $R_q = R/qR$ (with q large enough to support functionality). The secret encoding procedure encodes elements of a quotient ring $R/\mathcal{I}$, where $\mathcal{I}$ is a principal ideal $\mathcal{I} = \langle g \rangle \subset R$, generated by g. In the composite order setting, g is equal to a product of several (say t) "short" ring elements $g_1, g_2, \ldots, g_t$. These ring elements are chosen such that the norms $N(g_i) = |R/\langle g_i \rangle|$ are equal to "large" (exponential in λ) primes p_i for each g_i. By the Chinese Remainder Theorem (CRT for short) one can observe that the following isomorphism $R/\mathcal{I} \cong R/\mathcal{I}_1 \times \ldots \times R/\mathcal{I}_t$ for ideals $\mathcal{I}_i = \langle g_i \rangle$ holds. Hence each element $e \in R/\mathcal{I}$ has an equivalent CRT representation in $R/\mathcal{I}_1 \times \ldots \times R/\mathcal{I}_t$ that is denoted by $(e[\![1]\!], \ldots, e[\![t]\!])$. Recall that, in this representation it holds that $e \equiv e[\![i]\!] \bmod \mathcal{I}_i$ and $e[\![i]\!]$ is called the value of e in the i-th *slot*; moreover, any arithmetic operation over $R/\mathcal{I}$ can be done "slot-wise." The short generator g (and all g_i) is kept secret, and no "good" description of $\mathcal{I}$ (or of $\mathcal{I}_i$) is made public.

Let $\mathbb{U}$ denote the universe such that $\mathbb{U} = [\ell]$.[8] To enforce the restricted multilinear structure (a.k.a. straddling sets) secrets $z_1, \ldots, z_\ell$ are sampled randomly from R_q (and hence, they are "not short"). The sets $\mathbf{v} \subseteq \mathbb{U}$ are called the levels. An encoding of an element $a \in R/\mathcal{I}$ at a level $\mathbf{v}$ is given by $e = [\tilde{a}/\prod_{i \in \mathbf{v}} z_i]_q \in R_q$ where $\tilde{a}$ is a "short element" in the coset $a + \mathcal{I}$ sampled via a specific procedure.[9] The quantity $\|\tilde{a}\|$ is called the noise of the encoding e and is denoted by $\mathsf{noise}(e)$. Rigorous calculation from the sampling procedure (c.f. Appendix A of the full version [19]) shows that $\mathsf{noise}(e) = O(\exp(t, |\mathbf{v}|))$. Note that in Lin's construction [39] as well as our construction, t will be a constant, but $|\mathbf{v}|$ is not.

The arithmetic computations are restricted by the levels of the encodings: addition is allowed between encodings in the same level whereas multiplication is allowed at levels $\mathbf{v}$ and $\mathbf{v}'$ when $\mathbf{v} \cap \mathbf{v}' = \emptyset$.[10] Furthermore, the GGH map provides a public zero-testing mechanism to check if any given encoding at level $\mathbb{U}$ is an encoding of an element that is equal to $0 \bmod g$ (equivalently $0 \bmod g_i$ in the i-th slot for all $i \in [t]$). Notice that since the map allows κ-degree computations, the noise in the top-level encoding resulting after such a computation can be at most $O(\exp(\kappa, t, \ell))$.

Reducing Noise in GGH. We first elaborate on the GGH sampling procedure [24, Sect. 6.4] as follows: To encode at level $\mathbf{v} \subseteq \mathbb{U}$, the encoding procedure

---

[8] In the actual construction the structure of the elements of $\mathbb{U}$ are much involved. But for simplicity here we just assume $\mathbb{U} = [\ell]$ that suffices to convey the main idea.

[9] We use the notation $[\cdot]_q$ to denote operations in R_q.

[10] Note that in the actual construction we use different restriction for multiplication due to difference in the structure of the straddling levels and the universe.

samples a ring element from the fractional ideal $\langle g/z_{\mathbf{v}} \rangle$, where $z_{\mathbf{v}} = \prod_{i \in \mathbf{v}} z_i$.[11]
Hence, the amount of noise generated by the encoding procedure depends on
the size of the generator $g/z_{\mathbf{v}}$, which is in turn dominated by the size of $1/z_{\mathbf{v}}$.
Generally, following [25], one can sample *atoms* z_i such that their inverse $1/z_i$
is short in K, say n^2/q (where K is the quotient field of R). Now, expressing
$z_{\mathbf{v}}$ as $z_{\mathbf{v}} = \prod_{i \in \mathbf{v}} z_i$ we obtain $\|1/z_{\mathbf{v}}\| = O(\exp(|\mathbf{v}|)/q^{|\mathbf{v}|})$. We show in the full
version [19] (Appendix A) that the noise of the fresh encoding is dominated by
$\|z_{\mathbf{v}}\| \cdot \|1/z_{\mathbf{v}}\|$ which grows exponentially with $|\mathbf{v}|$, i.e., the cardinality of $\mathbf{v}$. As
mentioned earlier, in Lin's construction, some elements are encoded at levels $\mathbf{v}$
of cardinality polynomial in λ resulting in fresh encodings of noise $O(\exp(\lambda))$.

To handle the noise in encodings more carefully, we provide two possible
techniques specific to our construction. The first technique is fairly simple and
works by choosing the degree n of the ring R sufficiently large (larger than the size
of the circuit we obfuscate). With this parameter choice we can guarantee with
probability close to 1 that for all levels at which we encode and the zero-testing
level that $\|z_{\mathbf{v}}\| \cdot \|1/z_{\mathbf{v}}\| = \mathsf{poly}(n)$. This comes at the expense of making the
degree n of the ring R grow with the size of the circuit (which is still polynomial
in the security parameter).

The second technique follows a different strategy and avoids the dependence
of n on the size of the circuit. We first observe that, in our obfuscation construction many combinations of $\prod_i z_i$ (i.e. many subsets of $[\ell]$) terms actually
never arise. We illustrate our main idea with a toy example. Assume that we
only need to encode in levels $\hat{\mathbf{v}}_i = [\ell] \setminus \{i\}$ and $\mathbf{v}_i = \{i\}$ for all $i \in [\ell - 1]$ (and
not at the level $\{\ell\}$). Now, if we were to follow the above sampling procedure
then clearly we will end up with $\|z_{\hat{\mathbf{v}}_i}\| \cdot \|1/z_{\hat{\mathbf{v}}_i}\| = O(\exp(\ell - 1))$. Instead, we
actually follow a different strategy, namely we sample all the z_i for $i \in [\ell - 1]$
except the last z_ℓ term "as usual", i.e. such that $1/z_i$ is "short" in K. However
for the one remaining term (i.e. z_ℓ) we instead sample another value $z^\star$, such
that $1/z^\star$ is "short" in K and then set

$$z_\ell := \left[\frac{z^\star}{\left(\prod_{i \in [\ell-1]} z_i \right)} \right]_q .$$

Furthermore, we require that for $i \in [\ell - 1]$, $1/[z_i^{-1}]_q$ is also short in K. We can
now compute a value $z_{\hat{\mathbf{v}}_i} := z^\star \cdot [z_i^{-1}]_q$.[12] Observe that it holds that $[z_{\hat{\mathbf{v}}_i}]_q =$
$[\prod_{i \in [\ell] \setminus \{i\}} z_i]_q$ as desired. Moreover, $1/z_{\hat{\mathbf{v}}_i}$ is now short in K:

$$\|1/z_{\hat{\mathbf{v}}_i}\| = \|1/(z^\star \cdot [z_i^{-1}]_q)\| \leq \sqrt{n} \cdot \|1/z^\star\| \cdot \|1/[z_i^{-1}]_q\|,$$

which is "short". The cost incurred by this modification is that $1/z_\ell$ may *not*
be "short" in K. However, this will not pose a problem as z_ℓ is not used to

---

[11] Notice that $g/z_{\mathbf{v}}$ is in K. We generally use $a/b \in K$ to denote "division" in the
quotient field K of R for $a, b \in R$. This is not to be confused with the notation
$a \cdot [b^{-1}]_q \in R$ which is a multiplication operation in R where the inverse $[b^{-1}]_q$ is
in R_q.

[12] Notice that, the inverse is in R_q but the product is in R.

sample encodings anyway (recall that we do *not* require to sample at level $\{\ell\}$). We show that such modification brings down the noise bound of fresh encodings to $O(\mathsf{poly}(\ell)\mathsf{exp}(t))$ and maximum noise bound (in any encoding produced in our construction) to $O(\mathsf{poly}(\ell)\mathsf{exp}(\kappa, t))$. In our scheme, both κ and t will be constants.

Our Security Model: The Weak Multilinear Map Model. Typically, obfuscation candidates[13] were proven secure in the so-called *ideal graded encoding* model. In contrast, we prove security of our construction in the *weak multilinear map model* [44], a model that captures all currently known vulnerabilities of multilinear maps. This model is similar to the ideal multilinear map model (a.k.a., the ideal graded encoding model). However, it additionally allows for computation on ring elements resulting from a zero-test performed on encodings of 0. The security definition requires that the adversary can not come up with a polynomial which evaluates to 0 over these post-zero ring elements. In the composite order setting we require that the adversary can not come up with a polynomial which evaluates to 0 in *any* of the slots. Unlike the ideal model, this model is *not entirely agnostic* about the underlying multilinear map instantiation. In particular, our weak multilinear map model is based on the composite-order GGH multilinear maps and captures *all* the attack directions investigated in our cryptanalysis.

Self-fortification from Constant-Degree Multilinear Map. To prove security of our obfuscation candidate in the weak multilinear map model, we make another modification to Lin's obfuscation scheme for C_{seed} using a self-fortification technique similar to [27].

Recall that multilinear maps allow for testing of zero-encodings at the universe set (a.k.a. the top level). All known attacks against multilinear map candidates exploit the "sensitive information" leaked upon a successful zero-test. To protect against these attacks, the idea of [27] is to render this "sensitive leakage" useless by "masking" it with a PRF output. Similarly, we achieve this by augmenting the given circuit C with a *parallel* PRF computation, the output of which is used to mask the leakage from the real computation. More care is required so that the PRF computation does *not* affect the actual computation of C and "comes alive" only after a successful zero-test.

Before we describe our transformation, let us first describe the techniques of obfuscating circuits directly of [6,51], also used in [39]. At a high level, consider a universal circuit $\mathcal{U}$ that takes as input the circuit (to obfuscate) C and the input x to C and outputs $C(x)$. The obfuscation consists of a collection of values in $R/\mathcal{I}$ encoded at carefully chosen levels (i.e., straddling sets). Multiple slots are used where w.l.o.g. the first slot is used for actual computation and a bunch of other

---

[13] There are some works e.g. [7,11] that prove security of their constructions in slightly stronger models than the ideal graded encoding model which captures some attacks on multilinear maps.

slots are added with random values. These random values along with the choice of straddling sets ensure that the random values are nullified only with a correct (and consistent) evaluation corresponding to some input x. More precisely, a correct evaluation leads to an encoding of $(\mathcal{U}(\mathcal{C},x) \bmod g_1, 0 \bmod g_2, \ldots, 0 \bmod g_t)$ at the highest level $\mathbb{U}$; zero-testing of which would reveal the output. On the other hand, any incorrect computation would not cancel out all random values, and hence would result in a non-zero value in $\bmod\, g$ with all but negligible probability.

Our idea is to add an extra slot (say the second slot) for PRF computation such that a correct computation produces an encoding of $(\mathcal{U}(\mathcal{C},x) \bmod g_1, g_2 \cdot \mathcal{U}(\mathcal{C}^{\mathsf{PRF}_\psi}, x), 0 \bmod g_3, \ldots, 0 \bmod g_{t+1})$[14] at the top level.[15] Notice that due to a g_2 multiplier in the second slot, the computation is not affected by the PRF output as the value in the second slot is still $0 \bmod g_2$. Nonetheless, we show that a successful zero-test returns a ring element (say $\boldsymbol{f}$) in $R/\mathcal{I}$ that has an additive blinding factor $\boldsymbol{\alpha} \cdot \mathcal{C}^{\mathsf{PRF}_\psi}(x)$ for some $\boldsymbol{\alpha} \in R/\mathcal{I}$. Furthermore, we are able to show that as long as $\boldsymbol{\alpha}$ is invertible in (the composite order quotient ring) $R/\mathcal{I}$ the CRT representation of $\boldsymbol{f}$ given by $(\boldsymbol{f}[\![1]\!], \ldots, \boldsymbol{f}[\![t]\!])$ is "somewhat random" in each slot (formally, $\boldsymbol{f}[\![i]\!]$ has high min-entropy).

Cryptanalysis. In the full version [19] (Appendix C.2), we discuss our change to the composite order generators g from a cryptanalytic perspective. In a nutshell, existing lattice attacks, such as attacks against overstretched NTRU assumptions [3,35], do not exploit the specific distribution of instances, but rather geometric properties (i.e. noise terms being short). Thus, our construction resists currently known lattice attacks and there is no reason to believe choosing composite generators $g = \prod_i g_i$ leads to less secure schemes than choosing primes ones. However, we do know that top-level encodings of zero, with correlated randomness, can be dangerous. This is especially the case if they can be used to obtain an element in the ideal $\langle g \rangle$. In the composite order setting, we expect potential attacks if an element in the ideal $\langle g_i \rangle$ for any i can be computed. However, all these potential attacks are captured by the weak multilinear map model that we consider. At a high level, our proof in the weak multilinear map model guarantees that no element in the ideal $\langle g_i \rangle$ for any i can be computed.

In the full version [19] (Appendix C.3), we discuss reasons for the believed security of our variants of the GGH sampling procedure.

Roadmap. The rest of the paper is organized as follows. After providing notations and basic preliminaries on lattices in Sect. 2, in Sect. 3 we briefly summarize Lin's bootstrapping theorem and a few related definitions. Our main $i\mathcal{O}$-construction is provided in Sect. 4. We defer the rest to the full version [19], in which we provide a composite-order GGH multilinear map candidate

---

[14] $\mathcal{C}^{\mathsf{PRF}_\psi}$ is a circuit for computing PRF with the key ψ.

[15] In the construction this is implemented by canceling out the PRF value by multiplying with an appropriate encoding that encodes a value which is $0 \bmod g_2$ in the second slot.

(Appendix A); our our modifications on the composite-order GGH multilinear map to achieve low noise (Appendix B); a cryptanalytic discussion of our modifications to the asymmetric GGH multilinear maps (Appendix C); the formal description of weak multilinear map model (Appendix F); some additional preliminaries on number fields and ideal lattices (Appendix E).

2   Preliminaries

Notations. The natural security parameter throughout this paper is λ, and all other quantities are implicitly assumed to be functions of λ. We use standard big-O notation to classify the growth of functions, and say that $f(\lambda) = \tilde{O}(g(\lambda))$ if $f(\lambda) = O(g(\lambda) \cdot log^c \lambda)$ for some fixed constant c. We let $\mathsf{poly}(\lambda)$ denote an unspecified function $f(\lambda) = O(\lambda^c)$ for some constant c. A *negligible* function, denoted generically by $\mathsf{negl}(\lambda)$, is an $f(\lambda)$ such that $f(\lambda) = o(\lambda^{-c})$ for every fixed constant c. We say that a function is *overwhelming* if it is $1 - \mathsf{negl}(\lambda)$.

The *statistical distance* between two distributions X and Y over a domain D is defined to be $\frac{1}{2}\sum_{d \in D} |\Pr[X = d] - \Pr[Y = d]|$. We say that two ensembles of distributions $\{X_\lambda\}$ and $\{Y_\lambda\}$ are *statistically indistinguishable* if for every λ the statistical distance between X_λ and Y_λ is negligible in λ.

Two ensembles of distributions $\{X_\lambda\}$ and $\{Y_\lambda\}$ are *computationally indistinguishable* if for every probabilistic poly-time non-uniform (in λ) machine $\mathcal{A}$, $|\Pr[\mathcal{A}(1^\lambda, X_\lambda) = 1] - \Pr[\mathcal{A}(1^\lambda, Y_\lambda) = 1]|$ is negligible in λ. The definition is extended to non-uniform families of poly-sized circuits in the standard way.

Lemma 1 (Schwarz-Zippel Lemma). *Let $\mathbb{F}$ be a finite field and let $p \in \mathbb{F}[x_1, \ldots, x_n]$ be a multivariate polynomial of degree at most d. Further let $X_1, \ldots, X_n$ be independently distributed random variables on $\mathbb{F}$ such that $H_\infty(X_i) \geq k$ for all i. Then it holds that*

$$\Pr[p(X_1, \ldots, X_n) = 0] \leq \frac{d}{2^k},$$

where the probability runs over the random choices of $X_1, \ldots, X_n$.

2.1   Lattices

We denote set of complex number by $\mathbb{C}$, real numbers by $\mathbb{R}$, the rationals by $\mathbb{Q}$ and the integers by $\mathbb{Z}$. For a positive integer n, $[n]$ denotes the set $\{1, \ldots, n\}$. By convention, vectors are assumed to be in column form and are written using bold lower-case letters, e.g. $\boldsymbol{x}$. The ith component of $\boldsymbol{x}$ will be denoted by x_i. We will use $\boldsymbol{x}^T$ to denotes the transpose of $\boldsymbol{x}$. For a vector $\boldsymbol{x}$ in $\mathbb{R}^n$ or $\mathbb{C}^n$ and $p \in [1, \infty]$, we define the ℓ_p norm as $\|\boldsymbol{x}\|_p = \left(\sum_{i \in [n]} |x_i|^p\right)^{1/p}$ where $p < \infty$, and $\|\boldsymbol{x}\|_\infty = \max_{i \in [n]} |x_i|$ where $p = \infty$. Whenever p is not specified, $\|\boldsymbol{x}\|$ is assumed to represent the ℓ_2 norm (also referred to as the Euclidean norm).

Matrices are written as bold capital letters, e.g. X, and the ith column vector of a matrix X is denoted x_i. Finally we will denote the transpose and the inverse (if it exists) of a matrix X with X^T and X^{-1} respectively.

A lattice Λ is an additive discrete sub-group of $\mathbb{R}^n$, i.e., it is a subset $\Lambda \subset \mathbb{R}^n$ satisfying the following properties:

(subgroup) Λ is closed under addition and subtraction,
(discrete) there is a real $\varepsilon > 0$ such that any two distinct lattice points $x \neq y \in \Lambda$ are at distance at least $\|x - y\| \geq \varepsilon$.

Let $B = \{b_1, \ldots, b_k\} \subset \mathbb{R}^n$ consist of k linearly independent vectors in $\mathbb{R}^n$. The lattice generated by the B is the set

$$\mathcal{L}(B) = \{Bz = \sum_{i=1}^{k} z_i b_i : z \in \mathbb{Z}^k\},$$

of all the integer linear combinations of the columns of B. The matrix B is called a *basis* for the lattice $\mathcal{L}(B)$. The integers n and k are called the *dimension* and *rank* of the lattice. If $n = k$ then $\mathcal{L}(B)$ is called a *full-rank* lattice. We will only be concerned with full-rank lattices, hence unless otherwise mentioned we will assume that the lattice considered is full-rank.

For lattices $\Lambda' \subseteq \Lambda$, the quotient group Λ/Λ' (also written as $\Lambda \bmod \Lambda'$) is well-defined as the additive group of distinct *cosets* $v + \Lambda'$ for $v \in \Lambda$, with addition of cosets defined in the usual way.

2.2   Gaussians on Lattices

Review of Gaussian measure over lattices presented here follows the development by prior works [1,2,32,43,48]. For any real $s > 0$, define the (spherical) *Gaussian function* $\rho_s : \mathbb{R}^n \to (0,1]$ with parameter s as:

$$\forall x \in \mathbb{R}^n, \rho_s(x) = \exp(-\pi \langle x, x \rangle / s^2) = \exp(-\pi \|x\|^2 / s^2).$$

For any real $s > 0$, any n-dimensional lattice Λ and any vector $c \in \mathbb{R}^n$, define the (spherical) *discrete Gaussian distribution* over the coset $\Lambda + c$ as:

$$\forall x \in \Lambda + c, D_{\Lambda+c,s}(x) = \frac{\rho_s(x)}{\rho_s(\Lambda + c)}.$$

Klein [36] and Gentry, Peikert and Vaikuntanathan [32] provide an efficient algorithm to sample from a discrete gaussian given a *good* basis. We will use a version of this algorithm due to Peikert [47] which directly samples from a coset of a lattice.

Theorem 1 ([47], **Theorem 4.2**). *There exists an efficient algorithm* SampleD, *which given a basis B of an n-dimensional lattice Λ and a parameter $s \geq \|B\| \cdot \omega(\sqrt{\log(n)})$ and any vector c efficiently samples a distribution within negligible distance of $D_{\Lambda+c,s}$.*

Smoothing Parameter. Micciancio and Regev [43] introduced a lattice quantity called the *smoothing parameter*, and related it other lattice parameters.

Definition 1 (Smoothing Parameter, [43, Definition 3.1]**).** *For an n-dimensional lattice Λ, and positive real $\varepsilon > 0$, we define its smoothing parameter denoted $\eta_\varepsilon(\Lambda)$, to be the smallest s such that $\rho_{1/s}(\Lambda^* \setminus \{0\}) \leq \varepsilon$.*

Intuitively, for a small enough ε, the number $\eta_\varepsilon(\Lambda)$ is sufficiently larger than a fundamental parallelepiped of Λ so that sampling from the corresponding Gaussian "wipes out the internal structure" of Λ. The following Lemma 2 formally provide this claim. Finally Lemma 3 provides bounds on the length of a vector sampled from a Gaussian.

Lemma 2 ([32, Corollary 2.8]). *Let Λ, Λ' be n-dimensional lattices, with $\Lambda' \subseteq \Lambda$. Then for any $\varepsilon \in (0, \frac{1}{2})$, any $s \geq \eta_\varepsilon(\Lambda')$, the distribution of $(D_{\Lambda,s} \pmod{\Lambda'})$ is within a statistical distance at most 2ε of uniform over $(\Lambda \pmod{\Lambda'})$.*

Lemma 3 ([43, Lemma 4.4] **and** [9, Proposition 4.7]). *For any n-dimensional lattice Λ, an $s \geq \eta_\varepsilon(\Lambda)$ for some negligible ε, any vector c and any constant $\delta > 0$ we have*

$$\Pr_{x \leftarrow D_{\Lambda+c,s}} \left[(1-\delta)s\sqrt{\frac{n}{2\pi}} \leq \|x\| \leq (1+\delta)s\sqrt{\frac{n}{2\pi}} \right] \geq 1 - \mathsf{negl}(n).$$

Invertibility of Ring Elements. Let R denote the $2n^{th}$ cyclotomic ring and let R_q denote R/qR for a prime q. We note that R_q is also a ring and not all elements in it are invertible. Let $R_q^\times$ denote the set of elements in R_q that are invertible. We next provide a lemma of Stehlé and Steinfeld that points out that a (large enough) random element is R_q is also in $R_q^\times$ with large probability.

Lemma 4 ([50, Lemma 4.1]). *Let $n \geq 8$ be a power of 2 such that $X^n + 1$ splits into n linear factors modulo $q \geq 5$. Let $\sigma \geq \sqrt{n \ln(2n(1 + 1/\delta))/\pi} \cdot q^{1/n}$, for an arbitrary $\delta \in (0, 1/2)$. Then*

$$\Pr_{f \leftarrow D_{\mathbb{Z}^n, \sigma}} [f \pmod{q} \notin R_q^\times] \leq n(1/q + 2\delta).$$

We will use the following simple lemma to lower bound the length of the shortest vector in an ideal lattice via its norm.

Lemma 5 *Let $\mathcal{I} \subset R$ be an ideal lattice. Then it holds that $\lambda_1(\mathcal{I}) \geq \sqrt{n} \cdot N(\mathcal{I})^{1/n}$.*

Babai's Roundoff Algorithm. We will need to compute short representatives of residual classes $x \bmod \mathcal{I} \in R/\mathcal{I}$ for ideals $\mathcal{I} = \langle g \rangle$. A simple algorithm for this task is Babai's roundoff algorithm. Given an $x \in R$, we can find a small representative $\hat{x}$ of $x \bmod \mathcal{I}$ by computing

$$\hat{x} = x - \lfloor x \cdot g^{-1} \rceil \cdot g,$$

where the $\lfloor \cdot \rceil$ operation round each component to the nearest integer. Clearly, it holds that $\hat{x} \equiv x \bmod \mathcal{I}$ and

$$\|\hat{x}\| = \|x - \lfloor x \cdot g^{-1} \rceil \cdot g\| = \|(x \cdot g^{-1} - \lfloor x \cdot g^{-1} \rceil) \cdot g\|$$
$$\leq \sqrt{n} \cdot \|x \cdot g^{-1} - \lfloor x \cdot g^{-1} \rceil\| \cdot \|g\| \leq \frac{n}{2} \cdot \|g\|,$$

as $x \cdot g^{-1} - \lfloor x \cdot g^{-1} \rceil \in K$ is a field element with coefficients of size at most $1/2$. Therefore, if g is short then so is $\hat{x}$.

3 Bootstrapping $i\mathcal{O}$ for Special Purpose Circuits

In this section, we state the main results from [39] relevant to our work.

Theorem 2 (Bootstrapping $i\mathcal{O}$ for constant degree circuits, [39], Theorem 5]). *Assume sub-exponential hardness of LWE, and the existence of a sub-exponentially secure constant-degree PRG. There exist a family of circuit classes of constant degree, such that $i\mathcal{O}$ for that family with universal efficiency can be bootstrapped into $i\mathcal{O}$ for P/poly.*

Universal efficiency means the following: $i\mathcal{O}$ for constant degree circuits has universal efficiency if the run-time of the obfuscator is independent of the degree of the computation. More precisely, there is a universal polynomial p such that for every circuit C of degree d, obfuscating C takes time $p(1^\lambda, |C|)$, for a sufficiently large λ.

Moreover, in Lin's $i\mathcal{O}$ construction, it does not suffice that the circuits of the seed class are of a constant degree. In fact, the degree of multilinearity required of multilinear maps grows with the *type degree* and *input types* of the special circuits used for bootstrapping in the above theorem.

One of the main contributions of [39] is to prove that the seed class of circuits indeed has constant number of input types as well as constant type degree. For the purpose of being self-contained, we define the input types and type degree first.

Definition 2 (Type Function, [39], Definition 18). *Let Σ be any alphabet where every symbol in Σ is represented as a binary string of length $\ell \in \mathbb{N}$. Let $\mathcal{U}(\star, \star)$ be an arithmetic circuit over domain $\Sigma^c \times \{0,1\}^m$ with some $m, c \in \mathbb{N}$. We say that $\mathcal{U}$ has c input types and assign every wire $w \in \mathcal{U}$ with a type $t_w \in \mathbb{N}^{c+1}$ through the following recursively defined function $t_w = \mathsf{type}(\mathcal{U}, w)$.*

- **Base Case:** *If w is the ith input wire,*
  - *If $i \in [(k-1)\ell + 1, k\ell]$ for some $k \in [c]$ (meaning that w describes x^k), assign type $t_w = \mathbf{1}_k$ (a vector with one at position k and zeros everywhere else).*
  - *If $i \in [c\ell + 1, c\ell + m]$ (meaning that w describes the circuit C), assign type $t_w = \mathbf{1}_{c+1}$.*

- **Recursion:** *If w is the output wire of gate g with input wires u, v of types $t_u = \text{type}\,(\mathcal{U}, u)$ and $t_v = \text{type}\,(\mathcal{U}, v)$ respectively.*
  - *If g is an addition/subtraction gate and $t_u = t_v$, then assign type $t_w = t_u$.*
  - *Otherwise (i.e., g is a multiplication gate or $t_u \neq t_v$), then assign $t_w = t_u + t_v$.*

Definition 3 (Type Degree). *We define the type degree of the following objects:*

- *The type degree of a wire w of $\mathcal{U}$ is $\text{tdeg}\,(\mathcal{U}, w) = |\text{type}\,(\mathcal{U}, w)|_1$.*
- *The type degree of $\mathcal{U}$ is $\text{tdeg}\,(\mathcal{U}) = \max_{w \in \mathcal{U}}\,(\text{tdeg}\,(\mathcal{U}, w))$.*

The fact that the seed class of [39] has constant input types and constant type degree is summarized in the following lemma.

Lemma 6 (The Special-Purpose Circuits Have Constant Type-Degree, [39], Lemma 5). *The class of special purpose circuits $\{\mathcal{P}_\lambda^{T,n}\}$ has universal arithmetic circuits $\{U_\lambda\}$ of constant $c^{T,n}$ input-types, constant type degree $\text{tdeg}^{T,n}$, and size $u(1^\lambda, n, \log T)$, for a universal polynomial u independent of T, n.*

Given the above lemma, [39] gives an $i\mathcal{O}$ construction in the ideal graded encoding model, where the oracle has degree $d = O(\text{tdeg} + c)$, i.e. a constant. In our work, we give an $i\mathcal{O}$ construction that improves upon the construction of [39] in two ways. We show that our construction is secure against all known attacks including annihilation attacks [44] and has only a polynomial noise growth as mentioned in Sect. 1.

4   Construction of the Obfuscator

In this section, we give our $i\mathcal{O}$ construction for the seed class of circuits from [39] and prove security in the weak multilinear map model. We build on the construction of [39] in composite-order ideal graded encoding model, and use new ideas to achieve security in the weak multilinear map model and constant noise growth.

[39] gives a construction for obfuscation which obfuscates circuits with multi-bit outputs directly. The reason stated in [39] is the following: Direct conversion from a multi-bit output circuit C to a single-bit output circuit $\bar{C}$ by taking an additional input for an index of the output wire as $\bar{C}(x, i) = C(x)_i$ might not preserve constant type degree of C (which is crucial for the construction). This is because the multiplexer circuit that chooses the ith output depending on input i might not have constant type degree. In this work, we observe that obfuscating one-bit output circuits suffices if we give out a different obfuscation per-output bit of the circuit. Let $C_i = C(x)_i$ denote the circuit that that outputs the ith bit of the circuit. We can easily construct C_i by removing some gates of C that do not contribute to i^{th} output wire. This transformation does not increase the type-degree. Hence, for simplicity, we only focus on obfuscating Lin's seed class of circuits for one bit outputs.

Construction Overview. Let C be a circuit that has a single bit output and which can be computed by a universal arithmetic circuit $\mathcal{U}(x, C)$. Recall that $x \in \Sigma^c$ and each input wire takes in a symbol from Σ as input. At a high level, in Lin's [39] construction, for every input wire and every symbol, encodings are given per symbol bit. Also, encodings are given per description bit of the circuit C. Then given an input x, an evaluator can simply pick the encodings corresponding to x, C and homomorphically evaluate $\mathcal{U}$ on the encodings of x and C to obtain an encoding of $\mathcal{U}(x, C)$, which can then be zero-tested. This basic idea is not secure and [6,39] need a composite ring with many primes to make it secure. The actual computation happens in one of the sub-rings and computation on random elements happen in other sub-rings to protect against input-mixing attacks as well as low-level zeroes. Moreover, they also need carefully chosen straddling sets (to encode the elements) to ensure input consistency.

In our case, the goal is to prove security against post-zeroizing computations as well. For this, as already mentioned in the introduction, the main idea is the following: We add one more sub-ring where a PRF is computed.[16] The key idea is that though the PRF is being computed in only one of the sub-rings, after zero-testing it yields a random ring element in all the sub-rings, in particular, a random element in $R \bmod \mathcal{I}$, where $\mathcal{I} = \langle g \rangle$ (c.f. Appendix A of the full version [19] for definitions of R and $\mathcal{I}$). So we start by computing a one-bit PRF on input x in one of the sub-rings.

To argue security, we need that the PRF output has sufficient min-entropy. But since the PRF has a one-bit output similar to $\mathcal{U}$, it does not have enough min-entropy. So the final idea is to compute multiple PRFs in parallel and combine them to get a ring element. In doing this, we need to use an unbounded addition gate and need to take care that it does not blow up the type-degree of the computation. For this, we ensure that, before being added, all PRF outputs are at the same type-degree or straddling set and also have the same El-Gamal randomness of the encodings. Recall that [6,39] use El-Gamal encodings to encode elements and to be able to add two encodings without increasing the type-degree, it is important that they have the same randomness r term.

Finally, the straddling sets are matrices of polynomial size and as detailed in Appendix A.2 of the full version [19] if we pick a z_{ij} corresponding to each entry in the matrix, the noise of encodings would be too high. We explain in Appendix B of the full version, how we change the GGH instantiation of Appendix A (of the full version) to control the noise growth.

4.1 Setting and Parameters

Consider an arbitrary circuit class $\{C_\lambda\}$ with universal circuits $\{\mathcal{U}_\lambda\}$. The universal circuit $\mathcal{U} = \mathcal{U}_\lambda$ has the following parameters:

- alphabet Σ with $|\Sigma|$ symbols, each of length ℓ, both $|\Sigma|$ and ℓ being $\mathsf{poly}(\lambda)$,

---

[16] We note that such a PRF can be computed using constant input types and constant type degree. See more details in Appendix G.2 of the full version [19].

- domain $\Sigma^c \times \{0,1\}^m$, that is, every circuit $C \in C_\lambda$ has input $x = x^1, \cdots, x^c$ where $x^k \in \Sigma$ for every $k \in [c]$ and can be described by an m-bit string,
- degree of the universal circuit is $d = \deg(\mathcal{U})$,
- an output wire o, denoted by $\mathbf{t} = \mathsf{type}(\mathcal{U}, o) \in \mathbb{N}^{c+1}$ the type of the output wire (see Definition 2). Note that $\mathbf{t}[k]$ denotes the type degree of x^k in the output wire.

Recall the ring $R = \mathbb{Z}[X]/(X^n + 1)$ defined in the composite-order GGH graded encoding scheme (see Appendix A.1 of the full version [19]). In our construction, we will use PRF circuits with 1 bit output. Our construction uses n independent PRFs, where n is the dimension of R. Let $C^{\mathsf{PRF}^t} : \Sigma^c \to \{0,1\}$ be a PRF for all $t \in [n]$. As already shown in [39], these circuits also satisfy the constraints for constant input types and constant type degree as the seed-class (c.f. Lemma G.9 of the full version [19]). More precisely, $C^{\mathsf{PRF}^t}(x)$ is a circuit computing 1 bit for every $t \in [n]$, and each circuit can be described by an m-bit string.

Encoding Levels: We specify the levels used in the $i\mathcal{O}$ construction in Fig. 1. All levels are represented as a $(|\Sigma| + 1) \times (c + 2)$ matrices over $\mathbb{N}$.

Notation: In the following construction, we abuse the notations $0/1$ to refer to both bits $0/1$ and ring elements $\mathbf{0}/\mathbf{1}$.

4.2 Our Obfuscator

Input: Security parameter λ, program description $C \in C_\lambda$.
Output: Obfuscated program with the same functionality as C.
Algorithm: Our obfuscator proceeds as follows:

1. Instantiate a $(c + 3)$-composite graded encoding scheme (params, sparams, $\mathbf{p}_{zt}$) $\leftarrow$ InstGen($1^\lambda, 1^{c+3}, \mathbf{v}_{zt}$), and receive a ring $R \cong R_1 \times R_2 \times \cdots \times R_{c+3}$. Note that $R_i \cong \mathbb{Z}_{p_i}$ for some prime p_i for all $i \in [c + 3]$. Hence, given sparams it is easy to sample a uniform element in any of the sub-rings.
2. Compute encoding $Z^* = [w^*]_{\mathbf{v}^*}$ for $w^* = (1, 1, 1, \rho_1^*, \cdots, \rho_c^*)$ where $\rho_k^* \xleftarrow{\$} R_{k+3}$ for $\forall k \in [c]$.
3. **Encode the input symbol.** For $\forall k \in [c]$, encode the k-th input symbol:
   - For every symbol $s \in \Sigma$, sample $r_s^k \xleftarrow{\$} R$ and compute $R_s^k = [r_s^k]_{\mathbf{v}_s^k}$.
   - For $\forall j \in [\ell]$, sample $y_j^k \xleftarrow{\$} R_1$.
   - For every symbol $s \in \Sigma$, and every j-th bit s_j, compute encoding $Z_{s,j}^k = [r_s^k \cdot w_{s,j}^k]_{\mathbf{v}_s^k + \mathbf{v}^*}$ for $w_{s,j}^k = (y_j^k, s_j, s_j, \rho_{s,j,1}^k, \cdots, \rho_{s,j,c}^k)$ where $(\rho_{s,j,1}^k, \cdots, \rho_{s,j,c}^k) \xleftarrow{\$} R_4 \times \cdots \times R_{c+3}$.

$$\forall k \in [c], s \in \Sigma, \qquad \mathbf{v}_s^k = \begin{bmatrix} & & (k) & & \\ 0 & \cdots & 0 & \cdots & 0 & 0 \\ \vdots & \ddots & \vdots & \ddots & \vdots & \vdots \\ (s) \; 0 & \cdots & 1 & \cdots & 0 & 0 \\ \vdots & \ddots & \vdots & \ddots & \vdots & \vdots \\ 0 & \cdots & 0 & \cdots & 0 & 0 \\ \hline 0 & \cdots & 0 & \cdots & 0 & 0 \end{bmatrix}$$

$$\forall k \in [c], s \in \Sigma, \qquad \hat{\mathbf{v}}_s^k = \begin{bmatrix} & & (k) & & \\ 0 & \cdots & \mathbf{t}[k] & \cdots & 0 & 0 \\ \vdots & \ddots & \vdots & \ddots & \vdots & \vdots \\ (s) \; 0 & \cdots & 0 & \cdots & 0 & 0 \\ \vdots & \ddots & \vdots & \ddots & \vdots & \vdots \\ 0 & \cdots & \mathbf{t}[k] & \cdots & 0 & 0 \\ \hline 0 & \cdots & 1 & \cdots & 0 & 0 \end{bmatrix}$$

$$\mathbf{v}^* = \begin{bmatrix} 0 & \cdots & 0 & 0 \\ \vdots & \ddots & \vdots & \vdots \\ 0 & \cdots & 0 & 0 \\ \hline 0 & \cdots & 0 & 1 \end{bmatrix} \qquad \mathbf{v}^{c+1} = \begin{bmatrix} 0 & \cdots & 0 & 1 & 0 \\ \vdots & \ddots & \vdots & \vdots & \vdots \\ 0 & \cdots & 0 & 1 & 0 \\ \hline 0 & \cdots & 0 & 0 & 0 \end{bmatrix}$$

$$\tilde{\mathbf{v}} = \begin{bmatrix} 0 & \cdots & 0 & 0 & 0 \\ \vdots & \ddots & \vdots & \vdots & \vdots \\ 0 & \cdots & 0 & 0 & 0 \\ \hline 0 & \cdots & 0 & 1 & 0 \end{bmatrix} \qquad \bar{\mathbf{v}} = \begin{bmatrix} 0 & \cdots & 0 & 1 \\ \vdots & \ddots & \vdots & \vdots \\ 0 & \cdots & 0 & 1 \\ \hline 0 & \cdots & 0 & 0 \end{bmatrix}$$

$$\mathbf{v}_{zt} = \begin{bmatrix} \mathbf{t}[1] & \cdots & \mathbf{t}[c] & \mathbf{t}[c+1]+1 & 1 \\ \vdots & \ddots & \vdots & \vdots & \vdots \\ \mathbf{t}[1] & \cdots & \mathbf{t}[c] & \mathbf{t}[c+1]+1 & 1 \\ \hline 1 & \cdots & 1 & 1 & D \end{bmatrix} \qquad \text{where } D = d+c+2$$

Fig. 1. Levels used in the obfuscation.

4. **Encode the circuit and PRFs.** Compute encoding $R^{c+1} = \left[r^{c+1} \right]_{\mathbf{v}^{c+1}}$ where $r^{c+1} \overset{\$}{\leftarrow} \mathcal{R}$. For $\forall t \in [n]$, generate the following encodings for program description: We will encode the circuit $\mathcal{C}$ in $\mathcal{R}_2$ and circuit $\mathcal{C}^{\mathsf{PRF}}$ in $\mathcal{R}_3$.

   (a) For $\forall j \in [m]$, compute encoding $Z_{t,j}^{c+1} = \left[r^{c+1} \cdot w_{t,j}^{c+1} \right]_{\mathbf{v}^{c+1}+\mathbf{v}^*}$
   for $w_{t,j}^{c+1} = \left(y_{t,j}^{c+1}, \mathcal{C}_j, \mathcal{C}_j^{\mathsf{PRF}^t}, \rho_{t,j,1}^{c+1}, \cdots, \rho_{t,j,c}^{c+1} \right)$ where $y_{t,j}^{c+1} \overset{\$}{\leftarrow} \mathcal{R}_1$ and $\left(\rho_{t,j,1}^{c+1}, \cdots, \rho_{t,j,c}^{c+1} \right) \overset{\$}{\leftarrow} \mathcal{R}_4 \times \cdots \times \mathcal{R}_{c+3}$.

   (b) Compute encoding $Z_{t,m+1}^{c+1} = \left[r^{c+1} \cdot w_{t,m+1}^{c+1} \right]_{\mathbf{v}^{c+1}+\mathbf{v}^*}$
   for $w_{t,m+1}^{c+1} = \left(y_{t,m+1}^{c+1}, 1, \mathbf{e}^t, \rho_{t,m+1,1}^{c+1}, \cdots, \rho_{t,m+1,c}^{c+1} \right)$
   where $\mathbf{e}^t$ is an element in the ring R of the composite order GGH graded

encoding scheme (see Appendix A of the full version [19]),[17] $y_{t,m+1}^{c+1} \xleftarrow{\$} \mathcal{R}_1$ and $\left(\rho_{t,m+1,1}^{c+1}, \cdots, \rho_{t,m+1,c}^{c+1}\right) \xleftarrow{\$} \mathcal{R}_4 \times \cdots \times \mathcal{R}_{c+3}$. During computation, these encodings will be used to combine the n one-bit PRF computations into a ring element.

5. Encode c elements for the purpose of canceling ρ in the last c slots: For $\forall k \in [c]$ sample $\hat{w}^k = \left(\hat{y}^k, \hat{\beta}^k, \hat{\alpha}^k, \hat{\rho}_1^k, \cdots, \hat{\rho}_c^k\right)$ where $\hat{y}^k, \hat{\beta}^k, \hat{\alpha}^k, \hat{\rho}_1^k, \cdots, \rho_c^k$ are all uniformly random except that $\hat{\rho}_k^k = 0$ and generate the following encodings:

   For all $s \in \Sigma$, sample $\hat{r}_s^k \xleftarrow{\$} \mathcal{R}$ and compute encodings $\hat{R}_s^k = \left[\hat{r}_s^k\right]_{\hat{\mathbf{v}}_s^k}$ and $\hat{Z}_s^k = \left[\hat{r}_s^k \cdot \hat{w}^k\right]_{\hat{\mathbf{v}}_s^k + \mathbf{v}^*}$.

   For the following: denote $\hat{y} = \prod_{k=1}^c \hat{y}^k, \hat{\beta} = \prod_{k=1}^c \hat{\beta}^k, \hat{\alpha} = \prod_{k=1}^c \hat{\alpha}^k, \hat{w} = \prod_{k=1}^c \hat{w}^k = \left(\hat{y}, \hat{\beta}, \hat{\alpha}, 0, \cdots, 0\right)$.

6. Encode an element to cancel out the PRF computation in the 3rd slot: Compute encodings $\widetilde{R} = [\widetilde{r}]_{\widetilde{\mathbf{v}}}$ and $\widetilde{Z} = [\widetilde{r} \cdot \widetilde{w}]_{\widetilde{\mathbf{v}} + \mathbf{v}^*}$ for $\widetilde{r} \xleftarrow{\$} \mathcal{R}$ and $\widetilde{w} = \left(\widetilde{y}, \widetilde{\beta}, 0, \widetilde{\rho}_1, \cdots, \widetilde{\rho}_c\right)$ where $\widetilde{y}, \widetilde{\beta}, \widetilde{\rho}_1, \cdots, \widetilde{\rho}_c$ are all uniformly random in respective sub-rings.

7. Encode an element for the purpose of authentication of computation: Compute encodings $\bar{R} = [\bar{r}]_{\bar{\mathbf{v}}}$ and $\bar{Z} = [\bar{r} \cdot \bar{w}]_{\bar{\mathbf{v}} + D\mathbf{v}^*}$, where $D = d + c + 2$, for $\bar{r} \xleftarrow{\$} \mathcal{R}$ and $\bar{w} = \hat{w} \cdot \widetilde{w} \cdot (\bar{y}, n, 0, 0, \cdots, 0)$, where $\bar{y} = \sum_{t=1}^n \left(\bar{y}_t \cdot y_{t,m+1}^{c+1}\right)$ for $\bar{y}_t = \mathcal{U}\left(\left\{y_j^1\right\}_{j \in [\ell]}, \cdots, \left\{y_j^c\right\}_{j \in [\ell]}; \left\{y_{t,j}^{c+1}\right\}_{j \in [m]}\right)$.

8. **The obfuscation.** The obfuscated program consists of the following:
   - The evaluation parameters params, $\boldsymbol{p}_{zt}$.
   - The encoding Z^*.
   - For $\forall k \in [c], \forall s \in \Sigma$, the encodings $R_s^k, \hat{R}_s^k, \hat{Z}_s^k$, and for $\forall j \in [\ell]$, $Z_{s,j}^k$.
   - R^{c+1}, and for $\forall t \in [n], \forall j \in [m+1]$, $Z_{t,j}^{c+1}$.
   - The encodings $\widetilde{R}, \widetilde{Z}, \bar{R}, \bar{Z}$.

Efficiency: It is easy to see that the number of encodings in the obfuscated program is bounded by $\mathsf{poly}(1^\lambda, S(\lambda))$, where $S(\lambda)$ is the size of $\mathcal{U}_\lambda$. The size of each encoding and ℓ_1-norm of $\mathbf{v}_{zt}$ are also bounded by $\mathsf{poly}(1^\lambda, S(\lambda))$. It is easy to check that all poly above are fixed universal polynomials. Therefore the size of obfuscation is bounded by $p(1^\lambda, S(\lambda))$ for a universal polynomial, which satisfies the universal efficiency requirement in Sect. 3.

---

[17] The values of $\mathbf{e}^t$ is specified in the proof of Theorem G.20 of the full version [19], which is crucial for proving post zeroizing security, but does not affect the correctness of the obfuscator.

Evaluation: To evaluate the program on an input $x = x^1, \ldots, x^c \in \Sigma^c$, we will use the following encodings:

$$\left\{\left(R^k_{x^k}, Z^k_{x^k, j}\right)\right\}_{k \in [c], j \in [\ell]}, \quad \left\{\left(R^{c+1}, Z^{c+1}_{t,j}\right)\right\}_{t \in [n], j \in [m+1]},$$

$$\left\{\left(\hat{R}^k_{x^k}, \hat{Z}^k_{x^k}\right)\right\}_{k \in [c]}, \quad \left(\tilde{R}, \tilde{Z}\right), \left(\bar{R}, \bar{Z}\right), Z^*.$$

We in-line the analysis of *correctness* in the description of the evaluation below.

Input: Two pairs of encodings $\left(R_\alpha = [r_\alpha]_{\mathbf{v}_\alpha}, Z_\alpha = [r_\alpha \cdot w_\alpha]_{\mathbf{v}_\alpha + d_\alpha \mathbf{v}^*}\right)$ and $\left(R_\beta = [r_\beta]_{\mathbf{v}_\beta}, Z_\beta = [r_\beta \cdot w_\beta]_{\mathbf{v}_\beta + d_\beta \mathbf{v}^*}\right)$, encoding $Z^* = [w^*]_{\mathbf{v}^*}$, and an operator op,

Output: A pair of encodings $\left(R_\sigma = [r_\sigma]_{\mathbf{v}_\sigma}, Z_\sigma = [r_\sigma \cdot w_\sigma]_{\mathbf{v}_\sigma + d_\sigma \mathbf{v}^*}\right)$

Algorithm:
   i. Permute the operands to ensure that $\delta = d_\beta - d_\alpha \geq 0$.
   ii. Consider the operator op:
      – **Multiplication:** If op $= \times$, then $R_\sigma = R_\alpha \times R_\beta$ and $Z_\sigma = Z_\alpha \times Z_\beta$.
      $(r_\sigma = r_\alpha \cdot r_\beta, \mathbf{v}_\sigma = \mathbf{v}_\alpha + \mathbf{v}_\beta, \text{ and } d_\sigma = d_\alpha + d_\beta.)$
      – **Addition/Subtraction:** If op $= +/-$ and $\mathbf{v}_\alpha \neq \mathbf{v}_\beta$, then $R_\sigma = R_\alpha \times R_\beta$ and $Z_\sigma = Z_\alpha \times R_\beta \times (Z^*)^\delta + / - Z_\beta \times R_\alpha$.
      $(r_\sigma = r_\alpha \cdot r_\beta, \mathbf{v}_\sigma = \mathbf{v}_\alpha + \mathbf{v}_\beta, \text{ and } d_\sigma = d_\beta.)$
      – **Constrained Addition/Subtraction:** If op $= +/-$ and $\mathbf{v}_\alpha = \mathbf{v}_\beta = \mathbf{v}$ (by induction it is guaranteed that $r_\alpha = r_\beta = r$), then $R_\sigma = R_\alpha$ and $Z_\sigma = Z_\alpha \times (Z^*)^\delta + / - Z_\beta$.
      $(r_\sigma = r, \mathbf{v}_\sigma = \mathbf{v}, \text{ and } d_\sigma = d_\beta.)$

Fig. 2. Computation over El-Gamal encodings

1. For every $t \in [n]$, do the following:
   (a) Consider the encodings $\left(R^k_{x^k}, Z^k_{x^k, j}\right)$ for $k \in [c], j \in [\ell]$, and $\left(R^{c+1}, Z^{c+1}_{t,j}\right)$ for $j \in [m]$. Apply the circuit $\mathcal{U}$ on these pairs of encodings. More specifically, we recursively associate every wire α in $\mathcal{U}$ with a pair of encodings $\left(R_\alpha = [r_\alpha]_{\mathbf{v}_\alpha}, Z_\alpha = [r_\alpha \cdot w_\alpha]_{\mathbf{v}_\alpha + d_\alpha \mathbf{v}^*}\right)$ in El-Gamal form as follows:
      – **Base Case:** For every $k \in [c]$ and every $j \in [\ell]$, the jth input wire of x^k is associated with pair $\left(R^k_{x^k}, Z^k_{x^k, j}\right)$. For every $j \in [m]$, the jth program bit is associated with $\left(R^{c+1}, Z^{c+1}_{t,j}\right)$.
      – **Recursion:** For every gate $g \in \mathcal{U}$ with input wires α, β and output wire σ, apply the computation as described in Fig. 2, over the encodings $Z^*, (R_\alpha, Z_\alpha), (R_\beta, Z_\beta)$ and the operator of g.

A pair of encodings for the output wire o is obtained:

$$\left(R_{\mathcal{U}} = [r_{\mathcal{U}}]_{\mathbf{v}_{\mathcal{U}}}, Z_{t,\mathcal{U}} = [r_{\mathcal{U}} \cdot w_{t,\mathcal{U}}]_{\mathbf{v}_{\mathcal{U}}+d\mathbf{v}^*}\right),$$

where (let $\mathbf{1}$ denote an all-one vector, $\mathbf{0}$ an all-zero vector, and let $\mathbf{1}_i$ denote a vector with one at position i and zeros everywhere else)

$$\mathbf{v}_{\mathcal{U}} = \left[\begin{array}{cccc|c} \mathbf{t}[1] \cdot \mathbf{1}_{x^1} \cdots \mathbf{t}[c] \cdot \mathbf{1}_{x^c} & \mathbf{t}[c+1] \cdot \mathbf{1} & \mathbf{0} \\ \hline 0 \quad \cdots \quad 0 & 0 & 0 \end{array}\right],$$

$$w_{t,\mathcal{U}} = \left(\mathcal{U}\left(\{y_j^1\}_{j\in[\ell]}, \cdots, \{y_j^c\}_{j\in[\ell]}, \{y_{t,j}^{c+1}\}_{j\in[m]}\right), \mathcal{U}(x, \mathcal{C}), \mathcal{U}\left(x, \mathcal{C}^{\mathsf{PRF}^t}\right),\right.$$
$$\left.\star, \cdots, \star\right)$$
$$= \left(\bar{y}_t, \mathcal{C}(x), \mathcal{C}^{\mathsf{PRF}^t}(x), \star, \cdots, \star\right).$$

In the above, the values denoted by $\star$ do not matter for correctness, and hence are not mentioned explicitly.

(b) Take the product of $(R_{\mathcal{U}}, Z_{t,\mathcal{U}})$ with $\left(R^{c+1}, Z_{t,m+1}^{c+1}\right)$ and obtain a pair of encodings (computation done as in Fig. 2):

$$\left(\ddot{R}_{\mathcal{U}} = [\ddot{r}_{\mathcal{U}}]_{\ddot{\mathbf{v}}_{\mathcal{U}}}, \ddot{Z}_{t,\mathcal{U}} = [\ddot{r}_{\mathcal{U}} \cdot \ddot{w}_{t,\mathcal{U}}]_{\ddot{\mathbf{v}}_{\mathcal{U}}+(d+1)\mathbf{v}^*}\right), \text{ where}$$

$$\ddot{\mathbf{v}}_{\mathcal{U}} = \left[\begin{array}{cccc|c} \mathbf{t}[1] \cdot \mathbf{1}_{x^1} \cdots \mathbf{t}[c] \cdot \mathbf{1}_{x^c} & (\mathbf{t}[c+1]+1) \cdot \mathbf{1} & \mathbf{0} \\ \hline 0 \quad \cdots \quad 0 & 0 & 0 \end{array}\right],$$

$$\ddot{w}_{t,\mathcal{U}} = w_{t,\mathcal{U}} \cdot w_{t,m+1}^{c+1} = \left(\bar{y}_t \cdot y_{t,m+1}^{c+1}, \mathcal{C}(x), \mathcal{C}^{\mathsf{PRF}^t}(x) \cdot \mathbf{e}^t, \star, \cdots, \star\right).$$

Remark 1. Note that our construction ensures that $\left(\ddot{R}_{\mathcal{U}}, \ddot{Z}_{t,\mathcal{U}}\right)$ has the same level and same $\ddot{r}_{\mathcal{U}}$ for every $t \in [n]$. This is crucial to do the next step of addition of n terms using constrained addition. This ensures that the addition does not grow the levels of multilinearity needed.

2. Take the sum of $\left\{\left(\ddot{R}_{\mathcal{U}}, \ddot{Z}_{t,\mathcal{U}}\right)\right\}_{t\in[n]}$ and obtain a pair of encodings:

$$\left(\ddot{R}_{\mathcal{U}} = [\ddot{r}_{\mathcal{U}}]_{\ddot{\mathbf{v}}_{\mathcal{U}}}, \ddot{Z}_{\mathcal{U}} = [\ddot{r}_{\mathcal{U}} \cdot \ddot{w}_{\mathcal{U}}]_{\ddot{\mathbf{v}}_{\mathcal{U}}+(d+1)\mathbf{v}^*}\right), \text{ where}$$

$$\ddot{w}_{\mathcal{U}} = \sum_{t=1}^{n} \ddot{w}_{t,\mathcal{U}} = \left(\bar{y}, n \cdot \mathcal{C}(x), \mathcal{C}^{\mathsf{PRF}}(x), \star, \cdots, \star\right),$$

where $\mathcal{C}^{\mathsf{PRF}}(x) = \sum_{t\in[n]} \mathbf{e}^t \mathcal{C}^{\mathsf{PRF}^t}(x)$.

3. Take the product of $\left(\ddot{R}_{\mathcal{U}}, \ddot{Z}_{\mathcal{U}}\right)$ with the product of $\left\{\left(\hat{R}_{x^k}^k, \hat{Z}_{x^k}^k\right)\right\}_{k\in[c]}$ and obtain a pair:

$$\left(\hat{R}_{\mathcal{U}} = [\hat{r}_{\mathcal{U}}]_{\hat{\mathbf{v}}_{\mathcal{U}}}, \hat{Z}_{\mathcal{U}} = [\hat{r}_{\mathcal{U}} \cdot \hat{w}_{\mathcal{U}}]_{\hat{\mathbf{v}}_{\mathcal{U}}+(d+1+c)\mathbf{v}^*}\right), \text{ where}$$

$$\hat{\mathbf{v}}_{\mathcal{U}} = \left[\begin{array}{cccc|c} \mathbf{t}[1] \cdot \mathbf{1} \cdots \mathbf{t}[c] \cdot \mathbf{1} & (\mathbf{t}[c+1]+1) \cdot \mathbf{1} & \mathbf{0} \\ \hline 1 \quad \cdots \quad 1 & 0 & 0 \end{array}\right],$$

$$\hat{w}_{\mathcal{U}} = \hat{w} \cdot \ddot{w}_{\mathcal{U}} = \left(\hat{y}\bar{y}, \hat{\beta}n \cdot \mathcal{C}(x), \hat{\alpha} \cdot \mathcal{C}^{\mathsf{PRF}}(x), 0, \cdots, 0\right).$$

4. Take the product of $\left(\hat{R}_\mathcal{U}, \hat{Z}_\mathcal{U}\right)$ with $\left(\widetilde{R}, \widetilde{Z}\right)$ and obtain a pair:

$$\left(\widetilde{R}_\mathcal{U} = [\widetilde{r}_\mathcal{U}]_{\widetilde{\mathbf{v}}_\mathcal{U}}, \widetilde{Z}_\mathcal{U} = [\widetilde{r}_\mathcal{U} \cdot \widetilde{w}_\mathcal{U}]_{\widetilde{\mathbf{v}}_\mathcal{U} + D\mathbf{v}^*}\right), \text{ where}$$

$$\widetilde{\mathbf{v}}_\mathcal{U} = \left[\begin{array}{cccc|c} \mathbf{t}[1] \cdot \mathbf{1} \cdots \mathbf{t}[c] \cdot \mathbf{1} & (\mathbf{t}[c+1] + 1) \cdot \mathbf{1} & \mathbf{0} \\ 1 \cdots 1 & 1 & 0 \end{array}\right],$$

$$\widetilde{w}_\mathcal{U} = \widetilde{w} \cdot \hat{w}_\mathcal{U} = \left(\widetilde{y}\hat{y}\bar{y}, \widetilde{\beta}\hat{\beta}n \cdot \mathcal{C}(x), 0, 0, \cdots, 0\right).$$

5. Subtract the pair $\left(\bar{R}, \bar{Z}\right)$ from $\left(\widetilde{R}_\mathcal{U}, \widetilde{Z}_\mathcal{U}\right)$ and obtain the pair:

$$\left(\bar{R}_\mathcal{U} = [\bar{r}_\mathcal{U}]_{\bar{\mathbf{v}}_\mathcal{U}}, \bar{Z}_\mathcal{U} = [\bar{r}_\mathcal{U} \cdot \bar{w}_\mathcal{U}]_{\bar{\mathbf{v}}_\mathcal{U} + D\mathbf{v}^*}\right), \text{ where}$$

$$\bar{\mathbf{v}}_\mathcal{U} = \left[\begin{array}{cccc|c} \mathbf{t}[1] \cdot \mathbf{1} \cdots \mathbf{t}[c] \cdot \mathbf{1} & (\mathbf{t}[c+1] + 1) \cdot \mathbf{1} & \mathbf{1} \\ 1 \cdots 1 & 1 & 0 \end{array}\right],$$

$$\bar{w}_\mathcal{U} = \left(0, \widetilde{\beta}\hat{\beta}n \cdot (\mathcal{C}(x) - 1), 0, 0, \cdots, 0\right).$$

6. Finally, apply zero testing on $\bar{Z}_\mathcal{U}$. If isZero(params, $\mathbf{p}_{zt}$, $\bar{Z}_\mathcal{U}$) = 1 then output 1, otherwise output 0.

   As analyzed above, in an honest evaluation, $\bar{Z}_\mathcal{U}$ is an encoding of 0 under $\mathbf{v}_{zt}$ iff $\mathcal{C}(x) = 1$ with high probability over choice of $\widetilde{\beta}, \hat{\beta}$. Hence the correctness of the evaluation procedure follows.

References

1. Agrawal, S., Gentry, C., Halevi, S., Sahai, A.: Discrete Gaussian leftover hash lemma over infinite domains. Cryptology ePrint Archive, Report 2012/714 (2012)
2. Aharonov, D., Regev, O.: Lattice problems in Np cap coNp. J. ACM **52**(5), 749–765 (2005)
3. Albrecht, M., Bai, S., Ducas, L.: A subfield lattice attack on overstretched NTRU assumptions. In: Robshaw, M., Katz, J. (eds.) CRYPTO 2016. LNCS, vol. 9814, pp. 153–178. Springer, Heidelberg (2016). https://doi.org/10.1007/978-3-662-53018-4_6
4. Ananth, P., Sahai, A.: Projective arithmetic functional encryption and indistinguishability obfuscation from degree-5 multilinear maps. In: Coron, J.-S., Nielsen, J.B. (eds.) EUROCRYPT 2017. LNCS, vol. 10210, pp. 152–181. Springer, Cham (2017). https://doi.org/10.1007/978-3-319-56620-7_6
5. Ananth, P.V., Gupta, D., Ishai, Y., Sahai, A.: Optimizing obfuscation: avoiding Barrington's theorem. In: Ahn, G.-J., Yung, M., Li, N. (eds.) ACM CCS 14, pp. 646–658. ACM Press, November 2014
6. Applebaum, B., Brakerski, Z.: Obfuscating circuits via composite-order graded encoding. In: Dodis, Y., Nielsen, J.B. (eds.) TCC 2015. LNCS, vol. 9015, pp. 528–556. Springer, Heidelberg (2015). https://doi.org/10.1007/978-3-662-46497-7_21
7. Badrinarayanan, S., Miles, E., Sahai, A., Zhandry, M.: Post-zeroizing obfuscation: new mathematical tools, and the case of evasive circuits. In: Fischlin, M., Coron, J.-S. (eds.) EUROCRYPT 2016. LNCS, vol. 9666, pp. 764–791. Springer, Heidelberg (2016). https://doi.org/10.1007/978-3-662-49896-5_27

8. Barak, B., Garg, S., Kalai, Y.T., Paneth, O., Sahai, A.: Protecting obfuscation against algebraic attacks. In: Nguyen, P.Q., Oswald, E. (eds.) EUROCRYPT 2014. LNCS, vol. 8441, pp. 221–238. Springer, Heidelberg (2014). https://doi.org/10.1007/978-3-642-55220-5_13

9. Boneh, D., Freeman, D.M.: Linearly homomorphic signatures over binary fields and new tools for lattice-based signatures. In: Catalano, D., Fazio, N., Gennaro, R., Nicolosi, A. (eds.) PKC 2011. LNCS, vol. 6571, pp. 1–16. Springer, Heidelberg (2011). https://doi.org/10.1007/978-3-642-19379-8_1

10. Boneh, D., Silverberg, A.: Applications of multilinear forms to cryptography. Cryptology ePrint Archive, Report 2002/080 (2002). http://eprint.iacr.org/2002/080

11. Brakerski, Z., Dagmi, O.: Shorter circuit obfuscation in challenging security models. Cryptology ePrint Archive, Report 2016/418 (2016). http://eprint.iacr.org/2016/418

12. Brakerski, Z., Rothblum, G.N.: Virtual black-box obfuscation for all circuits via generic graded encoding. In: Lindell, Y. (ed.) TCC 2014. LNCS, vol. 8349, pp. 1–25. Springer, Heidelberg (2014). https://doi.org/10.1007/978-3-642-54242-8_1

13. Canetti, R., Lin, H., Tessaro, S., Vaikuntanathan, V.: Obfuscation of probabilistic circuits and applications. In: Dodis, Y., Nielsen, J.B. (eds.) TCC 2015. LNCS, vol. 9015, pp. 468–497. Springer, Heidelberg (2015). https://doi.org/10.1007/978-3-662-46497-7_19

14. Cheon, J.H., Han, K., Lee, C., Ryu, H., Stehlé, D.: Cryptanalysis of the multilinear map over the integers. In: Oswald, E., Fischlin, M. (eds.) EUROCRYPT 2015. LNCS, vol. 9056, pp. 3–12. Springer, Heidelberg (2015). https://doi.org/10.1007/978-3-662-46800-5_1

15. Coron, J.-S., et al.: Zeroizing without low-level zeroes: new MMAP attacks and their limitations. In: Gennaro, R., Robshaw, M. (eds.) CRYPTO 2015. LNCS, vol. 9215, pp. 247–266. Springer, Heidelberg (2015). https://doi.org/10.1007/978-3-662-47989-6_12

16. Coron, J.-S., Lee, M.S., Lepoint, T., Tibouchi, M.: Cryptanalysis of GGH15 multilinear maps. In: Robshaw, M., Katz, J. (eds.) CRYPTO 2016. LNCS, vol. 9815, pp. 607–628. Springer, Heidelberg (2016). https://doi.org/10.1007/978-3-662-53008-5_21

17. Coron, J.-S., Lepoint, T., Tibouchi, M.: Practical multilinear maps over the integers. In: Canetti, R., Garay, J.A. (eds.) CRYPTO 2013. LNCS, vol. 8042, pp. 476–493. Springer, Heidelberg (2013). https://doi.org/10.1007/978-3-642-40041-4_26

18. Coron, J.-S., Lepoint, T., Tibouchi, M.: New multilinear maps over the integers. In: Gennaro, R., Robshaw, M. (eds.) CRYPTO 2015. LNCS, vol. 9215, pp. 267–286. Springer, Heidelberg (2015). https://doi.org/10.1007/978-3-662-47989-6_13

19. Döttling, N., Garg, S., Gupta, D., Miao, P., Mukherjee, P.: Obfuscation from low noise multilinear maps. Cryptology ePrint Archive, Report 2016/599 (2016). http://eprint.iacr.org/2016/599

20. Ducas, L., Nguyen, P.Q.: Learning a zonotope and more: cryptanalysis of NTRUSign countermeasures. In: Wang, X., Sako, K. (eds.) ASIACRYPT 2012. LNCS, vol. 7658, pp. 433–450. Springer, Heidelberg (2012). https://doi.org/10.1007/978-3-642-34961-4_27

21. Ducas, L., Pellet-Mary, A.: On the statistical leak of the GGH13 multilinear map and some variants. IACR Cryptology ePrint Archive, 2017:482 (2017). http://eprint.iacr.org/2017/482

22. Gama, N., Nguyen, P.Q., Regev, O.: Lattice enumeration using extreme pruning. In: Gilbert, H. (ed.) EUROCRYPT 2010. LNCS, vol. 6110, pp. 257–278. Springer, Heidelberg (2010). https://doi.org/10.1007/978-3-642-13190-5_13

23. Sanjam Garg. Candidate Multilinear Maps. Association for Computing Machinery and Morgan & Claypool, New York, NY, USA (2015)

24. Garg, S., Gentry, C., Halevi, S.: Candidate multilinear maps from ideal lattices. Cryptology ePrint Archive, Report 2012/610 (2012). http://eprint.iacr.org/2012/610

25. Garg, S., Gentry, C., Halevi, S.: Candidate multilinear maps from ideal lattices. In: Johansson, T., Nguyen, P.Q. (eds.) EUROCRYPT 2013. LNCS, vol. 7881, pp. 1–17. Springer, Heidelberg (2013). https://doi.org/10.1007/978-3-642-38348-9_1

26. Garg, S., Gentry, C., Halevi, S., Raykova, M., Sahai, A., Waters, B.: Candidate indistinguishability obfuscation and functional encryption for all circuits. In: 54th FOCS, pp. 40–49. IEEE Computer Society Press, October 2013

27. Garg, S., Miles, E., Mukherjee, P., Sahai, A., Srinivasan, A., Zhandry, M.: Secure obfuscation in a weak multilinear map model. In: Hirt, M., Smith, A. (eds.) TCC 2016. LNCS, vol. 9986, pp. 241–268. Springer, Heidelberg (2016). https://doi.org/10.1007/978-3-662-53644-5_10

28. Gentry, C., Gorbunov, S., Halevi, S.: Graph-induced multilinear maps from lattices. In: Dodis, Y., Nielsen, J.B. (eds.) TCC 2015. LNCS, vol. 9015, pp. 498–527. Springer, Heidelberg (2015). https://doi.org/10.1007/978-3-662-46497-7_20

29. Gentry, C., Lewko, A., Waters, B.: Witness encryption from instance independent assumptions. In: Garay, J.A., Gennaro, R. (eds.) CRYPTO 2014. LNCS, vol. 8616, pp. 426–443. Springer, Heidelberg (2014). https://doi.org/10.1007/978-3-662-44371-2_24

30. Gentry, C., Lewko, A.B., Sahai, A., Waters, B.: Indistinguishability obfuscation from the multilinear subgroup elimination assumption. In: Guruswami, V. (ed.) 56th FOCS, pp. 151–170. IEEE Computer Society Press, October 2015. https://doi.org/10.1109/FOCS.2015.19

31. Gentry, C., Lewko, A.B., Waters, B.: Witness encryption from instance independent assumptions. Cryptology ePrint Archive, Report 2014/273 (2014). http://eprint.iacr.org/2014/273

32. Gentry, C., Peikert, C., Vaikuntanathan, V.: Trapdoors for hard lattices and new cryptographic constructions. In: STOC, pp. 197–206 (2008)

33. Gentry, C., Szydlo, M.: Cryptanalysis of the revised NTRU signature scheme. In: Knudsen, L.R. (ed.) EUROCRYPT 2002. LNCS, vol. 2332, pp. 299–320. Springer, Heidelberg (2002). https://doi.org/10.1007/3-540-46035-7_20

34. Hu, Y., Jia, H.: Cryptanalysis of GGH map. In: Fischlin, M., Coron, J.-S. (eds.) EUROCRYPT 2016. LNCS, vol. 9665, pp. 537–565. Springer, Heidelberg (2016). https://doi.org/10.1007/978-3-662-49890-3_21

35. Kirchner, P., Fouque, P.-A.: Comparison between subfield and straightforward attacks on NTRU. IACR Cryptology ePrint Archive 2016:717 (2016)

36. Klein, P.N.: Finding the closest lattice vector when it's unusually close. In: Shmoys, D.B. (ed.) 11th SODA, pp. 937–941. ACM-SIAM, January 2000

37. Langlois, A., Stehlé, D., Steinfeld, R.: GGHLite: more efficient multilinear maps from ideal lattices. In: Nguyen, P.Q., Oswald, E. (eds.) EUROCRYPT 2014. LNCS, vol. 8441, pp. 239–256. Springer, Heidelberg (2014). https://doi.org/10.1007/978-3-642-55220-5_14

38. Lenstra, A.K., Lenstra, H.W., Lovász, L.: Factoring polynomials with rational coefficients. Mathe. Ann. **261**(4), 515–534 (1982)

39. Lin, H.: Indistinguishability obfuscation from constant-degree graded encoding schemes. In: Fischlin, M., Coron, J.-S. (eds.) EUROCRYPT 2016. LNCS, vol. 9665, pp. 28–57. Springer, Heidelberg (2016). https://doi.org/10.1007/978-3-662-49890-3_2

40. Lin, H.: Indistinguishability obfuscation from SXDH on 5-linear maps and locality-5 PRGs. In: Katz, J., Shacham, H. (eds.) CRYPTO 2017. LNCS, vol. 10401, pp. 599–629. Springer, Cham (2017). https://doi.org/10.1007/978-3-319-63688-7_20

41. Lin, H., Tessaro, S.: Indistinguishability obfuscation from trilinear maps and block-wise local PRGs. In: Katz, J., Shacham, H. (eds.) CRYPTO 2017. LNCS, vol. 10401, pp. 630–660. Springer, Cham (2017). https://doi.org/10.1007/978-3-319-63688-7_21

42. Lin, H., Vaikuntanathan, V.: Indistinguishability obfuscation from DDH-like assumptions on constant-degree graded encodings. Cryptology ePrint Archive, Report 2016/795 (2016). http://eprint.iacr.org/2016/795

43. Micciancio, D., Regev, O.: Worst-case to average-case reductions based on Gaussian measures. In: 45th FOCS, pp. 372–381. IEEE Computer Society Press, October 2004

44. Miles, E., Sahai, A., Zhandry, M.: Annihilation attacks for multilinear maps: cryptanalysis of indistinguishability obfuscation over GGH13. In: Robshaw, M., Katz, J. (eds.) CRYPTO 2016. LNCS, vol. 9815, pp. 629–658. Springer, Heidelberg (2016). https://doi.org/10.1007/978-3-662-53008-5_22

45. Nguyen, P.Q., Regev, O.: Learning a parallelepiped: cryptanalysis of GGH and NTRU signatures. In: Vaudenay, S. (ed.) EUROCRYPT 2006. LNCS, vol. 4004, pp. 271–288. Springer, Heidelberg (2006). https://doi.org/10.1007/11761679_17

46. Pass, R., Seth, K., Telang, S.: Indistinguishability obfuscation from semantically-secure multilinear encodings. In: Garay, J.A., Gennaro, R. (eds.) CRYPTO 2014. LNCS, vol. 8616, pp. 500–517. Springer, Heidelberg (2014). https://doi.org/10.1007/978-3-662-44371-2_28

47. Peikert, C.: An efficient and parallel gaussian sampler for lattices. In: Rabin, T. (ed.) CRYPTO 2010. LNCS, vol. 6223, pp. 80–97. Springer, Heidelberg (2010). https://doi.org/10.1007/978-3-642-14623-7_5

48. Regev, O.: New lattice-based cryptographic constructions. J. ACM **51**(6), 899–942 (2004)

49. Schnorr, C.P., Euchner, M.: Lattice basis reduction: improved practical algorithms and solving subset sum problems. Math. Program., 181–191 (1993)

50. Stehlé, D., Steinfeld, R.: Making NTRU as secure as worst-case problems over ideal lattices. In: Paterson, K.G. (ed.) EUROCRYPT 2011. LNCS, vol. 6632, pp. 27–47. Springer, Heidelberg (2011). https://doi.org/10.1007/978-3-642-20465-4_4

51. Zimmerman, J.: How to obfuscate programs directly. In: Oswald, E., Fischlin, M. (eds.) EUROCRYPT 2015. LNCS, vol. 9057, pp. 439–467. Springer, Heidelberg (2015). https://doi.org/10.1007/978-3-662-46803-6_15

Secure Computations and Protocols

Non-Interactive and Fully Output Expressive Private Comparison

Yu Ishimaki[✉] and Hayato Yamana

Waseda University, Tokyo, Japan
{yuishi,yamana}@yama.info.waseda.ac.jp

Abstract. Private comparison protocols are fundamental to the field of secure computation. Recently, Lu et al. (ASIACCS 2018) proposed a new protocol, XCMP, which is based on a ring-based fully homomorphic encryption (FHE) scheme. In that scheme, two μ-bit integers a and b are compared in encrypted form without revealing the plaintext to an evaluator. The protocol outputs a bit in encrypted form, which indicates whether $a > b$. XCMP has the following three advantages: the output can be reused for further processing, the evaluation is performed without any interactions with a decryptor having a secret key, and the required multiplicative depth is only 1. However, XCMP has two potential disadvantages. First, the protocol result preserves both additive and multiplicative homomorphisms over $\mathbb{Z}_t$ only, whereas the underlying FHE scheme can support a much larger plaintext space of $\mathbb{Z}_t[X]/(X^N + 1)$ for a prime t and a power-of-two N; this restricts the functionality of applications using the comparison result. Second, the bit length μ of the integers to be compared is no more than $\log N$ (typically 16 bits, at most). Thus, it is difficult for XCMP to handle larger integers. In this paper, we propose a non-interactive private comparison protocol that solves the aforementioned problems and outputs an additively and multiplicatively reusable comparison result over the ring without adding an extremely large computational overhead over XCMP. Moreover, by regarding a μ (> 16)-bit integer as a sequence of *chunks*, we show that the multiplicative depth required for our comparison protocol is logarithmic in the number of *chunks*. This value is much smaller than the naïve solution with a multiplicative depth of $\log \mu$. Experiment results demonstrate that our protocol introduces a subtle overhead over XCMP. Remarkably, we experimentally demonstrate that our protocol for a larger domain is comparable to the construction given by one of the state-of-the-art bitwise FHE schemes.

Keywords: Homomorphic encryption · Secure computation · Non-interactive private comparison

1 Introduction

Private comparison is a fundamental protocol in the field of secure computation. The protocol takes two private μ-bit integers a and b as input and outputs a bit

© Springer Nature Switzerland AG 2018
D. Chakraborty and T. Iwata (Eds.): INDOCRYPT 2018, LNCS 11356, pp. 355–374, 2018.
https://doi.org/10.1007/978-3-030-05378-9_19

that indicates whether $a > b$. While there exist variants of private comparison protocols, we focus on two-party private comparison protocol since it is already adopted in many applications [4,6,17]. The protocol can be constructed from an interactive or non-interactive setting [15,19,24,27]. The interactive setting requires the parties to be online during the protocol evaluation, which generates a burden to the parties, i.e., they have to consume bandwidth and turn on their devices (machines) until the protocol finishes. In order to mitigate such burden, one can construct a non-interactive protocol where a third party (evaluator) receives two private integers from the two parties and runs private comparison protocol on his machine. Then, the two parties are able to go offline during the evaluation. Similarly, we can mitigate a burden for one of the two parties by letting the other party being the evaluator.

The non-interactive comparison protocol can be constructed through homomorphic encryption [12,15,22,24,25], where the protocol supposes a decryptor having a secret key and an evaluator. The evaluator takes two encrypted integers as input and the evaluation is performed on his machine without interaction with the decryptor. The protocol output is either an indicator that reveals whether the Boolean expression $a > b$ holds [15,24] or the result of a computation result based on the indicator [22]. The former output can be achieved by simply subtracting the difference of the encrypted integers from all possible integers in the integer domain, followed by decryption and determination of whether a zero is present in the resultant list [15,24,25]. Thus, it is difficult to reuse the result in encrypted form as the indicator can only be obtained after the decryption. In contrast, the latter approach typically requires that a plaintext integer is encoded as a bit string [12]. With fully homomorphic encryption (FHE), we can trivially obtain both additive and multiplicative homomorphisms for the comparison result over the native plaintext space at the cost of a large multiplicative depth of $O(\log \mu)$. Therefore, both time and space complexities of FHE scheme increase. Thus, there is no efficient non-interactive comparison where the output of the method preserves both additive and multiplicative homomorphisms.

Recently, this problem was noted by Lu et al. [22]. They partially addressed this issue by introducing a new private comparison protocol using ring-based FHE, called XCMP. Since it relies on a new integer encoding method, XCMP allows to evaluate the comparison with a multiplicative depth of only 1 for $\mu \leq 16$. The protocol takes two encrypted μ-bit integers as input and privately compare them on the evaluator, with no interactions with the decryptor. After the evaluation, the encrypted result represents the indicator. This protocol can efficiently evaluate the comparison with multiplicative depth only one for relatively small domain, and the comparison result can be used for subsequent computations that require additive homomorphism (i.e., the approach is *output expressive* [22]).

However, XCMP has two main disadvantages. First, the comparison result provides additive and multiplicative homomorphisms over $\mathbb{Z}_t$ only, which restricts the functionality of subsequent computations. In contrast, the encryption scheme provides much larger space, i.e., a polynomial ring $\mathbb{Z}_t[X]/(X^N + 1)$

for a prime t and a power-of-two N. Second, XCMP does not scale in the larger message domain.

In this paper, we continue to study the same setting as Lu et al. [22], but target a more practical and a more efficient result. More precisely, we make the following contributions:

- We propose a non-interactive and **fully** output expressive private comparison protocol providing a comparison result that is fully reusable over the polynomial ring, with no interactions with the decryptor. To the best of our knowledge, our protocol is the first non-interactive comparison protocol where the comparison result preserves both additive and multiplicative homomorphisms over the polynomial ring without encoding integers into bit strings. Our construction is performed by introducing a constant term extraction technique over the ring.
- We propose a more efficient non-interactive private comparison protocol for a large domain. The required multiplicative depth is $\lceil \log \lceil \frac{\mu}{\log N} \rceil \rceil + 1$, which is less than the naïve solution with a depth of $\log \mu$. We discuss the outcome that the multiplicative depth of the proposed protocol is significantly reduced from the theoretical construction given by Lu et al. [22] for $\mu \leq 2 \log N$. The constant term extraction also contributes to the construction. Moreover, we revisit the definition of a non-interactive and output-expressive protocol given in [22], to define a more practical metric.

The remainder of this paper is organized as follows. We first introduce the necessary background information and definitions in Sect. 2. Then, we present our proposed protocol in Sect. 3 and provide evaluation results in Sect. 4. We describe related work in Sect. 5 and, finally, conclude the paper in Sect. 6.

2 Preliminaries

2.1 Notations

All notation is summarized in Table 1.

Homomorphic addition and multiplication can take two ciphertexts or a single ciphertext and a single plaintext as operands. We distinguish the term "domain size" from "plaintext space." The domain size indicates the range of an integer to be compared (i.e., $[0, 2^\mu - 1]$), whereas the plaintext space specifies the message space that the homomorphic encryption scheme can natively use (i.e., R_t). We assume that an integer over R_t represents a constant polynomial; i.e., a polynomial with all coefficients being zero except its constant term (the integer). All logarithms are base 2.

2.2 Ring-Based Homomorphic Encryption Scheme

The proposed protocol can be implemented using a ring-based FHE scheme. In this study, we have implemented the protocol using Brakerski/Fan-Vercauteran

Table 1. Notation and definitions

Notation	Definition
μ	Domain bit size
N	Ring dimension (power of two)
R	Ring $\mathbb{Z}[X]/(X^N + 1)$
R_t	Ring defining plaintext space $R/tR = \mathbb{Z}_t[X]/(X^N + 1)$
R_q	Ring defining ciphertext space $R/qR = \mathbb{Z}_q[X]/(X^N + 1)$
$\boxplus$	Homomorphic addition
$\boxdot$	Homomorphic multiplication
$\sigma_i(\cdot)$	Automorphism by index $i \in \mathbb{Z}_{2N}^*$ on a polynomial or a vector of polynomials
$\mathsf{Enc}(\cdot)$	Encryption of an element over R_t

(BFV) scheme [7,18], which is one of the most well-used among the various homomorphic encryption schemes provided in open-source libraries (e.g., [1,11, 23]). Thus, in this subsection, we describe the core operations supported in ring-based FHE schemes, that are required for our method. Let $R_t = \mathbb{Z}_t[X]/(X^N+1)$ be a quotient ring for some positive integer t. A plaintext $p \in R_t$ is encrypted into a ciphertext $\mathsf{Enc}(p) \in R_q^2$, where $R_q = \mathbb{Z}_q[X]/(X^N + 1)$ and $q >> t$. Thus, a ciphertext can be viewed as a two-dimensional vector of polynomials in R_q.

Note that each ciphertext contains a noise term to guarantee the security of the encryption scheme. The inherent noise increases with every homomorphic operation, and the noise growth depends on the computation to be performed. When the noise exceeds some threshold, a correct decryption result can no longer be obtained. To avoid a decryption error, the parameter q, which determines the maximum noise that a single ciphertext can contain, is determined based on the performed operations.

The following operations provided by the ring-based FHE scheme are used:

- Homomorphic Addition
  Two ciphertexts can be added through component-wise addition of their polynomials, which involves additive noise growth.
- Homomorphic Multiplication
  Homomorphic multiplication is an expensive operation which is orders of magnitude slower than homomorphic addition and automorphism [3]. The noise growth involved in homomorphic multiplication is multiplicative. Note that ciphertext-plaintext multiplication is more efficient in terms of computational cost. The noise growth is also multiplicative, which depends on the number of non-zero coefficients and the value of each coefficient.
- Automorphism
  Applying automorphism over a ciphertext that encrypts $p \in R_t$ yields an

encryption of $p(X^i)$ for some $i \in \mathbb{Z}_{2N}^*$ [20]. After any automorphism on a ciphertext, the form of the corresponding secret key is changed. This operation is required when the associated secret keys of the two ciphertexts to be evaluated are different, while the original secret keys remain the same. To perform further evaluation over the ciphertexs, where the associated secret keys of the two ciphertexts are different, we must perform a key-switching operation which restores the current corresponding secret key to the original secret key associated with the original ciphertext. In this study, we assume that an appropriate key-switching operation is always performed after each automorphism.

One of the challenging aspects of homomorphic encryption is setting the minimum possible parameter size. In this study, we do not use *bootstrapping*, as that approach necessitates a large parameter size for the FHE scheme to homomorphically evaluate the decryption circuit. Thus, we set the smallest possible values for the parameter set to avoid costly *bootstrapping*, and limit the number of computations over the encrypted data without decryption error. Note that our proposed method can be evaluated on FHE with a smaller plaintext space using *bootstrapping*. Further, our proposed method is applicable to other ring-based homomorphic encryption schemes that support the above functions, such as BGV [8] and YASHE [5].

2.3 XCMP

We recall the non-interactive and output expressive private comparison protocol XCMP presented by Lu et al. [22], which can homomorphically evaluate the validity of $a \geq b$ for two $\log N$-bit integers a and b, with a multiplicative depth of 1 over R_t. The underlying concept is comparison of two integers over a degree of a monomial; hence, it is easier to obtain a comparison result as a bit on the coefficient over R_t. First, XCMP encodes two integers into two monomials. Specifically, two μ ($= \log N$)-bit integers $a, b \in [0, N-1]$ are encoded into X^a and X^b over R_t followed by encryption. Homomorphically negating the degree of one of those monomials (e.g., we obtain $\mathsf{Enc}(X^{-b})$ by homomorphically negating the degree of $\mathsf{Enc}(X^b)$ via an automorphism σ_{2N-1}), multiplication between $\mathsf{Enc}(X^a)$ and $\mathsf{Enc}(X^{-b})$ yields $\mathsf{Enc}(X^{a-b})$. Note that X^{-b} is congruent to $-X^{N-b}$ mod $X^N + 1$. As $a - b \in [-(N-1), N-1]$, we obtain an encryption of a polynomial $p(X) = dX^e$, where $d = \pm 1$ and $e \in [0, N-1]$. Then, $\mathsf{Enc}(p(X))$ is multiplied by a plaintext polynomial $T(X) = \sum_{i=0}^{N-1} X^i$ and we obtain an encryption of $p'(X) = p(X)T(X)$.[1]

The interesting property of $p'(X)$ is that it contains 1 at a constant term if $a \leq b$, and -1 otherwise. This is because the degree of the monomial $p(X)$ (i.e., the difference between the two integers) is some integer in $[-(N-1), N-1]$.

---

[1] In the original study [22], it is assumed that $T(X) = \alpha \sum_{i=0}^{N-1} X^i$ for some $\alpha \in \mathbb{Z}_t$, to render the protocol parameterizable. In this study, we define $T(X) = \sum_{i=0}^{N-1} X^i$, i.e. $\alpha = 1$ for simplicity.

There is a one-to-one mapping between the set of all possible differences of two integers $\{-(N-1), -(N-2), \ldots, -1, 0, 1, \ldots, N-1\}$ and a set of monomials $\{-X, -X^2, \ldots, -X^{N-1}, 1, X, \ldots, X^{N-1}\}$ with sign ± 1 over R. Therefore, multiplying $p(X)$ by $T(X)$ moves a coefficient of $p(X)$ to the constant term of $p'(X)$, iff. $a - b \in [-(N-1), 0]$; i.e., $a \leq b$. Otherwise $p'(X)$ contains -1 as the constant term $(a > b)$; i.e., $a - b \in [1, N-1]$.

However, because of the polynomial $T(X)$, the result $p'(X)$ consists of not only ± 1 as the constant term, but values for the higher terms. These values for the higher terms reveal the exact difference between the two compared integers. To hide the undesired information and to change the constant term from ± 1 to $\{0, 1\}$ simultaneously, a plaintext polynomial $R(X) = 1 + \sum_{i=1}^{N-1} r_i X^i$ (where each of the r_i terms represents a random value uniformly sampled from $\mathbb{Z}_t$) is added to $\mathsf{Enc}(p'(X))$. Finally, multiplying by an inverse of two, we obtain an encryption of a polynomial that contains 1 as the constant term, iff $a < b$, and 0 otherwise. The plaintext modulus t must be large enough in order to guarantee the correctness of a subsequent computation.

Recall that the result is an encryption of a polynomial that contains (random) values for the higher terms, except for the constant term, which is a bit. Thus, the result given by XCMP retains additive and multiplicative homomorphisms on the constant term (i.e., $\mathbb{Z}_t$) only, however, the encryption scheme itself supports a much larger plaintext space (i.e., R_t).

2.4   Efficient Private Comparison from Homomorphic Encryption

In order to clarify our goal, we define the term *efficient non-interactive and fully output expressive private comparison* below.

Definition 1. *Let a and b be two μ-bit integers. A private comparison protocol that takes encryptions of a and b as input can be said to be efficient non-interactive and fully output expressive iff it satisfies all the following properties:*

1. The comparison protocol does not require any communication with a party having a secret key during the evaluation (*non-interactive*);
2. The multiplicative depth required for the protocol is less than $\log \mu$ and the native plaintext space exceeds $\mathbb{Z}_2$ (*efficient*);
3. The comparison result is an encryption of a bit over the native plaintext space that preserves both the additive and multiplicative homomorphism over the native plaintext space (*fully output expressive*).

The terms *non-interactive* and *output expressive* are borrowed from [22]. Our definition differs from those of [22] in two aspects: we append the metric "*efficient*" and add the prefix "*fully*" to the last metric. Note that XCMP does not satisfy the second (for $\mu > \log N$) and third properties.

3 Proposed Protocol

In this section, we describe our proposed private comparison protocol that overcomes the disadvantages underlying XCMP. As mentioned in Sect. 2.4, XCMP suffers from the following two problems. First, the comparison result provides both additive and multiplicative homomorphisms over $\mathbb{Z}_t$ only, while the underlying FHE scheme supports much larger plaintext space R_t. Second, XCMP cannot efficiently deal with message domain of size more than $\log N$ bits. In order to solve the aforementioned problems and achieve *efficient non-interactive and fully output expressiveness*, we introduce a new *constant term extraction method* in Sect. 3.1. Then, in Sect. 3.2, we demonstrate how to integrate it into XCMP without changing the multiplicative depth of it for domain size of μ ($\leq \log N$) bits. Finally, in Sect. 3.3 we describe how to achieve the private comparison for $\mu > \log N$ by utilizing our constant term extraction method.

In addition to XCMP, a bitwise FHE scheme also supports non-interactive private comparison protocol. In Table 2, we demonstrate the advantages of our method compared with those methods.

Table 2. Comparison with existing non-interactive private comparison protocols using homomorphic encryption. Here, μ denotes the bit size of the input domain. The plaintext space of ring-based homomorphic encryption is $R_t = \mathbb{Z}_t[X]/(X^N + 1)$, where an element of R_t is a polynomial of degree $(N - 1)$ and its coefficient is modulo t. Being dependent on N (power of two) and t, d is determined by $\arg\min_d(t^d \equiv 1 \mod N)$.

Method	Message domain $[0, 2^\mu - 1]$	Multiplicative depth	Plaintext space	Homomorphism after comparison
Proposed method	$\mu \leq \log N$ $\mu > \log \mathbf{N}$ **(any)**	1 $\lceil \log\lceil \frac{\mu}{\log N}\rceil\rceil + 1$	$\mathbf{R_t}$	**Additive / multiplicative over $\mathbf{R_t}$**
XCMP [22]	$\mu \leq \log N$ $\log N < \mu \leq 2\log N$	1 $\log t + \log d + 2$	$\mathbf{R_t}$	Additive / multiplicative over $\mathbb{Z}_t$
Bitwise FHE	**any**	–	$\mathbb{Z}_2$	Additive / multiplicative over $\mathbb{Z}_2$

Recall that XCMP uses a single polynomial to represent a single integer, as the integer is encoded as a monomial degree. Thus, the upper limit of the message domain is the ring dimension ($\log N$ bits) of R. Lu et al. [22] concluded that their method is feasible in practice for domain sizes of up to $\log N$ bits. Although XCMP provides a means of extending the domain size to $2\log N$ bits, the multiplicative depth is extremely large when the plaintext modulus t is large, because of the use of Fermat's little theorem over R. The concrete multiplicative depth required for the protocol is $\log t + \log d + 2$ for $\mu \leq 2 \log N$, where d is a constant such that $t^d \equiv 1 \mod N$. Therefore, implementation of XCMP for a larger domain size was not provided. In contrast, our proposed method works with any size μ and a multiplicative depth of $\log \lceil \frac{\mu}{\log N} \rceil + 1$. For concrete comparison, we consider the case of $\mu = 2 \log N$. The multiplicative depth required for our method is, $1 + \log \frac{2 \log N}{\log N} = 2$. This is much more efficient than XCMP [22], as the multiplicative depth required for our method is independent of t.

For a bitwise FHE scheme, it is assumed that recent constructions [13,14,16] are used, which provide fast *bootstrapping* for an encryption of a single-bit message. The *bootstrapping* is performed on every homomorphic gate operation, such as AND-gate and XOR-gate operations. Thus, an unlimited number of computations without decryption errors for any function are offered. Therefore, we ignore the multiplicative depth for the comparison in the FHE. Our proposed protocol is able to preserve the homomorphism after the comparison over R_t; this feature is missing from the existing techniques. Specifically, for concrete evaluation, we compare the running time of our method with that of TFHE-based comparison in Sect. 4.2.

3.1   Constant Term Extraction

In this subsection, we describe our new *constant term extraction method* which allows to homomorphically extract a constant term of a polynomial. The output of XCMP is given on the constant term of a polynomial in encrypted form, however, there exists several random coefficients on the higher terms. Those random values restricts the functionalities of the subsequent computations based on the comparison result. Namely, (an encryption of) an element over $\mathbb{Z}_t$ can only be evaluated over the comparison result while the FHE scheme natively supports much larger plaintext space R_t. Thus, in order to leverage the native plaintext space after the comparison, the constant term p_0 must be homomorphically extracted from the encryption of the polynomial $p(X) = \sum_{i=0}^{N-1} p_i X^i \in R_t$.

Our proposed method is shown in Algorithm 1.

Algorithm 1. Constant Term Extraction

1: **Input**: an encryption of $p = \sum_{i=0}^{N-1} p_i X^i \in R_t$
2: **Output**: an encryption of $p_0 \in R_t$
3: $c \leftarrow \sigma_{N+1}(\mathsf{Enc}(p))$
4: $c \leftarrow \mathsf{Enc}(p) \boxplus c$
5: **for** $k = 1$ **to** $\log N - 1$ **do**
6: $\quad c' \leftarrow \sigma_{\frac{N}{2^k}+1}(c)$
7: $\quad c \leftarrow c \boxplus c'$
8: **end for**
9: $d \leftarrow c \boxdot (2^{\log N})^{-1}$
10: **return** d

The constant term extraction method is performed without using costly ciphertext-ciphertext multiplications, and primarily uses automorphism and ciphertext-ciphertext addition. The automorphism $\sigma_{\frac{N}{2^k}+1}$ maps $X \to X^{N+2^k}$, where X^{N+2^k} is congruent to $-X^{2^k} \mod X^N + 1$. When $k = 0$, the automorphism negates the odd coefficients of the input polynomial while keeping the even coefficients unchanged. Thus, we are able to discard the odd coefficients via $\sigma_{N+1}(\text{Enc}(p(X))) + \text{Enc}(p(X))$ for $p \in R_t$, while the even coefficients are doubled. Then, we recursively apply $\sigma_{N/2^k+1}$ followed by ciphertext-ciphertext addition for $k \in [1, \mu-1]$. The final result is a polynomial with the constant term multiplied by $2^{\log N}$, while the other coefficients are all zero. To obtain the exact constant term, we require multiplication by $(2^{\log N})^{-1} \mod t$. Our method is inspired by the optimization for linear transformation procedure in the context of *bootstrapping* proposed by Chen and Han (see Coefficient Selection [10, A.2]), which we adopt the technique in the context of private comparison.

Recall that an automorphism requires a key-switching operation; thus, a special key must be stored for every index subjected to automorphism. In our proposed method, we must perform $\log N$ automorphisms for distinct indices. Thus, our method contains both additional computational overhead and additional storage overhead compared to XCMP. As regards the computational cost, we must additionally perform $\log N$ automorphisms, $\log N$ ciphertext-ciphertext addition, and a single ciphertext-plaintext multiplication. Nevertheless we experimentally show that the additional computational overhead is acceptable. The storage overhead is more expensive than XCMP by a factor of $\log N$, as we require additional $\log N$ keys for automorphism.

3.2 Private Comparison for Small Domain ($\mu \leq \log N$)

In this subsection, we demonstrate our private comparison protocol for two μ-bit integers a and b, where $\mu \leq \log N$, using a single polynomial to represent a single integer. Our protocol is constructed from ring-based FHE scheme. We first describe private comparison protocol. Then, we also demonstrate construction of the private equality check protocol since it is required in order to construct the comparison protocol for a large domain.

Proposed Private Comparison Protocol. The private comparison protocol evaluates whether the Boolean expression $a > b$ holds in a privacy-preserving manner. The protocol is described in Algorithm 2. Our protocol is almost identical to XCMP. The difference is that constant term extraction (Sect. 3.1), which homomorphically extracts a constant term of a polynomial, is performed in our protocol. Thus, our protocol does not perform adding a random polynomial with a constant term of zero to the multiplication result. Moreover, we note that preparation for a plaintext polynomial $T(X)$ has two effects on the protocol.

The first is reduction of both the computational cost and noise growth. Recall that XCMP uses $T(X) = \sum_i^{N-1} X^i$ to compare $\log N$-bit integers over the exponents of X by a single ciphertext-plaintext multiplication. The caveat is that a wrap-around mod $X^N + 1$ (line 4 of Algorithm 2) is involved. Thus, the result appears as the constant term with the value being ± 1 depending on the comparison result. This requires adjustment of ± 1 to $\{0, 1\}$ (line 5,6 of Algorithm 2). These two operations can be removed by setting $\mu = \mu' < \log N$. That is, we set the plaintext polynomial $T(X) = \sum_{i=0}^{2^{\mu'}-1} X^i$ by setting $\mu = \mu' < \log N$. This allows to constant term have $\{0, 1\}$ after the multiplication $\mathsf{Enc}(X^{a-b}) \boxdot T(X)$, without multiplying by 2^{-1}. Disadvantage is that we must (at least) halve the original domain size.

The second affect is that parameter selection becomes easier in the protocol. As described in Sect. 2.2, the security level is determined by the choice of N and q. Then, q depends on a computation to be performed. Meanwhile, in our protocol, N is determined by μ and q for the comparison and the security parameter, respectively. Thus, selection for both q and N totally depends on the subsequent computation. In our evaluation, we consistently chose $\mu' = 10$. This is because we found the minimum recommended value of N is 1024 by following the recent security recommendation in the homomorphic encryption standardization [9]. Thus, treating $\mu' = 10$ is reasonable for parameter selection. This is helpful for the case that $\mu > \log N$, as we describe in Sect. 3.3.

Note that the multiplicative depth required for our protocol is identical to that of XCMP as we need to perform a single ciphertext-ciphertext multiplication and two ciphertext-plaintext multiplications. We also note that an integer encoded in the protocol is just a monomial, i.e., there is only one non-zero coefficient and the coefficient is ± 1. This indicates that there is no noise growth during the homomorphic multiplication of two monomials (line 3 of Algorithm 2) when one of the two monomials is not encrypted. In addition, it is possible to homomorphically convert X^b to X^{-b} via σ_{2N-1} as applied in XCMP. Here, we omit the procedure for simplicity.

By adopting our constant term extraction method, the output of the protocol is an encryption of a bit over R_t. Thus, our private comparison protocol is *efficient non-interactive and fully output expressive*.

Algorithm 2. Fully Output Expressive Private Comparison for Small Domain

---

1: **Input**: encryptions of two integers $a, b \in \mathbb{Z}_{2^\mu}$
2: **Output**: an encryption of a bit
3: $c \leftarrow \mathsf{Enc}(X^a) \boxdot \mathsf{Enc}(X^{-b})$
4: $c' \leftarrow c \boxdot T(X)$
5: $c' \leftarrow c' \boxplus 1$
6: $c' \leftarrow c' \boxdot 2^{-1}$
7: $d \leftarrow \mathsf{ConstantTermExtract}(c')$
8: **return** d

Private Equality Check. We here demonstrate the construction of private equality check protocol to handle a larger domain size for $\mu > \log N$. We can simply perform the private equality check protocol by applying constant term extraction on $\mathsf{Enc}(X^{a-b})$. This is because 1 appears as the constant term iff $a = b$, and 0 appears otherwise. We can trivially confirm that our private equality check protocol is *efficient non-interactive and fully output expressive*.

3.3   Private Comparison for Large Domain ($\mu > \log N$)

The private comparison described in the previous section assumes that domain size is $\log N$ bits at most. Although larger domain size can be handled by enlarging N by a factor of a power of two, this induces significant performance loss, as every operation and every object size in the ring-based homomorphic encryption becomes exponentially larger. In this subsection, we demonstrate management of the private comparison for a larger-bit integer without enlarging N, while retaining the *efficient non-interactive and fully output expressive* properties. We realize the private comparison protocol by regarding $\log N$-bit integers as one *chunk*. Namely, we split two μ-bit integers a and b by $\log N$ bits. That is, we define $a = \sum_{i=0}^{\lceil \frac{\mu}{\log N} \rceil - 1} a_i 2^{(\log N)^i}$ and $b = \sum_{i=0}^{\lceil \frac{\mu}{\log N} \rceil - 1} b_i 2^{(\log N)^i}$, with $a_i, b_i \in \mathbb{Z}_N$. We call each of a_i and b_i components i-th *chunks*. The comparison can be performed on every pair of i-th *chunks* individually. Therefore, we need a set of ciphertexts for a single integer by generating a ciphertext on each *chunk*.

The concept underlying our private comparison technique is similar to the standard solution in the literature [12]; that is, the private comparison is recursively computed from the most significant bit down to the least significant bit. In contrast to the literature, however, we can adopt the same concept while using a *chunk*-by-*chunk* approach rather than a bit-by-bit method.

Let K be the number of *chunks*, i.e., $\lceil \frac{\mu}{\log N} \rceil$, and let EQ_i and GT_i be the result (an encryption of a bit) of an equality check and a comparison applied on the i-th *chunk*, respectively. In addition, let $\overline{\mathsf{EQ}}_i$ be the flipped bit result of EQ_i, which can be computed by subtraction of EQ_i from 1 when $t > 2$. Recall that we can evaluate private comparison and private equality check for every *chunk* by using the method described in Sect. 3.2. Thus, the private comparison for large domain can be performed by recursively evaluating the following expression:

$$c_{K-1} = (\overline{\mathsf{EQ}}_{K-1} \boxdot \mathsf{GT}_{K-1}) \boxplus (\mathsf{EQ}_{K-1} \boxdot c_{K-2}),$$

where $c_0 = \mathsf{GT}_0$ and $c_i = (\overline{\mathsf{EQ}}_i \boxdot \mathsf{GT}_i) \boxplus (\mathsf{EQ}_i \boxdot c_{i-1})$ for $i \in [1, K-1]$. By recursively evaluating c_{K-2}, we can obtain the following:

$$c_{K-1} = (\overline{\mathsf{EQ}}_{K-1} \boxdot \mathsf{GT}_{K-1}) \boxplus_{i=1}^{K-2} (\overline{\mathsf{EQ}}_i \boxdot \mathsf{GT}_i \boxdot_{j=i+1}^{K-1} \mathsf{EQ}_j) \boxplus (\mathsf{GT}_0 \boxdot_{i=1}^{K-1} \mathsf{EQ}_i).$$

It is apparent that the maximum multiplicative depth required for the above expression is dominated by the evaluation of $\mathsf{GT}_0 \boxdot_{i=1}^{K-1} \mathsf{EQ}_i$, which is $\lceil \log \lceil \frac{\mu}{\log N} \rceil \rceil$. Note that, without expanding the expression, the depth required for the c_{K-1} evaluation is $\lceil \frac{\mu}{\log N} \rceil - 1$. The depth for evaluating a private comparison on every *chunk* is 1. Hence, the overall multiplicative depth to evaluate private comparison is $\lceil \log \lceil \frac{\mu}{\log N} \rceil \rceil + 1$. This is significantly smaller than that of the naïve method of $\log \mu$ since $\log N \geq 10$ in general. As the overall protocol consists of private comparison and private equality check on every *chunk*, where they are *efficient non-interactive and fully output expressive*, the result of the overall private comparison for large domain is also an encryption of $\{0, 1\}$ over R_t. Thus, we can say that our comparison protocol is *efficient non-interactive and fully output expressive* even for large domain. Our method can be viewed as a generalization of the XCMP approach proposed by Lu et al. [22] but in more efficient manner. Note that XCMP only provided a theoretical construction to extend the domain size to $2 \log N$-bit space with a large multiplicative depth, because it employed Fermat's little theorem over R.

Recall that we can control the domain size for each *chunk* by setting the domain range $[0, 2^{\mu'} - 1]$ to which a single *chunk* corresponds. Specifically, we set $\mu' < \log N$ and split the original large domain size μ by μ' rather than $\log N$. Hence, the number of *chunks* is $\lceil \frac{\mu}{\mu'} \rceil < \lceil \frac{\mu}{\log N} \rceil$, and we can retain easier parameter selection while also reducing the computational cost. In addition, the computation is easily parallelizable. We can reuse the multiplication result $\mathsf{Enc}(x^{a_i - b_i})$ for every *chunk* to evaluate EQ_i and GT_i.

3.4   Optimization

In our protocol, there are several (large) constant multiplications which increases the inherent noise by a factor of the same constant value. Thus, we do not wish to implement this multiplication in order to retain the smallest possible parameter size. As the constant term extraction multiplies the coefficient by $2^{\log N}$, multiplication by its inverse is required to obtain 1, if the coefficient is $2^{\log N}$ and 0 otherwise. On the other hand, the multiplication by $(2^{\log N})^{-1}$ (Line 9 in Algorithm 1) also multiplies the inherent noise. To avoid this multiplication, we set the plaintext modulus t to be $2^{\log N} - 1$. Then, we can treat $2^{\log N} \equiv 1 \mod (2^{\log N} - 1)$, thereby avoiding multiplication by a large constant. In addition, one can use plaintext space reduction technique for free when the underlying scheme is BFV with the plaintext modulus $t = 2^r$ for some $r > \log N$ [3]. Note that, in those cases, t must not be prime.

3.5   Secure Private Comparison Protocol

Now, we can construct a secure private comparison protocol in the semi-honest model, assuming the same stake-holders discussed in [22], i.e., encryptor, decryptor, and evaluator.

The evaluator receives encryptions of two μ-bit integers from two different encryptors. The evaluation is performed on the evaluator without any communications with the decryptor, while the comparison result offers both additive and multiplicative homomorphisms over R_t rather than $\mathbb{Z}_t$. Finally, the decryptor obtains the result by decrypting the data. We assume that the evaluator is semi-honest, i.e., he/she attempts to acquire meaningful information while correctly following the protocol. The security proof is identical to that of XCMP (thus, we refer the reader to [22] for more detail).

3.6 Applications and Limitation

The structure of plaintext R_t underlying ring-based FHE schemes enables to pack a set of integers into a single plaintext polynomial [26]. Since the computation to the plaintext performed element-wisely on each element in the vectors, computational cost is amortized. In contrast, our method requires an entire single polynomial to represent a single integer of up to $\log N$ bits, or requires multiple polynomials to represent a single integer of more than $\log N$ bits. Thus, our proposed method is best suited for applications in which the latency of the comparison is more important than the throughput.

4 Performance Result

In order to confirm the performance of our protocol, we evaluated runtime of our protocol. We implemented our protocol with C++ using the g++-7. We ran all experiments on a machine equipped with an Intel Xeon E5-1620 v4 with a 3.50 GHz CPU having four cores and 32 GB RAM running Ubuntu 16.04. All experiments were performed with a single thread setting. The protocol runtime performance was measured by the C++ Chrono library, and was averaged over 30 experiments.

4.1 Performance Comparison with XCMP

Recall that our proposed protocol introduces an additional operation, i.e., constant term extraction over XCMP, to render the scheme *fully output expressive*. Thus, there exists some additional overhead in our protocol. To determine the performance loss, we experimentally compared our protocol with XCMP by varying the bit length μ from 10 to 16.

Since XCMP was implemented with both SEAL v2.3.0.4 and HElib, we selected SEAL library for the implementation of XCMP for efficiency reason. Note that SEAL v2.3.0.4 does not support automorphism from user API. Thus, we used the more recent release of SEAL (version 2.3.1) which supports automorphisms from user API. Then, we implemented our protocol over XCMP.

In the experiments, we used almost the same FHE parameter set for both XCMP and our protocol for fair comparison. The only difference is the ciphertext modulus q due to the noise growth involved in the constant term extraction in our protocol. The ciphertext modulus q was determined in order to support multiplicative depth of 1 after the comparison. We set the ring dimension N equal to 2^μ. As smaller N requires smaller q to guarantee the security level, we chose a different parameter set according to μ. We chose the parameter set so that at least 80-bit security level is achieved against known attacks. The security level was estimated using an open-source implementation[2] of Learning With Errors estimator by Albrecht et al. [2]. The parameter set is shown in Table 3. Here, we chose $\log N = 11$ for $\mu = 10$ instead of choosing $\log N = 10$, in order to guarantee the correctness of the computation.

Table 3. FHE Parameters

μ		10	11	12	13	14	15	16
$\log N$		11	11	12	13	14	15	16
t		2,047	2,047	4,095	8,191	16,383	32,767	65,535
$\log q$	XCMP	120	120	120	120	120	120	180
	Ours						180	

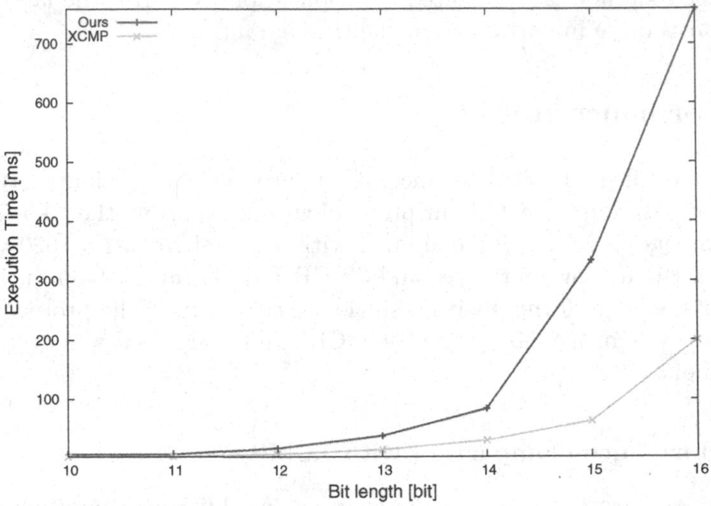

Fig. 1. Runtime comparison of proposed protocol with XCMP. After the comparison, both the protocol could perform a single ciphertext-ciphertext multiplication.

Figure 1 shows the runtime performance of the proposed protocol and XCMP. It is apparent that our protocol is slower than XCMP, with a maximum performance gap of a factor of six when $\mu = 15$. Our protocol involves a large computational overhead over XCMP, however, the runtime is still less than 1 s for $\mu \leq 16$. Note that our protocol has larger computational capability after the comparison than XCMP. Thus, the additional overhead associated with the proposed protocol is allowable in terms of a trade-off between functionality and efficiency.

4.2   Performance Comparison with Bitwise FHE-based Solution

In Sect. 3.3, we demonstrated extension of the message domain size from $\log N$-bit to a bit size more than $\log N$ by combining constant term extraction method and private comparison of multiple *chunks*. As XCMP does not facilitate implementation for a larger domain size, we cannot compare it with our protocol

---

[2]   https://bitbucket.org/malb/lwe-estimator commit:2094ada.

in experiment. Therefore, we compared our protocol with a construction by a bitwise FHE scheme instead. Specifically, we chose TFHE [13,14], a state-of-the-art FHE scheme with fast *bootstrapping* for an encryption of a single bit message, as a baseline. For the implementation, we used the TFHE open-source library[3]. Our protocol was implemented by using the PALISADE [23] lattice cryptography library, which provides implementations of several homomorphic encryption schemes. In particular, we chose the BFV [18] variant of Halevi-Polyakov-Shoup scheme [21]. Note that current version of TFHE provides neither multi-threading nor an asymmetric cryptosystem. Thus, we implemented our protocol with single thread setting to conduct fair comparison as much as possible, however, we used an asymmetric cryptosystem which is computationally more expensive than an symmetric one since PALISADE does not support symmetric one. In the experiment, we varied μ from 2 to 130.

For $\mu \in [2, 10]$, we represent a single integer by a single polynomial. Then, for $\mu \in [11, 130]$, we split a single μ-bit integer by 10 bits into $\lceil \frac{\mu}{10} \rceil$ *chunks*. We set $t = 2^{\log N} - 1$ to obtain $\{0, 1\}$ for the comparison in each *chunk* without multiplying the inverse of $2^{\log N}$. Figure 2 shows runtime performance for evaluation of the comparison ranging over the remaining multiplicative depth after the comparison from 0 to 2. Note that the comparison protocol is not *fully output expressive* when the remaining depth is 0; this functionality is exactly identical to that of XCMP, for which only several homomorphic additions can be performed over R_t.

Recall that our protocol is parameterized by N and q following the multiplicative depth of $\lceil \log \lceil \frac{\mu}{\log N} \rceil \rceil + 1$ (in our experiment, it was $\lceil \log \lceil \frac{\mu}{10} \rceil \rceil + 1$) and the security level. Thus, the parameter size increases as μ increases, which deteriorates the overall performance. From 2 to 130, the multiplicative depth changes when μ exceeds 10, 20, 40 and 80. Thus, there exist some performance gap when μ exceeds each of those values, which are due to the increase in the FHE parameter q. On the other hand, TFHE relies on *bootstrapping* after every gate operation. Hence, the parameter size does not change depending on the computation. As a result, the complexity of TFHE-based protocol is linear in μ with no concern for the parameter growth due to homomorphic multiplications. In contrast, the complexity of our proposed protocol is linear in the number of *chunks* with the increase of q and N at some value.

It is apparent that our method is comparable to TFHE. Specifically, for $\mu \in [3, 40]$ (resp. $\in [3, 20]$) our proposed protocol is faster than TFHE-based protocol by a factor of 2–3 with the setting of the remaining depth to 0 (resp. 1). For the other parameter sets, our proposed protocol is slower than TFHE-based protocol. For example, when $\mu = 130$ our protocol with remaining depth 2 is slower than TFHE-based protocol by a factor of approximately four. We remark that the performance gap seen on the comparison protocol can be absorbed depending on the computation after the comparison since the amortized cost per operation for TFHE that encrypts a bit is much smaller than other ring-based HE that encrypts a polynomial. Although there exists a trade-off between

---

[3] https://github.com/tfhe/tfhecommit:6297bc7.

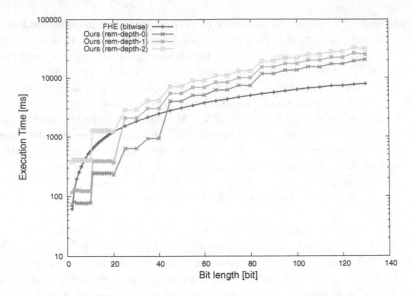

Fig. 2. Runtime comparison for proposed method and method based on bitwise FHE scheme. We ranged the remaining multiplicative depth after the comparison from 0 to 2 for our proposed method. In the figure, rem-depth-i represents that the scheme is able to handle multiplicative depth-i after the comparison. The bitwise FHE scheme has no multiplicative depth limitation.

the performance and the functionality, this is an allowable trade-off up to a few remaining depth in our case.

The concrete parameter set, which satisfies 80-bit security against known attacks, for our protocol is presented in Table 4. The difference in t is due to the constant term extraction. Recall that constant term extraction must be performed on all coefficients on a polynomial. Further, the degree of the polynomial is determined by the security level and q. In our protocol, q depends on μ as it determines the multiplicative depth. Thus, N varies based on μ due to the security reason.

5   Related Work

Many studies on private comparison protocols have been conducted. One well-known protocol was developed by Damgård et al. [15], i.e., the so-called DGK protocol. The DGK protocol can perform private comparison in a non-interactive manner, when one of two input integers are plaintext, however, the protocol does not offer multiplicative and additive homomorphisms over the comparison result since zero check after decryption is required.

Gentry et al. [19] constructed a private comparison protocol from a lattice-based somewhat homomorphic encryption in an interactive manner. While the comparison result is reusable over R_{2^r} for some r, it requires $O(\log \mu)$ rounds of

Table 4. FHE parameters for our proposed method

Remaining Depth	μ	$\log q$	$\log N$	t
0	[2, 20]	89.97	12	4,095
	[25, 40]	119.96	12	4,095
	[45, 80]	149.93	13	8,191
	[85, 130]	179.91	13	8,191
1	[2, 20]	119.96	12	4,095
	[25, 40]	149.92	13	8,191
	[45, 80]	179.91	13	8,191
	[85, 130]	209.89	13	8,191
2	[2, 20]	149.92	13	8,191
	[25, 40]	179.91	13	8,191
	[45, 80]	209.89	13	8,191
	[85, 130]	239.87	13	8,191

communication between two parties. Cheon et al. [12] developed a set of non-interactive homomorphic circuit evaluation primitives including a comparison protocol and equality check protocol for database query purposes. Although the suggested protocols can handle several integers by packing a set of bits into a single polynomial using a well-known batching technique [26], it does not overcome the multiplicative depth $O(\log \mu)$ complexity problem.

Saha and Koshiba [24] suggested a non-interactive private comparison protocol. Specifically, they proposed a new encoding method that enables efficient evaluation of multiple hamming distances of a set of two binary vectors and recursive addition of those hamming distances. The protocol outputs a vector that indicates whether $a > b$ holds in a similar manner to the DGK protocol, that is the comparison is performed by checking if there exists a zero in the result vector after decryption. Thus, it is difficult to non-interactively reuse the comparison result. The follow-up work [25] improved the efficiency by modifying the packing method in such a manner as to encode a vector of base-β integers into a single polynomial rather than to encode a binary vector, thereby handling much more integers in a single polynomial. The comparison result is not non-interactively reusable as well as the protocol of Saha and Koshiba [24].

From another technical perspective, one can use the constant term extraction method introduced by Ducas and Micciancio [16], which is used in the context of *bootstrapping* [13,14]. However, the result is an encryption of an element of $\mathbb{Z}_t$, not R_t, and homomorphically converting an element over $\mathbb{Z}_t$ back to an element over R_t requires *bootstrapping* that works in binary message space only. In contrast, the method proposed in this work does not use binary message space. Thus, it cannot be used directly for further processing over R_t.

6   Conclusion

In this paper, we improved upon the previous non-interactive private comparison protocol in two ways. First, we achieved a non-interactive and fully output expressive comparison protocol. Second, our protocol exhibits functionality comparable to the bitwise FHE solution with support for a few multiplicative depths, even for larger integers. Those two functionalities can be achieved by introducing a constant term extraction technique. The proposed protocol can be applied with a larger plaintext space in a similar manner to XCMP [22]. In addition, our method allows comparison of larger bits (>16), which were previously expensive in XCMP. The runtime performance of our protocol is comparable to a state-of-the-art bitwise FHE scheme. Further, the method developed in this work has the advantages that several addition and multiplication operations can be performed over R_t after the comparison has been conducted.

We hope that our protocol will be adopted as a sub-protocol for various secure computation protocols.

Acknowledgment. This work was supported by JST CREST Grant Number JPMJCR1503, Japan and Japan-US Network Opportunity 2 by the Commissioned Research of National Institute of Information and Communications Technology (NICT), JAPAN. The authors would like to thank Kurt Rohloff and Yuriy Polyakov for their supports for PALISADE library.

References

1. Aguilar-Melchor, C., Barrier, J., Guelton, S., Guinet, A., Killijian, M.-O., Lepoint, T.: NFLLIB: NTT-Based fast lattice library. In: Sako, K. (ed.) CT-RSA 2016. LNCS, vol. 9610, pp. 341–356. Springer, Cham (2016). https://doi.org/10.1007/978-3-319-29485-8_20
2. Albrecht, M., Player, R., Scott, S.: On the concrete hardness of learning with Errors. J. Math. Cryptol. **9**(3), 169–203 (2015)
3. Angel, S., Chen, H., Laine, K., Setty, S.: PIR with compressed queries and amortized query processing. In: Proceedings of the 2018 IEEE Symposium on Security and Privacy (SP), pp. 962–979 (2018)
4. Barni, M., et al.: Privacy-preserving fingercode authentication. In: Proceedings of the 12th ACM Workshop on Multimedia and Security (MM& Sec 2010), pp. 231–240 (2010)
5. Bos, J.W., Lauter, K., Loftus, J., Naehrig, M.: Improved security for a ring-based fully homomorphic encryption scheme. In: Stam, M. (ed.) IMACC 2013. LNCS, vol. 8308, pp. 45–64. Springer, Heidelberg (2013). https://doi.org/10.1007/978-3-642-45239-0_4
6. Bost, R., Popa, R.A., Tu, S., Goldwasser, S.: Machine learning classification over encrypted data. In: Proceedings of NDSS 2015 (2015)
7. Brakerski, Z.: Fully homomorphic encryption without modulus switching from classical GapSVP. In: Safavi-Naini, R., Canetti, R. (eds.) CRYPTO 2012. LNCS, vol. 7417, pp. 868–886. Springer, Heidelberg (2012). https://doi.org/10.1007/978-3-642-32009-5_50

8. Brakerski, Z., Gentry, C., Vaikuntanathan, V.: (Leveled) fully homomorphic encryption without bootstrapping. In: Proceedings of ITCS 2012, pp. 309–325 (2012)

9. Chase, M., et al.: Security of Homomorphic Encryption. Technical report (2017). HomomorphicEncryption.org

10. Chen, H., Han, K.: Homomorphic lower digits removal and improved FHE bootstrapping. In: Nielsen, J.B., Rijmen, V. (eds.) EUROCRYPT 2018. LNCS, vol. 10820, pp. 315–337. Springer, Cham (2018). https://doi.org/10.1007/978-3-319-78381-9_12

11. Chen, H., Han, K., Huang, Z., Jalali, A., Laine, K.: Simple Encrypted Arithmetic Library v2.3.0. Technical report (2017). https://www.microsoft.com/en-us/research/publication/simple-encrypted-arithmetic-library-v2-3-0/

12. Cheon, J.H., Kim, M., Kim, M.: Optimized search-and-compute circuits and their application to query evaluation on encrypted data. IEEE Trans. Inf. Forensics Secur. **11**(1), 188–199 (2016)

13. Chillotti, I., Gama, N., Georgieva, M., Izabachène, M.: Faster fully homomorphic encryption: bootstrapping in less than 0.1 seconds. In: Cheon, J.H., Takagi, T. (eds.) ASIACRYPT 2016. LNCS, vol. 10031, pp. 3–33. Springer, Heidelberg (2016). https://doi.org/10.1007/978-3-662-53887-6_1

14. Chillotti, I., Gama, N., Georgieva, M., Izabachène, M.: Faster packed homomorphic operations and efficient circuit bootstrapping for TFHE. In: Takagi, T., Peyrin, T. (eds.) ASIACRYPT 2017. LNCS, vol. 10624, pp. 377–408. Springer, Cham (2017). https://doi.org/10.1007/978-3-319-70694-8_14

15. Damgard, I., Geisler, M., Kroigard, M.: Homomorphic encryption and secure comparison. Int. J. Appl. Cryptol. **1**(1), 22–31 (2008)

16. Ducas, L., Micciancio, D.: FHEW: Bootstrapping homomorphic encryption in less than a second. In: Oswald, E., Fischlin, M. (eds.) EUROCRYPT 2015. LNCS, vol. 9056, pp. 617–640. Springer, Heidelberg (2015). https://doi.org/10.1007/978-3-662-46800-5_24

17. Erkin, Z., Franz, M., Guajardo, J., Katzenbeisser, S., Lagendijk, I., Toft, T.: Privacy-preserving face recognition. In: Goldberg, I., Atallah, M.J. (eds.) PETS 2009. LNCS, vol. 5672, pp. 235–253. Springer, Heidelberg (2009). https://doi.org/10.1007/978-3-642-03168-7_14

18. Fan, J., Vercauteren, F.: Somewhat practical fully homomorphic encryption. Cryptology ePrint Archive, Report 2012/144 (2012)

19. Gentry, C., Halevi, S., Jutla, C., Raykova, M.: Private database access with HE-over-ORAM architecture. In: Malkin, T., Kolesnikov, V., Lewko, A.B., Polychronakis, M. (eds.) ACNS 2015. LNCS, vol. 9092, pp. 172–191. Springer, Cham (2015). https://doi.org/10.1007/978-3-319-28166-7_9

20. Gentry, C., Halevi, S., Smart, N.P.: Fully homomorphic encryption with polylog overhead. In: Pointcheval, D., Johansson, T. (eds.) EUROCRYPT 2012. LNCS, vol. 7237, pp. 465–482. Springer, Heidelberg (2012). https://doi.org/10.1007/978-3-642-29011-4_28

21. Halevi, S., Polyakov, Y., Shoup, V.: An improved RNS variant of the BFV Homomorphic encryption scheme. Cryptology ePrint Archive, Report 2018/117 (2018)

22. Lu, W., Zhou, J., Sakuma, J.: Non-interactive and output expressive private comparison from homomorphic encryption. In: Proceedings of the 2018 on Asia Conference on Computer and Communications Security (ASIACCS 2018), pp. 67–74 (2018)

23. Polyakov, Y., Rohloff, K., Ryan, G.W.: PALISADE Lattice Cryptography Library User Manual (v1.2.0). Technical report (2018). https://git.njit.edu/palisade/PALISADE/blob/PALISADE-v1.2/doc/palisade_manual.pdf

24. Saha, T.K., Koshiba, T.: An efficient privacy-preserving comparison protocol. In: Barolli, L., Enokido, T., Takizawa, M. (eds.) NBiS 2017. LNDECT, vol. 7, pp. 553–565. Springer, Cham (2018). https://doi.org/10.1007/978-3-319-65521-5_48

25. Saha, T.K., Deevashwer, D., Koshiba, T.: Private comparison protocol and its application to range queries. In: Fortino, G., Ali, A., Pathan, M., Guerrieri, A., Di Fatta, G. (eds.) IDCS 2017. LNCS, vol. 10794, pp. 128–141. Springer, Cham (2018). https://doi.org/10.1007/978-3-319-97795-9_12

26. Smart, N.P., Vercauteren, F.: Fully homomorphic SIMD operations. Des. Codes Cryptogr. **71**(1), 57–81 (2014)

27. Yao, A.C.: How to generate and exchange secrets. In: Proceedings of 27th Annual Symposium on Foundations of Computer Science (SFCS 1986), pp. 162–167 (1986)

Secure Computation with Constant Communication Overhead Using Multiplication Embeddings

Alexander R. Block[✉], Hemanta K. Maji, and Hai H. Nguyen

Department of Computer Science, Purdue University, West Lafayette, IN, USA
{block9,hmaji,nguye245}@purdue.edu

Abstract. Secure multi-party computation (MPC) allows mutually distrusting parties to compute securely over their private data. The hardness of MPC, essentially, lies in performing secure multiplications over suitable algebras.

There are several cryptographic resources that help securely compute one multiplication over a large finite field, say $GF[2^n]$, with linear communication complexity. For example, the computational hardness assumption like noisy Reed-Solomon codewords are pseudorandom. However, it is not known if we can securely compute, say, a linear number of AND-gates from such resources, i.e., a linear number of multiplications over the base field $GF[2]$. Before our work, we could only perform $o(n)$ secure AND-evaluations.

Technically, we construct a perfectly secure protocol that realizes a linear number of multiplication gates over the base field using one multiplication gate over a degree-n extension field. This construction relies on the toolkit provided by algebraic function fields.

Using this construction, we obtain the following results. We provide the first construction that computes a linear number of oblivious transfers with linear communication complexity from the computational hardness assumptions like noisy Reed-Solomon codewords are pseudorandom, or arithmetic-analogues of LPN-style assumptions. Next, we highlight the potential of our result for other applications to MPC by constructing the first correlation extractor that has 1/2 resilience and produces a linear number of oblivious transfers.

Keywords: Secure computation · Multiplication embeddings
Oblivious transfer · Basis-independent circuit compututation
Leakage-resilient cryptography · Randomness extractors

1 Introduction

Secure multi-party computation [30,53] (MPC) allows mutually distrusting parties to compute securely over their private data. Even when parties follow

The research effort is supported in part by an NSF CRII Award CNS-1566499, an NSF SMALL Award CNS-1618822, and an REU CNS-1724673.

D. Chakraborty and T. Iwata (Eds.): INDOCRYPT 2018, LNCS 11356, pp. 375–398, 2018.
https://doi.org/10.1007/978-3-030-05378-9_20

the protocols honestly, but are curious to find additional information about other parties' private inputs, most functionalities cannot be securely computed [7,33,41,44]. So, we rely on diverse forms of cryptographic resources to help parties perform computations over their private data. These cryptographic resources can either be computational hardness assumptions [30,38] or physical resources like noisy channels [8,22,42,43], correlated private randomness [43,52], trusted resources [14,38,39], and tamper-proof hardware [18,24,40,46].

In this paper, for the simplicity of exposition of the key ideas, we consider 2-party secure computation against honest-but-curious adversaries. Suppose two parties are interested in securely computing a boolean circuit C that uses AND, and XOR, and represent the input, output, and the intermediate values of the computation in binary. Parties can use the oblivious transfer (OT) functionality to securely compute C (with perfect security and linear communication complexity) using the GMW protocol [30]. The OT functionality takes as input a pair of bits (x_0, x_1) from the sender and a choice bit b from the receiver, and outputs the bit x_b to the receiver. Notice Alice does not know Bob's choice bit b, and Bob does not know Alice's other bit x_{1-b}. Parties perform m calls to the OT functionality to securely compute circuits that have m AND gates (and an arbitrary number of XOR gates) with $\Theta(m)$ communication complexity. In this work, we consider secure computation protocols that have communication complexity proportional to the size of the circuit C.[1]

Parties can also compute arithmetic circuits that use MUL and ADD gates over large fields by emulating the arithmetic gates using finite fields. In particular, using efficient bilinear multiplication algorithms [21], parties can securely compute one multiplication over the finite field $\mathbb{GF}[2^n]$ by performing m OT calls and linear communication complexity, where $n = \Theta(m)$. In general, using m OT calls, parties can securely compute any circuit C that has m_i arithmetic gates over $\mathbb{GF}[2^{n_i}]$, for $i \in \mathbb{N}$, such that $\sum_i m_i \cdot n_i = \Theta(m)$, which measures the *size of C*. Intuitively, the size of the arithmetic circuit C refers to the cumulative size of representing the elements of the (multiplication) gates in the circuit.

Summarizing this discussion, we conclude that m OT calls help the parties securely compute arithmetic circuits (over characteristic 2 fields) of size $\Theta(m)$ with communication complexity $\Theta(m)$. Several cryptographic resources can implement the m instances of the OT functionality using a linear communication complexity. For example, there are instantiations based on polynomial-stretch local pseudorandom generators [35], the Phi-hiding assumption [36], LWE [26], DDH-hard groups [12], and noisy channels [34]. By composing these protocols, parties can use the corresponding cryptographic resources and securely compute linear-size circuits using only linear communication.

On the other hand, there are cryptographic resources that directly enable secure multiplication over a large extension field using communication that is proportional to the size of the field. For example, consider the constructions based on Paillier encryption [23,29,48], LWE [25,45], pseudorandomness of noisy

---

[1] Network latency considerations typically motivate the study of MPC protocols with linear communication complexity.

random Reed-Solomon codewords [39,47], and arithmetic analogues of well-studied cryptographic assumptions [1]. The key functionality in this context is a generalization of the OT functionality, namely the Oblivious Linear-function Evaluation [52] (OLE) over a field $\mathbb{K}$, say $\mathbb{K} = \mathrm{GF}[2^n]$. The OLE functionality takes as input a pair of field elements $(A, B) \in \mathbb{K}^2$ from the sender and an element $X \in \mathbb{K}$ from the receiver, and outputs the linear evaluation $Z = A \cdot X + B$ to the receiver. Note that, for $x_0, x_1, b \in \mathrm{GF}[2]$, we have $x_b = (x_0 + x_1)b + x_0$, i.e., OT is a particular instantiation of the OLE functionality. Using (the generalization of) the GMW protocol, parties can compute one multiplication over $\mathbb{K}$ with $\Theta(\lg |\mathbb{K}|)$ communication complexity. Note that the circuit with one MUL gate (over $\mathbb{K}$) has size $\lg |\mathbb{K}|$, so the communication complexity of the protocol is linear in the circuit size. However, using OLE over $\mathbb{K} = \mathrm{GF}[2^n]$, *can we securely compute boolean circuits such that the communication complexity is linear in the circuit size?*

The question motivated above with the illustrative example of $\mathbb{K} = \mathrm{GF}[2^n]$ and $\mathbb{F} = \mathrm{GF}[2]$ generalizes to any $\mathbb{K}$ that is an extension field of a constant-size base field $\mathbb{F}$. Before our work, the best solution securely evaluated size $m = o(n)$ boolean circuits using $\Theta(n) = \omega(m)$ communication complexity from one OLE over $\mathrm{GF}[2^n]$ (refer to Sect. 1.3 for the state-of-the-art construction). We present the first solution that securely evaluates size $m = \Theta(n)$ boolean circuits using one OLE over $\mathrm{GF}[2^n]$ and, thus, has communication complexity linear in the circuit size. Additionally, we found secure computation of size-m boolean circuits using linear communication from more diverse cryptographic resources. Because, any cryptographic resource that securely implements OLE over $\mathbb{K}$ with a linear communication complexity, also enables the secure computation of linear-size boolean circuits with a linear communication complexity. In particular, we provide the first linear communication protocols for m OTs from cryptographic hardness assumptions such as the pseudorandomness of noisy Reed-Solomon codewords [39,47] and arithmetic analogues of well-studied cryptographic assumptions [1].

1.1 Multiplication Embedding Problem

Our approach to the MPC problem begins with the following combinatorial embedding problem, which was originally introduced by Block, Maji, and Nguyen [10] in the context of leakage-resilient MPC. Let $\mathbb{F}$ be a finite field. Alice has private input $\mathbf{a} = (a_1, \ldots, a_m) \in \mathbb{F}^m$ and Bob has private input $\mathbf{b} = (b_1, \ldots, b_m) \in \mathbb{F}^m$. The two parties want Bob to receive the output $\mathbf{c} = (c_1, \ldots, c_m) \in \mathbb{F}^m$ such that $c_i = a_i \cdot b_i$, for all $i \in \{1, \ldots, m\}$.

Alice and Bob have access to an oracle that takes input $A \in \mathbb{K}$ from Alice and $B \in \mathbb{K}$ from Bob, where $\mathbb{K}$ is a degree-n extension of the field $\mathbb{F}$, and outputs $C = A \cdot B$ to Bob. Alice and Bob want to perform only one call to this oracle and enable Bob to compute $\mathbf{c}$. Note that Alice and Bob perform *no additional interactions*. Given a fixed value of n and a particular base field $\mathbb{F}$, how large can m be?

The prior work of Block et al. [10] constructed an embedding that achieved $m = n^{1-o(1)}$ using techniques from additive combinatorics. This paper, using algebraic function fields, provides an asymptotically optimal $m = \Theta(n)$ construction. Section 1.2 summarizes our results and a few of its consequences for MPC.

Recent Independent Work. Recently, in an independent work, Cascudo et al. [16] (CRYPTO-2018) also studied this embedding problem as reverse multiplication-friendly embeddings (RMFE), and provide a constant-rate construction. They use this result to achieve new amortization results in MPC.

1.2 Our Contributions

Given two vectors $\mathbf{a} = (a_1, \ldots, a_m) \in \mathbb{F}^m$ and $\mathbf{b} = (b_1, \ldots, b_m) \in \mathbb{F}^m$, we represent their Schur product as the vector $\mathbf{a} * \mathbf{b} = (a_1 \cdot b_1, \ldots, a_m \cdot b_m)$. We prove the following theorem.

Theorem 1 (Embedding Theorem). *Let $\mathbb{F}_q$ be a finite field of size q, a power of a prime. There exist constants $c_q^* \in \{1, 2, 3, 4, 6\}$, $c_q > 0$, and $n_0 \in \mathbb{N}$ such that for all $n \geqslant n_0$ where c_q^* divides n, there exist (linear) maps $E \colon \mathbb{F}_q^m \to \mathbb{K}$ and $D \colon \mathbb{K} \to \mathbb{F}_q^m$, where $\mathbb{K}$ is the degree-n extension of the field $\mathbb{F}_q$, such that the following constraints are satisfied.*

1. *We have $m \geqslant c_q n$, and*
2. *For all $\mathbf{a}, \mathbf{b} \in \mathbb{F}_q^m$, we have: $D\big(E(\mathbf{a}) \cdot E(\mathbf{b})\big) = \mathbf{a} * \mathbf{b}$.*

Intuitively, an oracle that implements one multiplication over a degree-n extension field $\mathbb{K}$ facilitates the computation of $m = \Theta(n)$ multiplications over the base field $\mathbb{F}$. For instance, assuming the base field $\mathbb{F} = \mathrm{GF}[2]$, our result shows that we can implement $m = \Theta(n)$ AND gates, which are equivalent to the MUL arithmetic gates over the $\mathrm{GF}[2]$, by performing only one call to the functionality that implements MUL over $\mathbb{K} = \mathrm{GF}[2^n]$. Section 1.3 presents a summary of the intuition that inspired our construction, and Sect. 2 provides the required technical background, and Sect. 2.2 presents the proof of Theorem 1.

Consequences for MPC. Recall that the OLE functionality over the field $\mathbb{K}$ takes as input (A, B) from the sender and X from the receiver, and outputs $Z = A \cdot X + B$ to the receiver. Essentially, OLE over the field $\mathbb{K}$ generates an additive secret share $(-B, Z)$ of the product $A \cdot X$. The embedding of Theorem 1 also helps Alice and Bob implement m independent OLEs over the base field $\mathbb{F}$, represented by the $\mathrm{OLE}\,(\mathbb{F})^m$ functionality, using one OLE over the extension field $\mathbb{K}$.

Theorem 2. *Let $\mathbb{F}$ be a finite field, and $\mathbb{K}$ be a degree-n extension of $\mathbb{F}$. There exists a 2-party semi-honest secure protocol for the $\mathrm{OLE}\,(\mathbb{F})^m$ functionality in the $\mathrm{OLE}\,(\mathbb{K})$-hybrid, where $m = \Theta(n)$, that performs only one call to the $\mathrm{OLE}\,(\mathbb{K})$ functionality (and no additional communication).*

Section 3 provides the proof of Theorem 2 in the semi-honest setting. Continuing our working example of $\mathbb{F} = \mathbb{GF}[2]$, we can implement $m = \Theta(n)$ independent OT functionalities by performing one call to the $\mathsf{OLE}\,(\mathbb{K})$ functionality.

Using Theorem 2, we can implement a linear number of OTs at a constant communication overhead based on computational hardness assumptions like the pseudorandomness of noisy Reed-Solomon codewords [39, 47] and arithmetic analogues of well-studied cryptographic assumptions [1], which help construct an OLE over large (but finite) fields. In general, if a cryptographic resource supports the generation of one OLE over $\mathbb{K}$ using $\Theta(\lg|\mathbb{K}|)$ communication complexity, then the following result also applies to that resource.

Corollary 1. *There exists a computationally secure protocol implementing m OTs using $\Theta(m)$ communication based on (any of) the following computational hardness assumptions.*

1. *Pseudorandomness of noisy random Reed-Solomon codewords [39,47],*
2. *Arithmetic analogues of "LPN-style assumptions" and the existence of polynomial-stretch local arithmetic PRGs [1].*

In fact, we can leverage efficient bilinear multiplication algorithms [21] that incur a constant communication overhead, to obtain the following result.

Corollary 2. *Let $\mathbb{F}$ be a finite field, and $\mathbb{K}$ be a degree-n extension of $\mathbb{F}$. Let $\mathbb{F}_1, \ldots, \mathbb{F}_k$ are finite fields such that $\mathbb{F}_i$ is a degree-n_i extension of the base field $\mathbb{F}$, for $i \in \{1, \ldots, k\}$. Let C be a circuit that uses m_i arithmetic gates over the field $\mathbb{F}_i$. If $m_1 n_1 + \cdots + m_k n_k \leq \Theta(n)$, then there exists a secure protocol for C in the $\mathsf{OLE}\,(\mathbb{K})$-hybrid that performs only one call to the $\mathsf{OLE}\,(\mathbb{K})$ functionality.*

Section 5 presents Corollary 2 and provides the outline of constant overhead secure computation of $\mathsf{OLE}\,(\mathbb{F}_i)$ by performing $\Theta(n_i)$ calls to the $\mathsf{OLE}\,(\mathbb{F})$ functionality, where $\mathbb{F}_i$ is a degree-n_i extension of the base field $\mathbb{F}$. We emphasize that Corollary 2 allows the flexibility to generate the (randomized version of the) $\mathsf{OLE}\,(\mathbb{K})$ in an offline phase of the computation without the necessity to fix the representation of the computation itself. We only fix the base field $\mathbb{F}$ and an upper-bound n estimating the size of the circuit C.

Finally, using our embedding, instead of the original multiplication embedding of [10], we obtain the following result for correlation extractors (cf., [36] for definitions and an introduction).

Corollary 3. *For every $1/2 \geq \varepsilon > 0$, there exists an n-bit correlated private randomness such that, despite $t = (1/2 - \varepsilon)n$ bits of leakage, we can securely construct $m = \Theta(\varepsilon n)$ independent OTs from this leaky correlation.*

Section 4 presents the details of the definition of correlation extractors and the proof of this corollary.

1.3 Technical Overview

To illustrate the underlying idea of our embedding, we use the example where $|\mathbb{F}| = 3n/2$, and $\mathbb{K}$ is a degree-n extension of $\mathbb{F}$. Note that in this intuition the size of the base field implicitly bounds the degree of the extension field $\mathbb{K}$ that we can consider. Ideally, our objective is to obtain multiplication embeddings for small constant-size $\mathbb{F}$ for infinitely many n, which our theorem provides. Nevertheless, we feel that the intuition presented in the sequel assists the reading of the details of Sect. 2.

Assume that n is even and $m := (n/2 - 1)$. We arbitrarily enumerate the elements in $\mathbb{F}$ as

$$\mathbb{F} = \{f_{-m}, \ldots, f_{-2}, f_{-1}, f_1, f_2, \ldots, f_{n-1}\}.$$

Suppose the field $\mathbb{K}$ is isomorphic to $\mathbb{F}[t]/\pi(t)$, where $\pi(t) \in \mathbb{F}[t]$ is an irreducible polynomial of degree n.

Recall that Alice and Bob have private inputs $\mathbf{a} = (a_1, \ldots, a_m) \in \mathbb{F}^m$ and $\mathbf{b} = (b_1, \ldots, b_m) \in \mathbb{F}^m$. Alice constructs the unique polynomial $A(t) \in \mathbb{F}[t]/\pi(t)$ of degree $< m$ such that $A(f_{-i}) = a_i$, for all $i \in \{1, \ldots, m\}$ using Lagrange interpolation. Similarly, Bob constructs the unique polynomial $B(t) \in \mathbb{F}[t]/\pi(t)$ of degree $< m$ such that $B(f_{-i}) = b_i$, for all $i \in \{1, \ldots, m\}$.

Suppose the two parties have access to an oracle that multiplies two elements of $\mathbb{K}$ and outputs the result to Bob. Upon receiving the inputs $A(t)$ and $B(t)$ from Alice and Bob, respectively, which correspond to elements in $\mathbb{K}$, the oracle outputs the result $C(t) = A(t) \cdot B(t)$ to Bob.[2] Note that $C(t)$ is the convolution of the two polynomials $A(t)$ and $B(t)$. Moreover, it has the property that $C(f_{-i}) = a_i \cdot b_i$, for all $i \in \{1, \ldots, m\}$. So, Bob can evaluate the polynomial $C(t)$ at appropriate places to obtain $\mathbf{c} = \mathbf{a} * \mathbf{b}$.

Note that this protocol crucially relies on the fact that the field $\mathbb{F}$ has sufficiently many *places* $\{f_{-1}, \ldots, f_{-m}\}$ to enable the encoding of $a_1, \ldots, a_m$ as the *evaluation* of polynomials at those respective places. For constant-size fields $\mathbb{F}$, this intuition fails to scale to large values of n. So, we use the toolkit of algebraic function fields for a more generalized and formal treatment of these intuitive concepts and construct these multiplication embeddings for *every* base field $\mathbb{F}$.

Prior Best Construction. [10] showed that $(\lg |\mathbb{F}|)^{1-o(1)}$ OTs could be embedded into one OLE over $\mathbb{F}$ if $\mathbb{F}$ has characteristic 2. Overall, this construction yields $s(\log s)^{-o(1)}$ OTs from one OLE over $\mathbb{K}$, where $s = \lg |\mathbb{K}|$.

Reduction of our Construction to Chen and Cramer [20]. Chen and Cramer [20] construct algebraic geometry codes/secret sharing schemes that have properties similar to the Reed-Solomon codes, except that these linear codes are over finite fields of appropriate size. It is not clear how to rely solely on the distance and independence properties of these codes to get our results. However, the algebraic geometric techniques underlying the construction of [20] and our construction have a significant overlap.

---

[2] Note that this is exact polynomial multiplication because the degree of $A(t)$ and $B(t)$ are both $< m$. So, the degree of $C(t)$ is $< 2m - 1 = n$. This observation, intuitively, implies that "mod $\pi(t)$" does not affect $C(t)$.

2   Embedding Multiplications

Our goal is to embed m multiplications over $\mathbb{F}_q$ using a single multiplication over $\mathbb{F}_{q^n}$ such that $m = \Theta(n)$. To do so, we use algebraic function fields over $\mathbb{F}_q$ with appropriate parameters.

2.1   Preliminaries

We introduce the basics of algebraic function fields necessary for our construction. For explicit details we refer the reader to the full version of our work [11]. Let $\mathbb{F}_q$ be a finite field of q elements, where q is a power of prime. Then an algebraic function field $K/\mathbb{F}_q$ of one variable over $\mathbb{F}_q$ is a finite algebraic extension of $\mathbb{F}_q(x)$ for some x transcendental over $\mathbb{F}_q$. Recall that $\mathbb{F}_q(x) = \{f(x)/g(x) \colon f, g \in \mathbb{F}_q[x], g \neq 0\}$. When clear from context, we write K in place of $K/\mathbb{F}_q$.

Every function field K has an infinite set of "points" called *places*, denoted by $P \in \mathbb{P}(K)$. Every place P has an associated *degree* $\deg P \in \mathbb{N}$, and for any $k \in \mathbb{N}$, the set $\mathbb{P}^{(k)}(K)$ denotes the set of places of degree k. In particular, for every $k \in \mathbb{N}$, this set is finite, and the set $\mathbb{P}^{(1)}(K)$ is called the set of *rational places*. For every element $f \in K$, and any place P, we can evaluate f at place P, denoted as $f(P)$. Then two cases occur: either f has a *pole* at P, which we denote as $f(P) = \infty$; or f is defined at P. For P which is not a pole of f and $\deg P = k$, we have that $f(P)$ is isomorphic to some element of $\mathbb{F}_{q^k}$. For any two functions f, g and place P such that P is not a pole of f or g, we have that $f(P) + g(P) = (f + g)(P)$, $f(P) \cdot g(P) = (f \cdot g)(P)$, and $xf(P) = (x \cdot f)(P)$ for any $x \in \mathbb{F}_q$. Every place P has an associated *valuation ring* $\mathcal{O}_P$. A valuation ring $\mathcal{O}$ of K is a ring such that $\mathbb{F}_q \subsetneq \mathcal{O} \subsetneq K$ and for every $z \in K$ either $z \in \mathcal{O}$ or $z^{-1} \in \mathcal{O}$.

A *divisor* D of K is a formal sum of places. Namely, $D = \sum_{P \in \mathbb{P}(K)} m_P P$ where $m_P \in \mathbb{Z}$ and $m_P = 0$ for all but finitely many places P. The set of places P where $m_P \neq 0$ is called the *support* of D and is denoted as $\mathsf{Supp}(D)$. Any divisor D also has associated degree $\deg D := \sum_{P \in \mathbb{P}(K)} m_P(\deg P) \in \mathbb{Z}$. Note that every place is also a divisor; namely $P = 1 \cdot P$. Such divisors are called *prime divisors*. For any two divisors $D = \sum m_P P$ and $D' = \sum n_P P$, we define $D + D' = \sum (m_P + n_P)P$. We say that $D \leq D'$ if $m_P \leq n_P$ for all places P.

For any $f \in K \setminus \{0\}$, the *principal divisor* associated to f is denoted as (f). Informally, the principal divisor $(f) = \sum a_P P$ for places P, where $a_P = 0$ if P is not a zero or a pole of f, $a_P > 0$ if P is a pole of f of order a_P, and $a_P < 0$ if P is a zero of f of order $|a_P|$. Given any divisor D, we can define the *Riemann-Roch space* associated to D. This space is defined as $\mathscr{L}(D) := \{f \in K : (f) + D \geqslant 0\} \cup \{0\}$. The Riemann-Roch space of any divisor D is a vector space over $\mathbb{F}_q$ and has dimension $\ell(D)$. This dimension is bounded by the degree of the divisor.

Imported Lemma 1 ([15, **Lemma 2.51**]). *For any $D \in \mathrm{Div}(K)$, we have $\ell(D) \leq \deg D + 1$. In particular, if $\deg D < 0$, then $\ell(D) = 0$.*

Every function field K has associated $g(K) \in \mathbb{N}$ called the *genus*. In particular, $g(K) := \max_D \deg D - \ell(D) + 1$, where the max is taken over all divisors D of K. When clear from context, we simply write $g := g(K)$.

2.2 Our Construction

In this section we present our construction that proves Theorem 1. We need three results to prove our result. First, we need Imported Lemma 1. The second needed result shows that there always exists a prime divisor of degree n for large enough n. This is given by the following lemma.

Imported Lemma 2 ([13, **Lemma 18.21**]). *Let $K/\mathbb{F}_q$ be an algebraic function field of one variable of genus g and degree at least n satisfying $n \geqslant 2 \log_q g + 6$. Then there exists a prime divisor of degree n of $K/\mathbb{F}_q$.*

Finally, we need the following result.

Lemma 1. *Let V be a subspace of dimension m of $\mathbb{F}_q^r$. Then there exists a linear mapping $\psi : \mathbb{F}_q^r \to \mathbb{F}_q^m$ such that ψ is a bijection from V to $\mathbb{F}_q^m$ and that $\psi(x) * \psi(y) = \psi(x * y)$ for every $x, y \in \mathbb{F}_q^r$.*

Proof. Let G be a generator matrix of $V \subseteq \mathbb{F}_q^r$, then $V = \{uG : u \in \mathbb{F}_q^m\}$ and for any $x \in V$ there exists a unique $z \in \mathbb{F}_q^m$ such that $x = zG$. Let G_T denote the columns of G indexed by set $T \subseteq \{1, \ldots, r\}$ and x_T denote the entries of $x \in \mathbb{F}_q^r$ indexed by T. Choose $S \subseteq \{1, \ldots r\}$ such that $G \setminus G_S$ has full rank. Note that $|S| = r - m$. Let $G' = G \setminus G_S$ and $S' = \{1, \ldots, r\} \setminus S$. Define $\psi : \mathbb{F}_q^r \to \mathbb{F}_q^m$ as $\psi(x) = x_{S'}$. Now, for any $x, y \in \mathbb{F}^r$, we have $\psi(x) * \psi(y) = x_{S'} * y_{S'} = (x * y)_{S'} = \psi(x * y)$. Finally, it follows that ψ is a bijection from V to $\mathbb{F}^m$ since G' has full rank and for any $x \in V$ there exists a unique $z \in \mathbb{F}^m$ such that $x = zG$. $\qquad\square$

At a high level, the proof of Theorem 1 follows from Fig. 1 and the intuition presented in Sect. 1.3, with the main difference being now the base field $\mathbb{F}$ here is of constant size.

Proof (Theorem 1). We consider two cases for the size q of the field: (1) q is an even power of a prime and $q \geqslant 49$, and (2) $q < 49$ or q is an odd power of a prime.

Case 1. Suppose $q \geqslant 49$ and q is an even power of a prime. In this case we choose $c_q^* = 1$. Suppose there exists an algebraic function field $K/\mathbb{F}_q$ of genus g and degree $n \geqslant \max\{2 \log_q g + 6, 6g\}$. Let P be a prime divisor of degree one of $K/\mathbb{F}_q$. By Imported Lemma 2 there exists a prime divisor Q of degree n. Let $s = \lfloor (n-1)/2 \rfloor$ and consider the Riemann-Roch space $\mathscr{L}(2sP) = \{z \in K/\mathbb{F}_q \mid (z) + 2sP \geqslant 0\}$ and the valuation ring $\mathcal{O}_Q$ of Q. The vector space $\mathscr{L}(2sP)$ is contained in $\mathcal{O}_Q$, which yields that the map $\kappa : \mathscr{L}(2sP) \to \mathbb{F}_{q^n}$ defined as $z \mapsto z(Q)$ is a ring homomorphism. The kernel of κ is $\mathscr{L}(2sP - Q)$, which has dimension 0 by Imported Lemma 1 (since $\deg(2sP - Q) = 2s - n < 0$). This implies that κ is injective. Since $\mathscr{L}(sP) \subseteq \mathscr{L}(2sP)$, the evaluation map κ

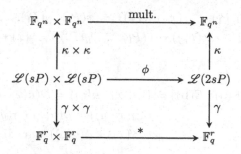

Fig. 1. Commutative diagram for performing r pointwise multiplications (the Schur product) over $\mathbb{F}_q$ using one multiplication over $\mathbb{F}_{q^n}$. The map ϕ represents polynomial multiplication.

restricted to $\mathscr{L}(sP)$, represented by $\kappa|_{\mathscr{L}(sP)}$, is a homomorphism from $\mathscr{L}(sP)$ to $\mathbb{F}_{q^n}$ and is injective.

Let $r > s$ and suppose $K/\mathbb{F}_q$ has at least $r+1$ distinct prime divisors of degree one. Let $P_1, P_2, \ldots, P_r$ be distinct prime divisors of degree one other than P. Consider the evaluation map $\gamma : \mathscr{L}(2sP) \to \mathbb{F}_q^r$ defined by $x \mapsto (x(P_1), x(P_2), \ldots, x(P_r))$. Since $\deg(sP - \sum P_i) = s - r < 0$, the kernel of $\gamma|_{\mathscr{L}(sP)}$ is $\mathscr{L}(sP - \sum P_i)$, which has dimension 0. Note that γ is a linear map, therefore by the rank-nullity theorem we have $\dim(\ker(\gamma|_{\mathscr{L}(sP)})) + \dim(\mathrm{Im}(\gamma|_{\mathscr{L}(sP)})) = \dim(\mathscr{L}(sP))$. So $\dim(\mathrm{Im}(\gamma|_{\mathscr{L}(sP)})) = \dim(\mathscr{L}(sP)) = s - g + 1$ since $\deg(sP) = s > 2g - 1$. Let $m = s - g + 1$ and $V = \mathrm{Im}(\gamma|_{\mathscr{L}(sP)})$. Then V is a vector subspace of $\mathbb{F}_q^r$ of dimension m. By Lemma 1, there exists a bijection $\psi : V \to \mathbb{F}_q^m$ such that it preserves the point-wise product operation; that is, $\psi(x) * \psi(y) = \psi(x*y)$ for every $x, y \in V$.

We define $E : \mathbb{F}_q^m \to K$ such that $E = \kappa \circ \gamma^{-1} \circ \psi^{-1}$, and $D : \mathrm{Im}(\kappa) \subseteq \mathbb{F}_{q^n} \to \mathbb{F}_q^m$ such that $D = \psi \circ \gamma \circ \kappa^{-1}$, where $\mathbb{K} = \mathbb{F}_{q^n}$.

Claim 1. *The maps E and D are well-defined.*

Proof. The definitions of E and D have inversion of functions and the fact is that not all functions have inverse functions. So we need to prove that we can always perform the inversions γ^{-1}, ψ^{-1}, and κ^{-1}. Since ψ is a bijection from V to $\mathbb{F}_q^m$ and γ is also a bijection from $\mathscr{L}(sP)$ to V, the mapping E is well-defined. Next, since κ is injective it is a bijection from $\mathscr{L}(2sP)$ to $\mathrm{Im}(\kappa)$. Thus, the mapping D is also well-defined.

Claim 2. *E and D are linear maps.*

This follows directly from the fact that ψ, κ, and γ are all linear maps. Next we will show that $D\big(E(\mathbf{a}) \cdot E(\mathbf{b})\big) = \mathbf{a} * \mathbf{b}$ for every $\mathbf{a}, \mathbf{b} \in \mathbb{F}_q^m$. Let $x, y \in \mathscr{L}(sP)$ such that $\mathbf{a} = \psi(x(P_1), x(P_2), \ldots, x(P_r)) = \psi(\gamma(x))$ and $\mathbf{b} = \psi(y(P_1), y(P_2), \ldots, y(P_r)) = \psi(\gamma(y))$ (such x and y always exist by properties of ψ and γ). Note that $(x \cdot y) \in \mathscr{L}(2sP)$ because $\mathscr{L}(sP) \cdot \mathscr{L}(sP) \subseteq \mathscr{L}(2sP)$, so γ has the following property.

$$\gamma(x \cdot y) = ((x \cdot y)(P_1), \ldots, (x \cdot y)(P_r)) = (x(P_1) \cdot y(P_1), \ldots, x(P_r) \cdot y(P_r))$$
$$= (x(P_1), \ldots, x(P_r)) * (y(P_1), \ldots, y(P_r)) = \gamma(x) * \gamma(y).$$

Therefore, we have

$$D(E(\mathbf{a}) \cdot E(\mathbf{b})) = D(\kappa(x) \cdot \kappa(y)) = D(\kappa(x \cdot y))$$
$$= \psi(\gamma(x \cdot y)) = \psi(\gamma(x) * \gamma(y))$$
$$= \psi(\gamma(x)) * \psi(\gamma(y)) = \mathbf{a} * \mathbf{b}.$$

Finally, since $s = \lfloor (n-1)/2 \rfloor$ and $n \geqslant 6g$, we have that $m = s-g+1 = \Theta(n)$. This completes the proof of Case 1.

Case 2. Suppose $q < 49$ is a power of prime or q is an odd power of a prime. Then Fig. 2 presents how to choose c_q^* such that $q^{c_q^*}$ is an even power of a prime and is at least 49.

											q		
			$q < 49$									$q \geqslant 49$	
	2	4	8	16	32	3	9	27	5	25	$q \geqslant 7$	$q = p^{2a+1}$	$q = p^{2a}$
c_q^*	6	3	2	2	2	4	2	2	4	2	2	2	1

Fig. 2. Table for our choices of c_q^* for Theorem 1. The value of c_q^* is chosen minimally such that $q^{c_q^*}$ is an even power of a prime and $q^{c_q^*} \geqslant 49$.

Let $q^* := q^{c_q^*}$. Suppose that n is sufficiently large and is divisible by c_q^*, and that $n/c_q^* \geqslant \max\{2\log_{q^*} g + 6, 6g\}$. Now q^* is an even power of a prime and $q^* \geqslant 49$, so we are in Case 1 with the following parameters. Let $n^* := n/c_q^*$, let $K/\mathbb{F}_{q^*}$ be an algebraic function field of genus g, and let Q be a prime divisor of degree n^*. Divisor Q exists since $n^* \geqslant 2\log_{q^*} g + 6$. Let $s = \lfloor (n^* - 1)/2 \rfloor$ and set $m = s - g + 1$.

Notice for every $x \in \mathbb{F}_q$, it holds that $x \in \mathbb{F}_{q^*}$ since $\mathbb{F}_q$ is a subfield of $\mathbb{F}_{q^*}$. Now consider any $\mathbf{a}, \mathbf{b} \in \mathbb{F}_q^m$. Again we have $\mathbf{a}, \mathbf{b} \in \mathbb{F}_{q^*}^m$. We define the maps of Case 1 with respect to q^* and n^*. In particular, we apply the algorithm from Case 1 with appropriate changes to q and n. Concretely, let $\kappa \colon \mathscr{L}(2sP) \to \mathbb{F}_{(q^*)^{n^*}}$, let $\gamma \colon \mathscr{L}(2sP) \to \mathbb{F}_{q^*}^r$, and let $V = \mathrm{Im}(\gamma|_{\mathscr{L}(sP)})$. Let $\psi \colon V \to \mathbb{F}_{q^*}^m$ be a bijection defined by Lemma 1. Let $E = \kappa \circ \gamma^{-1} \circ \psi^{-1}$ and $D = \psi \circ \gamma \circ \kappa^{-1}$. Consequently, we have $D(E(\mathbf{a}) \cdot E(\mathbf{b})) = \mathbf{a} * \mathbf{b}$.

We now have $s = \lfloor (n^* - 1)/2 \rfloor = \lfloor (n/c_q^* - 1)/2 \rfloor = \Theta(n)$ and $g = \Theta(n^*) = \Theta(n)$. Therefore, we have $m = s - g + 1 = \Theta(n)$. This completes the proof of case 2. Finally, Imported Theorem 1 gives concrete constructions of function fields with degree n prime divisors and $r + 1$ distinct rational places, which gives the result.   □

2.3 Function Field Instantiation Using Garcia-Stichtenoth Curves

In the proof of Theorem 1, we assume that there exists at least $r+1$ distinct places of degree one and there exists a prime divisor of degree n. We use appropriate Garcia-Stichtenoth curves to ensure this is indeed the case. Formally, we have the following theorem.

Imported Theorem 1 (Garcia-Stichtenoth [28]**).** *For every q that is an even power of a prime, there exists an infinite family of curves $\{C_u\}_{u\in\mathbb{N}}$ such that:*

1. *The number of rational places $\#C_u(\mathbb{F}_q) \geqslant q^{u/2}(\sqrt{q} - 1)$, and*
2. *The genus of the curve $g(C_u) \leq q^{u/2}$.*

For Theorem 1, we want the following conditions to be satisfied.

1. The number of distinct degree one places is at least $r + 1$
2. There exists a prime divisor of degree n.

Let $q \geqslant 49$ be an even power of a prime. Then for any $u \in \mathbb{N}$, we choose $n = q^{u/2}(\sqrt{q} - 1) \in \mathbb{N}$ and consider the function field given by the curve C_u. By Imported Theorem 1, we have that the number of rational points $\#C_u(\mathbb{F}_q) \geqslant q^{u/2}(\sqrt{q} - 1) = n$ and $g(C_u) \leq q^{u/2} = \frac{n}{\sqrt{q}-1}$. In particular, for $s = \lfloor \frac{n-1}{2} \rfloor$, we have $s < n$ and we can always choose r such that $s < r \leq n$. Setting $r = n - 1$, we have that the map γ in the proof of Theorem 1 defines a suitable Goppa code [31] over $\mathbb{F}_q$. With $r = n - 1$, we in fact have that there are at least $r + 1$ distinct prime divisors of degree one. Furthermore, we have $n \geqslant 6g$ and $g \leq \frac{n}{\sqrt{q}-1}$, so

$$2 \log_q g + 6 \leq 2 \log_q (n/(\sqrt{q} - 1)) + 6 \leq n.$$

So there exists a prime divisor of degree n by Imported Lemma 2. Finally we have

$$m = s - g + 1 \geqslant \lfloor (n-1)/2 \rfloor - n/(\sqrt{q} - 1) + 6 = \Theta(n).$$

Note that $g \geqslant 0$, so we also have $m \leq \lfloor \frac{n-1}{2} \rfloor + 6 = \Theta(n)$.

2.4 Efficiency

There are efficient algorithms to generate the places on the Garcia-Stichtenoth curves in Imported Theorem 1. In particular, the evaluation and the interpolation algorithms are efficient. The existence of such algorithms is one of the primary motivations for using Garcia-Stichtenoth curves instead of other alternate constructions.

For example, the cost of creating the generator matrices for multiplication-friendly linear secret sharing schemes as introduced by the seminal work of Chen-Cramer [20] corresponds to the cost of the encoding in our construction. The result of Shum-Aleshnikov-Kumar-Stichtenoth-Deolalikar [51], for instance, provides such an efficient encoding algorithm. The reconstruction/decoding problem has an efficient algorithm using the Berlekamp-Massey-Sakata algorithm with Feng-Rao majority voting.

3   Realizing OLE $(\mathbb{F})^m$ Using One ROLE$(\mathbb{K})$

In this section, we show how to securely realize m independent copies of OLE $(\mathbb{F})$ using one sample of ROLE$(\mathbb{K})$ (Random-OLE), for field $\mathbb{F} = \mathbb{F}_q$ and $\mathbb{K}$ a degree n extension field of $\mathbb{F}$. Intuitively, the ROLE$(\mathbb{K})$ functionality is an inputless functionality that samples A, B, X uniformly and independently at random from $\mathbb{K}$, and outputs (A, B) to one party and (X, Z) to the other party. This secure realization is achieved by composing two steps. First, we securely realize one OLE $(\mathbb{K})$ from one ROLE$(\mathbb{K})$ using a standard protocol (cf. the randomized self-reducibility of the OLE functionality [52]). Then, we embed m copies of OLE $(\mathbb{F})$ into one OLE $(\mathbb{K})$. Formally, we have the following theorem.

Theorem 3 (Realizing multiple small OLE using one large ROLE). *Let $\mathbb{F}$ be a field of size q, a power of a prime. Let $\mathbb{K}$ be a degree n extension field of $\mathbb{F}$. There exists a perfectly secure protocol for OLE $(\mathbb{F})^m$ in the ROLE$(\mathbb{K})$-hybrid that performs only one call to the ROLE$(\mathbb{K})$ functionality, $m = \Theta(n)$, and has communication complexity $3 \lg |\mathbb{K}|$.*

3.1   Preliminaries

We introduce the functionalities we are interested in.

Oblivious Linear-function Evaluation. For a field $(\mathbb{F}, +, \cdot)$, oblivious linear-function evaluation over $\mathbb{F}$, represented by OLE $(\mathbb{F})$, is a two-party functionality that takes as input $(a, b) \in \mathbb{F}^2$ from Alice and $x \in \mathbb{F}$ from Bob and outputs $z = ax + b$ to Bob. In particular, OLE refers to the OLE $(\mathbb{GF}[2])$ functionality.

Random Oblivious Linear-function Evaluation. For a field $(\mathbb{F}, +, \cdot)$, random oblivious linear-function evaluation over $\mathbb{F}$, represented by ROLE$(\mathbb{F})$, is a correlation that samples $a, b, x \in \mathbb{F}$ uniformly and independently at random. It provides Alice the secret share $r_A = (a, b)$ and provides Bob the secret share $r_B = (x, z)$, where $z = ax + b$. In particular, ROLE refers to the ROLE$(\mathbb{GF}[2])$ correlation.

3.2   Securely Realizing OLE $(\mathbb{K})$ using one ROLE$(\mathbb{K})$

The protocol presented in Fig. 3 is the standard protocol that implements the OLE $(\mathbb{K})$ functionality in the ROLE$(\mathbb{K})$-hybrid with perfect semi-honest security (cf. [52]).

3.3   Securely Realizing OLE $(\mathbb{F})^m$ using one OLE $(\mathbb{K})$

This section presents the realization of Theorem 2. Our goal is to embed m independent copies of OLE $(\mathbb{F})$ into one OLE $(\mathbb{K})$, where $m = \Theta(n)$. More concretely, suppose we are given an oracle that takes as input $A^*, B^* \in \mathbb{K}$ from Alice and $X^* \in \mathbb{K}$ from Bob, and outputs $Z^* = A^* \cdot X^* + B^*$ to Bob. Our aim is to implement the following functionality. Alice has inputs $\mathbf{a} = (a_1, \ldots, a_m) \in \mathbb{F}_q^m$ and

---

Pseudocode of the OLE protocol $\rho(\mathbb{K}, A^*, B^*, X^*)$

Given. Alice has $(\widetilde{A}_0, \widetilde{B}_0)$ and Bob has $(\widetilde{X}_0, \widetilde{Z}_0)$, where $\widetilde{A}_0, \widetilde{B}_0, \widetilde{X}_0$ are random elements in $\mathbb{K}$ and $\widetilde{Z}_0 = \widetilde{A}_0 \widetilde{X}_0 + \widetilde{B}_0$.

Private Inputs. Alice has private input $(A^*, B^*) \in \mathbb{K}^2$ and Bob has $X^* \in \mathbb{K}$.

Hybrid. Parties are in the ROLE($\mathbb{K}$)-hybrid.

Interactive Protocol.

1. **First Round.** Bob sends $M = \widetilde{X}_0 - X^*$ to Alice.
2. **Second Round.** Alice sends $\alpha = \widetilde{A}_0 + A^*$ and $\beta = \widetilde{A}_0 M + B^* + \widetilde{B}_0$.

Output Computation. Bob outputs $Z^* = \alpha X^* + \beta - \widetilde{Z}_0$.

Fig. 3. Perfectly secure protocol realizing OLE ($\mathbb{K}$) in the ROLE($\mathbb{K}$) correlation hybrid.

$\mathbf{b} = (b_1, \ldots, b_m) \in \mathbb{F}_q^m$, and Bob has input $\mathbf{x} = (x_1, \ldots, x_m) \in \mathbb{F}_q^m$. We want Bob to obtain $\mathbf{z} = (z_1, \ldots, z_m)$, where $\mathbf{z} = \mathbf{a} * \mathbf{x} + \mathbf{b}$, in other words, $z_i = a_i \cdot x_i + b_i$ for every $i \in [m]$. To do that, we extend our multiplication embedding with addition using a standard technique like in [10,32]. We define a randomized encoding function E_2 needed for our protocol as the following.

Definition of the (randomized) encoding function E_2

$E_2 : \mathbb{F}_q^m \to \mathbb{F}_{q^n}$. $E_2(\mathbf{b})$ returns a uniformly random $B \in \mathrm{Im}(\kappa) \subseteq \mathbb{F}_{q^n}$ such that $D(B) = \mathbf{b}$; that is, $\psi(\gamma(\kappa^{-1}(B))) = \mathbf{b}$.

We show that the protocol presented in Fig. 4 achieves $m = \Theta(n)$ and realizes Theorem 2.

Correctness. We argue the correctness of the protocol by showing that $D(Z^*) = \mathbf{a} * \mathbf{x} + \mathbf{b}$. In the protocol, Alice creates $A^* = E(\mathbf{a})$ and $B^* = E_2(\mathbf{b})$, and Bob creates $X^* = E(\mathbf{x})$. Calling the OLE ($\mathbb{K}$) functionality, Bob receives

---

Given. Linear maps E and D as in Theorem 1, and the linear map E_2 defined above.

Private input. Alice has private inputs $\mathbf{a} = (a_1, \ldots, a_m) \in \mathbb{F}_q^m$ and $\mathbf{b} = (b_1, \ldots, b_m) \in \mathbb{F}_q^m$. Bob has private input $\mathbf{x} = (x_1, \ldots, x_m) \in \mathbb{F}_q^m$.

Hybrid. Parties are in the OLE ($\mathbb{K}$)-hybrid.

Private Input Construction.

1. Alice creates private inputs $A^* = E(\mathbf{a})$ and $B^* = E_2(\mathbf{b})$.
2. Bob creates private input $X^* = E(\mathbf{x})$.
3. Both parties invoke the OLE ($\mathbb{K}$) functionality with respective Alice input (A^*, B^*) and Bob input X^*. Bob receives $Z^* = A^* X^* + B^* = E(\mathbf{a}) \cdot E(\mathbf{x}) + E_2(\mathbf{b})$.

Output Decoding. Bob outputs $\mathbf{z} = D(Z^*) = D(E(\mathbf{a}) \cdot E(\mathbf{x}) + E_2(\mathbf{b}))$.

Fig. 4. Protocol for embedding m copies of OLE ($\mathbb{F}$) into one OLE ($\mathbb{K}$), where $\mathbb{K}$ is a degree n extension field of $\mathbb{F}$.

$Z^* = A^* \cdot X^* + B^*$. In particular, Bob receives $Z^* = E(\mathbf{a}) \cdot E(\mathbf{x}) + E_2(\mathbf{b})$. Then Bob computes $D(Z^*)$. Since D is a linear map and by Theorem 1, we have $m = \Theta(n)$ and the following.

$$D(Z^*) = D\left(E(\mathbf{a}) \cdot E(\mathbf{x}) + E_2(\mathbf{b})\right) = D\left(E(\mathbf{a}) \cdot E(\mathbf{x})\right) + D(E_2(\mathbf{b})) = \mathbf{a} * \mathbf{x} + \mathbf{b}$$

Security. We argue the security for our protocol. The security relies on the observation that $E(\mathbf{a}) \cdot E(\mathbf{x}) + E_2(\mathbf{b})$ is uniformly distributed over the set

$$\{Z \colon Z \in \mathrm{Im}(\kappa) \text{ and } D(Z) = \mathbf{z}\},$$

where $\mathbf{z} = \mathbf{a} * \mathbf{b} + \mathbf{c}$. Note that Alice does not receive any message, so the simulation of semi-honest corrupt Alice is trivial.

Consider the case that Bob is semi-honest corrupt. In this case, the simulator receives $\mathbf{x}$ from the environment, sends $\mathbf{x}$ to the external functionality, and receives $\mathbf{z}$ as output. It samples $Z^* = E_2(\mathbf{z})$, and sends $(X^* = E(\mathbf{x}), Z^*, \mathbf{z})$ as the view of Bob to the environment.

We shall show that this simulation is perfect. Note that $E(\mathbf{a}) \cdot E(\mathbf{x}) \in \mathrm{Im}(\kappa)$. Observe that $E_2(\mathbf{b})$ is a uniform distribution over a coset of $E_2(0^m)$. Now, $E(\mathbf{a}) \cdot E(\mathbf{x}) + E_2(\mathbf{b})$ is a uniform distribution over the coset $\{Z \colon Z \in \mathrm{Im}(\kappa) \text{ and } D(Z) = \mathbf{z}\}$, where $\mathbf{z} = \mathbf{a} * \mathbf{x} + \mathbf{b}$. That is, the distribution of $E(\mathbf{a}) \cdot E(\mathbf{x}) + E_2(\mathbf{b})$ is identical to the distribution of $E_2(\mathbf{z})$.

3.4   Realization of OLE $(\mathbb{F})^m$ in the ROLE$(\mathbb{K})$-hybrid

The protocol that realizes Theorem 3 is the parallel composition of the protocols presented in Figs. 3 and 4 (Theorem 2). The composition of these protocols in parallel gives an optimal two-round protocol for realizing OLE $(\mathbb{F})^m$ in the ROLE$(\mathbb{K})$-hybrid with perfect security and $m = \Theta(n)$ by Theorem 1, as desired.

4   Linear Production Correlation Extractors in the High Resilience Setting

This section provides the necessary background of correlation extractors and proves Corollary 3. In particular, Corollary 3 is achieved by the construction of a suitable correlation extractor. A *correlated private randomness*, or *correlation* in short, is a joint distribution (R_A, R_B) which samples shares (r_A, r_B) according to the distribution and sends secret shares r_A to Alice and r_B to Bob. Correlations are given to parties in an offline preprocessing phase. Parties then use their respective secret shares in an online phase in an interactive protocol to securely compute an intended functionality. Correlation extractors take *leaky* shares of correlations and distill them into *fresh* randomness to be used to securely compute the intended functionality. Formally, we define a correlation extractor below.

Definition 1 (Correlation Extractor [36]). *Let (R_A, R_B) be a correlated private randomness such that the secret share size of each party is n'-bits. An (n', m, t, ε)-correlation extractor for (R_A, R_B) is a two-party interactive protocol in the $(R_A, R_B)^{[t]}$-hybrid that securely implements m copies of the OT functionality against information-theoretic semi-honest adversaries with ε-simulation error.*

Using this definition we restate Corollary 3 as follows.

Theorem 4 (Half Resilience, Linear Production Correlation Extractor). *For all constants $0 < \delta < g \leq 1/2$, there exists a correlation (R_A, R_B), where each party gets n'-bit secret shares, such that there exists a two-round (n', m, t, ε)-correlation extractor for (R_A, R_B), where $m = \Theta(n')$, $t = (1/2 - g)n'$, and $\varepsilon = 2^{-(g-\delta)n'/2}$.*

The construction of this correlation extractor achieves linear production $m = \Theta(n')$ and $1/2$ leakage resilience by composing our embedding (Theorems 1 and 2) with the correlation extractor of Block, Maji, and Nguyen (BMN) [10]. Prior correlation extractors either achieved sub-linear production, (significantly) less than $1/2$ resilience, or were not round-optimal.

4.1 Preliminaries

We introduce some useful functionalities and correlations.

Random Oblivious Transfer Correlation. Random oblivious transfer, represented by ROT, is a correlation that samples x_0, x_1, b uniformly and independently at random. It provides Alice the secret share $r_A = (x_0, x_1)$ and provides Bob the secret share $r_B = (b, x_b)$.

Recall also the **Oblivious Linear-function Evaluation** and **Random Oblivious Linear-function Evaluation** functionalities from Sect. 3.1. We denote ROLE($\mathbb{GF}[2]$) by ROLE. Note that ROT and ROLE are identical (functionally equivalent) correlations.

Inner-product Correlation. For a field $(\mathbb{K}, +, \cdot)$ and $n' \in \mathbb{N}$, inner-product correlation over $\mathbb{K}$ of size n', represented by IP$(\mathbb{K}^{n'})$, is a correlation that samples random $r_A = (x_0, \ldots, x_{n'-1}) \in \mathbb{K}^{n'}$ and $r_B = (y_0, \ldots, y_{n'-1}) \in \mathbb{K}^{n'}$ subject to the constraint that $x_0 + y_0 = \sum_{i=1}^{n'-1} x_i y_i$. The secret shares of Alice and Bob are, respectively, r_A and r_B.

4.2 Realizing Theorem 4

The realization of Theorem 4 is the parallel composition of two protocols. First, we utilize the BMN ROLE($\mathbb{K}$) extraction protocol [10, Fig. 7]. Informally, the BMN extraction protocol takes leaky shares of the inner-product correlation over the field $\mathbb{K}$, and securely extracts one sample of ROLE($\mathbb{K}$). In particular, the BMN extraction protocol is resilient to $t = (1/2 - g)n'$ bits of leakage, for any $g \in (0, 1/2]$.

Second, we utilize our new embedding protocol of Theorem 3 which produces m copies of OLE ($\mathbb{F}$) from one ROLE($\mathbb{K}$), and compose it in parallel with the BMN extraction protocol for ROLE($\mathbb{K}$). Previously, the BMN embedding achieved $m = (n')^{1-o(1)}$ production, whereas with Theorem 3 we achieve $m = \Theta(n')$ production with the following parameters. We take $\mathbb{F} = \mathrm{GF}[2]$ and $\mathbb{K} = \mathrm{GF}[2^{\delta n'}]$, where n' and δ are given, $\eta := \frac{1}{\delta} - 1$, and $n := \frac{n'}{(\eta+1)}$. In particular, $\mathbb{K}$ is a degree-n extension of $\mathbb{F}$, and n here corresponds to the n of Corollary 3. So $m = \Theta(n') = \Theta(n)$. We then take $(R_A, R_B) = \mathsf{IP}(\mathbb{K}^{1/\delta})$ to be the input correlation for the BMN extraction protocol.

The BMN extraction protocol is a perfectly secure semi-honest protocol for extracting one ROLE($\mathrm{GF}[2^{\delta n'}]$) in the $\left(\mathsf{IP}\left(\mathrm{GF}[2^{\delta n'}]^{1/\delta}\right)\right)^{[t]}$-hybrid which is resilient to $t = (1/2 - g)n'$ bits of leakage, for all $0 < \delta < g \leq 1/2$ (cf. [10, Theorem 1]). Then the parallel composition of the protocols of Figs. 3 and 4 is a perfectly secure semi-honest protocol for realizing m copies of OLE ($\mathrm{GF}[2]$) in the ROLE($\mathrm{GF}[2^{\delta n'}]$)-hybrid, and $m = \Theta(n') = \Theta(n)$. This proves Theorem 4, and thus Corollary 3.

4.3   Comparison with Prior Works

Correlation extractors were introduced by Ishai, Kushilevitz, Ostrovsky, and Sahai [36] as a natural generalization of privacy amplification and randomness extraction. Since the initial feasibility result of [36], there have been significant qualitative and quantitative improvements in correlation extractor constructions. Figure 5 summarizes the current state-of-the-art of correlation extractors.

Prior to our work, the Block, Gupta, Maji, and Nguyen (BGMN) correlation extractors [9] achieve the best qualitative and quantitative parameters. For example, starting with $n/2$ independent samples of the ROT correlation, they construct the first round-optimal correlation extractor that produces $m = \Theta(n)$

	Correlation Description	Message Complexity	Number of OTs Produced ($m/2$)	Number of Leakage bits (t)	Simulation Error (ε)
IKOS [36]	$\mathsf{ROT}^{n/2}$	4	$\Theta(n)$	$\Theta(n)$	$2^{-\Theta(n)}$
GIMS [32]	$\mathsf{ROT}^{n/2}$	2	$n/\operatorname{poly}\lg n$	$(1/4 - g)n$	$2^{-gn/m}$
	$\mathsf{IP}(\mathbb{K}^{n/\lg|\mathbb{K}|})$	2	1	$(1/2 - g)n$	2^{-gn}
BMN [10]	$\mathsf{IP}(\mathbb{K}^{n/\lg|\mathbb{K}|})$	2	$n^{1-o(1)}$	$(1/2 - g)n$	2^{-gn}
BGMN [9]	$\mathsf{ROT}^{n/2}$	2	$\Theta(n)$	$\Theta(n)$	$2^{-\Theta(n)}$
	$\mathsf{ROLE}(\mathbb{F})^{n/2\lg|\mathbb{F}|}$	2	$\Theta(n)$	$\Theta(n)$	$2^{-\Theta(n)}$
Our Results	$\mathsf{IP}(\mathbb{K}^{n/\lg|\mathbb{K}|})$	2	$\Theta(n)$	$(1/2 - g)n$	2^{-gn}

Fig. 5. A qualitative summary of prior relevant works in correlation extractors and a comparison to our correlation extractor construction. Here $\mathbb{K}$ is a finite field and $\mathbb{F}$ is a finite field of constant size. All correlations have been normalized so that each party gets an n-bit secret share. The parameter g is defined as the gap to leakage resilience s.t. $t \geqslant 0$.

secure ROT samples despite $t = (1/4 - \varepsilon)n$ bits of leakage, for any $\varepsilon > 0$. Note that any correlation extractor for $n/2$ ROT samples can have at most $t = n/4$ resilience [37].

Our correlation extractor is also round optimal. However, the BMN [10] correlation extractor and our correlation extractor have resilience in the range $t/n \in [1/4, 1/2)$. Intuitively, our correlation extractor is ideal where high resilience is necessary. Our correlation extractor needs a large correlation, for example, the inner-product correlation over large fields. Contrast this with the case of BGMN extractor that uses multiple samples of the ROT correlation. To achieve $t = (1/2 - g)n$ resilience, where $g \in (0, 1/4]$, we use the inner-product correlation over fields of size (roughly) 2^{gn}. Using the multiplication embedding in Theorem 1, our work demonstrates the feasibility of extracting $m = \Theta(gn)$ independent ROT samples when the fractional resilience is in the range $t/n \in [1/4, 1/2)$.

5   Chudnovsky-Chudnovsky Bilinear Multiplication

We discuss the reverse problems of Theorems 1 and 3. We assume familiarity with Sect. 2. First we consider the problem of computing one large field multiplications using many small field multiplications. This is given by the following theorem.

Theorem 5 (Field Extension Multiplication via Pointwise Base Field Multiplication). *Let $\mathbb{F}$ be a finite field of size q, a power of a prime. For sufficiently large n, there exists a constant $c' > 0$ and (linear) maps $E' \colon \mathbb{K} \to \mathbb{F}^m$ and $D' \colon \mathbb{F}^m \to \mathbb{K}$, where $\mathbb{K}$ is the degree-n extension of the field $\mathbb{F}$, such that the following constraints are satisfied.*

1. *We have $m \geqslant c'n$, and*
2. *For all $A, B \in \mathbb{K}$, the following identity holds*

$$D' (E'(A) * E'(B)) = A \cdot B$$

where "$*$" is pointwise multiplication over $\mathbb{F}^m$.

Note that since the maps E' and D' are linear, the following holds.

Corollary 4. *For all $A, B, C \in \mathbb{K}$, we have*

$$D' (E'(A) * E'(B) + E'(C)) = D' (E'(A) * E'(B)) + D' (E'(C)).$$

Theorem 5 follows from the results of Chudnovsky-Chudnovsk [21]. In particular, they show that the rank of bilinear multiplication is $\Theta(n)$.

Imported Theorem 2 (Chudnovsky and Chudnovsky [13, Theorem 18.20]). *For every power of a prime q there exists a constant c_q such that $R(\mathbb{F}_{q^n}/\mathbb{F}_q) \leq c_q n$, where R is the rank of the $\mathbb{F}_q$-bilinear map that is multiplication over $\mathbb{F}_{q^n}$.*

The theorem states that if $\mathbb{K}$ is a degree n extension of $\mathbb{F}_q$, then the bilinear complexity of multiplication over $\mathbb{K}$ is $\Theta(n)$. This result is due to the Chudnovsky-Chudnovsky interpolation algorithm (cf. Imported Lemma 2) and the following result of Garcia and Stichtenoth.

Imported Theorem 3 (Garcia and Stichtenoth [27], [13, **Theorem 18.24]).** *Let p be a power of prime, X_1 be an indeterminate over $\mathbb{F}_{p^2}$, and $K_1 := \mathbb{F}_{p^2}(X_1)$. For $i \geqslant 1$ let $K_{i+1} := K_i(Z_{i+1})$, where Z_{i+1} satisfies the Artin-Schreier equation $Z_{i+1}^p + Z_{i+1} = X_m^{p+1}$ and $X_i := Z_i/X_{i-1} \in K_i$ (for $i \geqslant 2$). Then K_i/F_{p^2} has genus g_i given by*

$$g_i = \begin{cases} p^i + p^{i-1} - p^{\frac{i+1}{2}} - 2p^{\frac{i-1}{2}} + 1 & \text{if } i \equiv 1 \bmod 2, \\ p^i + p^{i-1} - \frac{1}{2}p^{\frac{i}{2}+1} - \frac{3}{2}p^{\frac{i}{2}} - p^{\frac{i}{2}-1} + 1 & \text{if } i \equiv 0 \bmod 2, \end{cases}$$

and $|\mathbb{P}^{(1)}(K_i/\mathbb{F}_{p^2})| \geqslant (p^2-1)p^{i-1} + 2p \geqslant (p-1)g_i$.

Imported Lemma 3 (Chudnovsky-Chudnovsky Interpolation Algorithm [13, Proposition 18.22]). *Let $K/\mathbb{F}_q$ be an algebraic function field of one variable of genus g, $n \geqslant 2\log_q g + 6$, and assume that there exist at least $4g+2n$ prime divisors of degree one of $K/\mathbb{F}_q$. Then we have $R(\mathbb{F}_{q^n}/\mathbb{F}_q) \leq 3g + 2n - 1$.*

Imported Lemma 3 gives rise to the commutative diagram of Fig. 6 which defines the interpolation method. This interpolation method implements multiplication over $\mathbb{F}_{q^n}$ using r' pointwise multiplications over $\mathbb{F}_q^{r'}$. This gives that $r' = 3g + 2n - 1 = \Theta(n)$. Setting $m = r'$ and setting E' and D' according to the interpolation algorithm directly yields Theorem 5. Concretely, we have the maps E' and D' defined as follows.

$$E' := \kappa' \circ (\gamma')^{-1} \qquad\qquad D' := \gamma' \circ (\kappa')^{-1}.$$

Note both κ' and γ' are linear maps, so E' and D' are also linear maps.

Given E' and D' of Theorem 5, we compute the reverse problem of Theorem 3. That is, we can use multiple small ROLE to realize one large OLE.

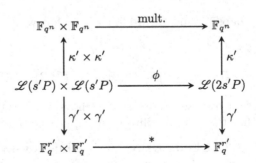

Fig. 6. Chudnovsky-Chudnovsky interpolation algorithm for performing multiplication over $\mathbb{F}_{q^n}$ using r' pointwise multiplications over $\mathbb{F}_q^{r'}$, where $r' = \Theta(n)$ and $s' = n+2g-1$.

Theorem 6 (Realizing one large OLE using multiple small ROLE). *Let* $\mathbb{F}$ *be a field of size* q, *a power of a prime. Let* $\mathbb{K}$ *be a degree* n *extension field of* $\mathbb{F}$. *There exists a perfectly secure protocol for* OLE $(\mathbb{K})$ *in the* ROLE $(\mathbb{F})^m$-*hybrid that performs only one call to the* ROLE $(\mathbb{F})^m$ *functionality,* $m = \Theta(n)$, *and has communication complexity* $3m \lg |\mathbb{F}|$.

To realize Theorem 6, we compose two steps in parallel. First we securely realize OLE $(\mathbb{F})^m$ from ROLE $(\mathbb{F})^m$ using a standard protocol. Then we use m copies of OLE $(\mathbb{F})$ to implement a single OLE $(\mathbb{K})$.

5.1 Securely Realizing OLE $(\mathbb{F})^m$ Using ROLE $(\mathbb{F})^m$

The protocol presented in Fig. 7 is an extension of the standard protocol that implements the OLE $(\mathbb{F})$ functionality in the ROLE $(\mathbb{F})$-hybrid with perfect semi-honest security. In particular, it is the m parallel composition of the OLE $(\mathbb{F})$ functionality in the ROLE $(\mathbb{F})$-hybrid.

Pseudocode of the OLE $(\mathbb{F})^m$ protocol

Given. Alice has $(\mathbf{a}', \mathbf{b}')$ and Bob has $(\mathbf{x}', \mathbf{z}')$, where $\mathbf{a}', \mathbf{b}', \mathbf{x}'$ are random elements in $\mathbb{F}^m$ and $\mathbf{z}' = \mathbf{a}' * \mathbf{x}' + \mathbf{b}'$.

Private Inputs. Alice has private input $(\mathbf{a}^*, \mathbf{b}^*) \in \mathbb{F}^{2m}$ and Bob has $\mathbf{x}^* \in \mathbb{F}^m$.

Hybrid. Parties are in ROLE $(\mathbb{F})^m$-hybrid.

Interactive Protocol.

1. **First Round.** Bob sends $\mathbf{m} = \mathbf{x}' - \mathbf{x}^*$ to Alice.
2. **Second Round.** Alice sends $\alpha = \mathbf{a}' + \mathbf{a}^*$ and $\beta = \mathbf{a}' * \mathbf{m} + \mathbf{b}^* + \mathbf{b}'$.

Output Computation. Bob outputs $\mathbf{z}^* = \alpha * \mathbf{x}^* + \beta - \mathbf{z}'$.

Fig. 7. Perfectly secure protocol realizing OLE $(\mathbb{F})^m$ in the ROLE $(\mathbb{F})^m$ correlation hybrid.

5.2 Securely Realizing OLE $(\mathbb{K})$ from OLE $(\mathbb{F})^m$

The goal is to use m copies of OLE $(\mathbb{F})$ to compute one OLE $(\mathbb{K})$, where $m = \Theta(n)$. Concretely, suppose we are given an oracle which takes as input $\mathbf{a}, \mathbf{b} \in \mathbb{F}^m$ from Alice and $\mathbf{x} \in \mathbb{F}^m$ from Bob, and outputs $\mathbf{z} = \mathbf{a} * \mathbf{x} + \mathbf{b}$ to Bob. Our aim is to implement the following functionality. Alice has private inputs $A \in \mathbb{K}$ and $B \in \mathbb{K}$, and Bob has input $X \in \mathbb{K}$. We want Bob to obtain $Z = AX + B \in \mathbb{K}$. We show that if Alice and Bob use the protocol presented in Fig. 8, we can achieve $m = \Theta(n)$. More formally, we have the following lemma.

Lemma 2 (Performing one large OLE using multiple small OLE). *Let* $\mathbb{K}$ *be an extension field of* $\mathbb{F}$ *of degree* n. *There exists a perfectly secure protocol for* OLE $(\mathbb{K})$ *in the* OLE $(\mathbb{F})^m$-*hybrid that performs only one call to the* OLE $(\mathbb{F})^m$ *functionality and* $m = \Theta(n)$.

Given. Two linear maps E' and D' as in Theorem 5.
Private input. Alice has private inputs $A \in \mathbb{K}$ and $B \in \mathbb{K}$. Bob has private input $X \in \mathbb{K}$.
Hybrid. Parties are in the $\mathsf{OLE}(\mathbb{F})^m$-hybrid.
Private Input Construction.

1. Alice creates private inputs $\mathbf{a} = E'(A)$ and $\mathbf{b} = E'(B)$.
2. Bob creates private inputs $\mathbf{x} = E'(X)$.
3. Both parties invoke the the $\mathsf{OLE}(\mathbb{F})^m$ functionality with respective Alice input $(\mathbf{a}, \mathbf{b})$ and Bob input $\mathbf{x}$. Bob receives $\mathbf{z} = \mathbf{a} * \mathbf{x} + \mathbf{b} = E'(A) * E'(X) + E'(B)$.

Output Decoding. Bob outputs $Z = D'(\mathbf{z}) = D'(E'(\mathbf{a}) * E'(\mathbf{x}) + E'(\mathbf{b})) = AX + B$.

Fig. 8. Protocol for computing one $\mathsf{OLE}(\mathbb{K})$ using m copies of $\mathsf{OLE}(\mathbb{F})$, where $\mathbb{K}$ is a degree n extension field of $\mathbb{F}$.

Figure 8 realizes Lemma 2. In the protocol, Alice creates $\mathbf{a} = E'(A)$ and $\mathbf{b} = E'(B)$, and Bob creates $\mathbf{x} = E'(X)$. Calling the $\mathsf{OLE}(\mathbb{F})^m$ functionality, Bob receives $\mathbf{z} = \mathbf{a} * \mathbf{x} + \mathbf{b}$. In particular, he receives $\mathbf{z} = E'(A) * E'(X) + E'(B)$. Bob then computes $D'(\mathbf{z})$. Since D' is a linear map and by Theorem 5, we have the following.

$$D'(\mathbf{z}) = D'(E'(A) * E'(X) + E'(B)) = D'(E'(A) * E'(X)) + D'(E'(B))$$
$$= AX + B$$

5.3   Proof of Theorem 6

The protocol which satisfies Theorem 6 is the parallel composition of the protocols presented in Figs. 7 and 8 (Lemma 2). The composition of these protocols in parallel gives an optimal two-round protocol for realizing $\mathsf{OLE}(\mathbb{K})$ in the $\mathsf{ROLE}(\mathbb{F})^m$-hybrid with perfect security and $m = \Theta(n)$ by Theorem 5, as desired.

5.4   Prior Work

Chudnovsky-Chudnovsky [21] gave the first feasibility result on the bilinear complexity of multiplication, showing $\Theta(n)$ multiplications in $\mathbb{F}_q$ suffice to perform one multiplication over $\mathbb{F}_{q^n}$. Since then there have been several works on explicit constructions and variants of the bilinear multiplication algorithms and improved the bounds on the bilinear complexity.

The works of [27, 28, 50] discuss the construction of appropriate function fields such that there is sufficient number of rational points for interpolation. Improvement on the bounds for the bilinear complexity of multiplication and generalizations of the Chudnovsky-Chudnovsky method appear in [3, 4, 6, 49]. Explicit construction of multiplication algorithms are discussed in [2, 4, 17], and in the particular case of function fields over elliptic curves in [5, 19].

References

1. Applebaum, B., Damgård, I., Ishai, Y., Nielsen, M., Zichron, L.: Secure arithmetic computation with constant computational overhead. In: Katz, J., Shacham, H. (eds.) CRYPTO 2017. LNCS, vol. 10401, pp. 223–254. Springer, Cham (2017). https://doi.org/10.1007/978-3-319-63688-7_8
2. Atighehchi, K., Ballet, S., Bonnecaze, A., Rolland, R.: On chudnovsky-based arithmetic algorithms in finite fields. CoRR **abs/1510.00090** (2015). http://arxiv.org/abs/1510.00090
3. Ballet, S., Rolland, R.: Multiplication algorithm in a finite field and tensor rank of the multiplication. J. Algebra **272**(1), 173–185 (2004). https://doi.org/10.1016/j.jalgebra.2003.09.031. http://www.sciencedirect.com/science/article/pii/S0021869303006951
4. Ballet, S., Baudru, N., Bonnecaze, A., Tukumuli, M.: On the construction of the asymmetric chudnovsky multiplication algorithm in finite fields without derivated evaluation. Comptes Rendus Mathematique **355**(7), 729–733 (2017). https://doi.org/10.1016/j.crma.2017.06.002. http://www.sciencedirect.com/science/article/pii/S1631073X17301577
5. Ballet, S., Bonnecaze, A., Tukumuli, M.: On the construction of elliptic chudnovsky-type algorithms for multiplication in large extensions of finite fields, March 2013
6. Ballet, S., Pieltant, J., Rambaud, M.: On some bounds for symmetric tensor rank of multiplication in finite fields. CoRR **abs/1601.00126** (2016). http://arxiv.org/abs/1601.00126
7. Beaver, D.: Perfect privacy for two-party protocols. In: Feigenbaum, J., Merritt, M. (eds.), vol. 2, pp. 65–77. American Mathematical Society, Providence (1989)
8. Beimel, A., Malkin, T., Micali, S.: The all-or-nothing nature of two-party secure computation. In: Wiener, M. (ed.) CRYPTO 1999. LNCS, vol. 1666, pp. 80–97. Springer, Heidelberg (1999). https://doi.org/10.1007/3-540-48405-1_6
9. Block, A.R., Gupta, D., Maji, H.K., Nguyen, H.H.: Secure computation using leaky correlations (asymptotically optimal constructions). Cryptology ePrintArchive, Report 2018/372 (2018). https://eprint.iacr.org/2018/372
10. Block, A.R., Maji, H.K., Nguyen, H.H.: Secure computation based on leaky correlations: high resilience setting. In: Katz, J., Shacham, H. (eds.) CRYPTO 2017. LNCS, vol. 10402, pp. 3–32. Springer, Cham (2017). https://doi.org/10.1007/978-3-319-63715-0_1
11. Block, A.R., Maji, H.K., Nguyen, H.H.: Secure computation with constant communication overhead using multiplication embeddings. Cryptology ePrint Archive, Report 2018/395 (2018). https://eprint.iacr.org/2018/395
12. Boyle, E., Gilboa, N., Ishai, Y.: Group-based secure computation: optimizing rounds, communication, and computation. In: Coron, J.-S., Nielsen, J.B. (eds.) EUROCRYPT 2017. LNCS, vol. 10211, pp. 163–193. Springer, Cham (2017). https://doi.org/10.1007/978-3-319-56614-6_6
13. Bürgisser, P., Clausen, M., Shokrollahi, M.A.: Algebraic Complexity Theory. Grundlehren der mathematischen Wissenschaften, vol. 315. Springer, NewYork (1997). https://doi.org/10.1007/978-3-662-03338-8
14. Canetti, R., Lindell, Y., Ostrovsky, R., Sahai, A.: Universally composable two-party and multi-party secure computation, pp. 494–503 (2002). https://doi.org/10.1145/509907.509980

15. Cascudo, I.: On asymptotically good strongly multiplicative linear secret sharing. Ph.D. thesis, Tesis doctoral, Universidad de Oviedo (2010)

16. Cascudo, I., Cramer, R., Xing, C., Yuan, C.: Amortized complexity of information-theoretically secure MPC revisited. In: Shacham, H., Boldyreva, A. (eds.) CRYPTO 2018. LNCS, vol. 10993, pp. 395–426. Springer, Cham (2018). https://doi.org/10.1007/978-3-319-96878-0_14

17. Cenk, M., Özbudak, F.: On multiplication in finite fields. J. Complex. 26(2), 172–186 (2010)

18. Chandran, N., Goyal, V., Sahai, A.: New constructions for UC secure computation using tamper-proof hardware. In: Smart, N. (ed.) EUROCRYPT 2008. LNCS, vol. 4965, pp. 545–562. Springer, Heidelberg (2008). https://doi.org/10.1007/978-3-540-78967-3_31

19. Chaumine, J.: Multiplication in small finite fields using elliptic curves, pp. 343–350 (2012). https://doi.org/10.1142/9789812793430_0018. https://www.worldscientific.com/doi/abs/10.1142/9789812793430_0018

20. Chen, H., Cramer, R.: Algebraic geometric secret sharing schemes and secure multi-party computations over small fields. In: Dwork, C. (ed.) CRYPTO 2006. LNCS, vol. 4117, pp. 521–536. Springer, Heidelberg (2006). https://doi.org/10.1007/11818175_31

21. Chudnovsky, D.V., Chudnovsky, G.V.: Algebraic complexities and algebraic curves over finite fields. Proc. Nat. Acad. Sci. 84(7), 1739–1743 (1987)

22. Crépeau, C., Kilian, J.: Achieving oblivious transfer using weakened security assumptions (extended abstract), pp. 42–52 (1988). https://doi.org/10.1109/SFCS.1988.21920

23. Damgård, I., Jurik, M.: A generalisation, a simpli.cation and some applications of paillier's probabilistic public-key system. In: Kim, K. (ed.) PKC 2001. LNCS, vol. 1992, pp. 119–136. Springer, Heidelberg (2001). https://doi.org/10.1007/3-540-44586-2_9

24. Damgård, I., Nielsen, J.B., Wichs, D.: Isolated proofs of knowledge and isolated zero knowledge. In: Smart, N. (ed.) EUROCRYPT 2008. LNCS, vol. 4965, pp. 509–526. Springer, Heidelberg (2008). https://doi.org/10.1007/978-3-540-78967-3_29

25. Damgård, I., Pastro, V., Smart, N., Zakarias, S.: Multiparty computation from somewhat homomorphic encryption. In: Safavi-Naini, R., Canetti, R. (eds.) CRYPTO 2012. LNCS, vol. 7417, pp. 643–662. Springer, Heidelberg (2012). https://doi.org/10.1007/978-3-642-32009-5_38

26. Dodis, Y., Halevi, S., Rothblum, R.D., Wichs, D.: Spooky encryption and its applications. In: Robshaw, M., Katz, J. (eds.) CRYPTO 2016. LNCS, vol. 9816, pp. 93–122. Springer, Heidelberg (2016). https://doi.org/10.1007/978-3-662-53015-3_4

27. Garcia, A., Stichtenoth, H.: A tower of artin-schreier extensions of function fields attaining the drinfeld-vladut bound. Inventiones Mathematicae 121(1), 211–222 (1995)

28. Garcia, A., Stichtenoth, H.: On the asymptotic behaviour of some towers of function fields over finite fields. J. Number Theory 61(2), 248–273 (1996)

29. Gilboa, N.: Two party RSA key generation. In: Wiener, M. (ed.) CRYPTO 1999. LNCS, vol. 1666, pp. 116–129. Springer, Heidelberg (1999). https://doi.org/10.1007/3-540-48405-1_8

30. Goldreich, O., Micali, S., Wigderson, A.: How to play any mental game or A completeness theorem for protocols with honest majority, pp. 218–229 (1987). https://doi.org/10.1145/28395.28420

31. Goppa, V.D.: Codes on algebraic curves. Soviet Math. Dokl. 24, 170–172 (1981)

32. Gupta, D., Ishai, Y., Maji, H.K., Sahai, A.: Secure computation from leaky correlated randomness. In: Gennaro, R., Robshaw, M. (eds.) CRYPTO 2015. LNCS, vol. 9216, pp. 701–720. Springer, Heidelberg (2015). https://doi.org/10.1007/978-3-662-48000-7_34

33. Impagliazzo, R., Luby, M.: One-way functions are essential for complexity based cryptography (extended abstract), pp. 230–235 (1989). https://doi.org/10.1109/SFCS.1989.63483

34. Ishai, Y., Kushilevitz, E., Ostrovsky, R., Prabhakaran, M., Sahai, A., Wullschleger, J.: Constant-rate oblivious transfer from noisy channels. In: Rogaway, P. (ed.) CRYPTO 2011. LNCS, vol. 6841, pp. 667–684. Springer, Heidelberg (2011). https://doi.org/10.1007/978-3-642-22792-9_38

35. Ishai, Y., Kushilevitz, E., Ostrovsky, R., Sahai, A.: Cryptography with constant computational overhead, pp. 433–442 (2008). https://doi.org/10.1145/1374376.1374438

36. Ishai, Y., Kushilevitz, E., Ostrovsky, R., Sahai, A.: Extracting correlations, pp. 261–270 (2009). https://doi.org/10.1109/FOCS.2009.56

37. Ishai, Y., Maji, H.K., Sahai, A., Wullschleger, J.: Single-use OT combiners with near-optimal resilience. In: 2014 IEEE International Symposium on Information Theory, Honolulu, HI, USA, June 29–July 4 2014, pp. 1544–1548. IEEE (2014). https://doi.org/10.1109/ISIT.2014.6875092

38. Ishai, Y., Prabhakaran, M., Sahai, A.: Founding cryptography on oblivious transfer – efficiently. In: Wagner, D. (ed.) CRYPTO 2008. LNCS, vol. 5157, pp. 572–591. Springer, Heidelberg (2008). https://doi.org/10.1007/978-3-540-85174-5_32

39. Ishai, Y., Prabhakaran, M., Sahai, A.: Secure arithmetic computation with no honest majority. In: Reingold, O. (ed.) TCC 2009. LNCS, vol. 5444, pp. 294–314. Springer, Heidelberg (2009). https://doi.org/10.1007/978-3-642-00457-5_18

40. Katz, J.: Universally composable multi-party computation using tamper-proof hardware. In: Naor, M. (ed.) EUROCRYPT 2007. LNCS, vol. 4515, pp. 115–128. Springer, Heidelberg (2007). https://doi.org/10.1007/978-3-540-72540-4_7

41. Kilian, J.: Founding cryptography on oblivious transfer, pp. 20–31 (1988). https://doi.org/10.1145/62212.62215

42. Kilian, J.: A general completeness theorem for two-party games, pp. 553–560 (1991). https://doi.org/10.1145/103418.103475

43. Kilian, J.: More general completeness theorems for secure two-party computation, pp. 316–324 (2000). https://doi.org/10.1145/335305.335342

44. Kushilevitz, E.: Privacy and communication complexity, pp. 416–421 (1989). https://doi.org/10.1109/SFCS.1989.63512

45. Lyubashevsky, V., Peikert, C., Regev, O.: On ideal lattices and learning with errors over rings. In: Gilbert, H. (ed.) EUROCRYPT 2010. LNCS, vol. 6110, pp. 1–23. Springer, Heidelberg (2010). https://doi.org/10.1007/978-3-642-13190-5_1

46. Moran, T., Segev, G.: David and goliath commitments: UC computation for asymmetric parties using tamper-proof hardware. In: Smart, N. (ed.) EUROCRYPT 2008. LNCS, vol. 4965, pp. 527–544. Springer, Heidelberg (2008). https://doi.org/10.1007/978-3-540-78967-3_30

47. Naor, M., Pinkas, B.: Oblivious polynomial evaluation. SIAM J. Comput. 35(5), 1254–1281 (2006)

48. Paillier, P.: Public-key cryptosystems based on composite degree residuosity classes. In: Stern, J. (ed.) EUROCRYPT 1999. LNCS, vol. 1592, pp. 223–238. Springer, Heidelberg (1999). https://doi.org/10.1007/3-540-48910-X_16

49. Randriambololona, H.: Bilinear complexity of algebras and the chudnovsky-chudnovsky interpolation method. J. Complex. 28(4), 489–517 (2012)

50. Shparlinski, I., Tsfasman, M., Vladut, S.: Curves with many points and multiplication in finite-fields. Lect. Notes Math. **1518**, 145–169 (1992)
51. Shum, K.W., Aleshnikov, I., Kumar, P.V., Stichtenoth, H., Deolalikar, V.: A low-complexity algorithm for the construction of algebraic-geometric codes better than the gilbert-varshamov bound. IEEE Trans. Inf. Theory **47**(6), 2225–2241 (2001)
52. Wolf, S., Wullschleger, J.: Oblivious transfer is symmetric. In: Vaudenay, S. (ed.) EUROCRYPT 2006. LNCS, vol. 4004, pp. 222–232. Springer, Heidelberg (2006). https://doi.org/10.1007/11761679_14
53. Yao, A.C.C.: Protocols for secure computations (extended abstract), pp. 160–164 (1982). https://doi.org/10.1109/SFCS.1982.38

Author Index

Printed in the United States
By Bookmasters